ENERGY:
ITS PHYSICAL
IMPACT ON THE
ENVIRONMENT

ENERGY: ITS PHYSICAL IMPACT ON THE ENVIRONMENT

DELBERT W. DEVINS

Professor of Physics
Indiana University

Robert E. Krieger Publishing Company
Malabar, Florida
1988

PHOTO CREDIT LIST

DEVINS, *ENERGY* PART OPENERS

Part I   Bill Gillette, E.P.A.-Documerica
Part II   Department of Energy
Part III   Department of Energy
Part IV   Arabian American Oil Company
Part V   Harry Wilks, Stock Boston

Original Edition 1982
Reprint Edition 1988

Printed and Published by
ROBERT E. KRIEGER PUBLISHING COMPANY, INC.
KRIEGER DRIVE
MALABAR, FLORIDA 32950

Copyright © 1982 by John Wiley & Sons, Inc.
Reprinted by Arrangement

Library of Congress Cataloging-in-Publication Data

Devins, D. W.
        Energy, its physical impact on the environment.

        Reprint. Originally published: New York : Wiley,
c1982.
        Includes bibliographies and index.
        1. Energy development--Environmental aspects.
2. Power resources. 3. Power (Mechanics) I. Title.
TD195.E49D48 1988   621.042   87-29764
ISBN 0-89464-271-5
                          68183

10   9   8   7   6   5   4   3   2

To Irene, for today,
and
Matthew, for tomorrow

PREFACE

In the early 1970s there was a surge of interest throughout the country in energy–environment problems, and many college courses in these topics were created. This textbook is an outgrowth of one such course. It was my desire to offer college students a course that would do more than skim the surface—one that would challenge them and allow them to use their newly formed scientific skills. In preparing to teach this physics course, however, I soon discovered that not only were there no suitable textbooks from which to choose but also that the information I needed was spread throughout a vast literature in several different fields. At the same time, it was very encouraging that there were many areas where physics could readily be applied to problems of current interest, and that *real* problems could be solved by beginning science students.

This book is the result of several years of accumulating data and ideas from many of the areas of energy and environment research where physics has been applied in the past and is being applied today. Diverse fields like combustion, meteorology, nuclear power, and solar cells are covered as well as others. Yet many interesting topics could not be included because of the sheer weight of their numbers. There is much more material here than can be covered in a one- or even two-semester course. This allows the course to be less structured than usual and even permits tailoring it to the current interests of the students and instructor.

In each chapter the physical fundamentals, applied science, and environmental ramifications are woven together in a coherent and logical fashion. A balance between formalism and application is not easy to achieve and, inevitably, my own prejudices must have appeared in the writing. The inclusion of calculus-based physics formalism will put some students at a disadvantage, but the larger body of information and the greater depth of understanding made possible by this level of mathematical analysis more than compensates. Many students have commented that this course material gave them their first inkling that textbook physics could actually be used in the real world!

Each chapter has two or more Group Projects; these can be used in making additional applications of textbook ideas to everyday life. There are a number of examples worked out within the text and a large number of homework exercises at the end of each chapter. These exercises can be most effective in bringing home the reality of physical analysis. Sometimes you will have to look up data in the library to solve a problem; sometimes you will not be able to find the information you need. Then you must do what you will have to do in the real world, namely, guess.

I hope this book will enable you to conclude that, although there are indeed many problems caused by the search for and use of energy in our country, there are also solutions to those problems. You may also discover that nothing is quite as simple as it might have seemed at one time; nothing is all bad or all good. For every action (or inaction), we must pay a price. The trick is to be able to decide whether the gain is worth the cost. This text will not always enable you to make that decision, but it will make you aware of the trade-offs involved.

No one writes a book like this without help. I have had a number of students over the past several years without whom it would not have been possible. They digested portions of the manuscript and labored over homework exercises. The feedback from this courageous group has been invaluable. I have also benefited from the encouragement of my colleagues, particularly G. D. Mahan, A. D. Bacher, D. B. Lichtenberg, and T. A. Ward. They critically read parts of the manuscript, acted as sounding boards, provided information, and gently pushed me on.

I thank Jewel Pierson, Kathy Klawitter, Sue Blue, Marlene Ellsworth, and Judy Chapple, who contributed their talents to typing various versions of the manuscript.

DELBERT W. DEVINS

CONTENTS

ENERGY: ITS PHYSICAL IMPACT ON THE ENVIRONMENT

PART ONE

SURVEY

CHAPTER ONE

THE ENERGY–ENVIRONMENT PROBLEM

1.1   THE NATURE OF THE PROBLEM

Conflicting Requirements

These days we seem to be confronted with conflicts at every turn. We read almost daily of shortages in fuels or crucial materials. We are told that hundreds of years of energy supply are available in coal and nuclear energy, but that there may be severe environmental consequences if these resources are used at an accelerated rate. We hear that arbitrary decisions made by foreign governments are responsible for our fuel supply shortages, but that higher pricing of fuels will lead to increased domestic production. We also hear that the maze of governmental regulations concerning clean air and water is driving the price of energy upward, but that without those regulations our general level of health would be substantially lowered. Life seems to be very complicated, indeed.

We have found ourselves suddenly in the "Age of Limits." We are on the threshold of recognizing that the earth's resources are finite, that it *is* a "Spaceship Earth," which must provide all the requirements of life for all its three to four billion inhabitants—forever. With this recognition has come the questioning of our traditional pattern of growth and con-

sumption, a national uneasiness about technology, and a search for environmentally clean energy sources. The public discussions of such "soft energy paths" have led to a heightened awareness of the impact of our society on its surroundings.

A coherent national energy–environment policy may soon emerge from these debates. Difficult, perhaps unpopular, decisions will be made by responsible leaders; an enlightened citizenry will accept those decisions. Our society will surely change, and we must accept change for the alternative is chaos. One impediment to the development of such a policy is an almost palpable lack of understanding by many people in and out of government of the physical principles that underlie energy generation and conversion processes. Information about resources, environmental impacts, alternative technologies, conservation, and other topics of current interest is spread throughout a variety of books, journals, and articles. It is difficult for the scientist, to say nothing of the average citizen, to obtain, collate, and digest the data needed to make rational decisions.

This textbook will provide some of the information and many of the tools required to bring to bear the power of physical insight on energy–environment problems. All of the problems cannot be studied in detail: society's interactions with its environment are too complex—no single textbook can adequately treat them. The plan of this textbook is to lay a foundation in each chapter, to demonstrate the application of physical principles where possible, to examine a few specific problems, to point out other perhaps unsolved or insoluble problems, and to start you thinking about them from a scientific point of view.

Trends and Projections

It is said that hindsight is always 20/20. Seeing the recent past so clearly may then help predict the immediate future. Although most efforts of this kind are singularly unsuccessful, we shall try to see what can be learned by studying trends.

As a first step let us examine the total consumption of energy in the United States over the past 200 years or so. In Figure 1.1 note that the energy consumption curve breaks away from the population growth curve around the turn of the century. Energy consumption per

capita has been increasing ever since (with the exception of the depression years). This tells us something about the past. Does it help with the future? In fact, this figure leads us to ask more questions than it helps to answer:

What have been the sources of this energy? Can these sources be maintained?
Is the growth in per capita energy consumption a necessary consequence of industrialization?
Have there been changes in the end-use distribution of energy in the United States in recent years?
Is there a similar curve for pollution problems?
How long can sustained per capita energy consumption growth and population growth be simultaneously maintained?

Some of these questions are easier to answer than others; some are not answerable at all. There has been a clear shift in our use of energy resources over the years, as seen in Figure 1.2. The development of engines powered by petroleum products pushed the

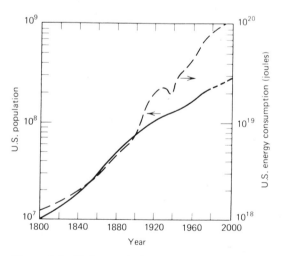

Figure 1.1 U.S. population versus time (left-hand vertical scale and solid line) and energy consumption versus time (right-hand vertical scale and dashed line) between the years 1800 and 2000.

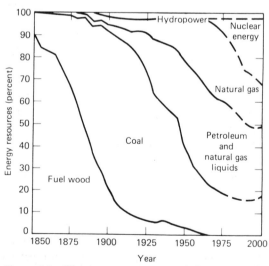

Figure 1.2 The energy resources mix between 1850 and 2000; dashed lines are estimates for the years after 1978.

demand for these products upward. Large ocean-going tankers made foreign oil cheap (until recently!). The completion of a nationwide network of pipelines in the 1930s made possible the widespread use of natural gas for home heating and electricity generation. The increasing difficulty in mining high-grade coal and the expense of transporting it to use sites helped to reduce the overall demand for coal. Petroleum products have been used in ever larger amounts for the production of petrochemicals—synthetic fabrics, ammonia, and pharmaceuticals, to name only a few.

The shift from coal and wood to oil and natural gas that took place after the turn of the century caused a great many social and environmental problems. Until then air and water pollution had been restricted primarily to the very large metropolitan areas, where coal burning for home heating, industry, and a relatively small number of central electricity-generating plants was the principal pollution source. Today home heating contributes little directly to metropolitan pollution problems, and generating plants can be made relatively clean. The major sources of pollution today are industry and the vast number of automobiles. Coal mine fatalities have been reduced significantly in the past 70 years, in part through better mine-safety laws and enforcement of these and in part from the decrease in subsurface mining and concurrent increase in aboveground strip mining—which has its own problems. Because of our reliance on petroleum we now have an increasing number of oil spills that affect us indirectly through the ocean food chain. As we now move to the end of the twentieth century, most analysts agree that the use of petroleum and natural gas for energy must begin to decline and the use of coal burning and nuclear power must increase.

It is not difficult to understand this conclusion. Petroleum is being consumed at a far greater rate than it is being produced by natural processes now. To be sure, the occurrence of oil is widespread, and many areas remain relatively unexplored—for example, the Atlantic coastal shelves. We will never "run out" of oil, but there will come a time when oil will be too expensive to warrant production and, obviously, too expensive to use! Just when this time will come is not easy to estimate (we shall discuss energy estimates later). We should note that a distinction is usually made between **resources** and **reserves**. A mineral resource is an estimate, sometimes only a guess, of the amount of a mineral present in a given area. A mineral reserve is that part of the resource that is believed to be recoverable. It usually is based on mapping or other means of more quantitative estimation. The figures for total oil reserves given in Table 1.1 are probably not too far off. Consider also that since oil has been produced commercially, starting at about the middle of the last century, almost one third of the original world estimated resources has already been pumped. The next third will be pumped before 1990. The end is definitely in sight.

TABLE 1.1 Estimated Proven World Oil Reserves

	Crude Reserves at End of 1979 ($\times 10^9$ bbl)[a]
Europe	13.758
Africa	57.361
Middle East	360.492
Asia, Far East	22.153
Oceania	16.016
USSR and satellites	61.501
United States	27.051
Latin America	66.989
TOTAL WORLD	625.822

Source: World Oil, August 15, 1980. Copyright © by Gulf Publishing Co.
[a]A barrel (bbl) is 42 gallons (gal) or 158.76 liters (ℓ).

For many of the energy end uses, substitutes for oil can be found. Oil can be prepared, at a high price, from coal. Nuclear fuels and coal can be used to produce the electricity now being generated from the heat obtained by burning oil. Additional electricity can be produced to supply the home heat and process heat now being obtained from oil. As we shall see, however, electricity represents energy of particularly high quality; it seems unreasonable to waste such quality on the wrong application. In any case, some end uses would be difficult to satisfy with an all-electric energy economy.

Many predictions for the U.S. energy picture of the future have been made. It is unlikely that any are correct, because it is difficult to take into account all the relevant factors, particularly the political and economic pressures.

It may be valuable to examine the energy resources and end-use distribution for 1970 and try to see how this distribution will have to change in the future. Refer to Figure 1.3. Clearly, the situation with petroleum has changed dramatically since 1970, when only 3.5×10^6 barrels of oil were imported per day. It is also clear that a continued increase in oil imports is neither desirable nor likely. Coal production is increasing at a rather slow rate. Geothermal, wind, and direct solar power presently contribute very little to the total energy picture, but could contribute more under the right conditions. Natural gas production fell every year from 1972 until 1979. At that time prices were decontrolled and more gas was produced, but a substantial improvement here is not realistic.

More and more electricity is being generated by nuclear plants, and, as a consequence, we will expect to see a higher percentage of rejected energy in the future. It seems there is a widening gap between our demand for energy, particularly in the form of petroleum, and our ability to supply it, either domestically or as imports. There is pressure from consumers to reduce the restrictions on the use of high-sulfur coal and in other ways to relax

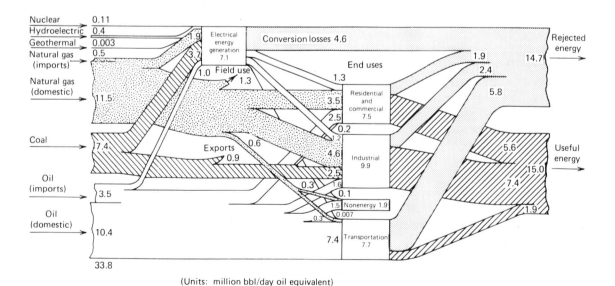

(Units: million bbl/day oil equivalent)

Figure 1.3 U.S. energy flow for 1970 in terms of million barrels of oil equivalent per day. (From W. Eaton, *Energy Storage.* Washington D.C.: U.S. Energy Research and Development Administration, 1975.)

environmental standards in order to boost energy supplies. Our choices are few, painful, and expensive:

Cut back on energy use (at the expense of economic growth?).

Develop exotic energy sources (at very high prices?).

Become more reliant on foreign suppliers (with significant political risks?).

Become substantially more efficient (yes, but how?).

All of the above.

Energy Definitions

The problems of energy supply and demand have been discussed thus far, and yet *energy* has not been defined. The energy concept is and has been very useful in physics; the Conservation of Energy Principle is a very powerful tool in the solution of problems. By defining energy appropriately it is possible to construct a self-consistent description of physical events in terms of energy transfer from one form to another. For example, when an automobile is brought to a stop, its kinetic energy is converted to thermal energy in the tires and roadway through friction. Some of the various forms of energy are:

■ Kinetic energy—the energy of relative motion.

■ Gravitational potential energy—the energy of position of mass in a gravitational field.

■ Electrical and magnetic energy—the energy of position of electric charges in electric and magnetic fields.

■ Electromagnetic energy—the radiant energy resulting from accelerations of electric charges.

■ Chemical energy—the energy of position of electrical charges in the electric fields of atoms.

■ Mass energy—the binding energy resulting from interactions of strong nuclear forces between nucleons.

■ Thermal energy—the kinetic energy of electrons, atoms, and molecules.

It is also possible to store energy in fields, commonly electric and magnetic fields. Finally, in atomic and nuclear processes the rest energy associated wih mass mc^2 must be taken into account. In these cases energy is not the consequence of the interaction between particles but, rather, is a static property of the particles or fields.

However, energy is often thought of as the result of an interaction between two or more objects. This picture is convenient because, although it is usually possible to calculate quite accurately the energy involved in these interactions, it is often not possible to calculate the interactions themselves to a satisfactory degree. Conversions between many of these forms of energy are possible; a large part of this textbook will be devoted to the study of these conversions, their efficiencies and by-products. But in order to make our studies quantitative so that they may be useful in the real world, it is necessary to have unit systems in which measurements can be expressed.

The unit system in which physicists work is the "Système International," or SI, system. In this system the unit of energy is the *joule*; 1 J equals 1 N m (newton-meter). Many of the calculations related to the environment are of an engineering nature, and the metric system is not always used. The unit of energy in the English system used by engineers is the foot-pound (ft-lb), but it is not largely used. Rather, the British thermal unit (Btu) is more common. Many of the calculations of interest will be connected with heat; both metric-system calculators and English-system workers habitually use the calorie (cal) as an energy unit for calculations involving heat, nutrition, electrical energy, and many others. It should be noted, however, that the calorie used in nutrition is 1000 times the size of the calorie used in heat calculations and, in fact, is often called the

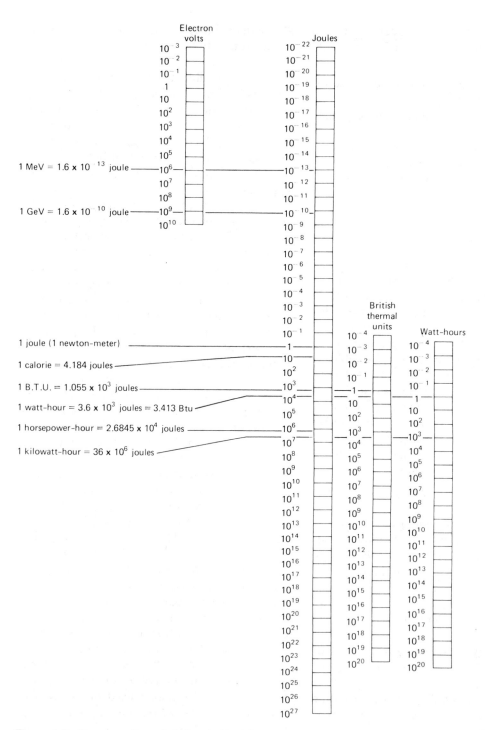

Figure 1.4 Energy units and their relationships.

kilocalorie (but just as often is called the Calorie with a capital "C," so confusion can abound). Another common practice is to express energy in terms of the units of power and time: in the MKS (meter-kilogram-second) system, the kilowatt-hour (kWh) and in the English system, the horsepower-hour (hp hr). A final energy unit is one used by atomic and nuclear physicists because of the smallness of energies in these realms: the electron volt (eV). The relationships between these various units are shown in Figure 1.4. See also Appendix A for a review of the various units. Contemporary practice in education, for engineering as well as physics, is to use the SI system. Except when quoting others, we will try to use SI units exclusively. We recommend their use; eventually the entire country will be metric.

In connection with energy resources, since the quantities of interest are so large, other units of measure have been devised:

$$1 \text{ quad} = 10^{15} \text{ Btu}$$

and

$$1 \text{ Q} = 10^{18} \text{ Btu}.$$

It is unfortunate that these terms have come into existence, not only because it is easy to mistake one for the other, but also because they are not SI units. Note that 1 Q is about the energy required to boil away Lake Michigan (the total U.S. energy use in 1970 was 0.07 Q). Note also that

$$10^{12} \text{ W yr} = 29.9 \times 10^{15} \text{ Btu}.$$

Doubling Time

Now that a system of measure for energy–environment studies has been introduced, let us take a detailed look at the energy consumption curve for the United States over the past few years. This is plotted in Figure 1.5. Except for the years immediately following the 1973–1974 Arab oil embargo, there has been a steady growth in the U.S. consumption of

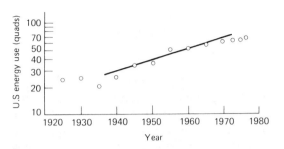

Figure 1.5 U.S. energy use between 1920 and 1980. [Adapted from Gibbons et al., "U.S. Energy Demands: Some Low Energy Futures," *Science* **200**, 142 (1978). Copyright © 1978 by the American Association for the Advancement of Science.]

energy. In fact, a straight line could be fitted to a portion of the curve, as indicated. Straight lines can be represented in a Cartesian coordinate system by the equation

$$y = mx + b. \tag{1.1}$$

In the case of Figure 1.5, since the vertical axis is a logarithmic scale, the equation would be

$$\log C = mt + B, \tag{1.2}$$

where C is the total energy consumption, m is the slope, t is the time (i.e., year), and B is the vertical axis intercept.

We could also write the equation in this form:

$$\log \left(\frac{C}{b} \right) = mt, \tag{1.3}$$

where $\log b = B$.

We have not specified the logarithm's base, although conventionally "log" is used to mean a base-10 logarithm. Any base could be used, however, and one particular base, base-e ($e = 2.718$), the "natural number," is used extensively in physics and mathematics. Base-e logarithms are usually indicated by "ln," so that Eq. 1.3 would be written

$$\ln \left(\frac{C}{b} \right) = mt, \tag{1.4}$$

where now $\ln b = B$. From this, by taking the antilogarithm of each side, we obtain

$$C = be^{mt}. \qquad (1.5)$$

We conclude that any data that yield a straight line when plotted on a semilogarithmic graph (i.e., one axis linear, one logarithmic) can be represented by an equation of the type of Eq. 1.5. In this equation m is a *constant*, the slope of the straight line through the data points.

You may also recognize Eq. 1.5 as the solution to this differential equation:

$$\frac{dC}{dt} = mC. \qquad (1.6)$$

This equation can be used to describe a wide variety of physical and biological phenomena.

Equation 1.6 emphasizes the fact that exponential change results when the rate of change of a variable is proportional to the amount of that variable present. This, of course, leads to *constant percentage change*.

Example 1.1 A culture of *E. coli* contains 200 mg/ml of bacteria at a certain time. Forty minutes later it is observed to contain 400 mg/ml. This growth is known to follow Eq. 1.4. What is the value of the slope m?

In Eq. 1.5, b must be the value C has when $t = 0$ ($e^{m \cdot 0} = 1$). Therefore, 400 mg/ml = (200 mg/ml)$e^{m \cdot 40}$ (i.e., let the time of the first observation be $t = 0$). Then

$$\frac{400 \text{ mg/ml}}{200 \text{ mg/ml}} = e^{m \cdot 40}.$$

Taking the natural logairthm of both sides,

$$\ln 2 = 40\, m.$$

So that

$$\boxed{m = 0.0173 \text{ or } 1.73 \text{ percent/min}}$$

How long could this growth continue?

A useful concept for phenomena that are characterized by constant percentage growth is **doubling time**. If Eq. 1.6 is solved for C,

$$C(t) = C(t = 0)e^{mt}, \qquad (1.7)$$

where we have indicated explicitly the time dependence of C. Doubling time is defined as the time required for the initial amount [in this case $C(t = 0)$] to double. Let T_2 be the doubling time; then

$$C(T_2) = C(0) \exp(mT_2), \qquad (1.8)$$

but

$$C(T_2) = 2C(0). \qquad (1.9)$$

So we have

$$2 = \exp(mT_2) \quad \text{or} \quad mT_2 = \ln 2. \quad (1.10)$$

Therefore,

$$T_2 = \frac{\ln 2}{m}$$

or

$$\boxed{T_2 = \frac{0.693}{m}.} \qquad (1.11)$$

Any phenomenon that is described by Eq. 1.7 is said to be characterized by **exponential growth**. Although many physical systems exhibit such behavior, they can do so only for a limited time. Indefinite exponential growth is clearly impossible, as can be easily seen.

Example 1.2 Assume the world population continues to grow at the rate of 2 percent per year. Start with a current population of about 3×10^9; calculate how long it will take for there to be one person for every square meter of land surface of the earth.

Rather than look up the land area of the earth, let us make a guess. It won't be too wrong. The earth's radius is 6400 km; therefore, its surface area is $5.15 \times 10^{14} \text{ m}^2$. Since

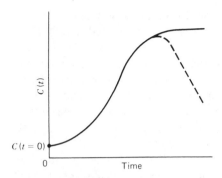

Figure 1.6 A plot of some variable
$C(t)$ as a function of time, illustrating
"Sigmoid" behavior.

about two thirds of this is water, the remainder must be land. Apply Eq. 1.6: $1.7 \times 10^{14} = 3 \times 10^9 e^{0.02T}$. Solve for T. $T = 547$ years!

This may sound like a long time, but it certainly isn't long on the cosmic scale!

An exponentially growing physical system must in time either level off at some asymptotic value or decline after reaching a maximum. Figure 1.6 illustrates this **Sigmoid** behavior. A bacteria colony will follow the dashed curve, as eventually the logistics of providing nutrients breaks down and the waste products from the bacteria begin to poison the colony. Clearly, the same phenomenon could occur with people; perhaps it has already begun.

1.2 CONTRIBUTING FACTORS

We have found that a variety of hostile forces have converged on us. We have forced ourselves into becoming almost totally dependent on commodities that are very limited in supply and that are beginning to show the first signs of a downturn in availability. In this section we examine some of the factors that are contributing to this squeeze and try to determine to what extent these factors can be eliminated or at least lessened by conscious action.

Population

The general trend of the earth's population has been upward, even though at times there have been short-term deviations and geographical differences from this general behavior; see Figure 1.7. It has only been since about 1750 that the rate of growth began to deviate from the almost steady 0.07 percent annually, a figure that has been estimated from the rather sparse population data available. This figure corresponds to a doubling time of about 1000 years; that is, every 1000 years up to 1750, the earth's population increased by a factor of 2. Around 1750 there was a relatively sharp transition to a growth rate of about 0.4 percent per year; and the average annual growth rate for each half-century thereafter has likewise increased, so that the present estimate for the growth rate for the last half of the twentieth century is 1.8 percent per year, corresponding to a doubling time of 39 years.

Demography, the study of populations, has become a complex and computerized science.

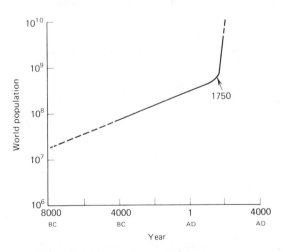

Figure 1.7 Estimated world population between 8000 BC and AD 2000. Note the rapid increase at about AD 1750, the beginning of the Industrial Revolution.

As a result population estimates made now are much more likely to be correct than those made earlier, even as short a time as 20 years ago. Projections made during and for about 10 years after World War II were woefully inadequate; the U.S. population predicted in 1943 for the year 2000 was about 154×10^6. Compare that to the current "correct" figure of about 220×10^6. The sustained postwar baby boom came as a complete surprise to the experts, as did the decline in birthrate in the United States in 1958. Demographers are now quite wary of making firm predictions; they would rather provide several alternative estimates covering a broad range of final populations.

But even conservative forecasting of this type may not be appropriate. The U.S. Census Bureau estimates the population and birthrate each year, based on the previous 10-year census figure and its understanding of demographic trends. In 1964, four birthrate trends were projected, Series A to D in Figure 1.8; but by 1967 the population and birthrate figures were so wrong that a new projection Series E was made. In 1972, Series F was added

because of the discrepancy between projected and actual figures. The estimated actual birthrate reached an all-time low in 1976 at 1.768 births per fertile woman. This is well below the 2.11 value needed to maintain zero population growth (ZPG). These numbers indicate that, *if nothing changes*, the population of the United States will peak somewhere near the year 2040 and then begin to decline.

Of course, *if nothing changes* is the key phrase. No one knows in specific terms what motivates families to have children. In certain countries religion and strong historical traditions prevail. In this country the rather large number of women in the work force, by choice of necessity, may militate against childbirth. The economic structure of our country and, to a much greater extent, other developed countries rewards childbirth: tax breaks for homeowners, large families, and family endowments; subsidies for milk, school aid, and scholarships are examples of childbearing encouragement. Nevertheless, the current generation of fertile women seem to be opting for fewer children, a different lifestyle.

A declining population, while having some obvious benefits in terms of relieving the energy–environment pressures, also has some obvious problems, such as the economic difficulties caused by the changing structure of society. Industries that rely on the growing consumption of a growing public will have to retrench—automobiles, housing, and so on are likely to be affected. The relationship of the United States to the rest of the world will also change greatly. Instead of being the fourth most populous country and the largest consumer of natural resources, we will soon become the sixth or seventh most populous country and conceivably third or fourth in terms of resource consumption. This change will be a difficult one to accept.

Although problems lie ahead, they are by no means insoluble. Many studies of the economic and social effects of ZPG and the

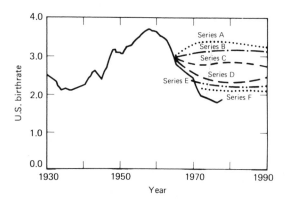

Figure 1.8 U.S. birth rate between 1930 and 1978 in live births per fertile woman. Series A through F represent several projections made by the U.S. Census Bureau. Zero Population Growth (ZPG) is estimated to result from a birthrate of 2.09.

related ZEG (zero economic growth) indicate that a satisfactory transition can be made. Of course, these conclusions are for the United States, but we do not exist in a vacuum. The energy–environment problem is worldwide, and the pressures of population are particularly evident outside the United States.

The world's annual population growth rate, currently 1.8 percent per year, is still increasing, even though the birthrate in many countries is declining. The direct relationship between birthrate and gross national product, which was valid earlier in the twentieth century, seems to have disappeared; in fact, no correlation between the two is now evident. In the developed countries, the United States, Japan, and most countries of Europe, for example, the population growth rate has become smaller and is expected to continue to decrease for the remainder of the century. In the underdeveloped or emerging countries, however, the growth rate is still increasing such that by the year 2000 the United Nations estimates a world population of between 5500 and 7000 million people with a still-increasing birthrate.

Needless to say, simply sustaining this population will require a minor miracle. It is difficult to believe that even the current world average per capita energy consumption level will ever become available to the large majority of the earth's population. This vast majority today does not accept their poor cousin status. What will happen tomorrow? There will always be upward mobility through economic strata so long as differences exist between people's living standards and so long as pathways between strata are available. Thus, either an increasing population or an increasing affluence—in fact, both are occurring—may present serious drains on the earth's resources and threaten to overtax its recuperative powers.

If, on the other hand, for reasons of price, commodity scarcity, strategic necessity, or whatever, pathways between strata wither away, then, if you listen carefully, you will hear the tick of what has been called the *population time bomb.*

Technology

We believe in an age of technology, and whether our high level of technology is a result of the availability of cheap energy or vice versa is quite irrelevant. While many of us may seriously question some technologies, no one would, after sober reflection, advocate a return to the days before X rays, penicillin, the Salk vaccine, adequate home heating, widespread literacy, longer life expectancies, and the myriad of other advantages that technology has provided.

At the same time, it is possible that we have allowed the narrow goals of certain technologies to blind us to the wider problems caused by them. We could examine a variety of technologies to illustrate this point, but we choose two: agriculture and air conditioning. This is not meant to imply that there is anything inherently wrong with either. It is simply that these two technologies represent the old and new and they demonstrate how undesirable side effects develop, even in cases where the primary goals are admirable.

Historically, fewer and fewer Americans have produced more and more food. This has come about because of four basic facts: mechanization, economies of scale, selective genetic breeding, and control of the growing environment. Mechanization of farming was an obvious application of petroleum-product-burning engines at a time when such products were cheap. This conversion from man power to machine power has meant that 1ℓ of gasoline in a 1-hp engine can now do the equivalent of 7 man-days of labor at the cost of about 12 cents per man-day (in 1979 prices). Machines are particularly efficient, and cost-effective when used a large part of each day. This has been one of the reasons large farms have flourished in recent years.

Genetic selection has made possible the development of very high-yield plant strains that respond to fertilizers, which, indeed, require heavy fertilization. In 1945 corn was fertilized with nitrogen, phosphorous, and potassium at about the rate of 7, 7, and 5 lb/acre, respectively. But by 1970, the corresponding respective figures were 112, 31, and 60 lb/acre. Not only that, but these newer varieties require extensive pesticide treatment, herbicides, and irrigation. Many types require a longer growing season, which means the cereal grains have too high a moisture content on harvesting to be stored safely. They must be dried using heat produced by burning a fossil fuel, usually propane.

The fossil fuel input required in 1945 to produce 1 kcal of corn was 0.25 kcal. By 1970 the input required had risen to 0.35 kcal. The fossil fuel input in 1970 was about 1.26 percent of the solar energy input per acre of corn. Only about 40 percent of the solar input is converted to the corn grain itself, so the fossil fuel requirement for ·the production of corn is about 11 percent of the total (assuming a yield of about 100 bushels per acre).

While this method of energy-intensive agriculture may be satisfactory for the United States, it is by no means clear that this same Green Revolution can be successfully transported to other, less-developed countries. In fact, it is becoming increasingly evident that expectations and realizations are growing further apart. The U.S. demand for cereal grains may be leveling off, although it is not obvious from Figure 1.9 that this is the case. In this country most of the cereals are used as feed for slaughter animals. In other countries the cereals are used directly in the diet, a rather more efficient utilization of energy.

In the rest of the world the demand for cereals is increasing. The United States has been the predominant exporter of cereals for

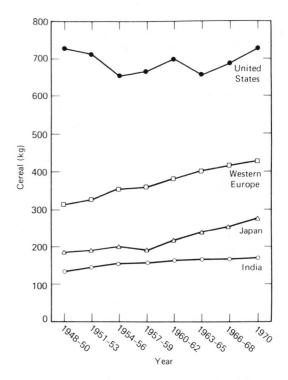

Figure 1.9 The per capita consumption of cereals in various parts of the world. [Reproduced, with permission, from H. Brown, "Energy in Our Future," *Annual Review of Energy* **1**, 1 (1976), copyright © 1976 by Annual Reviews Inc.]

the entire world. The worldwide demand has risen to the extent that the U.S. soil bank program, in which farmers were paid not to farm, has been canceled to provide for larger harvests. Other countries cannot afford the petrodollars required by the Green Revolution. We cannot feed everyone; they cannot feed themselves.

Another example of how a technology can change our energy dependence is the recent phenomenon of widespread air conditioning. Refrigeration has been with us since the development of steam engines and the understanding of the thermodynamics of real gases in the late nineteenth century. But the widespread application of refrigeration did not come about until the middle of the twentieth century.

The development of the vacuum tube in the 1920s brought on the science of electronics. As this technology grew, it became essential to provide a means of dissipating the large quantities of heat produced by these devices. This was especially true when large computers having thousands of tubes were built just after World War II. But refrigeration was a tricky business to have in the home or office, because the coolant in general use at that time was ammonia, a dangerous and toxic substance. In the late 1930s and early 1940s, fluorocarbon compounds were developed that were ideal as refrigerants: they were chemically inert,[1] nontoxic, and thermodynamically suited for the task. Home refrigeration (air conditioning) was then safe, economical, and, to most people at least, highly desirable. We do not mean to imply here that the development of electronics was solely responsible for the expansion of home air conditioning; but it clearly added a shove in that direction.

But summer air conditioning greatly taxes the abilities of some electric utilities to supply energy—witness the great blackouts and brownouts in New York City. So more plants are built to generate more electricity with more by-product emissions (smoke, SO_2, heat, etc.). What is more, many air conditioning units are very inefficient, especially, but not solely, the larger units; see Figure 1.10.

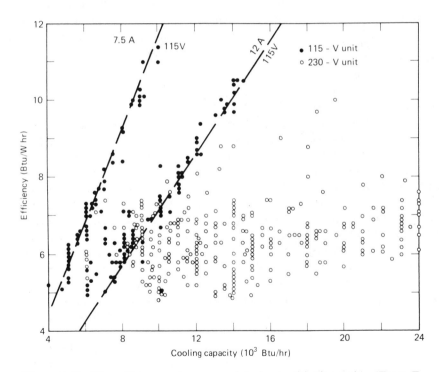

Figure 1.10 The efficiency of commercial air conditioning units. (From E. Hirst and J. Moyers, Oak Ridge National Laboratory Report ORNL-NSF-EP-59. Washington D.C.: National Science Foundation, 1973.)

[1]However, they are not inert in the stratosphere (see Chapter 12).

We should not conclude from this discussion that air conditioning is inherently wrong or immoral. Indeed, most of us, having grown accustomed to living and working in air-conditioned comfort, would find it very difficult to do without it now, particularly those of us who live east of the Rockies in the summer. We use this simply as an illustration of the fact that technology, *by its very nature*, creates more technology.

When we as a society, therefore, contemplate the addition of another technology perhaps we should ask some questions:

What are its life-cycle energy costs?
What are its life-cycle environmental costs?
What are its life-cycle economic costs?
Is it required?
Is there a better way to do it?

In this book we examine the various physical processes that are being used or could be used to extract energy from a primitive fuel and transform it to a useful form. We study the impacts of these transformations on the environment from the points of view given above. We invariably find there are trade-offs and compromises, gray areas and uncertainties. We find that nothing is free, that technical solutions almost always exist, but that they are often economically out of reach. However, we can go no farther without becoming equipped with some fundamentals, in terms of both the data concerning energy resources and in terms of the physics involved. Then, perhaps, the compromises and uncertainties will be understandable, if not resolvable.

SUMMARY

Energy consumption per capita has increased dramatically since the turn of the century.

Reliance on petroleum products has increased to the point that over half of our raw energy fuel today is crude oil.

This has caused multiple problems related both to the environmental disruption caused by the search for energy and by the pollution of the environment as a consequence of fossil fuel burning.

Metric units will be used in this text, although a variety of energy units are in use in the United States. Conversion factors are given.

Energy usage has increased exponentially; characterized by a constant percentage growth and a particular **doubling time**. Exponential growth can be found in many scientific fields.

Several factors have contributed to the growth of energy usage. The earth's population doubles every 39 years. Mechanization of many processes, for example, agriculture, increased energy intensification. Technological advances frequently substitute energy power for machines for human labor power.

REFERENCES

One characteristic that distinguishes human beings from animals is our ability to make a permanent record of our experiences so that others of our species might profit from them. This capability is a two-edged sword: valuable data in environmental physics are stored in a wide variety of journals, books, and magazines. The written record cannot be changed; rather, it is rewritten, reedited, and republished, creating a vast storehouse of literature. The first task facing a novice in this field is thus the development of a list of source material. Then these references must be sorted to distinguish the current ones from those that are out-of-date or otherwise not as useful. A few specific references will be given at the end of each chapter. The reader is urged to keep a constant eye on the literature for updates.

Abelson, P. H. (Ed.), *Energy: Use, Conservation and Supply*. Washington, D.C.:

American Association for the Advancement of Science, 1974.

Annual Review of Energy. Palo Alto: Annual Reviews. 1976.

Berkowitz, D. A., and A. M. Squires (Eds.), *Power Generation and Environmental Change.* Cambridge, Mass.: MIT Press, 1971.

Ehrlich, P. R., and Anne E. Ehrlich, *Population, Resources and Environment.* San Francisco: W. H. Freeman, 1970.

Energy and Environment, Readings from *Scientific American.* San Francisco: W. H. Freeman, 1980.

Energy and Power, Readings from *Scientific American.* San Francisco: W. H. Freeman, 1971.

Fisk, M. D., and W. W. Havens, Jr. (Eds.), *Physics and the Energy Problem—1974*, AIP Conference Proceedings No. 19. New York: American Institute of Physics, 1974.

Henry, Louis, *Population—Analysis and Models*, translated by Etienne van der Walle and Elise F. Jones. New York: Academic Press, 1976.

Holum, John R., *Topics and Terms in Environmental Problems.* New York: John Wiley, 1977.

Keyfitz, Nathan, "United States and World Populations," in *Resources and Man.* San Francisco: W. H. Freeman, 1969.

Romer, R. H., "Resource Letter ERPEE-1 on Energy, Resources, Production˙ and Environmental Effects," *Amer. J. Phys.* **40**, 805 (1972).

Turk, Amos, Jonathan Turk, and Janet T. Wittez, *Ecology, Pollution, Environment.* Philadelphia: W. B. Saunders, 1972.

GROUP PROJECTS

Project 1. Examine past enrollment records at your institution. Is any period characterized by exponential growth? doubling time? Is there a correlation with the growth of the general population? What is the current trend? Can you extrapolate to the year 2000? Discuss the implications.

Project 2. Determine the age distribution of class members and families. Include parents, aunts, uncles, siblings, and cousins. Compare with the general population. Do you expect a difference? Is the birthrate for this micropopulation group different from that of the general population? Discuss the result.

EXERCISES

One must learn
By doing the thing; for though you
  think you know it
You have no certainty until you try.

SOPHOCLES

Exercise 1. Update the figures for 1970 in Figure 1.3 to the present for energy sources. What are the trends?

Exercise 2.

(a) In mid-1975 the birthrates and death rates for several countries were as shown below. Net emigration was negligible in these countries. Determine for each country the natural rate of population growth and the

Country	Annual Birthrate (%)	Annual Death Rate (%)
Belgium	1.5	1.1
China	2.7	1.0
Mexico	4.2	0.9
New Zealand	2.2	0.8
Norway	1.7	1.0
Venezuela	3.6	0.7
West Germany	1.39	1.24
Zambia	5.1	2.0

number of years that would be required for the population to double. Do the countries appear to fall into any obvious categories?

(b) At the same time the United States had an annual birthrate of 1.62 percent, an annual death rate of 0.94 percent, and an annual rate of population growth of 0.90 percent. What was the natural rate of increase of the U.S. population? To what is the difference between the natural and actual population growth rates due? How long will it take the U.S. population to double?

Exercise 3. Comment on these statements:

(a) To maintain a constant human population no family should have more than two children.

(b) Improvements in medical diagnosis and treatment have contributed greatly to the population explosion by prolonging the lives of the elderly.

Exercise 4. To get a "feeling" for energy units express the following examples in joules, MeV, Btu, calories, and kilowatt-hours:

(a) The energy a 500-g ball gains in falling 2 m.

(b) The energy released from the fission of one ^{235}U atom.

(c) The heat generated by a 2-kW toaster turned on for 2 min.

(d) The energy required to air condition a single room for one day.

(e) The heat energy contained in 1 oz of bourbon.

Exercise 5. Express 1 MW yr in terms of each of the following: barrels of petroleum, tonnes of petroleum (1 t = 1000 kg = 7.4 barrels), thousands of cubic feet of natural gas, tonnes of coal (bituminous), and mQ.

Exercise 6. How long does it take a 1000-MW power plant to produce 1520 kWh of energy? The burning of 1000 lb of coal produces 1520 kWh of electrical energy. If a typical railroad car can transport 100 tonnes of coal, how many cars are required to supply the coal for one day's operation?

Answer: 5.47 s; 72 rail cars.

Exercise 7. An oil pipeline is to be built to supply a 1000-MW power plant ($\eta = 0.4$). If the flow is not to exceed 0.2 m/s, what should the diameter of the pipe be?

Exercise 8. In 1900 it took 20,000 Btu of heat energy to produce 1 kWh of electricity. Today it takes 9000 Btu. What are the corresponding efficiencies? (Be careful of units!) What would you expect is responsible for this twofold increase in efficiency?

Exercise 9. If you start with a large sheet of paper 75 μm ($\sim$ 3 mils) thick and fold it over successively 25 times (assuming you could), how thick will the folded stack be?

Exercise 10. If every 1000 MW of electric generating capacity requires about 4.047×10^5 m^2 (100 acres) of space, how much space will be required in 2072? (Assume the present rate of increase will remain constant.)

Exercise 11. At a reduced temperature a 200-mg/ml sample of *E. coli* after 40 min has grown to 225 mg/ml. What is the doubling time for this sample? Why is it different from that in Example 1.1?

Answer: 235.4 min.

Exercise 12. The earth receives radiant energy from the sun at a rate of 1.78×10^{17} W. Assume the present rate of increase of installed electrical generating capacity of 4 percent per year remains constant, and the 1978 generating capacity figure to be 579,000 MW. In what year will the United States generate as much power as the earth receives?

Answer: 2294.

Exercise 13. The decay of radioactive

materials can be represented by an equation of the form of Eq. 1.7,

$$N(t) = N(0)e^{-kt},$$

where $N(t)$ is the number of atoms present at a given time. If a certain nuclear reactor produces about 40 kg of ^{137}Cs (an isotope of cesium) in about a year of operation, how much of it is left after 100 years? (Nuclear decay rates are characterized by "half-life" τ_0, the time required for one half of the material to decay:

$$\frac{1}{2} = \frac{N(\tau_0)}{N(0)} = \exp(-k\tau_0).$$

Given τ_0, k can be determined.)

Exercise 14. About 30 g of ^{85}Kr is produced in one year of operation of a 1000-MW nuclear power plant. How long a time is required for this amount of krypton to decay to 0.3 g?

Exercise 15. Show that the time required to use totally a resource that has an initial amount Q_∞, if the present consumption rate is given by $C = C_0 e^{kt}$, can be written

$$T = \frac{1}{k} \ln \left(\frac{kQ_\infty}{C_0} + 1 \right).$$

At what value of k would the U.S. coal reserves last forever? What assumption does this formulation make about the shape of the dQ/dt versus time curve?

Exercise 16. Assume that the change in fossil fuel input for corn production is a linear function of time. In what year will the fossil fuel input equal the solar energy input per acre of corn? Could this happen? Explain.

CHAPTER TWO

ENERGY RESOURCES REVIEW

There are a number of energy fuels and techniques for converting those fuels to useful work. In this chapter, rather than discussing the conversion processes (that will come later), we review the current and potential energy sources, describe their general characteristics, determine the usable amount of such sources available to us, and attempt to determine the future utility of each.

2.1 THE EARTH'S ENERGY FLOW

Figure 2.1 is a schematic diagram of the energy flow system of the earth; the energy sources shown are the only ones available. Aside from a small amount of thermal energy provided by radioactivity in the earth's crust, stored thermal energy in the core, and gravitational energy provided by earth–moon and earth–sun interactions, the vast majority of the energy arriving at the surface of the earth comes from the sun. Even the fossil fuels in use today had origins in photosynthesis reactions in the primeval swamps. Not all the solar flux of about $1400 \, \text{W/m}^2$ that falls on the earth is utilized; about 30–40 percent is directly reflected. The amount of this reflected flux, called the **albedo**, depends on the surface features of the earth: sand, snow, ocean, clouds, and so on. Large changes in the reflecting properties of the

earth, caused by particulates in the atmosphere, for example, could have a serious effect on the energy budget of the earth. The part of the solar flux not reflected is absorbed, and part of this amount is reradiated. The reradiated flux, however, is at a longer wavelength than the absorbed radiation, as is the waste energy (heat) from electrical generation plants. The problem of supplying an increasing electrical demand would be easy up to some point, if it were possible to convert solar energy directly and efficiently into electrical energy. This conversion can and is being done at reduced efficiency, but to a trivial extent compared with the amount of energy liberated by fossil fuels, geophysical sources, and nuclear fission reactors. Supplies of these fuels may be limited, however, *and the sources of energy in Figure* 2.1 *are the only ones available to mankind, so far as we know.*

2.2 THE FOSSIL FUELS

Since the fossil fuels—coal, petroleum, and natural gas—account for the lion's share of our energy supply now and for the foreseeable future, we discuss them first. The fossil fuels are the result of ages of heat, pressure, and chemical and biological actions on plant and animal material deposited in all the geological

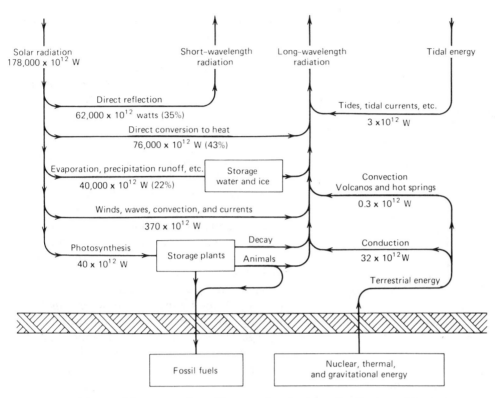

Figure 2.1 The earth's energy flow diagram showing the distribution of energy among the various sources and sinks on the earth. (From M. King Hubbert, "Man's Conquest of Energy: Its Ecological and Human Consequences," in *The Environmental and Ecological Forum 1971–1972.* Washington D.C.: U.S. Atomic Energy Commission Publication #TID-25857, 1972, Figure 2, p. 4.)

strata. These fuels are all carbon based; energy is released primarily by the formation of carbon dioxide.

The diagram in Figure 2.2 represents the flow of energy and matter in the combustion of fossil fuels to produce useful energy. The energy content of the various fossil fuels is listed in Table 2.1. There are several points of interaction with the environment apparent from Figure 2.2. There are some problems and one nonproblem. The latter has to do with oxygen. Some reports have worried about fossil fuel burning using up a significant amount of oxygen or that continued forest destruction would have the same effect. Measurements

over many decades have failed to detect any change in the oxygen level. This is not true for other gases.

The extraction of fossil minerals from the earth causes serious environmental problems. Those associated with the strip mining and subsurface mining of coal are familiar to virtually everyone. There are also problems with oil and natural gas extraction, primarily surface subsidence. The oil or gas trapped in porous rock provides a cushion of support for the overlying material. When the support is withdrawn, the surface sinks, as much as 30 feet in the case of Long Beach, California.

Many by-products result from the com-

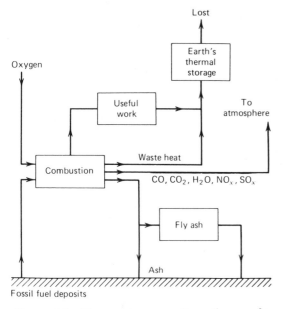

Figure 2.2 Mass and energy flow diagram for fossil fuel combustion.

bustion process. With coal, significant amounts of ash and fly ash are produced. Much of the fly ash can be controlled, but some cannot. All of the gases released are potentially harmful, even the water vapor and carbon dioxide. These gases absorb infrared radiation from the earth and reradiate part of it back, the **green-**

house effect. There may be global climatic changes if CO_2 levels continue to increase.

Combustion produces heat, part of which is transformed to useful work, sometimes in the form of electricity, sometimes in the form of motion as in an automobile. However, we should note that all the useful work produced eventually becomes waste heat as it is degraded through energy transformation processes, which are never 100 percent efficient. In a typical fossil-fueled plant the conversion efficiency is usually under 40 percent; in any case the efficiency is limited by the maximum and minimum steam temperatures. We will calculate later the theoretical maximum efficiency for power plants. Thermal storage indicates the waste heat may not be directly radiated to space but, rather, may be temporarily stored by the earth in water runoff, ice melting, atmospheric heating, and the like. This storage process may imply an appreciable increase in the earth's temperature if energy use continues to expand at the present rate; an increase in temperature could also have a profound effect on worldwide climate.

TABLE 2.1 Energy Content of Fossil Fuels

Fossil Fuel	Energy/(Unit Mass) or Energy/(Unit Volume)
Gasoline	4.6×10^7 J/kg
Crude oil	6×10^9 J/bbl
	4.3×10^7 J/kg
Natural gas	3.7×10^7 J/m^3
Natural gas liquids	4.9×10^9 J/bbl
	3.5×10^7 J/kg
Coal	$(3.0–5.5) \times 10^7$ J/kg

TABLE 2.2 Daily Emissions from a 1000-MW Power Plant[a]

Emission	Quantity
SO_x	1,100,000 kg[b]
N_xO_x	350,000 kg
CO_2	72,500,000 kg
CO	94,000 kg
Particulates	300,000 kg
Radioactivity	2.59×10^2 Bq[c]
Stack	1.35×10^{12} J
Heat	
Condenser	4.05×10^{12} J

[a]Assumes operation at full capacity at 40 percent efficiency.
[b]For a typical bituminous coal.
[c]Chiefly as ^{226,228}Ra. This figure is for coal; for oil it is about 50 times smaller. One bequerel (Bq) = one disintegration per second. See Chapter 14.

Effects just as drastic can result from a large influx of particulate matter into the atmosphere. Table 2.2 indicates the amounts of the various materials indicated on Figure 2.2 for a typical fossil-fuel-burning 1000-MW electrical generating plant. If the increase in power demand is to be met by fossil-fueled plants, it is easy to extrapolate the extent of pollution caused by these by-products. Likewise, the rate of fuel consumption is also easy to estimate; this leads immediately to the question of the sufficiency of reserves of the fossil fuels.

Petroleum Fuels

To estimate the reserves of these fuels it is convenient to divide them into three groups: liquids, gases and solids. The physical and chemical characteristics of these groups are quite different, and production statistics are more easily obtainable on this basis.

The crude oil pumped from wells is a complex mixture of hydrocarbons ranging from the volatile gasolines (not to be confused with the automotive fuel called "gasoline" in this country) to the very viscous tars. Petroleums are generally a mixture of molecules from three basic hydrocarbon families:

■ Paraffins.
■ Cycloparaffins or naphthenes.
■ Aromatics.

There are also usually small amounts of other elements chemically bonded into these molecules: sulfur (0 to 6 percent), oxygen (0 to 4 percent), nitrogen (0 to 1 percent), and some trace amounts of several metals. In addition to the basic hydrocarbon molecular formations, there are many much heavier molecular weight compounds formed by extensions of or cross connections between these basic building blocks. More than 300 different hydrocarbons have been identified in one crude oil sample from Oklahoma.

There is wide variation in crude oils from around the world, reflecting the wide variation in plant material origins of the oil. The paraffin content, sulfur content, viscosity, color, and many other characteristics are widely different. These variations have an effect on the availability of petroleum products.

The crude oil itself is of little utility. But it can be converted into extremely valuable substances by the process of *refining*. This is a general word covering three basic processes: physical separation, alteration, and purification. A schematic diagram of a petroleum–natural gas refining plant using these processes is shown in Figure 2.3. The various flow streams must be adjusted to account for product demand, seasonal variation in product composition, and variations in input feed stock composition. In particular, most refineries are constructed to use a particular feed stock, so that another, say, higher-sulfur-content crude input, would not be usable.

It is because of the wide range of products indicated or alluded to in Figure 2.3 that crude oil has become so highly prized in the past few decades. But even with the extremely wide range of nonenergy products made from petroleum on the market today, from nylon to paint to aspirin to plastics, these uses account for less than 3 percent of the total crude oil production. Most of the remainder is simply burned. It appears that for the near future this situation will continue, at least so long as the price per unit of energy obtained in this manner does not exceed that of energy obtained in some other fashion.

Petroleum liquids are found in sedimentary basins throughout the world. Since these types of geologic formations are well-known and the distribution of such formations is also well-known, it should be possible to estimate the earth's total oil reserves. Estimates of this kind vary between about 1300×10^9 barrels to 2100×10^9 barrels for the total petroleum resources of the earth. The total production of crude oil in the world from 1857, when it

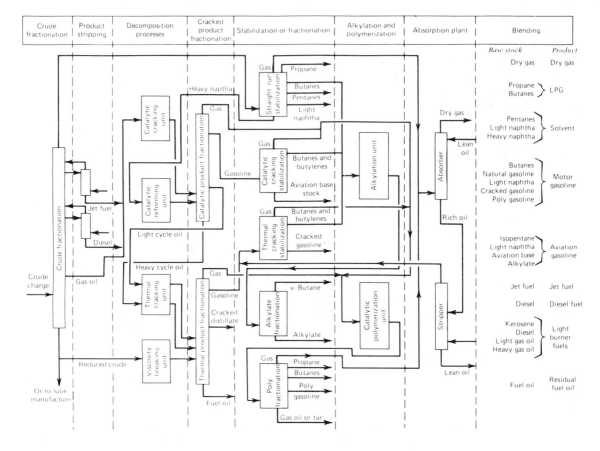

Figure 2.3 Schematic diagram of the refining process to produce light oil from crude oil. (From *McGraw-Hill Encyclopedia of Science and Technology.* New York: McGraw-Hill Book Co., 1977.)

began, to sometime in the future, when commercial exploitation ceases to be profitable, will probably equal some number between these approximate limits. If the crude oil production rate is given by dQ_p/dt and the ultimate recoverable amount by Q_∞, then,

$$Q_\infty = \int_{1857}^{T} \frac{dQ_p}{dt}\, dt. \qquad (2.1)$$

Unfortunately, we do not have an analytic expression for dQ_p/dt, nor do we know what this function will look like in the future. But we can tell what it will not look like!

Example 2.1 Suppose $Q_\infty = 2 \times 10^{12}$ barrels

and the production rate can be expressed as an exponential with a doubling time of 10 years and an initial value at $t = 0$ of 10^6 barrels. Starting with 1857 as $t = 0$, for what T will Eq. 2.1 be satisfied?

$$Q_\infty = \int_{t_0}^{T} f(t)\, dt = \int_{0}^{T} Q_0 e^{mt}\, dt = \frac{Q_0}{m} e^{mt} \Big|_{0}^{T}$$

$$= \frac{Q_0}{m} e^{mT} - \frac{Q_0}{m},$$

$$Q_\infty = \frac{Q_0}{m} (e^{mT} - 1);$$

$$\frac{1}{m} \ln\left(m\, \frac{Q_\infty}{Q_0} + 1\right) = T,$$

and
$$m \approx 0.07 \quad \text{(see Eq. 1.7)},$$

$$\boxed{T = 169 \text{ years!}}$$

An educated guess for this function can be made, based in part on past production trends. The worldwide production records show an almost exponential increase for the years 1900 to 1973; see Figure 2.4. If the area under this

dQ_p/dt versus t curve is to be finite, then clearly this curve must reach a maximum value and return to zero. Although it is not possible to predict the exact rate of falloff once the maximum is reached, it is not unreasonable to assume a symmetric shape for the curve. So if the maximum of the curve is adjusted to fit not only the available production data but also the predicted ultimate value Q_∞, curves as in Figure 2.5 result. Note that these curves predict 80 percent of the total world crude oil

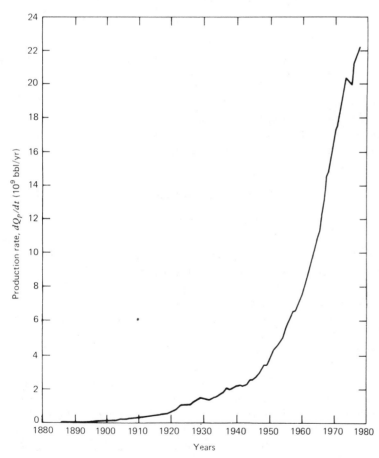

Figure 2.4 Worldwide crude oil production 1880 through 1977. (From M. King Hubbert, "Man's Conquest of Energy: Its Ecological and Human Consequences," in *The Environmental and Ecological Forum 1971–1972*. Washington D.C.: U.S. Atomic Energy Commission Publication #TID-25857, 1972.)

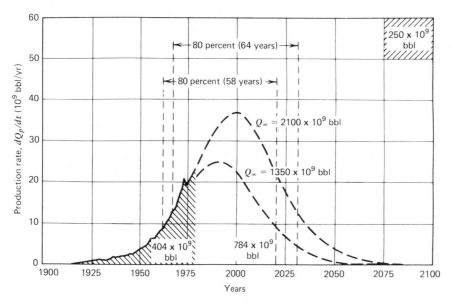

Figure 2.5 Complete cycles of world crude oil production for two values of Q_∞. The shaded area represents the total amount of crude oil produced through 1977. (From M. King Hubbert, "Man's Conquest of Energy: Its Ecological and Human Consequences," in *The Environmental and Ecological Forum 1971–1972*. Washington D.C.: U.S. Atomic Energy Commission Publication #TID-2587, 1972 and *Statistical Abstracts of the United States*. Washington D.C.: Department of Commerce, 1979.)

resource will be produced between 1962 and 2020—58 years for the lower estimate of Q_∞, and between 1968 and 2023—64 years for the higher. If the assumptions made in this analysis are correct, it would seem the world supply of crude oil is indeed small. The United States is also a large producer of petroleum products. How does the dQ_p/dt curve for the United States look?

A similar analysis for the United States yields a Q_∞ of about 200×10^9 barrels; the production curve for this value drawn to fit past production rates is shown in Figure 2.6. Since this situation is even worse than the worldwide production figures, it would seem logical that the United States should import oil as long as foreign suppliers are willing to sell it.

Of course, there are other considerations, on which we have already touched. The effect of oil imports on the economic stability of the country and the strategic effect of reliance on external energy sources are the most important. But crude oil is not the only fossil fuel.

Natural gas—principally methane—is found along with crude oil in the ratio of about 180 m^3 of gas per barrel of crude oil (averaged over all U.S. production). Some experts believe this estimation technique to be faulty, that natural gas and oil deposits are not correlated. In the absence of a better predictor, we shall use this one; hence, any projections of the ultimate production of natural gas must suffer from the same inadequacies as those for crude oil. An additional complication is that much of the natural gas produced in the past

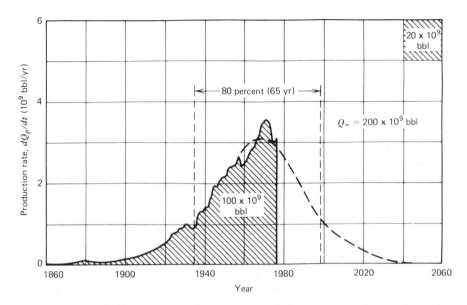

Figure 2.6 Complete cycle of U.S. crude oil production for $Q_\infty = 200 \times 10^9$ bbl. The shaded area represents the total U.S. crude oil production through 1977. (From M. King Hubbert, "Man's Conquest of Energy: Its Ecological and Human Consequences," in *The Environmental and Ecological Forum 1971–1972*. Washington D.C.: U.S. Atomic Energy Commission Publication #TID-2587, 1972 and *Statistical Abstracts of the United States*. Washington D.C.: Department of Commerce, 1979.)

(and presently in the Middle East) was simply burned at the wellhead for lack of an adequate market and transport system. In an attempt at releasing natural gas locked in semipermeable rock, an AEC-authorized underground nuclear explosion was detonated in Colorado in 1970. Gas was indeed released, but it was too radioactive to be utilized immediately. However, the project was determined to be a success and a second blast, this time three 30-kiloton nuclear devices, was detonated (May 1973). Recent trends in energy generation, along with the construction of large-diameter pipelines and large oceangoing vessels capable of maintaining hold temperatures low enough to liquefy the gas, will ensure the utilization of most of the available gas. Using the ratio given above and the world ultimate crude oil values, a range of between 235×10^{12}

and 380×10^{12} m^3 is obtained for the ultimate production Q_∞, whereas for the United States alone the corresponding figure is $Q_\infty = 30 \times 10^{12}$ m^3. The U.S. natural gas production curve peaked in 1972 and fell off each year until 1979, when production increased—a direct consequence of the deregulation of interstate gas prices.

The heavier components of natural gas, ethane, propane, butane, and others, are liquids at standard temperature and pressure (STP: 1 atm, 20°C). They are removed from the natural gas stream at the wellhead so that condensation will not cause difficulties in transmission of the gas; consequently their production has been listed separately from that of natural gas by the industry. When averaged over all U.S. production, these natural gas liquids (NGL) are found in the ratio 0.22 bbl

NGL per barrel of crude oil. Using this ratio and the Q_∞ for oil in the United States, one obtains for NGL $Q_\infty \approx 36.5 \times 10^9$ bbl. The production curve fitted to past production and the above Q_∞ peaks at 1979. The same analysis for world production of NGL yields $300 \times 10^9 \leq Q_\infty \leq 460 \times 10^9$ bbl with the peak year in the production curve falling between 1990 and 2010. Although there are uncertainties in this type of projection, it is apparent that the usual petroleum fuels will soon be in short supply. Before this happens, it may become economically desirable to exploit a more unusual form of petroleum.

Petroleum also exists naturally as a solid: kerogen; the difference between it and crude oil is that in it the long-chain hydrocarbon molecules are cross-linked, forming a more stable molecular configuration. In the Green River area of Utah, Wyoming, and Colorado, vast deposits of kerogen were laid down in an Eocene Epoch lake bed. These deposits must be considered in determining petroleum resources, since oil can be recovered from the shale by direct distillation at the rate of 10 to 30 bbl/tonne of shale. There are no technical difficulties to this process, but its application has not been widespread because of many factors. Principal among these is the high cost of the distilled petroleum compared to crude oil. Distilled petroleum has a higher nitrogen and sulfur content than normal crude and therefore presents more environmental problems when burned to provide heat; but the eventual difficulty in obtaining sufficient petroleum may make the distilled variety more desirable, even with the drawbacks. The U.S. Geological Survey estimates oil shale reserves of about 800×10^9 bbl in this one area of the western United States; estimates for the world are about $2,400 \times 10^9$ bbl. This quantity of oil could ease the pressure that current demand is applying if it could be economically recovered. In fact, if the cost of natural crude oil rises substantially, oil shale resources could dwarf the current estimates for easily obtained pumped oil reserves.

Potentially very large deposits of petroleum exist in the form of tar sands in the Athabaska region of Alberta Province in Canada. This is not strictly a petroleum solid, but rather a very viscous tar that acts as a binder between sand grains forming a hard, solid mass. The origin of these deposits is uncertain. The speculation is that oil from lower permeable formations seeped upward into the sands near the surface; conventional oil deposits are located within 200 miles south. The commercial exploitation of these deposits has been slow in developing; the costs of mining and transportation to a retorting plant make the end product more expensive than natural crude. Deriving the petroleum in place is made difficult by the necessity for providing hot water in large quantities in this rather remote location. One plan—rejected by the Canadian government—involved the use of a nuclear explosion to free the oil! Future supply and demand trends could make tar sands a valuable resource. Estimates of the reserves places them at about 300×10^9 bbl, between 15 and 25 percent of the world petroleum resources.

The figures presented so far for the ultimate production of petroleum products in the United States have excluded Alaska. The recent publicity on the Alaskan North Slope oil strike leads one to believe that vast amounts of petroleum underlie this area. Several estimates have been made; the most ambitious one predicts less than 50×10^9 bbl of crude oil from this reservoir. The other petroleum products scale according to the figures given earlier. Comparing these figures to those for the total U.S. Q_∞, we find the Alaskan oil cannot add significantly to the United States ultimate production.

All these estimates have been made on the basis of current production capabilities. Clearly this must be wrong; in the past 20 years the cost per foot of drilling deep oil wells

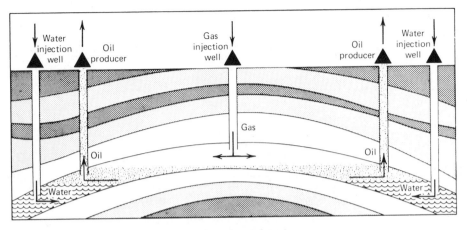

Figure 2.7 Several secondary petroleum recovery systems.

has risen only about as much as the wholesale price index, even though the maximum depth went from 4500 to more than 6000 m. Current oil production methods succeed in recovering only about 35 percent of the oil in a given field. Secondary recovery systems indicated in Figure 2.7, such as pressurized water or CO_2 and other methods, are being used to improve recovery. These techniques will certainly be improved and expanded. Drilling from offshore platforms will certainly increase in the next few years, although we have already experienced the environmental effects of poor safety controls in this type of operation. Cost reductions in the retorting of shale oil and tar sands will have a significant effect on domestic oil production. It may not be too speculative to predict that within the next three decades an improvement of a factor of 1.5 to 2 in oil recovery could be made; a corresponding improvement in petroleum reserves would result.

Most of the sudden concern for energy stems from the fact that the estimates for world crude oil reserves are small compared to our anticipated requirements. We should certainly ask how good these estimates are.

Since 1920 every estimate by petroleum experts of the remaining petroleum recover-

able by "current methods" has had to be revised upward within a few years, in fact about nine times the estimate made in 1920 for the total remaining petroleum has been recovered to date. It is interesting that a plot of the value of Q_x obtained by many estimators as a function of the year the estimate was made has a lot of scatter, but can be fitted without too much difficulty by a straight line with an upward slope. Also, the proven reserves between 1959 and 1976 seem to be increasing, although not with the same slope. Some of the estimates shown in Figure 2.8 were made by the same person at different times. In some instances the increased values are the result of new finds, new recovery technology, and new applications of geologic analogy. In others there is no known reason for the increase. There is an almost universal lack of agreement in the petroleum industry on the reliability of these estimates. This lack of agreement is a consequence of the difficulty in obtaining data. The method of predicting future productivity by geologic similarity to past productivity has been especially controversial; for example, Australia should be well endowed with petroleum because of its underlying sedimentary formations, but very little has

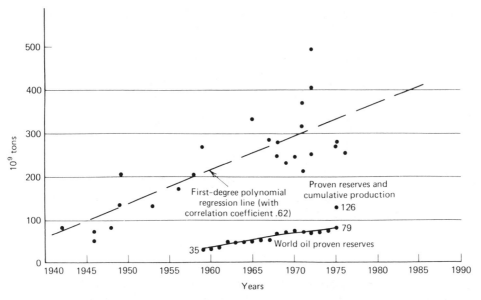

Figure 2.8 Petroleum reserve estimates as a function of time. [Reproduced, with permission, from M. Grenon, "Global Energy Resources," *Annual Review of Energy* **2**, 67 (1977), copyright © by Annual Reviews Inc.]

been found. Some authorities believe the only way to know the actual Q_∞ is to drill one well every square meter to a depth of 20,000 m throughout the earth. However, the hypothesis that continued drilling will result in continued discovery seems to be wrong. The rate of discovery of oil as a function of cumulative footage drilled has dropped from a high of about 280 bbl/ft in 1930 to about 40 bbl/ft in 1965.

Surely, uncertainties exist in these attempts at resource estimates, but if we have underestimated by as much as a factor of 10—and this is very unlikely—the peak in the dQ/dt versus t curve of Figure 2.5 would shift to about 2060. The earth's petroleum resources, whatever they may be, are finite.

Coal

Coal has a fundamentally different origin from oil. The latter is associated with saltwater sedimentation, whereas coal has resulted from the deposition of organic material in freshwater swamps. Coal is found in all geologic epochs from the Lower Paleozoic (350 million years ago) to the more recent Quaternary (1 million years ago). The various kinds of coal (peat, lignite and brown, subbituminous, bituminous, and anthracite coals) available today represent roughly a chronologic sequence from recently growing plant to the hardest, highest-carbon-content coal.

Coal can be classified many ways; for the purpose of this course we will use carbon content and thermal energy content—see Figure 2.9. A high energy content results from a high hydrogen content as well as the quantity of carbon present. Since hydrogen content is correlated—to a point—with carbon content, it is apparent that bacterial action reduces the complex hydrocarbon molecules of life, leaving the combustible hydrogen and carbon. Hence, the longer this action takes place, the higher the energy content is likely to be; in

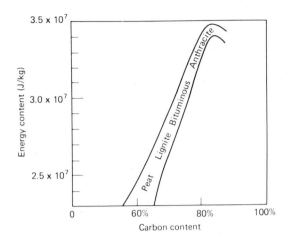

Figure 2.9 Classification of coal by energy and carbon content. Note that most of the U.S. anthracite coal has already been mined.

general, the older the coal, the better the coal (or the higher the *rank*, to use an industry term). These wide variations in energy content for various kinds of coal make assessments of

coal resources very difficult, for it is important to know not simply the quantity of coal available but, rather, the quantity of energy derivable from the coal that is available.

Coal has been mined for more than 1000 years; large-scale commercial exploitation is at least 200 years old. The location of coal-bearing strata is well-known. Because of the great lateral continuity of coal beds, prediction of mineable volumes is much easier than the corresponding problem with oil. However, as is the case with oil, the subsurface situation is not and can never be completely known. As a consequence, estimates of the remaining coal made over the years have had to be revised upward as more information becomes available. Figure 2.10 is an estimate of the world mineable coal resources. Table 2.3 summarizes the U.S. coal resources and reserves situation. This table also normalizes the estimates from tonnage to energy values and excludes that tonnage known to have a high sulfur content. We see that the United States' usable reserves

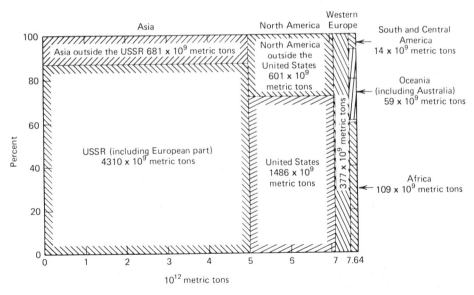

Figure 2.10 Estimate of world coal resources. (From M. King Hubbert, "Energy Resources," in *Resources and Man*. San Francisco: W. H. Freeman, 1969, with permission of the author and the National Academy of Sciences.)

TABLE 2.3 Summary Comparison of Estimated Low-Sulfur Coal Resources and Reserves in Millions of Short Tons

Coal Province	Conventional Estimates (Tonnage Only)		Standardized Estimates (Normalized for Energy/Sulfur)	
	Resources ($\times 10^6$ tons)	Reserves ($\times 10^6$ tons)	Resources ($\times 10^6$ tons)	Reserves ($\times 10^6$ tons)
Appalachian				
Bituminous	37,320	4,105	44,784	4,926
Anthracite	12,550	1,630	14,056	1,826
Interior				
Bituminous	445	70	472	74
Rockies				
Bituminous	45,215	4,585	48,581	4,925
Subbituminous	181,670	19,470	46,329	4,632
Lignite	344,620	37,905	0	0
West Coast				
Bituminous	900	80	855	76
Subbituminous	3,780	340	0	0
TOTAL	626,500	68,185	155,077	16,459

Adapted, with permission, from Richard A. Schmidt and George R. Hill, *Annual Review of Energy* **1**, 37 (1976), copyright © 1976 by Annual Reviews Inc.

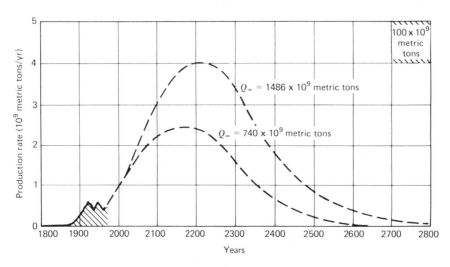

Figure 2.11 Complete cyles of U.S. coal production for two values of Q_∞. (From M. King Hubbert, "Man's Conquest of Energy: Its Ecological and Human Consequences," in *The Environmental and Ecological Forum 1971–1972.* Washington D.C.: U.S. Atomic Energy Commission Publication #TID-25857, 1972.)

are rather smaller than might have been expected.

The U.S. coal situation is summarized in Figure 2.11. The two values of Q_x are for total resources, not reserves, so these are clearly upper limits. However, if the energy/sulfur normalized numbers of Table 2.3 were used, a very different result would be obtained: the peak in the curve would come nearer the year 2000! There is yet another factor that may seriously affect the lifetime of our useful coal reserves.

Synthetic Petroleum Products

For any long-range program of independence from foreign energy sources to be successful, methods of producing crude oil synthetically and economically from coal and western oil shale must be found. We must also find a substitute for natural gas. Several processes have been developed for producing syncrude and SNG, as they are called.

These processes were developed in Germany during World War II as a matter of necessity; they are in use today in South Africa for the same reason. Such plants could be built in the United States except that the product would be too expensive to market. Even though the price of foreign oil continues to rise, it does not yet exceed the cost of synthetically produced petroleum. And since the actual cost of producing oil is far lower than the selling price,[1] and investor is faced with the prospect of being undersold at any time after having spent more than $2 billion on a syncrude or SNG plant.

A time will come when syncrude and SNG may compete economically with the natural product. From that time on, the rate of use of coal will increase dramatically. It is easy to

calculate, based on our current estimate of reserves, a production rate curve for coal, assuming all our oil and gas demands are met by synthesis. Figure 2.12 shows such a curve; it assumes use of coal by synthesis purposes for all petroleum products in the year 2000. The effect is so dramatic because of the very large amount of coal required: 4.8×10^6 tonnes of coal daily would be required to supply our SNG needs, but this is three times our coal production in 1978! The situation for syncrude is more startling: about 10 times our current daily coal production would be required just to satisfy current needs! But there is no way we can sustain that level of production. Indeed, we cannot even double the current tonnage. We do not have the capital, labor, transportation or distribution system capable of absorbing the increased demands. If this higher demand could be satisfied, then even coal, the most ubiquitous fossil fuel, may have a shorter useful lifetime than many of us have anticipated.

No energy source will be used commercially if a cheaper source is available. This is now often the situation with coal; although the cost at the mine head may be competitive, transportation costs rule out coal for many energy applications. Several factors may change this situation. It is possible to transport a coal–water slurry in a large-diameter pipeline directly from the mine to the consumers. The development of extra-high-voltage electrical transmission lines will allow generating stations to be located near coal mines, which in general are not located next to the population centers where the electricity is needed. Pressures for energy independence from foreign suppliers are placing more demands on coal. Production of substitute natural gas and oil from coal is technically feasible, if expensive; future price structures could make this process competitive. Restrictions placed on the use of high-sulfur fuels (generally cheaper than low-sulfur ones) by suddenly environmentally

[1]Petroleum costs are artificially inflated by taxes and royalties. The actual *production cost* of a barrel of Arabian Gulf oil at the dock is less than 35 cents per barrel. Compare this to the current *selling price*!

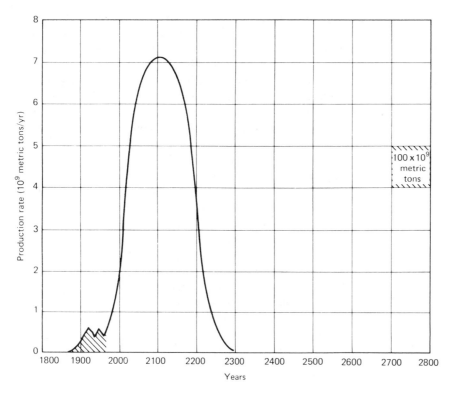

Figure 2.12 Projected coal production rate assuming coal is used starting in 1990 to produce SNG and syncrude. (Adapted from M. King Hubbert, "Man's Conquest of Energy: Its Ecological and Human Consequences," in *The Environmental and Ecological Forum 1971–1972.* Washington D.C.: U.S. Atomic Energy Commission Publication #TID-25857, 1972.)

aware governing bodies[2] are putting fossil fuels at a competitive disadvantage. Nuclear power plants have been built because in many situations the cost of electrical energy per kilowatt-hour may be cheaper with nuclear fuel than with any other.

There are no sure ways to calculate fossil fuel reserves; substantial uncertainties will always cast doubt on the reliability of estimates of reserves. What is more, to quote Putnam:

There is no such thing as an absolute reserve of coal, oil or gas. Reserves are relative. There is more coal, oil and gas in the earth's crust than will ever be used. It is not a question of emptying the bin. It is only a question of deciding how deep it is economical to dig.

The fossil minerals have required millions of years to create; the process continues today. But they are not being replaced at the rate they are being exhausted. The study of fossil fuels in time perspective is most easily illustrated by Figure 2.13. Sometime within the next two or three hundred years mankind will be forced to

[2]Many states require that native coal, often high in sulfur, be used in state institutions.

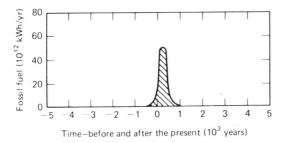

Figure 2.13 The fossil fuels in time perspective. (From M. King Hubbert, "Man's Conquest of Energy: Its Ecological and Human Consequences," in *The Environmental and Ecological Forum 1971–1972*. Washington D.C.: U.S. Atomic Energy Commission Publication #TID-25857, 1972.)

cease using all fossil fuels on a large scale to supply energy; alternative sources will have to have been found.

2.3 GEOPHYSICAL ENERGY

We must look at the earth's energy flow system, Figure 2.1, to find alternative energy sources. Let us first examine geothermal, gravitational, and solar energies; we will classify these as **geophysical energy sources**. The amount of energy available from these three is relatively easy to estimate, compared to fossil fuels. Let us briefly examine the methods by which geophysical energy can be converted to useful work, estimate the ultimate resource for each type, and discuss some of the environmental consequences of their use.

Hydroelectricity

The conversion of the potential energy of water stored in reservoirs to rotational energy to operate mills and other industries has been practiced since Roman times. The conversion to electrical energy had to await the physics and technology of the late nineteenth century. Around the turn of the century, large generat-

ing plants began to be installed; at the present time there are five hydroelectric generating plants in the United States with capacities exceeding 1000 MW.

The physical principles of this conversion process are really simple; the engineering details can be formidable. Briefly, water at high pressure from behind a dam is directed through a channel containing the turbine blades. The motion of the turbine turns a shaft that turns the generator armature, rotating in the magnetic field of the stator. The electrical power generated is the rate at which the stored potential energy of the water is converted to electrical energy.

The power available at a particular hydroelectric plant depends on the amount of drop the water makes between the reservoir surface and the generator; this distance is called the **head**. In a high-head reservoir the water has more potential energy than in a low-head system, so that for the former less water discharge is required to generate a given amount of electrical energy. This means smaller-diameter turbine blades are required and, therefore, a less expensive system can be built. But high heads are not always available; the terrain at the reservoir site obviously dictates a given amount of power. Hoover Dam in Arizona and the Aswan High Dam in Egypt are representative of projects involving high heads; whereas the Columbia River dams, such as Bonneville, and the TVA dams, such as the one at Kentucky Lake, are representative of dams having low-head reservoirs.

The current U.S. installed generating capacity (hydroelectricity) is on stream-flow reservoirs. The growth in installed capacity is shown in Figure 2.14, where P_∞ indicates the estimate by the Federal Power Commission for the ultimate generating capacity of stream flow in the United States. The corresponding estimate for the entire world is 2,857,000 MW; it is doubtful, however, that this figure will ever be reached.

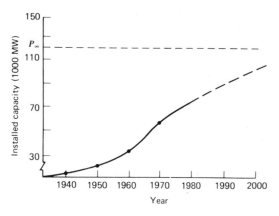

Figure 2.14 U.S. hydroelectric generating capacity. The dashed line is an estimate of the Federal Power Commission. (Adapted from *Hydroelectric Power Resources of the United States, Developed and Undeveloped, 1960.* Washington D.C.: Federal Power Commission, undated.)

Reservoirs created by dams may have adverse effects; they may destroy unique flora and fauna (e.g., Kentucky's Rough River), occupy otherwise useful land (e.g., Glen Canyon Project), reduce stream flow and seasonal flooding (e.g., the Aswan Project[3]), and are certainly aesthetically unpleasant in primitive areas. In addition, all streams carry silt that is deposited behind the reservoir, reducing its water-holding capacity. Hence, the utility of any dam site may be only 50 to 200 years—apparently to be much shorter in the case of the Aswan Dam. Many hydroelectric facilities have been built in arid regions. Creating large water surfaces in such areas must, over the long term, force climatological

changes, perhaps desirable, perhaps not. Finally, the establishment of a large reservoir places a very large weight load on the earth in that small area; stresses and strains are created which, if not relieved, could be a potential earthquake hazard. Countries or areas well endowed with high rainfall and consequent high stream flow are in general those that are least developed or lowest in population density: South America, Africa, and the Pacific Northwest of the United States and Canada. There is no point in developing hydroelectricity to its ultimate capacity in these areas; no one can benefit from the electrical energy. It is possible to transport electricity from remote generating sites to large urban loads, but there are limiting factors in this, too. We shall discuss them in Chapter 9.

The generation of electric energy by the conversion of potential energy of stored water requires only a reservoir for that water. We have considered only stream-flow reservoirs; another means exists to obtain the water for storage: tidal basins. In some areas of the world very large **tidal ranges** are found; that is, the difference between low tide and high tide is very large, as much as 10 m. If gates in a dam are opened between low and high tide to permit filling of the basin, then closed at high tide, the water can be directed through turbines at low tide to provide energy generation. What is even better, turbines can be built with reversible blades so that the water can be made to turn the turbines both in filling and in emptying the tidal basin. Obviously, electricity generated this way is available only during certain periods of the day; this would make the utilization of this energy by a large grid complicated.

The ultimate generating capacity for the world's tidal basins has been estimated to be 13,000 MW, very small compared with the estimated ultimate capacity for stream-flow hydroelectric generation. Of course, there may be serious errors in this estimate; but these are

[3]The problems created by the Aswan Project are staggering. The lack of seasonal flooding has allowed the encroachment of salinity from the Nile mouth and alkalinity from the desert areas into the farm lands served by the river. In addition, the lack of a dry season has contributed to the proliferation of a parasite, normally only a nuisance, but now found in epidemic quantities.

Figure 2.15   The tidal power system across the Rance River in France. (Photograph courtesy of Phototèque EDF.)

not likely to change the conclusion that tidal power cannot contribute significantly to mankind's total energy needs in the future. On the other hand, for isolated local use tidal power has clear advantages.

Tidal power produces no waste by-products and consumes no irreplaceable mineral resources. It does little damage (or possibly none) to the ecological or scenic environmemt. It would seem to be logical to develop tidal power in those regions where tidal, topo-graphical, and energy use factors make it practical.

A 240-MW generating plant is in operation now in the Rance estuary in the province of Brittany in France. The dam containing the generator is shown in Figure 2.15. A small (400-KW) plant is in operation, and several others are planned in the Soviet Union. The Canadian government signed an agreement in 1980 for the construction of a tidal power plant to be built on the Annapolis River in Nova

Scotia. Several studies have been made of Passamaquoddy Bay on the U.S.–Canadian border just off the Bay of Fundy, renowned for its high tides. But as yet the project has not been approved, principally because of the projected high cost.

Hydroelectric power, whether from streamflow or tidal reservoirs, has the potential of supplying virtually all the world's energy demands up to about AD 2000; however, because of the location of population and industrial densities this potential will never be realized. Instead, while the installed hydroelectric generating capacity is likely to rise in the next few decades, the percentage of total energy generated by this means will probably decrease. Hydroelectricity will be utilized primarily for peak-load relief of base generating plants running at near total capacity, except in those areas having unusually favorable sites for hydroelectric exploitation.

Wind Power

Another geophysical energy source that has a great deal in common with hydroelectricity is wind power. The energy contained in the wind has been used for some time to power ships, to turn mills, and to generate electricity. Most wind power generators are very small, having generating capacities of a few kilowatts and useful primarily in isolated locations, such as fire lookouts. One large-scale wind power generator, shown in Figure 2.16, was built during World War II at Grandpa's Knob, Vermont. This 1.25-MW generator operated successfully for several weeks, generating 61,780 kWh, after which one of the blades broke off during operation. Owing to wartime materials pressures and cost overruns, the generator was never repaired.

Since the "energy crisis" of 1973–1974, considerably increased effort has gone into wind power development. Several pilot plants of varying sizes have come into operation. The

Figure 2.16 The Smith–Putnam 1.25-MW generator at Grandpa's Knob, Vermont. (Photograph courtesy of John Wilbur, from MIT Museum and Historical Collection.)

cost of electricity produced by these devices is still quite high compared to fossil-fueled electricity, and some problems with electrical interference have been found. We should, nevertheless, consider wind power as an energy resource and attempt to discover its ultimate potential and undesirable environmental effects.

A wind power generator converts the kinetic energy of the wind into rotational energy of the propeller and generator shaft and thence to electrical energy. It is easy to show that the maximum rate of energy conversion, the power output, is proportional to the area of the intercepting blades and also to the third power of the wind velocity. Wind generators for producing high power, in the megawatt range, are of necessity very large, since wind speeds are, on the average, not very large.

It is this variability in the wind speed that poses one of the greatest problems to the widespread application of wind power generators. One cannot always count on the wind, even on mountain tops. What is more, electricity must be generated when the wind is blowing, not necessarily when it is needed. There is unfortunately no convenient method for storing large amounts of electricity efficiently and economically. Some proposals have been advanced to use electricity generated during off-peak hours to electrolyze water into hydrogen and oxygen, which would be stored. Then, later, these gases could be recombined in a fuel cell to provide electricity. The technology to do this is in its infancy but may become economically possible in a few years, in which case, wind power may be feasible on a large scale in the Midwestern plains area, as has been suggested.

One proposal would be to build 150,000 260-m towers with three-bladed propellers driving ~1.5-MW generators. These towers would be spaced about one per square mile and could provide about 225,000 MW of installed capacity, a substantial portion of the total U.S. capacity. These stations could probably not compete at present with other forms of electricity generation, however, particularly since the problem of off-peak storage has not been solved satisfactorily. In addition the engineering required to construct high-speed, variable-speed, heavy-duty generators, shafts, governors, and so on has not yet been perfected. There might well also be objections to this scheme from an aesthetic point of view. Some people believe oil well derricks to be bad enough, but 260-m towers are something else again!

However, any large-scale application of wind-interrupting machines in one localized area may have a profound effect on the climate of that region. Subtracting a substantial portion of the energy of a typical Midwestern storm ($\sim 4 \times 10^{12}$ J) could reduce the occurrence of severe storms and their effects. This might, of course, be advantageous—no one is advocating tornado production. On the other hand, other aspects of Midwestern climate may depend on the existence of these thunderstorms as they are now. Reducing the strength of thunderstorms could alter rainfall patterns to the extent that some of the eastern portions of the midwest might be lost to agriculture, or that irrigation might become a necessity there as it is in the Far West of the United States. The interactions of the various elements of the atmosphere are very complex and not at all understood. We must be very certain that any large-scale alteration of the natural processes of the earth is done with care and with a complete picture of the environmental ramifications involved.

In summary, whether wind power can be a convenient source for supplementary electricity generation is an open question. At present it is only marginally feasible. Advances in technology, particularly along the lines of off-peak electrical storage, may change this picture; but then one must worry about the more large-scale climatic effects of removing energy from the wind.

Geothermal Power

Geothermal power is being hailed in some quarters as the limitless, pollution-free power of the future. We need to examine the process carefully to see how well-founded these predictions may be. Figure 2.17 is a schematic diagram of the mass and energy flow in an advanced geothermal power plant. The actual generation of the electricity is the same as in a fossil fueled boiler: heat—in this case, from the earth—generates steam, which loses energy by turning the turbine, which in turn

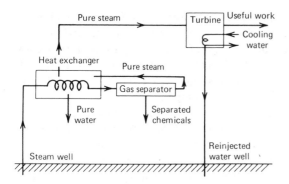

Figure 2.17 Matter and energy flow diagram for geothermal power.

provides the torque for the generator. The efficiency of energy conversion is lower than with fossil fuels because of the lower steam temperature. Additional complications result from the fact that the steam generated in the earth is not pure, but contains dissolved minerals, often in very large amounts. In the diagram a separator is shown that removes most of the unwanted chemicals from the steam and allows recovery of them on a commercial basis. Pure water is also recovered in the heat exchanger; this water may also be a saleable commodity in some locations, for example, California's Imperial Valley. External cooling—possibly via a cooling tower—is used to condense the spent steam to provide water for reinjection into the steam beds to provide for future use. In less sophisticated systems one or more of these extra components may be missing.

If the earth's crust, mantle, and core were uniform, the heat flux everywhere would be 7.3×10^{-6} J/cm^2 s; heat from the earth's core could not then be usable. The earth is not uniform, however, and the many areas of volcanic and hot spring activity are proof that magmatic intrusions exist near the surface of the earth. In some unique locations where magma is found close to water-bearing rocks that also are overlain by an impermeable cap

rock, conditions for useful steam generation exist. Wells drilled in the rock bearing the heated water will recover the steam, often at high temperatures, 100° to 300°C, if the depth and associated pressures are large enough. Sometimes natural cracks or fissures allow for the release of this steam in the form of geysers. This picture of the steam generation process is largely a result of conjecture by geologists, as adequate research into the relationships of the heat source and the reservoir has not been done. It is only in the last few years that serious interest in geothermal power has been shown by the energy corporations; techniques for extensive geothermal exploration and exploitation are only now being developed.

Geothermal power is currently being produced in several regions around the world; see Table 2.4. It is difficult to estimate geothermal resources; any number that we arrive at is likely to be wrong, but probably not wrong enough to make a serious difference in our conclusions. The method that has been used is to examine all the known areas of geothermal activity in the world and estimate the amount of heat stored under them to a depth of 19 km. This amounts to 4×10^{22} J. (The United States has about 10 percent of this

TABLE 2.4 World Geothermal Generating Capacity, 1977

Nation	Capacity (MW)
United States	504
Italy	405
New Zealand	202
Mexico	75
Japan	43
Soviet Union	6
Iceland	3.4
TOTAL	1238.4

From John C. Rowley, "Geothermal Energy Development," *Physics Today*, January 1977, p. 36.

total, mostly in the western states.) Assume that only 1 percent of this heat is recoverable and that the efficiency of conversion to electrical energy is 25 percent. This would yield 10^{20} J of electrical energy; if this were withdrawn over a period of 50 years, the resulting average power would be 60,000 MW. This is about 120 times the present installed geothermal generating capacity, but is of the same order of magnitude as the ultimate power available from tidal generation.

In order to capitalize on this small source of energy we must first overcome several technological and environmental problems. Geothermal power has to be competitive with other power to be exploited effectively. The major cost now is in drilling the well; steam or hot-water wells, although not as deep in general as petroleum wells, are large in diameter, up to 60 cm. The high mineral content of the subsurface water causes plugging of the well within a period of a few years of operation. The well must be unplugged or redrilled in another location, adding to the expense. Most geothermal wells produce hot water instead of steam; this means the electricity is produced at lower efficiency. Extraction of the heated fluid usually takes place at a more rapid rate than it is replaced. Eventually the temperatures go up and the moisture content goes down, signaling the end of the usefulness of the steam bed. To counteract this effect, water can be reinjected under high pressure; reinjection, especially in faulted regions, can be risky, however. Earth movements along fault lines have resulted from subsurface injection. A series of earthquakes in the Denver area a few years ago was correlated with subsurface waste disposal at the U.S. Army's Rocky Mountain Arsenal. In fact, research is under way to study the feasibility of using controlled injections to relieve stresses by triggering minor earth movements to ward off major slippages. A proposal has been made to detonate a nuclear weapon several

thousand meters below the surface in a geothermal region to create a chamber in which hot water could accumulate. This proposal has many environmental risks; the amount of power that can be realized from this procedure must be weighed carefully against the risks.

There are many unanswered questions in connection with the large-scale application of geothermal power. These questions should be examined before large investments have been made in capital equipment. If past history can be used as a guide, once these investments are made, exploitation is likely to occur regardless of the environmental consequences.

Solar Power

The geophysical energy sources discussed so far can supply only a small fraction of the energy demand in the next few decades and may be impractical for large-scale applications. The fossil fuels described earlier are nonrenewable and present environmental hazards in their use. We need an abundant, cheap, renewable, nonpolluting energy source. Fortunately, we have one: the sun. The sun radiates at the rate of about 3.8×10^{26} W spread throughout the electromagnetic spectrum, with the bulk of it in the ultraviolet, visible, and infrared. Of this amount a flux of 1.4 kW/m^2 is incident upon the earth's atmosphere with about 1 kW/m^2 on the earth's surface. Unfortunately, we do not yet know how to convert large quantities of this energy into electricity on an economically competitive way, although several techniques have been tried on a small scale.

The most obvious means of converting solar energy to useful work is to use it to replace the heat of combustion of a fossil fuel in a boiler and generate steam. But, as in the case of fossil fuels, the efficiency of conversion is limited by the temperature range of the working fluid, steam in this case. Since very large collecting systems for concentrating sunlight are difficult to fabricate and maintain, steam temperatures produced in this manner will generally be low. As a consequence, the efficiency of converting sunlight to electricity may be about 10 percent. In order to generate 1000 MW of electricity, 10,000 MW of solar power would be required. A collection area of 10×10^6 m^2 or a square about 3.15 km on a side would be necessary. There is no question of the technical feasibility of constructing such a collecting surface, the circulating fluid system, and other ancillary equipment. There is also little question that it would be prohibitively expensive compared with fossil-fueled or even nuclear-fueled plants. The Barstow solar tower pilot plant is estimated to produce electricity at a cost about 10 times higher than that from fossil-fueled plants.

Two other drawbacks to large solar-electric plants should be considered: energy storage and energy transportation. Some sort of storage system will be required to smooth out the roughly 12-hour day–night cycle; see Figure 2.18. One suggestion is to store heat in some compound as heat of fusion. If a suitable material having a high heat of fusion, low melting point, and ready availability could be found, then excess heat collected during the day could be stored to fire the boilers at night.

Another suggestion is to store the electricity. Storage batteries are a very uneconomic prospect, but pumped hydrostorage could be used. In this system (see Figure 2.19) excess electricity generated during the day would be used to pump water up to an elevated reservoir. Then, at night, the water would fall through a tailrace, turning a standard water turbine. Both of these storage systems have drawbacks: high cost and low efficiency. Pumped hydro is in use at several fossil-fueled plants now, however, so it is practical.

All energy conversion systems described so far can be called *indirect systems*, since energy in one form is transformed to electrical energy through one or more intermediate states in-

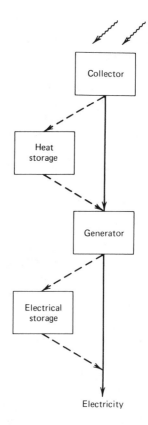

Figure 2.18 Solar-electric system with energy storage capability.

volving motion. Energy losses are inevitable in these devices because of friction. But it is possible to convert solar energy directly into electrical energy without moving parts or heat engines. The theoretical efficiency of such systems could be very high. Photons of electromagnetic energy can deposit their energy in a semiconductor crystal by collisions with atoms. This extra energy can release electrons in the crystal that can then move through the crystal into an external circuit, as if the crystal were a battery. We shall examine this process, called **photovoltaic conversion**, in detail in Chapter 5, along with other techniques for converting energy directly into electricity.

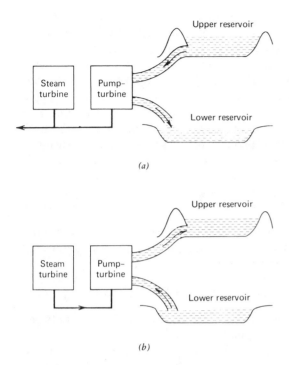

Figure 2.19 Pumped hydro storage: (*a*) daytime operation, (*b*) nighttime operation.

However, sunlight contains a large range of frequencies. Photons having low energies for a given material are not effective in producing photoelectrons. The probability of photon–atom collisions also diminishes as the frequency becomes high. What is more, the material must be thin to reduce noise and edge-capacitance effects; but the probability of freeing an electron goes down as the thickness of the material is reduced. The net result is that the conversion efficiencies obtainable by this effect are rather small—0.05 to 15 percent; in addition, the materials involved, primarily ultrapure silicon and gallium arsenide, are difficult to produce and fabricate and consequently are quite expensive.

Two schemes for major applications of photovoltaic conversion have appeared recently. One would use a system of earth-orbiting satellites in geosynchronous orbits with an array of photoelectric cells. The electrical energy thus generated would be converted to electromagnetic waves in the microwave frequency range and this energy beamed to earth. A receiving antenna array about 3 km square could supply about 3×10^7 kW with a power density of 1 kW/m². Since this is about the same as the energy density in solar radiation, no extensive damage would be done by an out-of-control microwave beam.[4] It must be noted that this microwave energy would eventually be degraded by energy transformations to waste heat and would cause an eventual increase in the earth's temperature. However, the conversion efficiency of presently available devices is much too low and the weight and cost are much too high to make this scheme practical in the immediate future.

In the other system, solar cell collecting panels would be arranged on earth in some sparsely populated and little used desert area. This system would not add to the earth's heat burden, since no solar radiation apart from that already incident on the earth is used. However, prolonged interception of solar radiation in a desert region could have a pronounced effect on the ecology of that area, since the local albedo would change. At present this system is still being studied for feasibility, and if conversion efficiencies as high as 30 percent could be obtained at a cost of less than $60/m², this system would be economical.

Other methods for generating large amounts of electricity with solar energy have been suggested. One of these would use the temperature gradient between the ocean surface and lower levels. We shall look at this in more

---

[4]The only biological effect of microwaves established with certainty so far is heating. A continuous exposure tolerance for humans of 10 MW/cm² has been established; this is about the level of power at the boundaries of the receiving array. There is a general agreement that additional research on the biological effects of narrow-band microwave exposure is necessary.

detail in Chapter 6. All other methods have one major failing, at least at present: they are probably far too expensive to apply on a large scale. Specialized small-scale uses for solar energy as in power for space satellites will continue. While solar energy is clearly abundant and renewable—at least for any foreseeable future—a technological breakthrough will be required to use it as a replacement for fossil or nuclear fuels in generating electricity.

Solar energy can and is being used in the home to supplement commercial heating. When we look again at this application later, we will discuss the advantages and disadvantages of home solar systems.

2.4 NUCLEAR ENERGY

Fossil fuels may be able to supply man's energy requirement for four or five decades, assuming a major changeover is made to coal. Solar energy may be able to become the major supplier of energy after that, perhaps not. A transitional energy source is thus needed almost immediately; this source should be abundant, cheap, renewable, and nonpolluting. Although nuclear energy does not meet all these requirements, it is available and is being developed rapidly. It very likely *will* be our transitional energy source for the simple reason that no other source appears likely to become available. Let us examine briefly both known nuclear processes—fission and fusion—in order to make reasonable estimates of their ultimate resources.

Fission

Nuclear reactors use the excess energy available from the fission of the isotope of uranium having a mass number 235:

$$^{235}_{92}U + n \rightarrow {}^{236}_{92}U \rightarrow {}^{A}_{Z}X + {}^{A'}_{Z'}Y + xn + E. \qquad (2.5)$$

This energy E is the mass difference between the original ^{236}U nucleus and the products resulting from the fission; the average value for ^{235}U is 210 MeV per fission (1 eV = 1.6×10^{-19} J). While reactors may seem to be complex, and indeed they are in terms of their

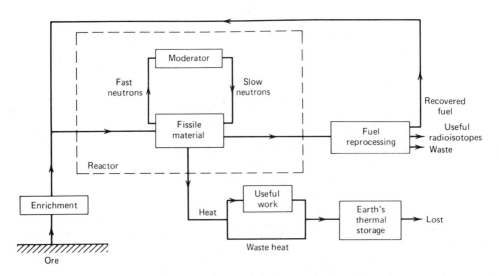

Figure 2.20 Energy and matter flow diagram in light water nuclear reactor (LWR).

engineering details, it is nevertheless true that they are really only boilers. The purpose of the reactor is to provide steam to turn a turbine.

Figure 2.20 is a diagram of matter and energy flow in a typical U.S. light water reactor (LWR). Fast neutrons (having an energy greater than about 1 MeV) from the fission reactions in what is called fissile material—so far only ^{235}U has been used commercially—are slowed in a **moderator**, giving up their kinetic energy to the moderator in the form of heat. This is the purpose of the moderator. The slowed neutrons are used to sustain the chain reaction in the fissile material. The fission products carry away kinetic energy, which is converted to heat in the fuel elements as these recoiling residual nuclei come to rest. The fuel temperatures normally exceed 1000°C. The water moderator of the typical U.S. reactor has a negative temperature coefficient of reactivity; that is, as the temperature of the water increases, its ability to moderate the neutron energies decreases and so the fission process in the fuel diminishes.

Two systems are now in use, as shown in Figure 2.21: the boiling water reactor (BWR) in which steam bubbles are allowed to form in the reactor core (2.21a) and the pressurized water reactor (PWR) in which the water is kept under high pressure so that steam does not form (2.21b). In the BWR the steam generated in the core is used to turn the turbines; in the PWR a heat exchanger is used so that the core coolant is not used in the turbines. Instead, a secondary steam loop is used. High temperatures are produced in the core by the

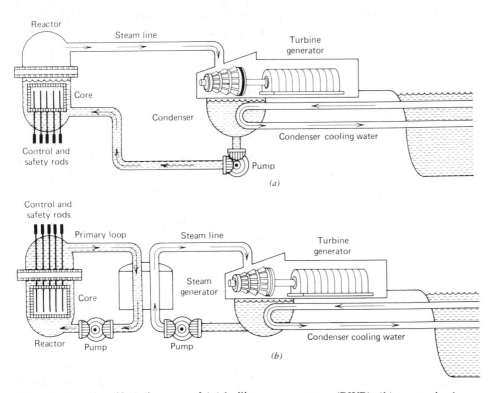

Figure 2.21 Simplified diagrams of (*a*) boiling water reactor (BWR), (*b*) pressurized water reactor (PWR).

fission products that lose their kinetic energy in the fuel elements. The nuclear reactor cannot produce as high a steam temperature as a fossil-fueled boiler because at coolant temperatures above about 300°C the moderation efficiency is too low. As a result the thermal efficiency is only about 30 percent, compared to the 40 percent efficiency achieved with coal, gas, or oil. This means that for the same *electrical* power output the nuclear reactor creates half again as much waste heat as the fossil-fueled generating plant. Both types of plants have potentially harmful by-products; in the case of the nuclear reactor it is the high-level radioactive fission products produced in the fuel, many of which have long half-lives. Table 2.5 lists the output of various items from a 1000-MWe BWR power generating plant. This table should be compared with Table 2.2 for the comparable output of a fossil-fueled plant.

The radioactive waste products contained in the spent reactor fuel constitute an unsolved problem in the development of nuclear power. Current plans require that the spent fuel be transported to reprocessing plants so that the valuable uranium and plutonium can be removed. The remainder must be isolated from the biosphere effectively for many years. These operations—transportation, reprocessing, and storage—present technical problems that have not been satisfactorily solved. No commercial reprocessing plants now exist. Nuclear power generation on the scale many predict for the future may present a risk of environmental contamination.

It is estimated by the Department of Energy that by the year 2000 nuclear power plants will account for approximately 27 percent of total U.S. electrical energy. Assuming the continued growth in energy demand at the present rate, this would amount to about 690,000 MW of installed capacity. A typical 1000-MW LWR would contain a fuel core of about 1.0×10^6 kg of uranium, 3 percent of which is ^{235}U. Of this, 3 kg of pure ^{235}U is consumed each day; and because of the low abundance of ^{235}U, 430 kg of natural uranium must be refined every day to fuel the reactor. This in turn means that 2,150 t of uranium ore must be mined *every day* for *each 1000-MW reactor*.

A 1968 Atomic Energy Commission (AEC) estimate of nuclear fuel requirements through 1980 including an eight-year forward reserve placed the requirements at 5.9×10^8 t of typical ore (5.9×10^5 t of the refined U_3O_8, "black oxide"). This very large projected rate of uranium use makes it imperative that reserves of this fissionable material be examined to see if they can support an industry of such magnitude.

Most inorganic salts of uranium are water-soluble; consequently, uranium is very widespread at low concentration throughout the surface of the earth. In most granites and shales its concentration varies between 10 and 100 parts per million (ppm). Concentrated ores, principally pitchblendes, carnotite, davidite, and conglomerates, occur in many areas. The Zaire—then Belgian Congo—pitchblende mines were the sources of uranium during World War II. Subsequent U.S. exploration

TABLE 2.5   Daily Emissions from a 1000-MWe Nuclear Reactor Power Generating Plant (Operating at Full Capacity at 32 Percent Efficiency)

Radioactivity	
To atmosphere[a]	~1.48×10^{13} Bq[b]
Contained liquids	~3.7×10^8 Bq
Fuel elements[c]	~1.11×10^{14} Bq
Heat	
Up the stack	None
Via a condenser	7.67×10^{12} J

[a]From a BWR without off-gas storage.
[b]The becquerel (Bq) is a measure of activity; it is one disintegration per second. See Chapter 14.
[c]Not emitted; stopped in fuel element.

resulted in discoveries of large deposits in the Colorado Plateau, the Wyoming Basins, and the Gulf Coastal Plains of Texas. The origin of these large deposits is still the subject of considerable discussion among geochemists; they seem to be associated with Paleozoic stream beds. Deposits of this nature in concentrations averaging 1 part per 1000 are currently being mined in the United States, Canada, South America, South Africa, and Europe.

Estimates of uranium reserves suffer from the same inadequacies as those for any underground mineral: lack of adequate knowledge of the subsurface geology. Nevertheless, it is imperative that estimates be made, even though it should be clearly understood that large uncertainties exist.

Uranium resource estimations are even more controversial than those for oil. This is because the usable amount of uranium ore in existence depends on how one uses it—in particular, on how the enrichment is done. We discuss this in detail in Chapter 7. It is clear, though, that there are problems with uranium just as there are with oil. Figure 2.22 shows the rate of discovery of the "cheapest" U_3O_8 ore as a function of cumulative footage drilled. (The cheapest was $17/kg in 1974, but by 1980 it was $66/kg and the price was still rising!) There has been a compelling drop in the discovery rate.

On the basis of the Q_∞ obtained from this figure, the complete cycle for "cheap" U_3O_8 is shown in Figure 2.23. This estimate projects a rather short lifetime for relatively inexpensive U_3O_8.

Breeder Reactors

It is clear that U.S. reserves of inexpensive uranium are perilously close to the anticipated demand for the next five or six years. Although low-concentration uranium ores are abundant, these more expensive ones will not be mined unless the cost of generating electricity by

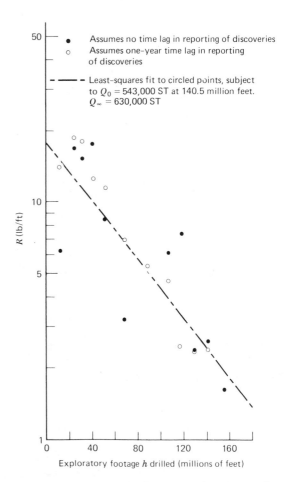

Figure 2.22 Rate of discovery of "cheapest" U_3O_8 versus exploratory footage drilled in short tons (ST). [From M. A. Lieberman, "U.S. Uranium Resources," *Science* **192**, 431 (1976). Copyright 1976 by the American Association for the Advancement of Science.]

other means, principally coal-fired boilers, becomes higher than it is now, thereby keeping nuclear power competitive. Even if the uranium reserves situation were not so critical, another factor must be taken into account; ^{235}U, the only naturally occurring fissionable isotope, is a nonrenewable resource. This isotope is not being created in nature; when all

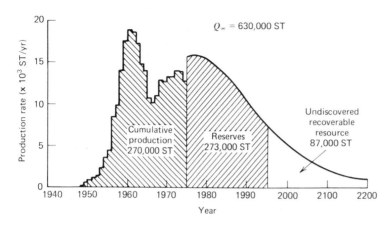

Figure 2.23 Complete cycle for U.S. uranium resources in short tons (ST). [From M. A. Lieberman, "U.S. Uranium Resources," *Science* **192**, 431 (1976). Copyright 1976 by the American Association for the Advancement of Science.]

the economically obtainable ^{235}U has been used in thermal reactors, it will be gone forever. A technique is needed to make use of the far more abundant isotope ^{238}U. This isotope does not undergo thermal neutron-induced fission, but it can be transmuted into an element that does.

The ^{238}U nucleus absorbs "fast" neutrons, that is, those with energies greater than a few keV; the resulting ^{239}U nucleus undergoes beta decay with a half-life of 23.5 min, becoming an element that does not appear in nature *neptunium*. This isotope also decays by emitting a beta particle, becoming *plutonium*; the half-life for this decay is 2.35 days. Symbolically we have

$$^{238}\text{U} + \text{n} \rightarrow \, ^{239}\text{U}$$
$$\rightarrow \, ^{239}\text{Np} + \beta^- + \bar{\nu}$$
$$\rightarrow \, ^{239}\text{Pu} + \beta^- + \bar{\nu}. \qquad (2.2)$$

The process represented by Eq. 2.2 is called **breeding**. This isotope of plutonium is relatively stable, having a half-life of more than 24,000 years; this is short enough, however, that plutonium as well is not found in nature.

Plutonium-239 undergoes thermal neutron fission even more readily than ^{235}U, with greater energy and a larger average number of neutrons released per fission. These properties of ^{239}Pu were discovered very early in the research on nuclear fission; a large effort was expended during World War II to produce it in kilogram quantities using reactors. The first nuclear explosion, the Trinity Bomb exploded on July 16, 1945, near Alamogordo, New Mexico, was generated by a plutonium fission device:

The physical properties of plutonium make it a very interesting and very dangerous material. It is interesting to the solid-state physicist because the pure metal has six distinct crystalline forms, *allotropes*, each having strikingly different physical properties. It is dangerous for several reasons. First, it undergoes fission very readily; a large accumulation of the pure metal will release so many neutrons from spontaneous fission that the chance of a sustained, runaway chain reaction is very large. The "critical mass" at which this is likely to occur is of the order of a few kilograms, depending on the geometry, state of the material, and other factors. Plutonium is also very toxic. Because of its radioactivity, very small amounts of the element ingested into the body can be very damaging. The maximum concentration of plutonium in air allowed by the Department of Energy (DOE) is 0.00003 μg/m^3. In addition, heated plutonium metal reacts very strongly with many gases, bursting into flame in an oxygen atmosphere. This characteristic, the fact that its radioactivity constantly generates heat in the metal, and its brittleness make it difficult to machine and handle. For these reasons and because of the worry about nuclear proliferation the U.S. government has not made a firm commitment to breeder reactors. We shall discuss the ideas behind them anyway, because while governments may change, our energy picture does not.

The operation of a breeder power reactor is shown schematically in Figure 2.24. Fast neutrons are produced by fissions in the fissile material, ^{239}Pu; the fission products produce heat in the fuel rods. The heat is then absorbed by a coolant and used to generate steam. The fast neutrons create more fissile material in a blanket of fertile material, ^{238}U. The plutonium thus created must be separated chemically from the blanket material. Since fewer neutrons from the fissile material are available to sustain the chain reaction, the concentration of fissile material in the fuel must be higher in the breeder than in a typical light water reactor, 30 percent compared to about 3 percent. The coolant in the breeder is not water because neutron moderation is not desired. Instead, liquid sodium is used in current designs. As a consequence there is a small negative temperature coefficient of reactivity. A sudden loss of coolant could result in a core meltdown, increase in fission reactivity, and criticality of mass. On the other hand, since water is not used as a coolant, the reactor can be operated at higher temperatures, thereby increasing the thermal efficiency of power generation to more than 40 percent, comparable to or better than that obtainable with fossil-fueled generators.

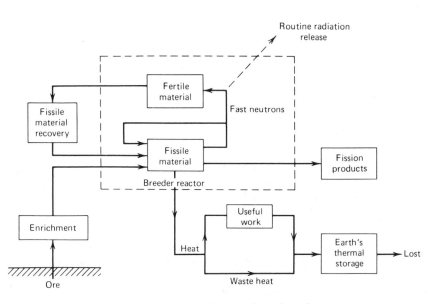

Figure 2.24 Energy and matter flow diagram for a breeder reactor.

Plutonium production goes on to some extent even in thermal light water reactors. But in this kind of reactor the number of fissile plutonium atoms produced in less than the number of ^{235}U atoms used.

Despite obvious drawbacks, the breeder reactor has the potential of multiplying our uranium reserves many times over. This could extend far enough into the future the time when energy becomes so expensive to produce that alternative forms of energy conversion—solar energy, for example—could become economically achievable.

Successful operation of breeder reactors could make practical the utilization of low-grade ores that at present cannot be considered. As an example, consider the Chattanooga Shale that underlies much of Tennessee, Kentucky, Ohio, Indiana, and Illinois. Assume a layer of minable rock having a density of $2.5 \, g/cm^3$ ($2.5 \, t$ per cubic meter) and a uranium content of $150 \, g/m^3$. If for each square meter of surface area there is about $5 \, m^3$ of ore, it would take only about $20 \, km^2$ to provide the rough equivalent of the ultimate U.S. oil resources, a region less than $5 \, km$ square. The corresponding number for the ultimate U.S. coal resources is about $30 \, km^2$. An even greater savings in resources would be realized if the thorium content of the rock were to be used in a breeder reactor to produce fissionable ^{233}U. The limiting factor to the ultimate energy production by breeder reactors would be the depth at which mining of fertile materials becomes impractical or uneconomic. So even though the fast breeder reactor can extend our resources into the future, it cannot be the eventual energy-conversion device for mankind; a system using an even more plentiful raw material is required.

Fusion

The most abundant element in the universe is hydrogen. On the earth, while very little hydrogen exists in the free state in the atmosphere, a very large amount is fixed in compounds on the earth's surface, principally in water. Three isotopes of hydrogen exist: ^{1}H, ^{2}H, and ^{3}H. The nucleus of the first is the proton; that of the second, the deuteron, is stable and is found in nature with an abundance of 0.015 percent. The third nucleus, called the triton, is unstable with a half-life of 12.26 years and can be produced easily by a variety of nuclear reactions. (Molecules formed of pairs of deuterons and tritons with electrons are called "deuterium" and "tritium," respectively.) These isotopes can be made to undergo nuclear reactions in which the products' masses are less than the reactants' masses. The excess mass appears, as in the fission case, as kinetic energy of the reaction products. The reactions of importance are as follows:

$$^2H + {}^2H \rightarrow {}^3He + n + 3.2 \, MeV, \qquad (2.3)$$

$$^2H + {}^2H \rightarrow {}^3H + p + 4.0 \, MeV, \qquad (2.4)$$

and

$$^2H + {}^3H \rightarrow {}^4He + n + 17.6 \, MeV. \qquad (2.5)$$

Nuclear reactions of this nature in which at least one of the product nuclei is more massive than either of the original nuclei are called **fusion reactions**.

Whereas the amount of energy available in a single fusion event is small compared to a fission event, the energy release per kilogram of material is comparable: $2.37 \times 10^{13} \, J$. This amount of energy could be obtained from less than $3 \, m^3$ of water by the fusion reactions described in Eqs. 2.3, 2.4, and 2.5. The fuel equivalent of $1 \, km^3$ of seawater would be $1.36 \times 10^9 \, bbl$ of crude oil, about 1/1000 of our estimate for the world's crude oil resources. The total volume of the oceans has been estimated to be about $1.5 \times 10^9 \, km^3$. The amount of energy potentially available, if fusion could be made to succeed, is truly incredible.

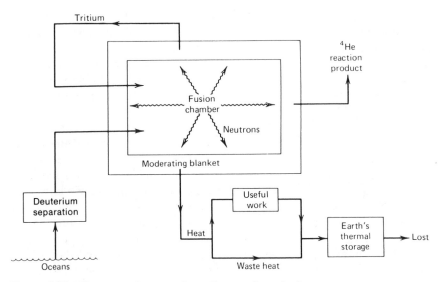

Figure 2.25 Energy and matter flow diagram for a fusion reactor.

Nuclear fusion is not simply a theoretical possibility; we believe it to be the mechanism of energy generation in stars. And, of course, the hydrogen bomb is an example of nuclear fusion at work. However, maintaining fusion over an extended period of time in an environment in which the heat can be withdrawn to produce steam is an entirely different problem. One might liken it to trying to find a box to hold a star! Figure 2.25 is the matter and energy flow diagram for a fusion reactor. High temperatures are required to force the fusing nuclei together; the mechanism for satisfactorily producing and maintaining these temperatures has not yet been devised. Much effort has gone into the development of magnetic confinement systems. Recent research using intense laser beams may also prove fruitful. Assuming that the fusion reaction can eventually be maintained, the problem then is to extract the excess energy, which appears mainly as kinetic energy of neutrons.

A coolant is required that will have a high heat capacity and at the same time will not be adversely affected by the intense neutron environment of the reactor. A good candidate is lithium metal; it has a high boiling point and has excellent heat-transfer characteristics. The two naturally occurring stable isotopes would react with neutrons as follows:

$$^6Li + n \rightarrow {}^3H + {}^4He + 4.8\,MeV, \qquad (2.6)$$

$$^7Li + n \rightarrow {}^3H + {}^4He + n - 2.5\,MeV. \qquad (2.7)$$

The recoil energy of the reaction products, the triton and helium, would be absorbed by collisions in the lithium metal producing heat, the usable end product of the reactor. Tritium (^3H) is quite dangerous in the environment because of the possibility of its being inhaled; but the production of tritium in the coolant is advantageous.

The deuterium–tritium fusion reaction of Eq. 2.5 produces the largest amount of excess energy per kilogram of reactants. But tritium is not normally found in nature, so it would be desirable to be able to produce the required tritium in the reactor. In this sense the fusion reactor would be a "breeder" reactor, and this aspect of the fusion reactor is the most dangerous to the environment. Estimates are that for a 5000-MW reactor there would be about

7×10^{18} Bq of tritium in the reactor at any given moment. This is comparable to the amount of the most dangerous radioactivity ^{131}I in a similarly sized fission reactor; but the biological effects of tritium compared to those of ^{131}I are substantially different. Estimates are that although the tritium problem would have to be considered carefully, it does not present the technical difficulties that maintaining the high-temperature plasma does.

Since the onset of the "energy crisis" fusion research has been pursued with vigor. Many fundamental and engineering difficulties remain, however, before fusion can contribute to our energy supply. Most estimates place commercial fusion well beyond the turn of the next century; it must be at least that far away.

Nuclear energy accounts for a relatively small proportion of the total energy being produced in the world today; but in the next 50 years it may produce more than 30 percent. This is in spite of the fact that many of the environmental problems associated with its use are not solved. The materials technology required to produce a safe and economical breeder reactor does not exist; and the government mandate to develop it has been withdrawn.[5] The much more difficult technical problem of contained fusion is being studied. Fusion has the potential of producing power efficiently, cleanly, and in a manner less injurious to the environment. There are many apparent contradictions in the nuclear energy story; none of these involves fundamentally technical issues. These problems arise because economic and political factors may be as important as any technical aspect in influencing decisions involving energy. But because the energy problem is of such significance, the development of alternative sources utilizing means that would have been called exotic a few years ago is being pursued.

Few industrial or governmental leaders are pointing to the other option society has: to reduce at least our rate of demand growth by a different level of consumption or a more efficient application of the energy currently in use, or both. Some of the means of doing this will be examined in Chapter 11. We must assume, for the time being, that the somewhat irrational pattern of consumption will continue.

The industrialized world exists because of the energy it can generate and consume. In the next four or five decades fossil fuels will become much more expensive and the cost of increased or even continued industrialization will be much higher than at present. The GNP gap between the industrialized and the developing nations continues to widen each year. A more expensive energy base will mean even more serious inequities may develop between the "haves" and the "have-nots." The recent awareness of the environmental repercussions of energy generation, highly mechanized agriculture, and automobile transportation will also add to the restrictions placed on these activities in the future. Different fuel sources are even now being developed. Nuclear power may be able to supply energy during this period at a reasonable cost; however, many serious environmental problems need to be solved before the construction of extremely large plants becomes too widespread. Power directly from the sun may become competitive in the next few decades, in which case the question of earth's resources for energy generation will become academic. A more serious question of the earth's heat balance then would have to be resolved. In any case, an understanding of the physics of energy conversion, transportation, storage, and utilization is an important foundation on which to begin discussions of our energy-induced environmental problems.

[5]Note that the French, British, and Soviet governments are proceeding with breeder development. However, their designs will not have to compete in the free market with other techniques. With the change in U.S. governments in 1980, there may also come a change in the U.S. approach to breeder development.

SUMMARY

There are only a few sources of energy available to mankind: direct solar; indirect solar, as in hydroelectricity, wind, tides, and the fossil fuels; and nuclear and geothermal.

The fossil fuels are used more than all the others, but they have many environmental problems. There are three types: **petroleum**, **natural gas**, and **coal**. All except coal are in rather short supply; Q_∞ for petroleum for the United States is estimated to be about 200×10^9 bbl. We are using the fossil fuels at an exponentially increasing rate. Petroleum and natural gas may not last beyond the early part of the next century. Coal may also be used up quickly if the production of synthetic oil and methane from coal becomes widespread.

Hydroelectricity can never account for more than a few percent of our energy demand. The sources and loads are too far apart. There are also environmental problems associated with hydroelectricity.

Tidal power also cannot contribute significantly except in a few locations where conditions are favorable.

Wind power has potential, but wind is diffuse and irregular. Large-scale applications are unlikely. Recent pilot projects indicate a rather high cost.

Geothermal power can be used in a few favorable locations where high-temperature steam is available. Widespread use seems unlikely.

Solar power can be used to heat homes, but large-scale generation of electricity seems far away because of costs. **Solar cells** are very expensive; only the space program can afford them. Also, power can be obtained only when the sun shines; no effective power storage system has been devised.

Nuclear power using fission accounts for in excess of 10 percent of our total electricity today. Two types of power reactors are used in the United States: pressurized water reactors (PWR) and boiling water reactors (BWR). Both use water as the **coolant** and the **moderator**. The thermal efficiency of nuclear reactors is lower than that of fossil-fueled plants. Radioactive by-products are problems that have not yet been solved. There is probably not enough low-cost **uranium ore** to fuel a significantly larger number of reactors.

Breeding fissile material out of nonfissile ^{238}U is a way to stretch our uranium resources. **Breeder reactors** have been built, but not commercially (at least, not successfully). Breeders would produce plutonium, a toxic and highly fissile material that could be diverted to nuclear weapons use. The U.S. government has not yet decided what to do about a U.S. breeder program.

Fusion would use a very abundant fuel for an almost unlimited lifetime. However, fusion has not yet been achieved, even under very carefully controlled laboratory conditions. The transition from laboratory success to operating commercial success will require many years.

REFERENCES

Barnea, Joseph, "Geothermal Power." *Sci. Amer.*, January (1972), p. 70.

Blair, I. M., B. D. Jones, and A. J. Van Horn (Eds.), *Aspects of Energy Conversion.* Oxford: Pergamon Press, 1976.

Culler, Floyd L., Jr., and William O. Harms, "Energy from Breeder Reactors," *Phys. Today*, May (1972), p. 28.

Daniels, Farrington, *Direct Use of the Sun's Energy.* New Haven: Yale University Press, 1964.

Enge, Harold A., *Introduction to Nuclear Physics.* Reading, Mass.: Addison-Wesley, 1966.

Fenner, David, and Joseph Klarmann, "Power from the Earth," *Environment* **13**(10), 19 (1971).

Fowler, John M., *Energy and the Environment.* New York: McGraw-Hill, 1975.

Ginsberg, D. M., "Resource Letter Scy-1 on Superconductivity," *Amer. J. Phys.* **32**, 1 (1964).

Ginsberg, D. M., "Resource Letter Scy-2 on Superconductivity," *Amer. J. Phys.* **38**, 949 (1970).

Gough, W. C., and J. Eastland, "The Prospects of Fusion Power," *Sci. Amer.*, February (1971), p. 50.

Hubbert, M. King, "Energy Resources," in *Resources and Man.* San Francisco: W. H. Freeman, 1969.

Landsberg, Hans H., Leonard L. Fischman, and Joseph L. Fisher, *Resources in America's Future.* Baltimore: Johns Hopkins Press, 1963.

McKelvey, Vincent E., "Contradictions in Energy Resource Estimates," in *Energy: Proceedings of the Seventh Biennial Gas Dynamics Symposium*, Lawrence B. Holmes (Ed.). Evanston: Northwestern University Press, 1968.

Meinel, Aden Baker, and Marjorie Pettit Meinel, "Physics Looks at Solar Energy," *Phys. Today*, February (1972), p. 44.

Michael, P., and R. I. Schermer, "Resource Letter Rea-1 on Reactors," *Amer. J. Phys.* **36**, 659 (1968).

Putnam, Palmer Cosseltt, *Energy in the Future.* New York: D. Van Nostrand, 1953.

Rau, Hans, *Solar Energy.* New York: Macmillan, 1964.

Rose, D. J., "Controlled Nuclear Fusion: Status and Outlook," *Science* **172**, 797 (1971).

Schurr, Sam H., *Energy Research Needs.* Washington, D.C.: U.S. Government Printing Office, Stock Number 0300-00360, 1971.

Snowden, Donald P., "Superconductors for Power Transmission," *Sci. Amer.*, April (1972), p. 84.

Stoker, H. Stephen, Spencer L. Seager, and Robert L. Capener, *Energy from Source to Use.* Glenview, Ill.: Scott, Foresman, 1975.

_____, *Energy in Transition: Final Report of the Committee on Nuclear and Alternative Energy Systems*, National Academy of Sciences. San Francisco: W. H. Freeman, 1980.

Journals

Energy Sources: An International Interdisciplinary Journal of Science and Technology. New York: Crane, Russak.

International Journal of Energy Research. New York: John Wiley.

Energy—The International Journal. Oxford: Pergamon Press.

GROUP PROJECTS

Project 1. Determine the heat supply for your institution. Is it a central plant or by individual building heat? What are the sources of fuel? What is the amount of fuel on hand (quantity as well as heating days)? What has been the change in price in each of the past 10 years? Plans for the future?

Project 2. Examine one of the sources of raw material nearest you—for example, coal, iron ore, timber, or stone. Estimated total reserves in this location? Fraction already used? Change in value in the past 10 years? Primary users of this material? Availability of lower grades?

Project 3. Determine the average amount of solar energy available in your locality. Take into account meteorological records, average available sun, rainy days, and so on.

EXERCISES

Exercise 1. If crude oil sells for \$40/bbl, at what price would natural gas have to sell so that its energy content would be priced the same as that in oil? Give your answer in dollars per thousand cubic feet. Compare with the current selling price in your area. Do the same for electricity; use appropriate units.

Exercise 2. The production rate versus time curve of Figure 2.6 can be approximated by a **Gaussian curve**:

$$\frac{dQ}{dt} = \frac{dQ}{dt}\bigg)_0 \exp[-a^2(t - t_0)^2],$$

where a is a measure of the width, t_0 is the time about which the curve is symmetric, and $[dQ/dt]_0$ is the value of the curve at t_0. The area Q_∞ under this curve can be found by integration:

$$Q_\infty = \int_{-\infty}^{+\infty} \frac{dQ}{dt} \, dt,$$

where a value of $a = 0.03$ approximates the curve in the figure. Find Q_∞ and show that the time span for which the curve is greater than $0.2[dQ/dt]_0$ is 85 yr.

Answer: 1.77×10^{11} bbl, 84.58 yr.

Exercise 3. Assume Q_∞ for U.S. oil production to be 1.65×10^{12} bbl, a factor of 10 higher than our estimate. Assume also that the maximum production rate $[dQ/dt]_0$ is 4×10^9 bbl/yr and a Gaussian curve of the type in Exercise 2, except that a is no longer 0.03. Find the year at which the curve reaches a maximum.

Exercise 4. The oil production rate curve can also be approximated as the difference between two logistics curves of the form

$$y = \frac{k}{1 + e^{a-bt}}.$$

The first curve would correspond to cumulative discovery with $k = 4.19 \times 10^9$ bbl/yr and the constants $a = 101.6$ and $b = 0.0523$. The second curve would correspond to cumulative production with $k = 4.19 \times 10^9$ bbl/yr, $a = 122.4$, and $b = 0.061$. Plot these two curves between 1900 and 2060. Plot the difference and estimate Q_∞ and the peak year.

Exercise 5. It is often difficult to estimate the lifetime of a finite resource as it depends critically on both the rate of consumption and the known reserves. For the case of petroleum consider the following way of estimating an absolute upper limit.

(a) Take the volume of petroleum in the earth to be equal to the volume of the earth.
(b) Use 7 percent per year for the growth rate of the world petroleum production. (This is the average rate for the period 1890 to 1970.)
(c) In 1970 the rate of production was 1.67×10^{10} bbl/yr.

How long will the oil last? Are you able to improve on your estimate by modifying any of the assumptions?

Exercise 6. A 1000-MWe generating plant of 40 percent thermal efficiency uses an oil-fired boiler. The oil is obtained by distillation of oil shale at the rate of 110 ℓ/tonne of shale. If the density of the shale after removal of the oil is 10^3 kg/m^3 and if these "tailings" are stacked in a rectangular object 50 m wide and 25 m high, how long would this object be after 100 days of operation?

Answer: 180 m.

Exercise 7. Coal having a carbon content of 80 percent and an energy content of 7.4×10^6 J/kg is burned in a 40 percent efficient generating plant producing 1000 MW of electric power. If the combustion process is 85 percent efficient (i.e., if 85 percent of the carbon is converted to CO_2), how much CO_2 is produced each day?

Exercise 8. Suppose that a windmill can convert the kinetic energy of the moving air into useful energy with perfect efficiency. Show that the output power of the windmill is given by

$$P = \frac{\rho_{air} A v^3}{2},$$

where A is the area covered by the blades and v is the wind speed. If $v = 16$ km/hr, what is the maximum possible putput power of a 3-m-diameter windmill?

Exercise 9. If the wind speed as a function of time in a particular location is given by

$$v = 20 \sin^2 \left(\pi \frac{t}{24} \right) \text{km/s},$$

where t is the time in hours with midnight being 0.0 hr, what is the maximum average electrical power obtainable from a wind generator at this location? Average over 24 hr.

Exercise 10. The average rate at which solar energy is incident on the southwestern United States is about 250 W/m². If solar cells have a direct conversion efficiency of 13 percent, how many square meters of collector would be required to construct a 1000-MW generating plant? What total cost is required if the solar cell material costs $10,000/m²? Compare with a coal generating plant of similar size that would cost about 1.5×10^9.

Answer: 3.08×10^7 m².

Exercise 11. In 1973 the total energy production in the United States amounted to 72.8 mQ. Of this total, 6.64 mQ was produced electrically.

(a) Suppose that 2 percent of the land area of Arizona were allocated as a Solar Power Preserve. Assuming an average of 250 W/m² and 13 percent efficiency, what fraction of the above quantities might be produced?

(b) According to one plan, 150,000 windmills each 260 m tall might be distributed throughout the Great Plains. If each unit were to produce on the average 1.5 MW, what fraction of the above quantities might be produced?

Exercise 12. The average rate at which the earth absorbs solar energy is one fourth the solar flux of 1.4 kW/m². Part of this energy is reflected, the albedo, and part is reradiated; the process can be described by

$$\Omega = 0.35\Omega + sT^4,$$

where Ω is the average absorption rate, 0.35Ω is the albedo, and the sT^4 term describes the radiation from an object at a temperature T in kelvins. For the earth, s is about 6×10^{-8} J/m² s K⁴. Find the equilibrium temperature of the earth, and compare this value to its average value of 20°C. Can you account for the difference?

Exercise 13. In an ocean thermal generating plant, heat is extracted from warm water at a rate of 3000 MW by cooling the water 2°C. Assume that the heat exchanger is 100 percent efficient. What flow of water must be established? Compare this water flow with the rate of water flow through a 100-MW hydroelectric plant with a water head of 30 m. (As we shall see in Chapter 3, about 3000 MW must be removed from the warm water to obtain 100 MW of useful work from the ocean thermal plan.)

Exercise 14. Suppose that we are to design a pumped-storage hydroelectric plant to provide 20 percent of the electrical energy demand of a region of western Massachusetts comprising a population of about 1.4×10^6 persons. If they use electrical energy at an *average* rate of 1 kW (as do most Americans), what volume of water is needed each day for a storage reservoir with a 240-m height difference (assume 100 percent conversion efficiency)? If the reservoir has an area of 130 ha, what is the required average depth?

Exercise 15. A 30-m head hydroelectric plant generates 500 MW. What must be the surface area of the reservoir so that the level does not drop more than 5 cm over the course of a day? Assume no water inflow and 90 percent efficiency of conversion.

Answer: $2.94 \times 10^9 \, m^2$.

Exercise 16. A certain tidal generating plant has a rectangular reservoir of $100 \, km^2$ area and a tidal range of 8 m. The water drains in over a 12-hr period. The efficiency of converting potential energy to electrical energy is 90 percent. The electrical voltage thus generated is stepped up by a transformer from 100 V to 500,000 V with an efficiency of 95 percent. Transmission lines carry current to a city a distance of 30 km and have a resistance of 0.0003 Ω/m. Another 95 percent efficient step-down transformer delivers electrical energy to the load at 100 V. What power is delivered to the load by this generator? How much energy is lost at each step? How does it appear? (Assume load + losses = stored energy.)

Exercise 17. If the heat flow from the earth were uniformly distributed, the heat flux would be $7.3 \times 10^{-6} \, J/cm^2 \, s$. Owing to nonuniformities, however, there are areas where this heat flow is strongly localized. At one such location, the Geysers in California, there are plants extracting this thermal energy and converting it to an electrical output of about 400 MW with an efficiency of 25 percent. Assuming that the source covers $30 \, km^2$ and that only 10 percent of the total has been tapped, how concentrated is the heat flow in this area?

Exercise 18. A geothermal well brings up water at 250°C to a heat exchanger. The water is rejected at 90°C. The generator produces 200 MW of electricity with an efficiency of 30 percent. If the geothermal water contains 20 g of dissolved minerals per liter that are extracted, what quantity of minerals is produced per day?

Exercise 19. Assume 80 percent of the energy contained in a thunderstorm comes from the latent heat of condensation of water vapor. How much water is contained in a typical thunderstorm? In grams? In gallons? In acre-feet? How does this compare with other estimates?

Exercise 20. How many joules per kilogram are available from the complete fission of ^{235}U? How does this compare with the thermal energy available from a pound of coal? (Give the ratio of the two values!)

Answer: $8.61 \times 10^{10} \, J$.

Exercise 21. A particular fossil-fueled plant produces 1000 MW of electrical power and is 42 percent efficient. A certain nuclear plant produces the same electrical power, but is 31 percent efficient. What is the ratio of the waste heat produced by the latter to that produced by the former?

Exercise 22.
(a) Assume that present reactors use only the ^{235}U isotope (0.7 percent of naturally occurring uranium) and that each fission yields about 150 MeV. How long will the estimated U.S. uranium reserves ($\approx 10^6$ tons) last if we continue to use electrical energy at the 1973 rate (6.64 mQ) and generate it all with nuclear power plants?
(b) The ^{238}U—^{239}Pu breeder would allow the same fission energy to be obtained from each uranium atom. How long would the same reserves last at the above rate? How long would they last if the electrical energy usage grew at a rate of 5 percent per year from 1973 on?

PART TWO

ENERGY CONVERSION PROCESSES

CHAPTER THREE

ELEMENTS OF THERMO-DYNAMICS

The interactions of real objects, either on the microscopic or macroscopic scale, are usually too complex to calculate exactly. This is one of the reasons the energy concept has proven to be so valuable. Many of the processes of interest in the study of energy–environment problems can be expressed in terms of a series of energy transfers from one system to another. In Figure 3.1 the fundamental processes (indicated by arrows) and energy states (indicated by circles) in the conversion of either fossil fuels (starting from the upper left) or fissile fuels (from the lower left) into electricity are shown.

These energy conversions are fundamental; we shall see all of them in the chapters that follow. The devices we consider employ gases in the energy-to-work conversion—either "real" gases like steam or "effective" gases such as the electrons in solids. We shall have to understand, in the physics sense, how large numbers of these particles interact with their surroundings during energy transfers.

The branch of physics called **thermodynamics** is used for this purpose. This discipline was developed in the nineteenth century, before the microscopic understanding of matter was achieved. Later, the connection between macroscopic and microscopic understanding came with the development of statistical mechanics and the quantum theory. We

do not need these more advanced topics here. Basically, we would like to find a way to make quantitative comparisons between alternative conversion processes. One simple standard for comparison is the efficiency. By this we mean

$$\text{efficiency} = \frac{\text{work out}}{\text{energy in}}. \quad (3.1)$$

Recall that in the previous chapter fossil-fueled power plants were said to be about 40 percent efficient and that nuclear plants were about 30 percent efficient. Is this because our technology is so poor? Not really; there are some real limitations imposed by nature, not technology, that fix energy conversion efficiencies under certain circumstances.

Our goals for this chapter are to discover what those limitations are and to build a firm foundation on which to base a more general study of energy conversion processes and devices. As a consequence, we introduce a number of new terms and ideas, but always try to use them in the context of energy–environment problems.

3.1 THE FIRST LAW OF THERMODYNAMICS

We shall spend some time discussing power plants, so let us devise a simple model of a power plant on which to begin our study of

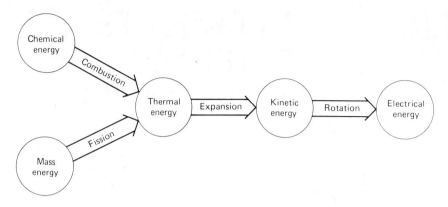

Figure 3.1 Energy conversion steps in electricity generating plants.

thermodynamics. Figure 3.2 illustrates such a model. Everything inside the dashed line we shall call the system; this includes the boilers, piping, turbines, generator, condenser, and pumps. At this stage we are not interested in the details of these various pieces of the system. Inside the system there are three main processes indicated: vaporization, expansion, and condensation. The heavy lines with arrows connecting these processes indicate the flow of what we shall call the **working fluid** between the various stages. In a fossil-fueled or nuclear power plant this working fluid is almost always water, although other fluids can be and have been used.

The operation of our model plant is simple: Energy input to the system is used to vaporize the working fluid. At the point B the fluid is characterized by a high temperature, pressure, and volume and is a vapor. The fluid (note "fluid" does not necessarily mean a liquid) expands, causing the turbine to turn, generating electricity that can be used to do work. At the point C the fluid is still a vapor, but it has a lower temperature, a very low pressure, and a large volume. The condenser is used to con-

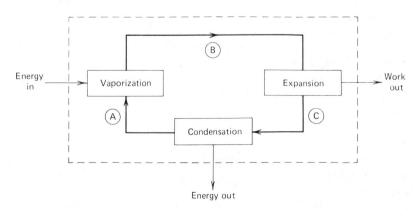

Figure 3.2 A model power plant consisting of three processes: vaporization, expansion, and condensation.

dense the working fluid from a vapor to a liquid and return the temperature, volume, and pressure to the same values that the fluid had at the point A. This requires that some energy be carried out of the system, usually by cooling water.

It is important to note that the working fluid is returned with no net change after completing the circuit from A to B to C and back to A.

Conservation of Energy

If we apply the conservation of energy principle to the system, we see that if there is no change in the system,

$$E_{in} = E_{out} + W_{out}. \qquad (3.2)$$

But this does not enable us to calculate the efficiency, because we do not know the magnitude of E_{out} for a given E_{in} and W. It is not even obvious that there must be an E_{out} in the most general situation. Clearly there must be in this case, as we can see by examining the energy input and output during each step. We must also note that energy can be used to change the character of the working fluid. It obviously requires energy to vaporize the fluid; that energy is stored in the fluid. We note the change in character of the fluid by means of energy storage as ΔU. Then Table 3.1 indicates the energy in, out, and stored for each step. The total energy in must equal the sum of the total energy out and total work out. The sums of all the ΔU's must

equal zero, because the fluid was returned to its original character—it could have no excess or deficiency of stored energy.

We can conclude that for our model of a power plant Eq. 3.2 must be correct, but we still cannot calculate the efficiency. We must go one step further and realize that the energy in or out in this case is in a particular form: thermal energy. We have used the concept of heat in previous chapters without ever relating it to other physical processes. Making this connection occupied many physicists in the nineteenth century, so it is rather well-established today. Let us make certain we have all the appropriate definitions. We have an intuitive feeling that hot things have a high heat content and cold things do not. We can do better than that, but we need a yardstick for measuring hot and cold.

Temperature

Over the years several temperature scales have been used; two are still in general use. In the Celsius scale, zero is the ice–water equilibrium point and 100° is the steam–water equilibrium point. Originally, in the Fahrenheit scale, 32° was the ice–water equilibrium point and 100° was to be body temperature. Unfortunately, body temperature varies considerably, even in a well individual. Today the Fahrenheit degree is defined as 1/180 of the

TABLE 3.1 Energy Transfers at Each Step in the Power Plant Model

Step	Energy Input	Work Output	Energy Output	Change in System
Vaporization	E_{in}	0	0	ΔU_1
Expansion	0	W_{out}	0	ΔU_2
Condensation	0	0	E_{out}	ΔU_3
Total	E_{in}	W_{out}	E_{out}	$\Delta U_1 + \Delta U_2 + \Delta U_3$

temperature difference between the ice point and the steam point.

In both the Celsius and Fahrenheit scales, negative temperatures are allowed, and the interpretation of temperature dependence is difficult. For example, if the volume occupied by a gas at a certain pressure is known to depend directly on the temperature, is it true that at 0°C the volume of the gas is zero? Obviously not. The zeros of the two common scales, although they may be based on some physical property, cannot have the same meaning as the zero of velocity or the zero of pressure or volume. The zeros of the Fahrenheit or Celsius scales are not absolute. An absolute zero can be determined, if a physical property of the constituents of matter is zero at some low temperature and if that property then depends in a reasonable fashion on the temperature.

There is such a property, the average kinetic energy:

$$KE_{av} = \frac{3}{2} kT. \qquad (3.3)$$

The average kinetic energy of a particular type of gas molecule is proportional to the temperature. The constant k is the Boltzmann constant, 1.38×10^{-23} J/K.

The temperature in Eq. 3.3 has a physical meaning at 0°, namely, that at that point $KE_{av} = 0$. But this is the only point on a temperature scale that can be established. By specifying the size of the degree, a new scale is defined. It is convenient to use the Celsius degree, since that would leave the ice and steam points 100 degrees apart. This new scale is called the Absolute or Kelvin scale. Note that 0.0 K $= -273.16$°C (the degree Kelvin is referred to as a *kelvin*). Note also that

$$T_F = \frac{9}{5} T_C + 32°. \qquad (3.4)$$

There is also an absolute scale based on the Fahrenheit degree. It is called the Rankine scale and has been used in U.S. engineering work. Note that

$$0.0°R = -459.67°F. \qquad (3.5)$$

Many textbooks place temperature measurements on a firmer theoretical and experimental base. This is not essential for our needs, so we refer you to the References for further information.

Heat

Now that we can be quantitative about hot and cold, we return to the question of heat. We know from experience that if two objects initially at different temperatures are joined by a heat conducting path, energy will transfer from the hotter to the cooler until both have the same temperature. We call this transfer heat. It is the exchange of average kinetic energy of atoms, molecules, and electrons in one substance to those in another. The transfer of heat into a system will be denoted by $+Q$; out of the system by $-Q$. Heat has the units of energy: joules or calories in the MKS system, British thermal units in the English system.

We return to the intuitive feeling: hot things have a high heat content. Is this true? If we walk barefoot at the beach on a sunny day, the sand will feel hot, but a paper plate lying in the sun will not. The sun shines equally on the plate and the same area of sand, yet the two differ in temperature. We conclude that there is a proportionality between the temperature change and the amount of heat received, but that the constant of proportionality depends upon the type of material receiving the heat:

$$Q = C \Delta T. \qquad (3.6)$$

This constant of proportionality is called the **heat capacity**. With this definition of heat capacity the value of C will depend on the physical dimensions of the particular system under study and not simply the composition. A more useful quantity is often the heat capacity

TABLE 3.2 Specific Heat at Constant Pressure for Certain Solids[a]

Solid	c_p (cal/g°C)[b]	c_p (cal/mol °C)
Aluminum	0.215	5.82
Carbon	0.122	1.46
Copper	0.0924	5.85
Ice	0.550	9.9
Iron	0.113	7.45
Lead	0.0315	6.32
Silver	0.0564	6.09
Tungsten	0.0321	5.92

[a] $T = 20°C$, $p = 1$ atm.
[b] The units here are not SI, but are those usually found in tabulations.

per unit mass or **specific heat** c, defined by

$$c = \frac{1}{m}\frac{\Delta Q}{\Delta T} \quad \text{J/g °C.} \tag{3.7}$$

Note that c is often quoted as molar heat capacity, which is the heat capacity per mole of substance. (Be sure of the units; some texts use pound-moles, some gram-moles, etc.[1])

[1] A gram-mole of a substance contains exactly Avogadro's number of molecules of the substance: 6.02×10^{23}. A gram-mole of water would have a mass of 18 g.

Specific heats for several solids are given in Table 3.2.

Example 3.1 10,000 calories of heat is added to each of two pans. Each pan has a mass of 1000 g; one is aluminum, one is iron. What is the temperature change of each?

$$c = \frac{1}{m}\frac{\Delta Q}{\Delta T}, \quad \Delta T = \frac{\Delta Q}{mc},$$

$$\left.\Delta T\right)_{Al} = \frac{10,000}{1000}\left(\frac{1}{c_{Al}}\right) = 44.5°C,$$

$$\left.\Delta T\right)_{Fe} = \frac{10,000}{1000}\left(\frac{1}{c_{Fe}}\right) = 88.5°C.$$

We conclude that hot things do not necessarily have a high heat content; they may simply have a small heat capacity. Note that the specific heat in Table 3.2 is given as c_p with subscript p. This indicates that the value is determined at a constant pressure. It is possible to make specific heat measurements with either a constant pressure or a constant volume. In the case of solids there is little volume change so that c_p is clearly implied when one says specific heat. For gases the differences can be quite large so that one must

TABLE 3.3 Specific Heats of Several Gases

Type of Gas	Gas	c_p (cal/mol K)	c_v (cal/mol K)	$c_p - c_v$	$\gamma = c_p/c_v$
Monatomic	He	4.97	2.98	1.99	1.67
	A	4.97	2.98	1.99	1.67
Diatomic	H_2	6.87	4.88	1.99	1.41
	O_2	7.03	5.03	2.00	1.40
	N_2	6.95	4.96	1.99	1.40
	Cl_2	8.29	6.15	2.14	1.35
Polyatomic	CO_2	8.83	6.80	2.03	1.30
	SO_2	9.65	7.50	2.15	1.29
	C_2H_6	12.35	10.30	2.05	1.20
	NH_3	8.80	6.65	2.15	1.31

be careful to differentiate between c_p and c_v. See Table 3.3 for the specific heats of several gases.

To return to our power plant model, we can now generalize the results of Table 3.1 for any step in the process:

$$Q_{in} - Q_{out} = \Delta U + W_{out}. \qquad (3.8)$$

In any given step one of these terms may be zero. This equation is called the **First Law of Thermodynamics**. It is simply an application of conservation of energy to a system that has thermal energy inputs or outputs. This may also be written using the notation of calculus:[2]

$$dQ = dU + dW. \qquad (3.9)$$

3.2 APPLICATIONS OF THE FIRST LAW

We now know how to apply conservation of energy to problems involving the exchange of thermal energy. There are a number of processes involving heat and work that are very important in the study of energy–environment problems. In this section we examine some of them.

Work

How does a thermodynamic system transfer energy by means of work? In Figure 3.2 it is presumed to be an expansion, but the details are not given. The implication is that the working fluid is a vapor and that the vapor volume increases. This is one of several ways to do work; electrical work can also be included in thermodynamic systems, as we see in Chapter 5. One of the simplest kinds of expansion is that of a gas against a piston. In

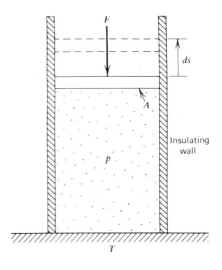

Figure 3.3 Work is done by the gas at pressure p as it expands against the piston of surface area A, moving it a distance ds. Heat may enter or leave the system via the constant temperature reservoir on which the cylinder rests.

Figure 3.3, the work done when the piston is moved a distance ds is given by

$$dW = pA\,ds \qquad (3.10)$$

or

$$dW = p\,dV, \qquad (3.11)$$

where $p = F/A$. In the case of a finite change from V_i to V_f the amount of work is given by

$$W = \int_{V_i}^{V_f} p(V)\,dV. \qquad (3.12)$$

If the functional form of $p(V)$ were known, Eq. 3.12 could be solved for W. This functional relationship describes the "path" of integration. If $V_f > V_i$, then W is positive; work is done *by* the system. If, however, $V_i > V_f$, then W is negative and work is done *on* the system.

Example 3.2 Assume $p = k/V$; find W be-

---

[2]This form implies that there are differentiable functions for Q, U, and W; this may not always be true. But for most of the applications of interest to us, there will be no difficulty if we use Eq. 3.9.

tween the limits V_i and V_f. Using Eq. 3.2,

$$W = \int_{V_i}^{V_f} \frac{k}{V} dV,$$

$$W = k \ln V \Big|_{V_i}^{V_f}$$

$$W = k \ln\left(\frac{V_f}{V_i}\right).$$

The natural logarithm of a number less than 1 is negative, so the convention that negative work is work done on the system is at least self-consistent.

Figure 3.4 illustrates two paths and a complete cycle. In Figure 3.4a the shaded area beneath the curve represents the area beneath the curve; it is also the integral indicated—the work done in going from the point i to the point f along the curve. In the center section a similar area is indicated; but this represents a negative work, since $V_i < V_f$. The integral along the complete path, as shown in Figure 3.4c, must be the sum of the other two:

$$\oint p\, dV = \int_{V_i}^{V_f} p_1(V)\, dV + \int_{V_f}^{V_i} p_2(V)\, dV. \quad (3.13)$$

The net work around this path is the area *inside* the path, as indicated. In general, the work depends not only on the end points V_i and V_f but also on the particular path chosen; this is the same as saying the integral in Eq. 3.12 cannot be performed unless the functional form of $p(V)$ is known. Diagrams of this type will be very helpful to us; this particular one is called a p–V diagram. Note that we cannot tell from it whether or not energy is being exchanged between the system and its surroundings by virtue of a temperature difference during a compression or expansion.

It may happen that the functional form of $p(V)$ is simple; for example, $p = $ const. In this case, Eq. 3.11 can be written

$$W = pV = p(V_f - V_i), \quad (3.14)$$

and considerable simplification results.

Example 3.3 The piston in Figure 3.3 has a diameter of 10.0 cm. If this piston moves a distance of 4.0 cm, what is the change in volume in the cylinder? If this displacement is caused by a constant 10^4 N force, how much work is done on the process? What is the pressure in the cylinder?

$$\Delta V = V_f - V_i = A\,\Delta s = \frac{\pi \cdot 10^2}{4}\, cm^2 \times 4\, cm$$

$$= 3.142 \times 10^{-4}\, m^3$$

$$W = F \cdot \Delta s = 10^4 \times 0.04 = 4.0 \times 10^2\, J$$

$$W = p\,\Delta V, \qquad p = \frac{W}{\Delta V}$$

$$\frac{4.0 \times 10^2}{3.142 \times 10^{-4}} = 1.3 \times 10^6\, N/m^2$$

As a check, note that $p = F/A$ as well.

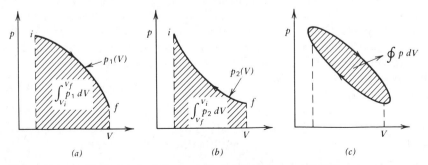

(a)   *(b)*   *(c)*

Figure 3.4 p–V diagrams: (a) expansion, (b) compression, (c) expansion and compression together, a complete cycle. The shaded area bounded by the curve between V_i and V_f is given by the integral indicated in each case.

A process for which $p = $ const is called an **isobaric process**; it has a unique signature in a p–V diagram—a horizontal line.

Internal Energy

It is possible to construct a system so well insulated that a compression can be performed without the transfer of heat out of the system. Any process for which $\Delta Q = 0$ is called an **adiabatic process**. A valuable property of adiabatic processes is that in these cases the work performed depends only upon the end points and not on the path; the path is then called an **adiabat**.

It is easy to see that the work is path-independent in an adiabatic process. From Eq. 3.9 if $Q_{in} = Q_{out} = 0$, then

$$W = -\int_{U_i}^{U_f} dU = U_i - U_f. \qquad (3.15)$$

Clearly, only the end points U_i and U_f are required to specify the work done. The nature of the path (in the p–V plane) is irrelevant, so long as it is an adiabat. Obviously, if the initial and final points are identical (closed path), $W = 0$.

We have been calling U the stored energy or the energy of change or something rather vague. We must of necessity be somewhat uncertain because it is not always clear precisely how the energy is manifested. It can be the change in temperature of the system, it can be a change in chemical composition, it can be a change in phase—liquid to solid, or many others. Obviously, ΔU is difficult to determine; the particular problem under study must be examined carefully for the exact nature of this internal energy change. In any case, however, we know that Eq. 3.9 will always apply and that Eq. 3.16 will apply in any adiabatic process. We will refer to U as the **internal energy**.

Example 3.4   A 1-kg sample of ice at 0°C absorbs 80,000 cal of heat. It is found that

afterward 1 kg of water at 0°C remains. The volume change is negligible. Explain.

Apply the First Law: $dQ = dU + dW$. Since there is little volume change the work term is zero. In this case the heat is totally absorbed by the ice as internal energy change. We conclude from the data that 80 cal/g is required simply to melt ice isothermally.

Let us examine the change in internal energy for a particular kind of process, one in which the volume remains constant—an **isovolumnic process**. In this case, $dV = 0$, so that the First Law may be written

$$dQ = dU \quad (dV = 0), \qquad (3.16)$$

and we change nothing if we divide both sides by dT:

$$\frac{dQ}{dT} = \frac{dU}{dT} \quad (dV = 0). \qquad (3.17)$$

But dQ/dT is the definition of heat capacity, and since we have started this assuming constant volume,

$$\frac{dQ}{dT} = C_v. \qquad (3.18)$$

We can then rewrite Eq. 3.17 in terms of C_v:

$$dU = C_v dT \quad (dV = 0). \qquad (3.19)$$

We have been careful to include at each step the fact that $dV = 0$ for this process. There are circumstances for which Eq. 3.19 could be correct, even if $dV \neq 0$. In fact, in most of the situations we will deal with, it is the case. So without proper justification as yet, we conclude that the First Law may be written:

$$dQ = C_v dT + p\, dV. \qquad (3.20)$$

This is really quite a simplification. Now, each term of the First Law is easily defined and, more importantly, easily measured.

Example 3.5   A container has 10 g of O_2 at atmospheric pressure. The gas absorbs 562 J of heat. It goes from 30° to 91.1°C. How much does its volume change?

Using Eq. 3.20:

$$dQ = C_v dT + p \, dV,$$

$$dV = \frac{1}{p}(dQ - nc_v dT)$$

(note $C_v = nc_v$, where n = number of moles)

$$dV = \frac{1}{1.013 \times 10^5}$$

$$\times \left[562 - \frac{10}{32} \times 5.03 \times (91.1 - 30) \times 4.186 \frac{J}{cal} \right]$$

$$dV = 157.9 \times 10^{-5} \text{ m}^3$$

$$\boxed{\Delta V \sim 1560 \text{ cm}^3.}$$

Ideal Gas

The data in Table 3.3 provide an interesting clue to the nature of gases. For each one of the three types of gases there, the difference $c_p - c_v$ is nearly a constant. This implies that the heat capacities do not depend on the exact makeup of the molecules of the gas, in which case it should also be true that interactions between molecules should also be independent of the identity of the gases. In the preceding section we implicitly assumed that c_v depended only on temperature and not volume or pressure. This will be the case only if gas molecules do not interact with each other at all, if they act as point particles. This clearly is an idealization; real gases do interact. But over a surprising range of pressure and temperatures most gases behave as if they were noninteracting **ideal gases** for which the internal energy is only a function of temperature.

Ideal gases obey exactly a relationship that has been found to be approximately true for real gases:

$$pV = nRT, \tag{3.21}$$

where n is the number of moles of the gas and R is a constant 8.317 J/mol K or 1.987 cal/mol K. Equation 3.21 is often called the *equation of state* of an ideal gas. Several state equations have been developed for real gases that accurately describe the p, V, T relationship over a wide range of state variables.

In fact, C_v and C_p are related, as can be shown (see the Exercises at the end of this chapter):

$$C_p = C_v + nR. \tag{3.22}$$

Equations 3.21 and 3.22 can be combined by using the definitions of heat capacities in terms of derivatives to provide a second formulation of the First Law:

$$dQ = C_p dT - V \, dp. \tag{3.23}$$

Do not try to identify the first term of the left-hand side with dU and the second with dW, because it is not correct to do so.

We now have most of the tools required to examine specific processes and to be able to calculate work, energy out, temperature change, and all the other quantities of interest in the energy–environment field. In the next few sections we apply these tools to see if there is anything yet that we need. If a system is allowed to change under the action of general forces, many different kinds of processes can occur. There are, however, four processes that are important to study, since they are fundamental to the study of complex devices, especially engines.

Four Processes

Adiabatic Process In an adiabatic process, $dQ = 0$, so Eqs. 3.20 and 3.23 could be solved simultaneously to eliminate dT. If γ is defined as the ratio C_p/C_v, then this procedure yields

$$pV^\gamma = \text{const.} \tag{3.24}$$

Two other equations relating p to T and T to v can likewise be obtained. (See the Exercises.)

These equations allow the calculation of thermodynamic variables during an adiabatic process. They will be particularly helpful in our study of heat engines.

The work done in an adiabatic process is obtained from the First Law in the form of Eq. 3.20:

$$0 = C_v\,dT + p\,dV, \qquad (3.25)$$

or

$$\Delta W = -C_v\,\Delta T. \qquad (3.26)$$

Example 3.6 Steam enters a low-pressure turbine at 260°C and exits at 35°C. If there is no heat lost to the surroundings, how much work can the steam do? The p–V diagram for this process is shown in Figure 3.5.

From Eq. 3.26

$$w = -0.48 \times 4.186 \times (35 - 260)$$

$$\boxed{= 452.1 \text{ J/g.}}$$

Note that this result is given in terms of the joules per gram of steam, since the heat capacity 0.48 is also a specific quantity, that is, calories per gram per degree Celsius.

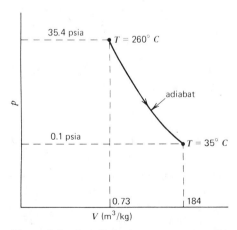

Figure 3.5 An adiabatic process on a p–V diagram. Note that the *specific volume* (volume per unit mass) is plotted.

Isovolumnic Process This kind of process, also called an isochoric process, is very simple, a straight vertical line on a p–V diagram. It is also simple to understand in terms of the First Law. Since $dV = 0$, there is no work done and from Eq. 3.20

$$dQ = C_v\,dT \qquad (3.27)$$

or

$$Q = C_v\,\Delta T. \qquad (3.28)$$

Knowing the heat input during this process enables the calculation of the pressure change using Eq. 3.23.

$$C_v\,dT = C_p\,dT - V\,dp, \qquad (3.29)$$

$$dp = \frac{nR}{V}\,dT. \qquad (3.30)$$

Example 3.7 The combustion of the fuel charge in an automobile cylinder is approximately isovolumnic. The temperature changes from 400° to 2180°C. What is the pressure change? $(V = 5 \text{ in.}^3)$

We assume the working fluid is air. Then from Eq. 3.21,

$$n = \frac{p_0 V}{R T_0}$$

$$\Delta p = \frac{p_0 V R}{R T_0}\frac{\Delta T}{V}$$

$$\Delta p = \frac{1 \text{ atm} \times (2453 - 673)}{673}$$

$$\boxed{\Delta p = 2.64 \text{ atm}}$$

Isothermal Processes A process that occurs at a constant temperature does not have a unique signature on a p–V diagram; furthermore, calculations of heat and work are more complicated than those we have so far encountered. From the First Law, Eq. 3.20,

$$dQ = p\,dV, \qquad (3.31)$$

$$Q = \int p\,dV. \qquad (3.32)$$

The integral of Eq. 3.32 cannot be evaluated unless the functional relationship between p and V is known. In real situations, this relationship may be very difficult to discern. In our model world we can assume that the working fluid is an ideal gas, so the integration can be done as in Example 3.2 and

$$Q = nRT \ln\left(\frac{V_f}{V_i}\right). \qquad (3.33)$$

A similar expression in terms of pressure can be obtained by starting with Eq. 3.23.

Example 3.8 It is desired to compress air in a large underground cavity as a means of storing energy, but to do so isothermally. If the beginning pressure is 1 atm and the final pressure is 25 atm, how much heat should be removed per mole? (Assume $T = 30°C$.)

The equation analogous to Eq. 3.33 is

$$Q = nRT \ln\left(\frac{P_i}{P_f}\right),$$

$$q = 2 \times 303 \times \ln\frac{1}{25},$$

$$\boxed{q = -1951 \text{ cal/mol.}}$$

The minus sign indicates the heat must be removed. Note that if this compressed gas were expanded adiabatically through a turbine to produce work, the temperature would drop significantly. The small q indicates that a molar quantity of heat is being calculated.

Isobaric Processes This process completes the list of those we shall study initially. There are many constant pressure processes in the real

world; most chemical reactions, for example, occur isobarically. We can easily calculate the heat and work involved. From the First Law as written in Eq. 3.23, since

$$dQ = C_p\,dT, \qquad (3.34)$$

the heat may be easily calculated:

$$Q = C_p\,\Delta T. \qquad (3.35)$$

The work may also be calculated using this result in Eq. 3.20:

$$C_p\,\Delta T = C_v\,\Delta T + p\,\Delta V. \qquad (3.36)$$

Example 3.9 A highly schematic diagram of a pressurized water nuclear reactor system is shown in Figure 3.6. The pressurizer is usually operated with about half its volume filled with steam. The water (and therefore the steam) in the pressurizer can be heated automatically to increase the steam pressure to compensate for small volume changes in the reactor caused by fuel changes, bubble formation, and the like. In this fashion the cooling fluid in the reactor is maintained at a constant pressure of about

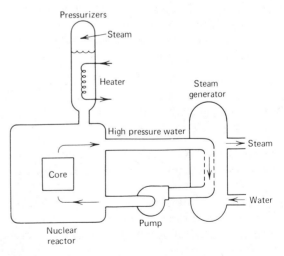

Figure 3.6 A schematic diagram of a nuclear power reactor showing the flow of cooling water through the core and steam generator.

2500 psia.[3] If the normal operating temperature is 370°C and the normal internal volume of the pressurizer is 50 m³, what temperature increase must be available to accommodate a 0.1 percent decrease in water volume?

First find the number of moles of steam in the pressurizer:

$$n = \frac{pV}{RT} = \frac{2500 \times 6.89 \times 10^3 \times 25}{8.317 \times (370 + 273)} = 8.05 \times 10^4.$$

From Eq. 3.36:

$$(8.05 \times 10^4) \times 2 \times 4.186 \times \Delta T$$

$$= 2500 \times 6.89 \times 10^3 \left(\frac{N/m^2}{lb/in.^2}\right) \times 0.025 \text{ m}^3,$$

$$\Delta T = 0.64 \text{ K or } 0.64°C.$$

We have discussed the processes that are the building blocks of the thermodynamic models of real-world devices; we examine many of these in subsequent chapters. There remains the question we started with: efficiency. Given a model for any cyclic set of processes, we can now calculate the efficiency. But we do not know if the efficiency we determine is the best that could be obtained by that device, or if another device could do the job better.

3.3 THE SECOND LAW OF THERMODYNAMICS

The highest possible efficiency would be 1.0; that is, all the input energy is converted to work. Is it possible? Yes, for a limited period of time. The isothermal expansion of a gas would be one example; the expansion would continue until atmospheric pressure were

reached. Is a set of processes executed cyclically possible such that $Q = W$ with $\Delta U = 0$? The first law of thermodynamics does not prohibit such a process. What would be the consequences of the existence of such a device?

If such a set of cyclic processes were available, it would be possible to use them to extract heat from any reservoir, say the ocean, and convert it to work, say to power ships across the ocean. This sounds very much like the "perpetual motion" machines of the nineteenth century; and it won't work for the same reason they didn't work. There is another apparent law that nature follows; we have come to call it the **Second Law of Thermodynamics**.

There are many ways of stating this law; one simple one is,

It is impossible to construct a device operating in a closed cycle which produces no effect other than the transfer of heat from a cooler to a hotter body.

(CLAUSIUS STATEMENT)

Another way of saying it would be,

It is impossible to construct a device operating in a closed cycle which produces no effect other than the conversion of heat from a reservoir to work.

(KELVIN STATEMENT)

The Second Law cannot be proved; we should note, however, that it has never been observed to be disobeyed in nature. It is a consequence of the random and microscopic nature of matter. It comes about because we have no mechanism to organize completely the actions of these random microscopic entities. We can control the average properties of a collection of molecules, but we cannot control all molecules. If we accept that the Second

---

[3]The unit psia is pounds per square inch absolute. This measure includes atmospheric pressure. The pressure over and above atmospheric is known as gauge pressure, psig.

Law must be true, then we can produce our long-desired efficiency relationship.

In any real set of cyclic processes there must be heat rejected at a lower temperature; otherwise the Kelvin statement of the Second Law would be violated. The work done must be the difference between the heat taken in at T_1 and the heat rejected at T_2:

$$W = Q_1 - Q_2. \tag{3.37}$$

The efficiency is then given by

$$\eta = \frac{Q_1 - Q_2}{Q_1}. \tag{3.38}$$

The efficiency is a maximum when $Q_2 = 0$; but this is not possible. What is the maximum value η can have for real processes?

Reversibility

We can calculate thermodynamic variables only for processes that are composed of a series of infinitesimal steps, each infinitesimally nearer **thermodynamic equilibrium**.

The value of a given variable for a given system has a meaning only when the system is in a particular kind of state. Because these variables are in a sense an average over some microscopic properties, we must be certain the statistical rules being used to do the averaging are applicable. They will generally not be if the system is undergoing a transition from one state to another; that is, if chemical changes are occurring, if accelerated or turbulent flow is in process, or if heat is being exchanged by friction with the surroundings. In these transition situations the precise composition, pressure, and temperature of the gas may vary from point to point and therefore cannot be determined for the entire system. Only if none of these three effects is occurring can our variables be defined; when this is true, the system is said to be in thermodynamic equilibrium.

Implicit in this definition was the idea that the system could go through these equilibrium states equally well in either direction, that it be thermodynamically **reversible**. This will not necessarily be true, if dissipative effects are present.

Consider a system that accepts Q_1 units of heat from a reservoir at T_1, converts part to work W, and rejects Q_2 units to the atmosphere. Could the system and its *surroundings* be restored without changes? In order for that to happen, Q_2 units of heat would have to be extracted from the atmosphere, converted to work W, which is isothermally transformed to heat at the reservoir T_2. This is impossible, as it violates the Clausius statement of the Second Law. In general, any dissipative effect—viscosity, friction, inelasticity or the transfer of heat between finite temperature differences—cannot be reversed without producing changes in the system or its surroundings.

In short, all real processes are irreversible, since they cannot be nondissipative. But the *concept* of reversibility is extremly useful in thermodynamics. Having laid this foundation we may now try to calculate the maximum efficiency of a cyclic device that has thermal energy as input.

The Carnot Cycle

This problem was first studied systematically by a French engineer, Sadi Carnot, in the early nineteenth century. He envisioned an ideal device initially in thermodynamic equilibrium with a cold reservoir at T_2. Then the following operations are performed: (1) a reversible adiabatic process in which the temperature of the working substance goes from T_2 to T_1, the temperature of a hot reservoir, a to b in Figure 3.7; (2) the working fluid is maintained in contact with the T_1 reservoir and allowed to expand reversibly and isothermally, b to c; (3) a reversible adiabatic process until the temperature is again at T_2, c to d; and (4) contact maintained at T_2 and the working material iso-

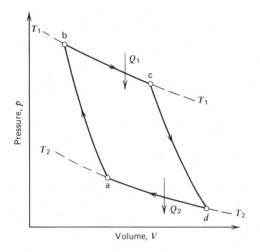

Figure 3.7 Carnot cycle of an ideal gas.

thermally and reversibly compressed until the system has its original internal energy, d to a.

Any device that operates using the processes outlined above, a Carnot cycle, is called a Carnot engine; there are no such devices in the real world. But the utility of the concept of a Carnot engine is widespread. We can calculate the efficiency of such a device and show that no other device can exceed it in efficiency under the same operating conditions. This can aid us in the allocation of effort and resources. For example, if we can calculate the ultimate efficiency of a given device, we will be in a position to decide whether additional work to improve the device's operation is warranted or whether research ought to be channeled in a new direction.

The efficiency of the Carnot engine, as for any heat engine, is given by Eq. 3.38. Since all the processes in the Carnot cycle are reversible, it should be possible to calculate Q_1 and Q_2 in terms of the variables of the system. But in the development of the Carnot cycle we did not mention the detailed nature of the system; the working substance was not even specified. In fact, the only variable that could be in-

volved is the temperature. It is easy to show (see the Exercises) that the Carnot efficiency must be given by

$$\eta_c = 1 - \frac{T_2}{T_1}. \qquad (3.39)$$

An efficiency of 100 percent could be obtained only if $T_2 = 0$, but we believe absolute zero to be unattainable. Any real engine operating between two temperatures must have an efficiency lower than that of a Carnot engine operating between the same temperatures.

Example 3.10 A propane-powered forklift truck has an operating efficiency of about 20 percent in terms of work out per unit of fuel energy in. If the propane flame temperature is 2000°C, how does this device compare to a Carnot engine operating between the same temperatures?

The Carnot engine would have

$$\eta_c = 1 - \frac{293}{2273},$$

$$\boxed{\eta_c = 87.1 \text{ percent.}}$$

The forklift does rather poorly in comparison!

To show that η_c is the maximum possible for a given set of temperature reservoirs consider the following. Let us use two reversible engines, I and R, operating between temperatures T_1 and T_2, where $T_1 > T_2$. These engines do not have to be identical. But since they are reversible, in the thermodynamic sense, we can reverse the operation of engine R, causing it to operate as a refrigerator. Refer to Figure 3.8; engine I removes Q_1' units of heat from the hot reservoir, does W units of work, and deposits $Q_1' - W$ units of heat at the cold reservoir.

The efficiency of this engine would be $\eta_I = W/Q_1'$. If we assume the first engine has a higher efficiency $\eta_I > \eta_R$ and that R uses the

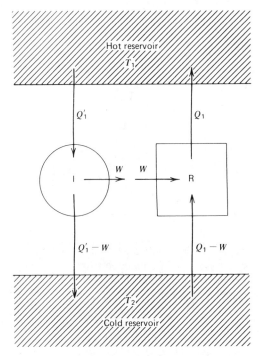

Figure 3.8 Reversible engine I driving a reversible refrigerator R.

from a cold to a hot reservoir. This phenomenon is a violation of the Clausius statement of the Second Law. So it cannot be true that $\eta_I > \eta_R$ for reversible engines. But a reversible engine is just a Carnot engine, so we can conclude that no engine can have an efficiency greater than that of a Carnot engine.

It is possible to formulate the Second Law in a more mathematical fashion in terms of a new thermodynamic variable.

Entropy

Any cyclic set of reversible processes can be approximated in terms of alternating adiabats and isotherms, as in Figure 3.9. Then the general cycle can be approximated by a sequence of Carnot cycles, for example, a–b–c–d–a in the figure. For this cycle, from Eqs. 3.38 and 3.39 it must be true that

$$\frac{Q_1}{T_1} + \frac{Q_2}{T_2} = 0 \qquad (3.43)$$

(remember Q_2 is a negative number).

Since this is true for all the Carnot cycles

work W produced by I to remove $Q_1 - W$ units of heat from the cold reservoir and deposit Q_1 units at the hot reservoir, then $\eta_R = W/Q_1$.

But if $\eta_R < \eta_I$, as we have assumed, then

$$\frac{W}{Q_1} < \frac{W}{Q_1'} \qquad (3.40)$$

and

$$\frac{1}{Q_1} < \frac{1}{Q_1'} \qquad (3.41)$$

or

$$Q_1 > Q_1'. \qquad (3.42)$$

But this implies that the hot reservoir gets hotter, which implies this arrangement of engines has the *net* effect of transferring energy

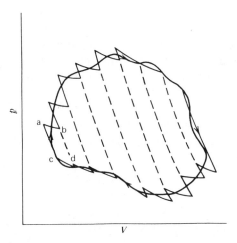

Figure 3.9 A general p–V contour approximated by a sequence of adiabats and isotherms.

that make up the general cycle

$$\sum \frac{dQ}{T} = 0, \qquad (3.44)$$

or if the steps are made infinitesimally small,

$$\oint \frac{dQ}{T} = 0. \qquad (3.45)$$

If we then make this definition

$$S_2 - S_1 \equiv \int_1^2 \frac{dQ}{T}, \qquad (3.46)$$

we will have a new variable that meets our requirements. This variable S is called **entropy**. Note that we have defined only the entropy *difference* between thermodynamic states. The entropy of a system in a given state can be calculated by arbitrarily defining the entropy of a "standard" system. In thermodynamics the entropy of a system is not of as much consequence as the *entropy change* during a process.

If the two points in Eq. 3.46 are made to be very close together, we can write

$$dS = \frac{dQ}{T}. \qquad (3.47)$$

This is often stated as the mathematical formulation of the Second Law. To understand this let us examine an irreversible process.

In Figure 3.10, the dashed path i–f represents an irreversible adiabatic process. Note that we cannot represent it by a line, since the state variables are not well-defined during an irreversible process. We now perform the following *reversible* processes starting at f:

- Adiabat from f to k ($\Delta Q = \Delta S = 0$).
- Along an isotherm from k to j [$Q = T(S_j - S_k)$ and some work W is done].
- Adiabat from j to i ($\Delta Q = \Delta S = 0$).

We wish to calculate $\Delta S_{IR} = S_f - S_i$ for the irreversible process. It must be the same as the total change for the set of reversible pro-

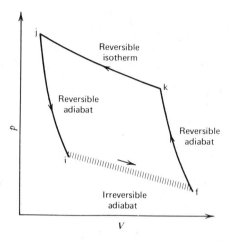

Figure 3.10 An *irreversible* adiabatic process between i and f compared to a sequence of *reversible* adiabats and isotherms.

cesses:

$$\Delta S_R = (S_k - S_f) + (S_j - S_k) + (S_i - S_j). \qquad (3.48)$$

But from the first and third items in the list above,

$$\Delta S_R = S_j - S_k = \Delta S_{IR}. \qquad (3.49)$$

It cannot be true that $Q > 0$ for the entire cycle. If it were, then the net result of this complete cycle would be conversion of Q units of heat to work W, forbidden by the Second Law. Then it must be that

$$T(S_j - S_k) \leq 0 \qquad (3.50)$$

or that

$$\Delta S_{IR} \geq 0. \qquad (3.51)$$

In fact, it must be true that

$$\Delta S_{IR} > 0; \qquad (3.52)$$

otherwise, the irreversible process would be exactly the same as a reversible process.

We see that for this *real* process the entropy change must be positive. It is a general truth that entropy must increase for irreversible processes. This increase in entropy means an increase in disorder on the molecular scale or a loss of opportunity to do work. The entropy change during a reaction can also be related to the level of order of the reacting systems.

By level of order we mean the degree of organization. A pure crystal of salt is highly organized, but water vapor diffusing into the atmosphere rapidly loses organization, and the disorder of the system—vapor plus atmosphere—increases. The entropy concept is wide-ranging and has applications in fields far removed from classical thermodynamics; see the References at the end of this chapter. For our purposes at present, we shall confine our discussion to a few rather simple applications. From the definition of entropy, we can write the First Law as follows:

$$T\,dS = dU + dW. \qquad (3.53)$$

For a cyclic process

$$T\,dS = dW. \qquad (3.54)$$

A T versus S plot has the same significance as a p versus V plot: the area enclosed by the curve, for a cyclic process, represents the net work done during the process, either by the system or on the system. The T–S diagram for the Carnot cycle takes a particularly simple form.

We use the entropy formulation for one simple application here but will encounter it many times again in later chapters.

Second Law Efficiency

We now know that any device that takes in heat to do work must reject some heat. The higher the temperature at which the energy transfer occurs, the higher will be the efficiency of that transfer. High-temperature heat is referred to as **high-quality heat** for this reason; it can do work more efficiently than low-temperature heat. We should keep in mind that high-quality heat has a premium value and

that efforts should be made to preserve that quality. This feature of heat transfers makes it of value to study not only the efficiency as we have defined it, the First Law efficiency, but also a more general measure of merit.

The First Law efficiency is defined for a specific device; it is often more important to determine efficiencies for a specific *task*. In energy-environment studies we frequently encounter questions of trade-offs and compromises. Is electric heat better than gas-fired? Is a heat pump better than both? An efficiency measure based on the task would give us the answer. Accordingly we define the **Second Law efficiency** to be

the heat or work usefully transferred by a given device or system *divided by* the maximum possible heat or work usefully transferable for the same function by any device or system using the same energy input as the given device or system. (3.55)

To make use of this definition we must understand what is meant by maximum possible heat or work. In the case of heat, it is apparent. In the case of work, it may not be. How much work can be done by the heat extracted from a reservoir? Certainly not an amount equal to the heat extracted, because that would violate the Second Law.

The work done by a system can be divided into two kinds of work: useful work ΔW_u, and work against the atmosphere ΔW_a. The First Law then becomes:

$$\Delta Q = \Delta U + \Delta W_u + \Delta W_a. \qquad (3.56)$$

Using the definition of entropy change, Eq. 3.46, we can write the heat transfer in terms of the entropy change of the atmosphere plus the system ΔS_{sa}, and the entropy change of the system only ΔS.

$$\Delta Q = T_0 \Delta S_{sa} - T_0 \Delta S. \qquad (3.57)$$

If we recall that $\Delta W_a = p_0 \Delta V$, then Eq. 3.56 can be written in terms of entropy changes using Eq. 3.57:

$$\Delta W_u = -T_0 \Delta S_{sa} - (\Delta U + p_0 \Delta V - T_0 \Delta S). \quad (3.58)$$

Now, define a quantity ΔB by

$$\Delta B \equiv \Delta U + p_0 \Delta V - T_0 \Delta S. \quad (3.59)$$

With this definition, Eq. 3.58 becomes

$$\Delta B = \Delta W_u + T_0 \Delta S_{sa}. \quad (3.60)$$

If we now assume that the useful work done by the system is absorbed by one or more other systems that do not interact otherwise with the system under study or the atmosphere, then by the Second Law it must be true that

$$S_{sa} \geqslant 0, \quad (3.61)$$

so that

$$\Delta W_u \leqslant \Delta B. \quad (3.62)$$

We see that the quantity ΔB can be identified with the change in available work. The available work is the maximum work that a system can do. It can be defined as

B = the maximum work that can be provided by a system (or by fuel) as it proceeds (by any path) ro a specific final state in thermodynamic equilibrium with the atmosphere. (3.63)

In the case of heat extraction from a reservoir,

$$B = Q_1' \left(1 - \frac{T_2}{T_1}\right). \quad (3.64)$$

The idea of available work demonstrates quite strikingly the effect of heat quality. The lower the temperature T_1 at which the thermal energy is transferred, the smaller the available work.

In terms of available work, the Second Law efficiency can be written

$$\epsilon = \frac{B_{min}}{B_{actual}} = \frac{W_u}{W_u + T_0 \Delta S_{sa}}. \quad (3.65)$$

With this result ϵ can be calculated for some processes.

We shall develop the ideas of available work

and Second Law efficiencies in more detail as we begin the study of specific devices and energy conversion processes. We have only briefly reviewed some introductory elements of thermodynamics. This branch of physics has been well-studied for many years and is widely documented. More thorough discussions can be found in any number of excellent texts, some of which are given in the References. This introduction is not complete; additional material will be developed as needed in the discussion of the various energy conversion processes, the first group of which are the heat engines.

SUMMARY

Thermodynamics is the study of large ensembles of molecules, their dynamics and their interactions.

The *First Law of Thermodynamics* is nothing more than conservation of energy applied to a system in which heat may be exchanged.

Heat is the transfer of energy by virtue of a temperature difference. Three temperature scales are in general use: Fahrenheit, Celsius, and Absolute. Absolute zero is $-273.16°C$ and $-459.67°F$.

Thermodynamic work is the integral of $p\,dV$. Work on or by a system can result in a loss or gain of heat and a change in the **internal energy** of the system.

Ideal gases are characterized by a particular equation of state, relating certain state variables.

Four processes are important in thermodynamics: **isovolumnic**, **isothermal**, **isobaric**, and **adiabatic**. Any arbitrary process can be modeled by a combination of these.

The Second Law of Thermodynamics is used to describe the direction that a given action will take.

Reversible processes are ideal ones in which the system does not undergo any dissipative processes in going from one state to another.

The Carnot cycle is a set of alternating adiabats and isotherms. This is the most efficient process that can be carried out cyclically between two temperature reservoirs.

Entropy is a thermodynamic variable that can be used to determine the direction of a process. For all *real* (nonreversible) processes entropy must increase during the process. Entropy can be related to the level of organization of the system.

The Second Law efficiency is a useful idea because it relates the work done by a *device* to the maximum work that can be done by any device under the same conditions.

REFERENCES

Granet, Irving, *Thermodynamics and Heat Power.* Reston, Va.: Reston Publishing, 1974.

Obert, E. R., and R. A. Gaggioli, *Thermodynamics.* New York: McGraw-Hill, 1963.

Prigogine, Ilya, Gregoire Nicolis, and Agnes Babloyantz, "Thermodynamics of Evolution," *Phys. Today,* November (1972), p. 23.

Van Ness, H. C., *Understanding Thermodynamics.* New York: McGraw-Hill, 1969.

Zemansky, Mark W., *Heat and Thermodynamics.* New York: McGraw-Hill, 1968.

_____, *Efficient Use of Energy.* New York: American Institute of Physics, 1975.

GROUP PROJECTS

Project 1. Estimate the heat capacity of your classroom. Include all heat sources when making measurements: lights, bodies, and so on. Estimate the heating requirements during the heating season. Compare with your institution's records.

Project 2. Do an experiment to answer an old question: If you want your coffee to cool to drinking temperature rapidly, should you put the cream in right away or wait until just before drinking? Can you explain your result?

EXERCISES

Exercise 1. A 10-kg aluminum mass falls 1 m onto a 5-kg iron block that is thermally isolated. What is the temperature increase of the iron?

Answer: $8.6 \times 10^{-3} \, °C$.

Exercise 2. Sunlight is incident on a glacier at the rate of 90 W/m^2. If the ice temperature is $-9°C$, how much time will be required to bring a 2.5-cm surface layer of ice up to a temperature of 30°C? Neglect loss of heat from the 2.5-cm layer.

Exercise 3. What was the initial volume of the gas in Example 3.5?

Exercise 4. Think of the interior of a house as an object having a certain heat capacity. A house having a surface area of 150 m^2 exposed to the sun absorbs solar radiation at an average rate of 400 W/m^2 over an 8-hr day. During this period the interior temperature goes from 20° to 26°C. What is the heat capacity for this interior? What have you neglected? A cube of iron of what dimension would have the same heat capacity?

Exercise 5. The specific heat of the rocks used in a solar heat storage system is 580 J/kg °C. What mass of rock is required to store one day's heating ($\sim$200 MJ/hr for 24 hr) provided the rock temperature varies between 21° and 66°C only?

Answer: 184 t.

Exercise 6. Show that the heat capacities for an ideal gas satisfy the equation

$$C_p = C_v + nR.$$

Hint: Consider two processes to get from T_1 to T_2: an isovolumnic and an isobaric. Compare ΔU in each case.

Exercise 7. Using the procedure outlined in the text, show that for an ideal gas undergoing an *adiabatic process, pV^γ = const.*

Exercise 8. For an ideal gas undergoing an adiabatic process, show that

$$TV^{\gamma-1} = \text{const} \quad \text{and} \quad Tp^{(1-\gamma)/\gamma} = \text{const}.$$

Exercise 9. Derive Eq. 3.23.

Exercise 10. If the steam flow rate corresponding to Example 3.6 is 10 t per hour and if the efficiency of conversion of turbine shaft work to electricity is 0.9, what electrical power could be generated? What volume condenser is required for one hour's steam?

Answer: 1.13×10^6 W, 1.84×10^6 m³.

Exercise 11. If in Example 3.7 the initial pressure is 8 atm, how many moles of air are in the cylinder? If the heat input is 5190 cal per stroke, how much gasoline is burned per cylinder per stroke?

Exercise 12. Compressed air in an underground storage cavern at 25 atm and 30°C is expanded adiabatically to 1 atm. What is its final temperature? (γ for air is about 1.4.)

Exercise 13. If the steam in the pressurizer of Example 3.9 is to be heated in 1 min by an

electric heater, how much power is required? If the turbine were down, where would this power come from?

Answer: 3.10×10^4 W.

Exercise 14. Plot the behavior of an ideal gas on a p–V diagram for the following processes:

(a) $V = \text{const}$ (isochoric process)
(b) $p = \text{const}$ (isobaric process)
(c) $\Delta T = 0$ (isothermal process)
(d) $\Delta Q = 0$ (adiabatic process)

Exercise 15. Calculate the work *done by* an ideal gas for the same four processes above.

Exercise 16. One mole of an ideal monatomic gas is taken around the cycle shown in Figure 3.11.

> Process $1 \to 2$ is at constant volume.
> Process $2 \to 3$ is adiabatic.
> Process $3 \to 1$ is at constant pressure.

Determine numerical values for the heat ΔQ, the change in internal energy ΔU, and the work done W for each of the three processes and for the cycle as a whole.

Exercise 17. A natural-gas-fired furnace is usually believed to be about 65 percent

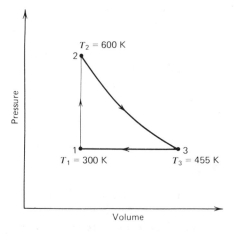

Figure 3.11 Diagram for Exercise 16.

efficient; that is, 65 percent of the fuel energy is converted to heat that is ducted into the house. But if the flame temperature is 2100°C, what is the ultimate efficiency of a device using this fuel?

Answer: 87.6 percent.

Exercise 18. In a Carnot cycle the isothermal expansion of the gas takes place at 400 K and the isothermal compression at 300 K. During the expansion, 500 cal of heat energy is transferred to the gas. Determine the following:

(a) The work performed by the gas during the isothermal expansion.
(b) The heat rejected from the gas during the isothermal compression.
(c) The work done on the gas during the isothermal compression.
(d) The efficiency of the cycle.

Exercise 19. If you had the choice of increasing the efficiency of a Carnot engine by decreasing the low temperature at which heat is rejected by an amount ΔT or by increasing the high temperature at which heat is accepted by the same amount, which would you choose? Use a general mathematical argument to justify your choice.

Exercise 20. For a Carnot cycle,

$$\eta = 1 - \frac{Q_2}{Q_1}.$$

Show that

$$\frac{Q_2}{Q_1} = \frac{T_2}{T_1}$$

and hence that

$$\eta = 1 - \frac{T_2}{T_1}.$$

(*Hint*: Refer to the Carnot cycle of Figure 3.7, $a \to b \to c \to d \to a$. For the isothermal processes, determine Q_1 and Q_2 from the expression

$$Q_1 = W_1 = \int_{V_b}^{V_c} p\, dV = nRT_1 \ln \frac{V_c}{V_b}.$$

You can now relate the volumes at a, b, c, and d using the expression $TV^{\gamma-1} = \text{const}$ for the adiabatic processes.)

Exercise 21. Using the First Law as written for an ideal gas, Eq. 3.20, and the definition of entropy, derive the following:

$$S = C_p \ln T - nR \ln p + S_0,$$

where $S_0 = S_s - C_p \ln T_s + nR \ln p_s$, the subscript s referring to the "standard" values.

Exercise 22. Show that heating a substance (1 mol with specific heat per mole of c) from T_a and T_b leads to an entropy change of the *material* given by

$$\Delta S_{a \to b} = c \ln \frac{T_b}{T_a}.$$

Exercise 23.

(a) Prove that the net work output of any closed cycle engine is given by $\oint T\, dS$.
(b) For the Carnot cycle, construct the appropriate T–S diagram and evaluate its area in terms of Q_1, Q_2, T_1, and T_2.

Exercise 24. Derive Eq. 3.64.

Exercise 25. How much available work is lost when a liter of gasoline is burned in a catalytic heater at 43°C compared to normal combustion ($T = 2200°C$, the flame temperature)?

Answer: 2.75×10^7 J.

CHAPTER FOUR

HEAT ENGINES

In this chapter we examine in detail several devices that use a high-temperature fluid in the energy conversion process. These belong to a general class of devices called heat engines. We shall use this term to mean anything that accepts heat at one temperature and converts it to or requires work and rejects heat at another temperature in a cyclic fashion.

4.1 INTERNAL COMBUSTION ENGINES

Passenger automobiles and trucks account for about 25 percent of the total energy consumed in the United States. For the most part these vehicles use **internal combustion engines (ICEs)**, either gasoline- or diesel-powered. Since such a large amount of energy is consumed by ICEs, we shall examine their operation carefully to understand their problems and to see what direction should be taken either to improve them or replace them.

Diesel-powered vehicles make up less than 1 percent of all U.S. vehicles and consume about 5 percent of the fuel. But because of the substantially different emissions from diesel engines, there is considerable interest in them for mass transportation. Consequently, we shall also examine the diesel engine and discuss emissions and possible control procedures.

The actual operation of a spark-ignited, reciprocating (gasoline powered), four-cycle ICE is familiar and is diagrammed schematically in Figure 4.1. The gasoline vapor–air mixture is compressed on an upward stroke of the piston followed by combustion, rapid expansion of the gas, and delivery of power to the crankshaft. The next upward stroke clears the cylinder of exhaust; this is followed by a downward stroke that loads the cylinder with another gas–air charge. Each one of these separate processes is irreversible; each has friction, turbulence, chemical change or all of these. The p–V diagram for a real engine with realistic values of operating parameters is shown in Figure 4.2. (The intake and exhaust lines are not shown.) Note that there are no straight lines, no adiabats, and no isotherms. Exact analysis of this cycle to extract work, heat values, and efficiency would be very difficult. What is needed is a simplified *model* having reversible steps that would approximate this cycle. Such a model was developed in the late nineteenth century and has been named after its developer: the Otto cycle.

The Otto Cycle

In the air-standard Otto cycle the following assumptions are made. The working substance

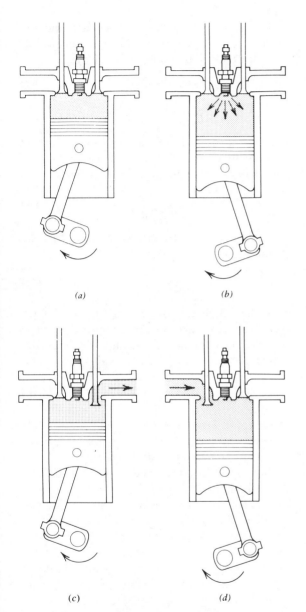

Figure 4.1 The four strokes of the spark-ignited internal combustion engine: (*a*) compression, (*b*) combustion, (*c*) exhaust, and (*d*) fuel-air intake.

is air (effectively an ideal gas), and it undergoes no chemical change throughout the cycle. There is no friction between the piston and the cylinder, and there is no turbulence or ac-

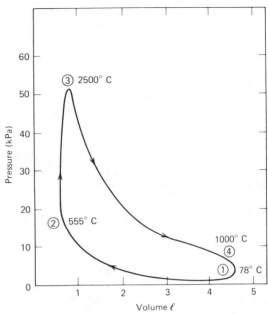

Figure 4.2 $p–V$ diagram for a real ICE.

celeration in the gas nor any other dissipative effect. The following processes are performed:

- A reversible adiabatic compression resulting in a temperature rise from T_a to T_b, path a to b in Figure 4.3.
- A reversible isovolumnic temperature and pressure increase, T_b to T_c and p_c to p_c, path b to c.
- A reversible adiabatic expansion with temperature decrease, T_c to T_d, path c to d.
- A reversible isovolumnic decrease in temperature and pressure, T_d to T_a and p_d to p_a, path d to a.

Line a to e represents the exhaust stroke at pressure p_0, normally atmospheric pressure, and line e to a represents the air–fuel intake stroke, also at p_0.

In comparing Figures 4.2 and 4.3, we can see a rough equivalence between the four processes outlined above and the real engine. Note that these four processes correspond to only

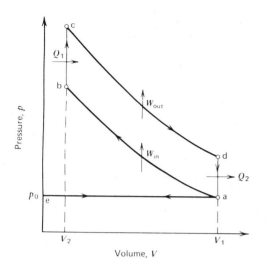

Figure 4.3 The air-standard Otto cycle. Compression: a→c; combustion: b→c and c→d; exhaust: d→a and a→e; air intake e→a.

two of the "strokes" of the real engine: compression and power delivery.

To calculate the efficiency of this cycle we need to know how the heat accepted Q_1 and rejected Q_2 depend on the parameters of the system. From Eq. 3.18 for a isovolumnic process:

$$Q_1 = C_v(T_c - T_b) \qquad (4.1)$$

and

$$Q_2 = C_v(T_a - T_d). \qquad (4.2)$$

The net work W is then

$$W = Q_1 + Q_2; \qquad (4.3)$$

note that $Q_2 < 0$. The efficiency is given by

$$\eta = \frac{W}{Q_1} = \frac{Q_1 + Q_2}{Q_1}, \qquad (4.4)$$

which can be written

$$\eta = 1 - \frac{T_d - T_a}{T_c - T_b}, \qquad (4.5)$$

assuming C_v is a constant throughout the

cycle. We next make use of the fact that the coordinates of two points located on an adiabat in a p–V diagram are related by

$$T_d V_1^{\gamma-1} = T_c V_2^{\gamma-1} \qquad (4.6)$$

and

$$T_a V_1^{\gamma-1} = T_b V_2^{\gamma-1}, \qquad (4.7)$$

where γ is the ratio of heat capacities for the working fluid, air in this case. These expressions can be combined finally to yield

$$\eta = 1 - \frac{1}{(V_1/V_2)^{\gamma-1}}. \qquad (4.8)$$

If we define the **compression ratio** for the engine to be

$$R_c = \frac{V_1}{V_2}, \qquad (4.9)$$

then the efficiency of the air-standard Otto cycle depends only on the compression ratio and the heat capacities of the working substance and no other parameters of the system:

$$\eta = 1 - \frac{1}{R_c^{\gamma-1}}. \qquad (4.10)$$

The compression ratio of a typical automobile is of the order of 9:1. Using $\gamma = 1.4$ for hot air, we obtain an efficiency of $\eta = 0.58$.

The actual performance in a real ICE is about half this value. The ideal processes assumed in the Otto cycle are not realized in practice. Experience indicates that about 33 percent more work is required for the a to b path than is calculated; that is, the idealized adiabatic process is in reality only about 75 percent adiabatic. For example, from a to b an amount of work W_{in} given by

$$W_{in} = C_v \, \Delta T \qquad (4.11)$$

is required by the Otto cycle. In a real engine

$$W_{real} = \frac{W_{in}}{0.75}. \qquad (4.12)$$

The difference between the actual work required and the work calculated must appear as heat loss during this "adiabatic" process. Similarly, the process c to d should be modified to take into account dissipative effects; about 25 percent less work is produced than is calculated.

Example 4.1 Calculate Q_1, W_{in}, W_{out}, Q_2 and η for an air-standard Otto engine having $T_a = 60°C$, $T_b = 400°C$, $T_c = 2180°C$, and $T_d = 855°C$. Consider 1 mol of ideal gas having $\gamma = 1.4$. Then recalculate these same quantities for a real engine as outlined above.

$$Q_1 = C_v(T_c - T_b) = (3 \text{ cal/K})(2453 \text{ K} - 673 \text{ K})$$
$$= 5340 \text{ cal},$$

$$Q_2 = C_v(T_a - T_d) = (3 \text{ cal/K})(333 \text{ K} - 1128 \text{ K})$$
$$= -2385 \text{ cal},$$

$$W_{in} = C_v(T_b - T_a) = (3 \text{ cal/K})(673 \text{ K} - 333 \text{ K})$$
$$= 1020 \text{ cal},$$

$$W_{out} = C_v(T_d - T_c) = (3 \text{ cal/K})(1128 \text{ K} - 2453 \text{ K})$$
$$= -3975 \text{ cal},$$

$$\eta = \frac{Q_1 - Q_2}{Q_1} = \frac{5340 - 2385}{5340} = 55 \text{ percent.}$$

But, in a real engine

$$W_{in} = W_{in} \times 1.33 = 1357 \text{ cal},$$

$$W_{out} = W_{out} \times 0.75 = -2981 \text{ cal.}$$

The heat input Q_1 will remain the same, but the heat lost will now include Q_2 *plus* the differences between the old and new values of W_{in} and W_{out}

$$Q_2' = -2385 + (1020 - 1357) + (-3975 - [2981]),$$
$$Q_2' = -3716 \text{ cal},$$

$$\eta = \frac{Q_1 + Q_2'}{Q_1} = \frac{5340 - 3716}{5340} = 30.4 \text{ percent.}$$

Example 4.2 Use the $p-V$ diagram of the air-standard Otto engine to describe the effect of a faulty spark plug (a one-cylinder engine). Discuss the engine performance.

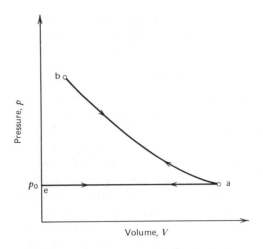

Figure 4.4 $p-V$ diagram for an ICE with no spark.

The $p-V$ diagram in this case is particularly simple, as shown in Figure 4.4. Air–fuel intake is along path e–a. Compression is path a–b, but since there is no ignition, there is no heat addition, and the expansion is back along path b–a. Exhaust follows line a–e. Since the work done is the net area around the path a–b–b–a, the work done is zero.

It would seem the simplest way to improve the performance of an ICE would be to increase the compression ratio R_c. There is, however, an upper limit to this ratio of about 10:1 imposed by the heating effect of the compression stroke. If the gasoline-vapor–air mixture in a real engine becomes too hot, it will ignite prematurely, causing engine knock and reducing performance. The means of combating this effect used extensively by the gasoline industry is to use additives in the gasoline to improve its **octane rating**.[1] But these

[1]The octane rating of a fuel is measured in a standard test engine by comparison with a standard fuel. For example, a rating of 95 octane means the fuel under test would require the same compression ratio as a mixture of 95 percent isooctane and 5 percent heptane to produce knocking in the test engine.

additives have a significant bearing on the environmental effects of spark-ignited ICEs.

Environmental Effects

All commercially available gasolines are mixtures of volatile hydrocarbons and specialized additives. One company advertises its gasoline as containing the following: butanes, isopentanes, isooctanes, toluene, and xylene as the basic fuel with polymeric amide imide as a detergent to reduce hydrocarbon emission, salicylidene propane diamine to reduce gum formation, oxygenated compounds to increase icing protection, phenylene dianine or alkyl phenol to reduce gum formation by oxidation, lead alkyls (tetraethyl lead and tetramethyl lead) to retard detonation and improve octane rating, and finally a rust inhibitor. Not stated as being in the gasoline, but surely there, are organic scavengers, ethylene dichloride and dibromide, which combine with the lead to form compounds that are exhausted.

A given brand of gasoline does not have a constant set of constituents. Changes are continually being made depending on season, geographic demand, and the spot market for the refinery inputs required for gasoline production: naphthas and gas oils. These spot markets fluctuate considerably, responding to international pressures and speculators.

In view of all this, it is not surprising that automobiles account for about 60 percent of the air pollution in the United States. Most of this pollution is a consequence of the operational characteristics of the engine.

Table 4.1 lists automotive emissions, their sources, and typical quantities for a pre-1968 car having no federal emissions standards. In Table 4.2 we show the federal standards for exhaust emissions and their evolution since the passage of the Clean Air Act of 1970. It is a sad comment on the technical expertise of the automotive industry or the political processes of our country or both that the standards for

TABLE 4.1 Automotive Emissions

Source	Pre-1968[a] (No Standards)
Evaporation from fuel tank and carburetor	
Hydrocarbons	210 ppm (2.4 g/veh. mile)
Crankcase blowby	
Hydrocarbons	275 ppm (3.2 g/veh. mile)
Exhaust	
Hydrocarbons	990 ppm (10.5 g/veh. mile)
Carbon monoxide	3.5% (77 g/veh. mile)
NO_x	1500 ppm (5.3 g/veh. mile)
Lead	2.4 g/gal (reg.)
Phosphorous	<0.92% by weight

[a]*Note*: These figures depend on engine speed and driving conditions. Average values are given.

automotive emissions first proposed in 1970 were not met by 1981.

The oil embargo of 1973–1974 and the realization in the late 1970s that energy supplies are indeed limited have brought considerable pressure on the government from energy companies to relax air pollution standards. Federal pressure to increase gasoline mileage has been quite successful, however, so that there is not as strong a movement to relax automotive standards as there is in favor of repealing strong standards for coal-burning electric power generators. Nevertheless, some attitudes change slowly; there will always be sentiment for relaxing or removing federal emissions standards for automobiles. The recent economic difficulties of the U.S. auto industry coupled with the 1980 change in federal administration may change considerably the climate for mandated performance, even in the face of considerable evidence of its efficacy.

Table 4.1 gives three sources of air pollution

TABLE 4.2 Evolution of Federal Automobile Emissions Standards (Grams/Vehicle Mile)

	1970–1971[a]	1972–1974[a]	1975[b]	1976[d]	1977[e]	1978[f]	1980[g]	1981[g]
HC	4.1	3.0	1.5 (0.9)[c]	1.5	1.5	[0.41]	0.41	0.41
CO	34.0	28.0	15.0 (9.0)	15.0	15.0	[3.4]	7.0	3.4
NO_x	None	3.1	3.1	3.1	2.0	[0.4]	2.0	1.0

[a]Imposed administratively by the Environmental Protection Agency (EPA).
[b]Statutory standards by the Clean Air Act of 1970; also set goals for 1977.
[c]California values given in parentheses.
[d]1974 amendment to the Clean Air Act; also suspended future goals.
[e]Interim standards imposed by EPA, except NO_x standard imposed by Congress (P. L. 93–319).
[f]Standards set by EPA, July 1, 1977; subsequently deferred, amended, or both.
[g]Standards set by the amendments of 1977.

associated with automobiles. Fuel evaporation contributes about 15 percent of the automotive hydrocarbon emissions. This can be reduced by controlling the volatility of the fuel by means of additives or by using totally different fuel. The former is being done in Los Angeles County and the latter seems quite unlikely to happen in the immediate future. For diesel fuel, evaporation is not much of a problem, since diesel fuel has a much higher boiling point than gasoline (175° to 400°C, compared to 30° to 200°C).

Unburned fuel can be vented to the atmosphere through the crankcase. During the compression stroke fuel can escape the cylinder into the crankcase past the piston ring seals. Venting the crankcase fumes into the carburetor, required on U.S. automobiles since the 1968 model year (required in California since the 1963 model year), has effectively eliminated this problem. Crankcase blowby is not a problem with diesel engines as no fuel is present in the cylinder during the compression stroke.

Exhaust gases account for about 65 percent of all automotive emissions; they consist of CO_2, H_2O, CO, NO_x, partially burned and unburned hydrocarbons, lead compounds, and traces of other materials. The CO_2 and H_2O would be the only products of complete combustion of the fuel; they are not ordinarily thought of as being pollutants. But in Chapter 12 we see that these two compounds in sufficient quantity can have significant global effects.

Carbon monoxide (CO) results from the combustion of the fuel with an insufficient amount of air. The CO emission versus the air to fuel ratio is shown in Figure 4.5; the CO percentage becomes very small near an air to fuel ratio of about 15:1. The **stoichiometric ratio** would be 14.5:1 determined by balancing this chemical equation for the various fuel components:

$$C_xH_y + (x + \tfrac{1}{4}y)O_2 = xCO_2 + \tfrac{1}{2}yH_2O. \quad (4.13)$$

Example 4.3 What is the stoichiometric air-fuel ratio for the complete combustion of methane (CH_4)?

Using Eq. 4.13 we have $x = 1$, $y = 4$: $CH_4 + 2O_2 = CO_2 + 2H_2O$. Therefore, 2 mol of oxygen is required for each mole of methane. Since air

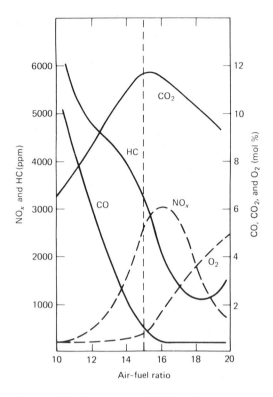

Figure 4.5 Summary of ICE emissions as a function of air to fuel ratio.

is about 20 percent O_2, the air to fuel ratio should be about 10:1.

Engines designed after 1968 have incorporated the correct air to fuel ratio without seriously degrading engine performance. Even with these improvements, it seems unreasonable to expect the ICE to operate with less than about 0.50 percent CO emission.[2] Carbon monoxide can also be removed from the exhaust by various attachments, filters, and so on.

The stoichiometric air–fuel mixture minimizes CO emissions, but enhances NO_x for-

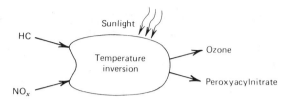

Figure 4.6 Simplified diagram of photochemical smog formation.

mation, as shown in Figure 4.5. Nitric oxide (NO) and nitrogen dioxide (NO_2) production in the ICE is a complex process, but we should try to understand it, as these compounds are significant in the formation of photochemical smog. It is the NO that is produced in high-temperature combustion, but all oxides of nitrogen convert to NO_2 in the presence of air.

Smog production, while not understood in all its detail, can be summarized by a single figure—Figure 4.6. We discuss this problem in more detail in Chapter 13. Without HCs or NO_x, principally from automobiles, smog would not be the very serious problem it is today.

Operating an ICE engine with an extremely lean fuel mixture, say an air to fuel ratio greater than 18:1, would improve the NO_x emissions while not seriously increasing HC emissions. However, it would be difficult to obtain vehicle performance with such operation; misfires would be common (a large percentage of emissions from older engines result from misfires). In addition to the air to fuel ratio the formation of NO_x also depends strongly on the combustion temperature in the cylinder; the higher the temperature, the more enhanced the NO_x production. Studies have shown that combustion in the cylinder does not take place uniformly.

Combustion propagates from the spark to the cylinder walls and piston as a flame front. The time of propagation apparently depends on engine speed, ranging from 10 to 50 ms for completion. The temperature of the flame front

---

[2]It should be noted that, even though an engine may be tuned at the factory to operate at the stoichiometric mixture, a few hours of driving can change its operational characteristics considerably.

itself is substantially higher than the bulk temperature of the gas, about 2700°C compared to 830°C. It is in this high-temperature flame-front region that NO_x is formed; the mechanism by which the high NO_x concentration is maintained after flame-front passage is not well understood and is the subject of research at present.

The only obvious means of reducing NO_x formation is to reduce the flame-front temperature; of course, this should be done without affecting engine performance. This means a certain amount of inert, noncombustible material should be added to the air-fuel mixture to serve as a heat sink, thereby reducing the flame-front temperature but not the bulk temperature of the gas in the cylinder. Experiments using up to 25 percent of the exhaust gases recycled to the intake manifold or injecting water at a rate of about 1.4 kg water to 1.0 kg fuel have yielded reductions in NO_x emissions of up to 80 percent. But these experiments also indicated problems in fuel economy and engine performance. Development of exhaust filters for NO_x removal has proved to be difficult and expensive; however, because of the mechanism for photochemical smog formation, these compounds are among the most important for removal.

Another solution for the NO_x problem in ICEs has been developed by the Japanese firm Honda Motors. The Honda system, called the

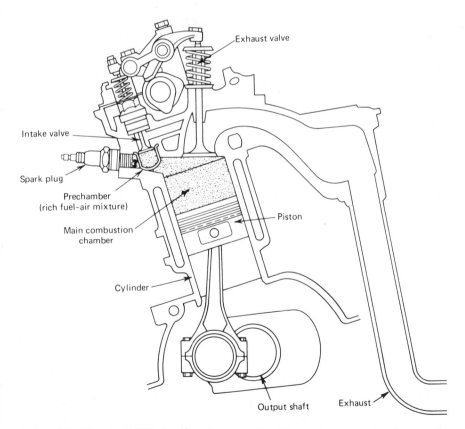

Figure 4.7 Honda CVCC stratified charge engine. The main chamber intake valve is is not shown; it is behind the exhaust valve.

stratified charge engine, uses two combustion chambers. In a small chamber a very rich fuel–air mixture is swirled in as a vortex; the flame front then expands into a second chamber, the regular piston chamber, which contains a very lean mixture. In each chamber the mixture is far away from the stoichiometric ratio; see Figure 4.7. Preparation of the flamefront in the smaller chamber improves combustion in the larger chamber so that reliable performance, free of misfires, is obtained. This engine is able to pass all current EPA emissions standards without additional external devices.

The CVCC (controlled vortex combustion chamber) engine was introduced and certified by the EPA the same year U.S. automakers were pleading to have the 1975 standards deferred as they were "unreachable with current technology." The Mazda rotary and Peugeot diesel engines introduced in the same year also met the 1975 EPA standards. The Ford Motor Company was developing a kind of stratified charge engine for its smaller cars called the Proco for "programmed combustion." It is not clear that the same two-chamber system will continue to function as the engine size is increased. It may be that reliably operating, lean-burning engines will eventually become available in the larger sizes. It is also not clear that Ford will place the Proco in production, since emissions and mile per gallon standards are being met without it.

We should note here that fuel injection is in a sense a stratified charge system, since a very rich region is created by the injector. The Ford Proco is a fuel-injection engine. Flame fronts start in the rich fuel-cloud region, then propagate to the leaner region of the chamber. However, fuel-injection engines, as a rule, have somewhat worse emissions characteristics than two-chamber engines of the same size.

The existence of hydrocarbons (HC) of low molecular weight in the exhaust is difficult to

explain; if the air–fuel mixture has been adjusted appropriately, complete combustion should result. Yet, studies indicate that up to 50 percent of the photochemically active HCs in exhaust is low-molecular-weight HCs not originally present in the fuel. The origin of these HCs was discovered in a series of experiments at General Motors. It was found that during flame-front passage the gas in the cylinder was separated into shells. Near the cylinder wall, out to about $25 \mu m$, there was pure, unburned fuel before and after the flame passage. From about 25 to $125 \mu m$ from the wall, "cracked" fuel products were found— HCs of low molecular weight. Beyond about $125 \mu m$, combustion was more or less complete. This region of partial burning has been called the **quench zone**; whether it exists because of the temperature difference between the cylinder wall and the gas or for some other reason is not known. Clearly, more research on this phenomenon is required. At present the only means of substantially reducing or eliminating HC emission from automobiles is by external attachments to the exhaust.

More than 80 percent of the lead content of the fuel is emitted in the exhaust as lead compounds—chiefly $PbClBr$ and forms of $NH_4Cl(PbCl)_2Br$. The remaining 20 percent is retained in the engine and exhaust system or suspended in the crankcase oil. These lead compounds account for a significant fraction of the particulate matter in automobile exhaust. There exists no technology for removing lead from the exhaust; in fact, the lead compounds effectively "poison" catalytic converters attached to the exhaust to remove CO and HC. The only solution is to remove the lead from the fuel.

The EPA had ruled in 1970 that the lead content of gasoline should be reduced to 0.5 g/gal by 1979 (compared to the 2.4 g/gal figure in Table 4.1). But in 1974 the District of Columbia Circuit Court of Appeals vacated the regulation in response to a suit brought by the

Ethyl Corporation, which claimed that the regulation was without scientific basis—this despite overwhelming evidence of the toxicity of lead and its existence in the environment from old paint and leaded gasoline, almost exclusively.

In 1976 the National Resources Defense Council won a suit to compel EPA to establish ambient air-quality standards for lead. These standards do not directly specify the permissible lead content of gasoline, but they do make it difficult for a community that relies heavily on leaded gasoline transportation to remain in compliance.

The biggest factor in the appearance of unleaded gasoline has been the requirement of catalytic converters on automobiles to achieve the reduced levels of permissible emissions. Catalytic converters are disabled or "poisoned" by lead compounds, and so substitutes for lead as antiknock agents had to be found. These early substitutes also produced environmental risks, as shown in visibility tests of effects from leaded and unleaded gasolines.

Catalytic Converters

To meet the 1975 standards (not yet enforced), U.S. automakers decided to avoid high-risk, low-cost alternatives, choosing instead the low-risk, high-cost catalytic converter. A catalyst is a substance that promotes a chemical reaction but does not enter into it. In essence, a catalyst increases the rate constant for the reaction.

In a chemical reaction the concentration of one of the reactants usually varies with time. Since molecules can react only when they are in reasonably close physical contact, the catalyst must somehow be able to bring reacting species together at a higher rate than would be the case in the absence of the catalyst. The exact mechanism by which this occurs is not well understood; as a consequence, the method to determine the best catalyst for a given purpose is largely one of trial and error.

Catalysts have received wide application in the petrochemical industry, both as promoters of cracking reactions and the reverse, polymerization. They are also used extensively in hydrogenation, desulfurization, and in reactions involving nitrogen.

Virtually every element in the periodic table has been examined as a potential catalyst for automobile exhaust. The catalysts in current devices consist of a support material that is covered with a thin layer of a promoter material. The support material is made up of small granules of a porous refractory inorganic oxide, such as alumina or silica, or of calcined clays. The promoter, usually a metal or metallic oxide, covers the support to a few molecules' depth. The thin layer is necessary so that the support material's pores do not become clogged. A high porosity is desirable as it increases the exposed promoter area, but a compromise between porosity and mechanical support is required. A 20-kg catalytic bed has an effective area in the neighborhood of $10^6 \, m^2$ (about 200 acres).

The catalytic converters in use today are oxidizers:

$$2CO + O_2 \rightarrow 2CO_2, \qquad (4.14)$$

$$4HC + 5O_2 \rightarrow 4CO_2 + 2H_2O. \qquad (4.15)$$

Clearly, an oxidizing reaction is of little use in controlling NO_x. But catalyzing reduction reactions is much more difficult, and has not been achieved in a readily marketable form.

Control of NO_x emissions today relies mainly on **exhaust gas recirculation** systems (EGR). In these systems part of the exhaust gas is diverted back into the cylinders along with the combustion air. This lowers the flame-front temperature, thereby reducing the production of NO_x. Future automobiles will need EGR, lean-burn engines, and reduction catalysts to achieve the 1981 NO_x standards.

Converters currently in use require a rather expensive metal, platinum.[3] Aside from that, there are other difficulties with catalysts. A minimum temperature is required for catalytic action; during engine warmup undesirable emissions escape into the atmosphere through the exhaust system. On the other hand, if the temperature becomes too high, the support material begins to lose surface area by pore growth and evaporation. Pores can also be clogged by particulate matter in the exhaust—especially the lead compounds. It is, then, especially important to remove lead from gasolines. Some early fears were that acid sulfur compound fixation would be enhanced by catalytic converters, but these have proven to be groundless.

Catalysts, in combination with EGR and other techniques, may be effective, at a high price, in controlling auto emissions at some arbitrary level. In the long range, one of two alternative solutions may be appropriate: replacement of gasoline with methane, propane, or other fuel, or replacement of the ICE in automobiles with a propulsion device less prone to undesirable emissions or one in which the emissions can be more easily controlled. Each of these alternatives represents a drastic departure from the current situation. Each would have enormous economic ramifications; for this reason alone these alternatives can only be considered as long range. But even if the economics were not an overriding factor, what alternative propulsion systems are there?

Diesel Engines

One alternative has been available for some time. The diesel engine was developed in the late nineteenth century as a replacement for

the steam engine. The rather poor efficiency—about 7 percent then—of the steam engine and the pressure of a burgeoning industrial economy provided the impetus for the development of new engines of all types. The diesel engine, the steam turbine, and even the gasoline-powered ICE as we know it all resulted from these early efforts.

The diesel engine has not been used very much in private automobiles because, until recently, it was much heavier and bulkier than spark-ignited gasoline-fueled ICEs of the same horsepower. But developments in materials technology have changed this picture. Automobiles with diesel engines are on the market. This engine may be an important alternative to the ICE, since automobiles with it meet the interim emissions standards *without* using catalytic converters. For this reason we shall consider the operation of the diesel engine in some detail.

In the diesel engine the chamber contains only air during the compression stroke. At the appropriate point, when the air has been heated to a high temperature by the compression, fuel is injected under pressure into the chamber. Spontaneous combustion takes place, forcing the piston downward on the power stroke. When the engine is first started, it may be too cold for spontaneous combustion. A glow plug is used under these circumstances; it is shown in Figure 4.8.

Except for the ignition by compression and the conditions under which combustion occurs, the Diesel cycle is similar to the Otto cycle. In the model cycle for the diesel engine, called the air-standard Diesel cycle, the combustion is assumed to occur isobarically. Thus, in Figure 4.9, b–c represents combustion, c–a the power stroke, a–e exhaust, e–a air intake, and a–b compression.

We can calculate the efficiency of this idealized cycle easily:

$$Q_1 = C_p(T_c - T_b) \qquad (4.16)$$

---

[3]A commercially available catalyst, PTX from Englehard Minerals and Chemicals, consists of 0.5 percent (by weight) platinum on silicon dioxide—aluminum dioxide in a honeycomb structure.

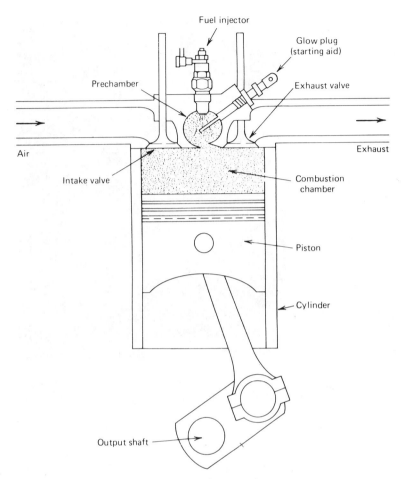

Figure 4.8 Cutaway drawing of a diesel engine.

and

$$Q_2 = C_v(T_a - T_d). \qquad (4.17)$$

$$\eta = \frac{Q_1 + Q_2}{Q_1}, \qquad (4.18)$$

but where $Q_2 < 0$.

So

$$\eta = 1 + \frac{Q_2}{Q_1} = 1 + \frac{C_v(T_a - T_d)}{C_p(T_c - T_b)} \qquad (4.19)$$

or

$$\eta = 1 - \frac{1}{\gamma} \frac{(1/r_E)^\gamma - (1/r_C)^\gamma}{(1/r_E) - (1/r_C)}, \qquad (4.20)$$

where $r_E = V_1/V_3$, the expansion ratio, and $r_C = V_1/V_2$, the compression ratio.

So again, as in the Otto cycle, the efficiency of the Diesel cycle is expressed in terms of compression ratios. But in this case, since only *air* is compressed and there is no danger of preignition, much higher compression ratios are possible. In theory, with $r_C \sim 15$, $r_E \sim 5$, and $\gamma = 1.5$, an efficiency of 64 percent should be possible. But, just as before, the many irreversible processes in practice lower this figure by about half. The diesel engine is nevertheless more efficient than its equivalent horsepower gasoline-powered counterpart.

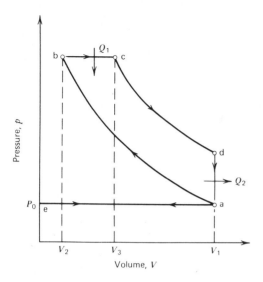

Figure 4.9 The air-standard Diesel cycle.

Figure 4.10 The world's largest diesel engine, rated at 47,300 hp. (Photograph courtesy of B&W Diesel A/S.)

Example 4.4 In a particular diesel engine, air is compressed from atmospheric pressure and room temperature to 1/15 its original volume. Assume a reversible adiabatic compression; find the final temperature.

Use $TV^{\gamma-1} = $ const or $T_1 V_1{}^{\gamma-1} = T_2 V_2{}^{\gamma-1}$.

$$T_2 = T_1 \left(\frac{V_1}{V_2}\right)^{\gamma-1} = 293(15)^{0.4} = 866 \; K.$$

Of course, in order to withstand the much higher compression ratio, the engine must be substantially heavier and beefier. For many applications—heavy-duty trucks, ships, and power-generating plants, for example—there are no particular weight penalties. Figure 4.10 is an example of a very large marine diesel engine. The high efficiency, cheaper fuel, and exceptional mechanical reliability have made the diesel an obvious choice. It is only recently that materials technology has allowed compact, high-efficiency diesel engines suitable for passenger automobiles to be built.

There are some difficulties with passenger diesel automobiles. Lubricating oils for diesels are particularly temperature-sensitive; as the climate changes seasonally, the oil must be replaced. Many service stations do not sell diesel fuel, and many of those that do are not open on Sundays, so let the buyer beware!

Diesel fuel is essentially the middle weight fraction of distillate fuel oils, with the volatile and heavier substances removed. A substance heavier than gasoline is required because the fuel must be injected into the cylinder under high pressure, more than 35 atm, forming small-diameter particles and igniting in a controlled fashion. Diesel fuel is rated by a *cetane rating*; this is a measure of the ease of ignition. Like the octane rating for gasoline, the cetane rating is obtained by comparing performance of a test engine with the fuel in question and

TABLE 4.3 Comparison of Diesel Fuel with Other Fuels

Fuel	Average (kg/ℓ)	Average (Btu/gal)	Average (J/ℓ)
Diesel	0.85	137,750	38.4×10^6
Gasoline	0.72	123,500	34.4×10^6
Propane	0.50	91,800	25.6×10^6
Butane	0.58	102,400	28.6×10^6

with a standard fuel of cetane and α-methyl-naphthalene, a poor igniter. Diesel oil is compared with other fuels in Table 4.3.

Because of the differences between diesel fuel and gasoline and the operation of compression-ignited and spark-ignited ICEs that they fuel, diesel-engine emission problems are substantially different. They have less than 0.1 the CO of gasoline ICEs, about the same HC, and perhaps slightly more NO$_x$. These emissions can be controlled by EGR. The major emission problems are those of smoke and odor.

Diesel smoke levels can be controlled; 1970 federal regulations require that static smoke emission from diesel engines not reduce transmission of a light beam by more than 20 percent. Barium additives can reduce the smoke level by 50 percent in concentrations of less than 0.25 percent. The chemical reactions taking place are not clearly understood, but the barium is exhausted primarily as $BaSO_4$. This compound is insoluble and cannot easily find its way into the human food chain. Soluble barium compounds are toxic and are present in diesel exhaust at levels that are thought to be harmless. One difficulty in obtaining smoke-free diesel operation is that proper maintenance is critically important. Adjustment of the diesel engine to obtain higher power by increasing the fuel comsumption is a common practice, but this also increases the smoke output.

The reason for the characteristic diesel odor is at present unknown, even though a substantial amount of research has been done to discover it. It is apparently caused by HCs of high molecular weight (greater than 10 carbon atoms) that are somehow produced during combustion. Catalytic devices may aid in reducing odor, but, clearly, much more effort needs to be expended in determining the origin of the odor before it can be effectively combated.

The argument against standard automotive engines can be made on two grounds: the undesirable emissions and the inefficiency of energy conversion, both chemical and mechanical. The standard engine is of the reciprocating type; that is, the linear motion of the piston is converted to rotary motion of the drive shaft through a system of linkages to the crankshaft. Energy loss at the many bearing surfaces is unaviodable. What is more, because of the unsymmetrical motion of the pistons, vibration is a serious problem, and the engine block and mounts must be heavy and rugged. The search for a better design began almost as soon as the limitations to the reciprocating ICE were realized, in the 1920s.

Rotary Engines

Rotary ICEs have two advantages over reciprocating ICEs: (1) they are virtually vibration-free and have no loss of power in complex linkages, and (2) they are substantially more compact for the same power output. Over the years, many rotary ICEs have been designed and built. Of the several classes of rotary engines, one type, the eccentric rotor engine, has been marketed in an automobile, the Mazda, produced by Toyo Kogyo of Japan. The Mazda was announced with some fanfare in the early 1970s, but is no longer available in the United States. The engine in this car is an improved version of the original design by Felix Wankel. Since it could have some significant advantages over the reciprocating

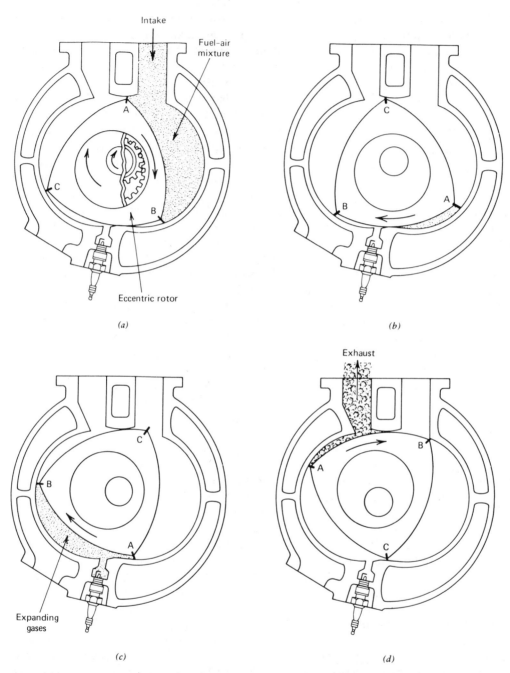

Figure 4.11 The Wankel rotary ICE. (a) Fuel–air intake between seals A and B, (b) compression, (c) power, and (d) exhaust.

engine, as well as some distinct disadvantages, we discuss it in more detail.

The operation of the Wankel engine is illustrated in Figure 4.11. In the Wankel the four strokes of the reciprocating engine are reproduced as the three-cornered rotor turns on the eccentric gear. In Figure 4.11a, the fuel–air mixture enters the combustion chamber in the region A to B. In 4.11b this mixture is compressed, while another charge enters in region C to A. In 4.11c, ignition has occurred, and the gases are beginning to expand to provide the power. In 4.11d the original section of the combustion chamber is being exhausted, the A to C region is providing power, and the fuel–air charge in region B to C is being compressed.

The Wankel engine is one third the size of a conventional reciprocating engine (V-8), delivering the same horsepower with half the weight and about one third as many moving parts. A typical American 195-hp V-8 has a mass of about 1000 kg, occupies about $1.6 \, m^3$, and has about 1000 parts, 390 of which are moving parts. A 185-hp Wankel would have a mass of about 500 kg, occupy about $0.5 \, m^3$, and have about 600 parts, 150 of which move. The Wankel is expected to be about 35 percent cheaper to manufacture than its reciprocating counterpart.

The Mazda Wankel is not an inherently cleaner engine than a comparable reciprocating type. It has a compression ratio of 9.4:1, but it operates on a cheap, low-octane fuel. The air–fuel mixture is rich, as reports indicate a somewhat worse fuel consumption, 12 to 16 miles/gal, than was expected from such a small engine. This was done on purpose, since a rich mixture tends to minimize NO_x at the expense of CO and HC. But both CO and HC are effectively controlled in the Mazda by thermal reactors. The small size of the Wankel is very important in this regard, since substantial under-the-hood space is made available for the thermal reactor.

A thermal reactor is simply a hold-up tank or "stove" with air injection in which the CO and HC emissions have time to oxidize to more benign species. This system can easily meet the EPA interim standards when used with smaller engines. It is not clear that sufficient reactor volume could be found to handle the emissions from larger engines.

The major problem with the Wankel rotary concerns the seals at the contact points of the rotor and the trochoidal chamber. These seals will have to be very durable in a highly corrosive environment. In the Mazda engine they are made of a carbide–aluminum alloy that was expected to last 60,000 to 100,000 miles before requiring replacement, a job demanding complete engine teardown. Early operating experience with the Mazda was not entirely satisfactory. The seals wore out quickly, causing reduced gas mileage and increased emissions. Improvement in materials and design may bring the rotary ICE back as a contender for the future. But for now it does not seem to be the answer.

The importance of the ICE in man's environment cannot be overstated; it is estimated that in Los Angeles alone in 1968, automobiles emitted 1700 tons/day of HC, 9500 tons/day of CO, and 620 tons/day of NO_x. The emissions problem has been approached on three fronts: fuel technology, engine technology, and exhaust conversion device technology. Perhaps a fourth should be added: adequate maintenance and inspection. Older, marginally operating automobiles must be removed from the highways if a low level of emissions is to be obtained. The emissions problem should be looked at as a systems problem with all parts being optimized. It may be possible to conclude in the future that further improvements of ICEs are not possible, and that alternative propulsion systems should be used—several automobile manufacturers have looked or are looking others now. The steam or gas turbine, the external combustion engine, and battery power are being considered.

4.2 TURBINES

Since their perfection in about 1890, turbines have become the mainstay of electrical energy generation and aircraft and ship propulsion. Turbines provide a very efficient means of converting the internal energy of a hot fluid into rotational energy of a shaft. Turbines have a low capital investment per shaft horsepower output, low maintenance cost, high conversion efficiency, and uniform angular velocity with vibration-free operation. The original turbines were small, a few hundred horsepower, and were built for military ships; one of the largest built to date was a unit to generate 1300 MW of electrical power for TVA. The automobile industry has examined turbines as possible propulsion units. Because of the widespread use of turbines, we examine their operation in general terms.

In a turbine a fluid enters and leaves in steady flow with an amount of mechanical work W_s being produced. The motion of the fluid in the turbine is complex; we shall have to understand some fundamental properties of fluid flow before we examine turbines in detail.

Fluid Flow

A moving fluid, gas or liquid, can carry momentum and kinetic energy. It can do work; it can be characterized by an internal energy. Heat can be transferred into or out of a flowing fluid. In the most general case, a wide variety of effects can occur.

In Figure 4.12, we indicate a generalized flow process. Before we proceed to discuss some specific types of processes, let us examine this generalization. In the figure, w' is the work output per unit mass of the device, and q is the heat input per unit mass. The quantities p_i and p_f are the (constant) inlet and

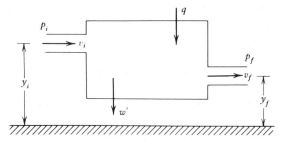

Figure 4.12 Generalized flow device.

outlet pressures, and the v's are the corresponding uniform speeds. The total energy of a mass of fluid at the inlet and outlet can be written

Inlet energy: $\quad E_i = U_i + p_i V_i + \frac{1}{2} m v_i^2 + mgy_i,$
$$\tag{4.21}$$

Outlet energy:

$$E_f = U_f + p_f V_f + \frac{1}{2} m v_f^2 + mgy_f. \tag{4.22}$$

The U's are internal energy functions; the pV terms represent work done on and by the fluid at the inlet and outlet, respectively. The remaining terms are obviously kinetic and potential energy. The difference between the total energy of this fluid at the inlet and the outlet must, by the first law, be equal to the net work and net heat transferred into or out of the fluid:

$$mq - mw' = E_f - E_i. \tag{4.23}$$

This can be written

$$Q = U_f - U_i + mg \left(y_f - y_i + \frac{v_f^2}{2g} - \frac{v_i^2}{2g} \right)$$
$$+ p_f V_f - p_i V_i + W_s, \tag{4.24}$$

where Q and W_s are related to q and w' in an obvious way.

This result is too general to illuminating. To develop a clearer quantitative understanding of fluid flow, let us examine three specific processes: the throttle, the nozzle, and the general turbine, as shown in Figure 4.13.

In the throttle, the process is assumed to be

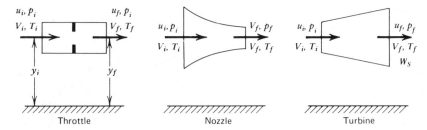

Figure 4.13 Three specific types of flow devices: the throttle, the nozzle, and the turbine.

adiabatic and reversible and the flow horizontal so the velocities of the fluid are equal. Applying these criteria to Eq. 4.24 yields

$$0 = U_f - U_i + p_f V_f - p_i V_i$$

or

$$U_i - p_i V_i = U_f + p_f V_f. \qquad (4.25)$$

In the case of a nozzle there is a change in speed of the fluid, because of a change in cross-sectional area of the pipe; we will assume that there is horizontal flow and no work is done, nor is there any transfer of heat. Equation 4.22 becomes

$$0 = U_f - U_i + mg \left(\frac{v_f^2}{2g} - \frac{v_i^2}{2g} \right) + p_f V_f - p_i V_i. \qquad (4.26)$$

In the turbine, of course, work is done, but we assume that there is horizontal flow and that the passage of the fluid is so rapid that there is no heat exchanged. Equation 4.22 becomes

$$0 = U_f - U_i + mg \left(\frac{v_f^2}{2g} - \frac{v_i^2}{2g} \right) + p_f V_f - p_i v_i + W_s. \qquad (4.27)$$

These three results and calculations concerning them could be simplified if a new thermodynamic variable were defined.

Enthalpy

In a constant-pressure process, the first law may be written

$$Q = U_2 - U_1 + p(V_2 - V_1). \qquad (4.28)$$

If a new thermodynamic variable, the **enthalpy** H, is defined:

$$H = U + pV, \qquad (4.29)$$

then the first law can be written

$$Q = H_2 - H_1. \qquad (4.30)$$

Enthalpy is a measure of the heat in an *isobaric* process. Some textbooks do not make it clear that enthalpy can be interpreted as heat only under this special condition. We have defined heat as the *transfer* of energy, so the phrase "heat content of a system" has no meaning. However, it is possible to calculate the enthalpy of a given system in a given state. Many calculations involving fluids used in energy conversion devices make use of this fact, and extensive tables of enthalpy covering a wide range of temperatures and pressures for various fluids such as steam have been prepared. Extracts of steam tables appear in Appendix B; we shall discuss their use shortly.

Using the definition of enthalpy, Eq. 4.25 for the throttle can be written

$$H_i = H_f. \qquad (4.31)$$

Flow through a throttle, then, is *isenthalpic*.

For the nozzle, Eq. 4.26 becomes

$$\tfrac{1}{2}mv_f^2 = \tfrac{1}{2}mv_i^2 + H_i - H_f. \qquad (4.32)$$

The equation of continuity assures us that in a nozzle the fluid velocity must increase. That being the case, the enthalpy must decrease.

In a turbine a stationary set of blades always acts as a nozzle directing the fluid onto the moving blades. The work done by this fluid in passing through the moving blades is given by Eq. 4.27 with the definition of enthalpy

$$W_s = H_i - H_f + \tfrac{1}{2}mv_i^2 - \tfrac{1}{2}mv_f^2. \qquad (4.33)$$

In most designs the velocity change of the fluid is small, so that Eq. 4.33 can be written

$$W_s = \Delta H. \qquad (4.34)$$

Example 4.5 Steam expanded through a particular nozzle goes from an initial specific enthalpy of 600 cal/g to a final specific enthalpy of 550 cal/g. If the initial steam velocity is 300 m/s, what is the exhaust steam speed?

Write Eq. 4.32 in terms of specific quantities:

$$\tfrac{1}{2}v_f^2 = \tfrac{1}{2}v_i^2 + (h_i - h_f),$$
$$v_f^2 = (300)^2 + 2 \cdot (600 - 550) = 316 \text{ m/s}.$$

Properties of Real Substances

In order to understand in detail the steam turbine system, we need to develop a little more formalism, since we are now studying a real gas, steam, for the first time. And real gases are different from ideal ones. Real substances can change **phase**, from solid to liquid to vapor. Figure 4.14 is the p–V diagram of a pure substance (note that lower-temperature isotherms are not shown, so the solid phase is not indicated). The dashed line in this figure is the **saturation curve**. This separates the purely liquid, partially liquid, and purely vapor portions of the diagram. Above the **critical isotherm** no

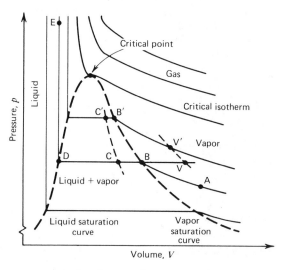

Figure 4.14 Isotherms of a pure substance.

combination of pressure and volume will result in liquid; vapor in this region is called a **gas**.

If a small amount of water is placed in a large container that has previously been evacuated, the condition of the result is indicated by the point A in Figure 4.14. If the resulting vapor is compressed slowly and isothermally, the pressure will rise until the point B is reached. At this point the vapor content is as high as it can be at that temperature; the vapor is said to be **saturated**. Continued compression will result in some condensation of vapor to liquid so that from B to D, constant pressure and constant temperature, an increasing fraction of vapor is condensed to liquid. At the point D there is only liquid, no vapor; the liquid is said to be saturated. Continued compression will have little effect on the volume, so the line D–E is almost vertical.

The same processes would occur along any isotherm below the critical isotherm. Along any two isobar–isotherms inside the **vapor dome**, there will always be two points having the same vapor fraction. Lines through such points, such as C–C′, are called curves of **constant quality**, as in Figure 4.14.

Points B and B′ in Figure 4.14 lie on the vapor saturation curve. If the vapor is isobarically heated from point B to point V, it is said to be **superheated**. The temperature difference between the two isotherms at B and V is called the number of degrees of superheat. Points V and V′, which lie an equal number of degrees of superheat away from the vapor saturation curve, lie on curves of constant superheat.

An ideal gas does not exhibit the properties shown in Figure 4.14. At the critical point the slope of the critical isotherm must be zero:

$$\left.\frac{dp}{dV}\right)_{T=T_c} = 0. \qquad (4.35)$$

There must also be an inflection point at p_c, V_c, T_c:

$$\left.\frac{d^2p}{dV^2}\right)_{T=T_c} = 0. \qquad (4.36)$$

For an ideal gas the only temperature for which Eq. 4.35 is satisfied is $T_c = 0°$. But there are several equations of state for real gases that have had varying degrees of success.

The Rankine Cycle

We needed to establish some of the properties of real gases because in the steam turbine the working fluid changes phase twice. The idealized model representing the steam turbine or steam engine is called the Rankine cycle. It is as approximate as the Otto or Diesel cycles and is as useful in calculating efficiency. The cycle, illustrated in Figure 4.15, consists of a series of reversible processes:

1. Fluid is pumped to the boiler, path a–c; this is assumed to occur in two steps. From a–b it is isovolumnic with W_{in} required; from b–c it is isobaric with some heating occurring.
2. From c–d isobaric, isothermal vaporization of the fluid occurs; an amount of heat Q_{in} is transferred.

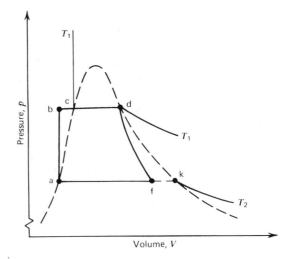

Figure 4.15 Curves of constant quality and constant superheat.

3. An adiabatic expansion with work output W_{out} occurs from d–f.
4. The fluid undergoes an isothermal condensation from f–a with the rejection of Q_{out} units of heat.

The efficiency of conversion in this case, as in all engines, is just the work output divided by the energy input or

$$\eta = \frac{W_{out}}{Q_{in}} = \frac{Q_{in} - Q_{out}}{Q_{in}}. \qquad (4.37)$$

Since processes b–d and f–a are isobaric, the energy transferred during them, Q_{in} and Q_{out}, can be written in terms of the enthalpies at the end points of those processes (from Eq. 4.30):

$$\eta = \frac{(H_d - H_b) - (H_f - H_a)}{H_d - H_b}. \qquad (4.38)$$

Extensive tabulations of specific enthalpy for steam and pure water (Appendix B) as well as other substances, as a function of temperature, have been available for many years. But these values, as well as specific volume, are tabulated only for the saturated states.

The enthalpy at point f would not be listed

since at this point a mixture of vapor and liquid exists. A means of determining the vapor fraction at point f is required. If another thermodynamic parameter existed, one that was constant during an adiabatic process and that could be easily calculated for pure substances, then the vapor fraction could be determined.

We have already defined such a thermodynamic variable: entropy. We see from Eq. 3.47 that if $dQ = 0$, $dS = 0$; an adiabatic process is also **isentropic**. Then, if the entropy for steam at d is S_{sd}, if the vapor fraction at f is y, and if the entropy of steam at f is S_{sf} and for water at f is S_{wf}, then

$$yS_{sf} + (1 - y)S_{wf} = S_{sd}. \qquad (4.39)$$

Once the vapor fraction y is known, the enthalpy of the combined steam–water system at f H_f can be calculated:

$$H_f = yH_{sf} + (1 - y)H_{wf}. \qquad (4.40)$$

The T–S diagram for the Rankine cycle is shown in Figure 4.16. Values of entropy are also tabulated, usually along with enthalpy in

steam tables, so the determination of the vapor phase at point f should be straightforward. The calculation of the efficiency along the lines outlined above yields, for reasonable values of T_a and T_b, a value of about 45 percent. In actual practice a figure near 30 percent would be obtained. The difference between these two efficiencies represents the extent to which the Rankine cycle is too much of an idealization. In a real system all the processes are irreversible and losses due to friction, radiation, and conduction must be taken into account. However, improvement in performance almost to the Rankine cycle limit can be made by carefully examining the p–V diagram and attempting to increase the area enclosed by the curve.

Example 4.7 Given the following operating conditions for a Rankine engine as in Figure 4.16:

$$p_d = 1555.099 \text{ kPa}, \qquad T_1 = 200°\text{C},$$
$$p_f = 1.7055 \text{ kPa}, \qquad T_2 = 15°\text{C},$$

find the liquid fraction at f.

From the steam tables, Appendix B:

$s_d = s_f = 6.4321 \text{ kJ/kg K}$ (saturated vapor),
$s_a = 0.2244 \text{ kJ/kg K}$ (saturated liquid),
$s_k = 8.78 \text{ kJ/kg K}$ (saturated vapor),
$s_d = s_f = X_f s_a + (1 - X_f)s_k$,
$X_f = $ liquid fraction at f,
$6.4321 = 0.2244 X_f + (1 - X_f)8.78$,

$$\boxed{X_f = 0.274.}$$

If point d of Figure 4.16 were extended to the right, far into the vapor region, the p–V and T–S diagrams would be as shown in Figure 4.17. This requires that the steam be superheated. There are limits to the temperature to which this process can be carried

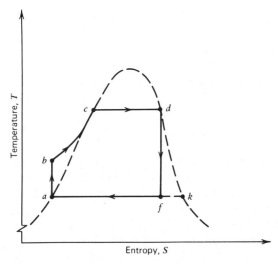

Figure 4.16 T–S diagram for the Rankine cycle.

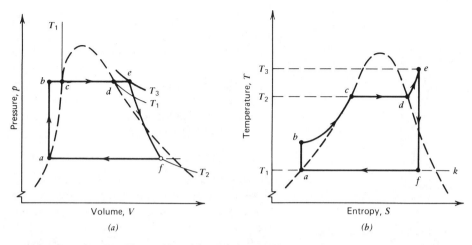

Figure 4.17 The Rankine cycle with superheat: (*a*) *p–V* diagram; (*b*) *T–S* diagram.

because of the limitations of the materials used to transport the steam from boiler turbine.

Superheat increases the efficiency of the cycle. Refer to Figure 4.18; the Rankine cycle with superheat shown can be thought of as three independent cycles. Cycle A is a pumping cycle added to the Carnot cycle B. Cycle C is the superheat cycle. By using the numbers indicated in the figure, it can be shown that the

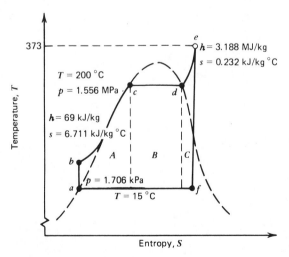

Figure 4.18 Some typical operating values for a Rankine cycle with superheat.

superheat cycle contributes about 12.7 percent of the total heat input and requires about 16.1 percent of the work input and that the overall efficiency is 40 percent. Without superheat the efficiency would be 38.4 percent.

In addition to improving the net efficiency slightly, superheating also helps to relieve a major problem encountered in steam turbine operation: that of wet steam corrosion. The steam–water mixture in the low-pressure end of the turbine, that is, along path h to f in Figure 4.17, is highly corrosive, causing pitting of the blades and pipes.

Another method of improving the performance of the system is to duct the steam back to be reheated after it has done some work in the high-pressure turbines. The *T–S* diagram corresponding to superheat with reheat is shown in Figure 4.19. The increase in efficiency with reheat may be very small, but the reduction in the steam–water path (i.e., h to k in Figure 4.19) makes the procedure more than worthwhile.

Example 4.8 In Figure 4.19 assume the *p*, *T*, *h*, *and s* values from Figure 4.18 for the points a to e with the added values of $s_h = 6.711$ kJ/kg K (assume h lies on the vapor

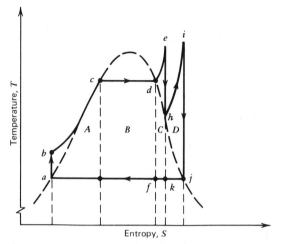

Figure 4.19 *T–S* diagram for a Rankine cycle with superheat and heat.

saturation line), $p_h = 650.4$ kPa, $T_h = 162°C$, and $h_h = 2.753$ MJ/kg. Assume also the values at the point i are $p_i = 650.4$ kPa, $h_i = 3.247$ MJ/kg, and $s_i = 7.628$ kJ/kg K. Find the overall efficiency of the total cycle A + B + C + D.

The heat added during the process $h \rightarrow i$ is given by $q = \Delta h = 495$ kJ/kg. The heat rejected during the process $j \rightarrow k$ is given by $q_r = T \Delta s$ ($i \rightarrow j$ is isentropic or adiabatic), $q_r = -2.647$ MJ/kg. The work done is $q + q_r = 104.7$ kJ/kg. These numbers added to those previously obtained make the overall efficiency $\eta = (6.72 \times 10^5)/(1.643 \times 10^6) = 0.409$.

If the efficiencies of the three cycles of Figure 4.18 were calculated separately, that of cycle C would be the highest and cycle A the lowest. Of course, this is because in cycle A heat is added at a low temperature. This situation can be improved by bleeding off steam at various temperatures from the turbine set and using it to preheat the condensate in stages before it is returned to the boiler. This process, known as regenerative heating, is diagramed in Figure 4.20. The correspondig *T–S* diagram is shown in Figure 4.21.

By the use of superheat, reheat, and regenerative heating, modern steam turbine

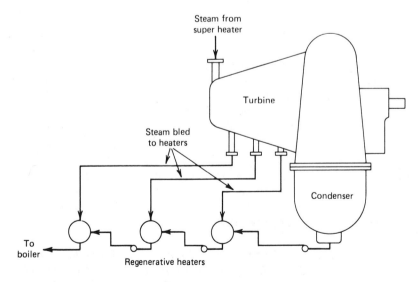

Figure 4.20 Steam is bled off from the turbine set at intermediate pressures to provide for regenerative heating of the condensate.

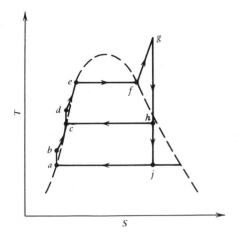

Figure 4.21 *T–S* diagram of regenerative heating with a Rankine cycle.

plants are able to achieve efficiencies near the Carnot maximum for the temperatures involved. Because of the low cost per kilowatt-hour, reliability of operation, and relatively high efficiency, steam turbine generators will probably continue to be the single most important method of generating electricity, at least for large, central-station installations. In circumstances where higher temperatures are desirable or unavoidable, however, fluids other than steam may be appropriate. Table 4.4 lists several fluids that have been used or studied for use in Rankine cycle systems.

TABLE 4.4 Fluids for Rankine Cycle

Material	Melting Point	Boiling Point
Water	0°C	100°C
Diphenyl	70°C	254°C
Aluminum bromide	98°C	264°C
Mercury	−39°C	357°C
Sulfur	113°C	480°C
Rubidium	38°C	700°C
Sodium	98°C	881°C

We have seen that Rankine cycle devices are relatively efficient in terms of the First Law. What about the Second Law efficiency, which we defined in Chapter 3? To calculate this value we need to know the minimum work required to perform the same process. The minimum work is surely just the amount of energy available from the fuel. In most power plants, more than 90 percent of that energy is used to produce steam, so we would expect steam turbine systems to have rather high Second Law efficiencies, virtually the same as the first law value. This is indeed the case. The only way to improve the energy utilization would be to find a method of converting the fuel energy directly to electricity. Such techniques do exist, and we examine some of them in Chapter 5.

Gas Turbines

The steam turbine was developed as an improvement over the reciprocating steam engine, which we have not discussed since it has largely disappeared from the scene, although some people would like to see it revived in automobiles. The turbine enabled designers to use much higher temperatures and therefore higher efficiencies and work outputs. For the industrialized world of the late nineteenth and early twentieth centuries, turbine engines revolutionized marine transport and electric power generation. A little later a new industry began to develop that also needed powerful, lightweight propulsion systems: aviation. The steam turbine was not the answer: it was too heavy, it required a large supply of water as well as fuel, the water needed to be condensed, and changes in the speed of rotation could not be made rapidly. These requirements and problems led to the development of the high-speed gas turbine for the aviation industry. Recently, attempts have been made to use a gas turbine for automobile propulsion. The operation of the gas turbine is

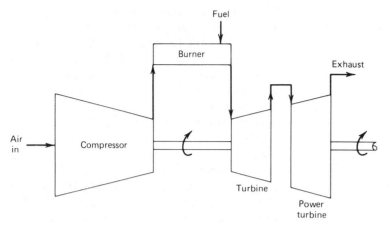

Figure 4.22 Schematic diagram of an open-cycle gas turbine engine.

substantially different from that of a steam turbine; we shall examine the thermodynamics of that operation and, later, the related environmental effects.

A typical gas turbine is shown schematically in Figure 4.22. Air is compressed by a compressor; it is then mixed with the fuel and ignited. The hot exhaust gases are expanded through one or more turbines. The physical design of these turbines is essentially the same as for steam turbines, differing only in size and construction material. The first turbine is used to provide mechanical power for the compressor. Additional turbines may be used to supply mechanical power for auxiliary electrical generation, or, in the case of the turboprop aviation engine, power for the propeller. In the turbojet engine the change in momentum between the air at the compressor input and the high-speed exhaust gases at the turbine outlet serves as the propulsion system.

To analyze the operation of a gas turbine it

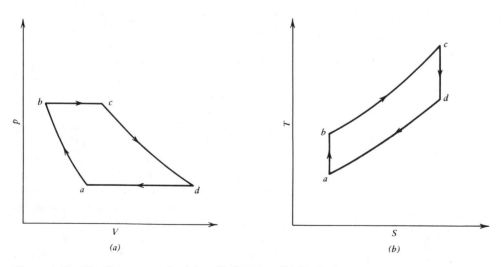

Figure 4.23 The Brayton cycle: (a) p–V diagram; (b) T–S diagram.

is necessary to have an idealization of the cycle, as for other heat engines. This idealization, called the **Brayton cycle**, is illustrated in Figure 4.23. Step a to b is an adiabatic (isentropic) compression followed by isobaric heating from b to c. The compressed hot gas expands through the turbine from c to d, at that point having a temperature higher than at the beginning, point a, but having the same pressure. In a closed system the gas would be cooled to T_a and the process repeated. In an open system the atmosphere serves as a heat sink for the isobaric process d to a.

The efficiency for this engine, as for all engines, is given by

$$\eta = \frac{Q_{in} - Q_{out}}{Q_{in}}. \tag{4.41}$$

Since the heat is added and rejected during isobaric processes, we could calculate the efficiency using enthalpy, as in the Rankine cycle. It is more instructive, however, to study the temperature dependence of the efficiency explicitly. Assuming a constant heat capacity C_p we have

$$Q_{in} = C_p(T_c - T_b) \tag{4.42}$$

and

$$Q_{out} = C_p(T_d - T_a). \tag{4.43}$$

In these terms we have for the efficiency

$$\eta = \frac{(T_c - T_b) - (T_d - T_a)}{(T_c - T_b)}. \tag{4.44}$$

The pressures at the four points are related:

$$\frac{p_a}{p_b} = \frac{p_d}{p_c}, \tag{4.45}$$

and, if we assume an ideal gas, the temperatures are also related:

$$\frac{T_a}{T_d} = \frac{T_b}{T_c}. \tag{4.46}$$

The efficiency can then be expressed in terms

of temperature alone:

$$\eta = \frac{T_c - T_d}{T_c}. \tag{4.47}$$

Ordinarily, T_d for a gas turbine and also the pressure ratio p_c/p_d are known. But, using the relationship between temperature and pressure for an ideal gas, T_c can be calculated.

For a specific case of an ideal gas for which $\gamma = 1.4$, $T_c = 820°C$, and a pressure ratio of 5:1, application of the above technique yields $T_d = 415°C$ and an efficiency of 37 percent. If it is assumed that the inlet temperature is 21°C, a Carnot engine operating between those temperature limits would have an efficiency of 73 percent. Hence, the Brayton cycle does not do very well in comparison to the ultimate efficiency; and even the idealized Brayton cycle does not take into account pressure and heat losses, inefficiencies in the compressor and turbine, and the effect of irreversible processes.

Gas turbine operation can be improved in much the same way that steam turbine operation is. In a process called regenerative heating part of the exhaust heat is used to preheat the compressed air just before mixing with the fuel. This has the effect of raising T_b and T_c in Figure 4.23 without altering T_d. The improvement in efficiency for a typical installation is substantial. If the assumption is made that T_b is made equal to T_d, and the figures for the above special case are used, the efficiency is increased to about 58 percent. Further improvement can be obtained by making use of the fact that less work is required to compress the air if it is at a lower temperature. In a system called intercooling, the compression is done in stages; after each stage the gas is cooled. The combination of regenerative heating and intercooling can produce efficiencies as high as 65 percent. The gas turbine can easily be made more efficient than the steam turbine of the same power output; this is because it is possible to utilize higher-temperature gases in the gas turbine. Clearly, the key to heat engine performance is *temperature.* In the ICE, steam turbine, and gas turbine the temperature limits are imposed by the available materials technology. Research in materials continues, and no doubt improvements in each of these engines will be made in the next few years.

Some work has been done on the feasibility of using turbines with automobiles. A major factor is the initial cost of a turbine-powered car. The ICE is as cheap as it is now because of more than 80 years of experience and development. Another problem with turbine engines is that they are most suited for constant, high-speed operation. Typical stop-and-go city driving would result in poor gas mileage and higher repair costs. Although Chrysler's experiment a few years back appears not to have been successful, the need for emission regulations may prompt a closer look at automotive gas turbine engines by the industry.

4.3 EXTERNAL COMBUSTION ENGINES

External combustion engines (ECEs) are those in which combustion of the fuel occurs outside the cylinder containing a closed-cycle gas system and power-transferring pistons. This type of engine is important today because, like the closed-cycle gas turbine that is an ECE, undesirable emissions resulting from the fuel combustion can be minimized, or even eliminated entirely. What is more, a theoretical efficiency equal to that of a Carnot engine can be obtained for a particular type of ECE, the **Stirling engine**.

The Stirling engine was invented in 1816, and, although very efficient, it was nevertheless large and heavy compared to other engines of the same power. As a consequence it could not compete with ICEs and turbines when they were developed. Several European firms have

shown a revived interest in Stirling engines since World War II. Advances in metallurgy and general technological developments have produced efficient, compact, and silent Stirling engines. Because of the current interest we should examine the physics of the Stirling engine and its applicability as a power source.

The Stirling cycle, the idealized model of the Stirling engine's operation, consists of an isothermal compression, a to b in Figure 4.24, followed by heat addition at constant volume along b to c. Then this is followed by an expansion at constant temperature from c to d, and another constant-volume process from d to a completes the cycle.

The efficiency of this cycle is the net work out divided by the net energy in, as for any cycle. In this case, heat energy is shown entering the system in two of the processes, a to b and c to d; likewise, heat is rejected during d to a and a to b. An amount of heat Q_R is shown for both processes. If an engine could be constructed such that the Q_R rejected were the same as the Q_R added, then this heat would be purely internal to the workings of the engine and would not have to be considered in the calculation of the efficiency. Let us assume this to be the case. We need calculate only Q_{in} and Q_{out}.

For an isothermal process the First Law becomes

$$dQ = p \, dV, \qquad (4.48)$$

and for an ideal gas, $pV = RT$ (molal quantities), so

$$dq = RT \frac{dV}{V}. \qquad (4.49)$$

For the two processes, then, we have

$$Q_{cd} = RT_c \ln \left(\frac{V_d}{V_c} \right), \qquad (4.50)$$

$$- Q_{ab} = RT_a \ln \left(\frac{V_a}{V_b} \right). \qquad (4.51)$$

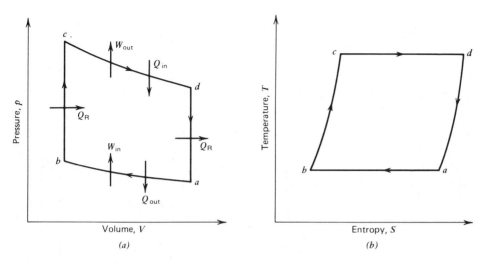

Figure 4.24 The Stirling cycle: (*a*) *p–V* diagram; (*b*) *T–S* diagram.

The efficiency of the process

$$\eta = \frac{Q_{in} - |Q_{out}|}{Q_{in}} \qquad (4.52)$$

can be written

$$\eta = \frac{T_c - T_a}{T_c}, \qquad (4.53)$$

which is identical to the efficiency of a Carnot engine operating between the same temperature extremes.

Of course, to obtain this theoretical efficiency, one needs to design a mechanical system that is described rather well by the idealized Stirling cycle. Such engines are being built. Figure 4.25 is a simplified diagram of a Stirling engine. The gas in the "hot space" of a cylinder expands through the regenerator, is cooled, and then enters the space between the pistons, where it forces the power piston down. (*Note*: This actual process has clearly been idealized by the reversible processes c–d and d–a; in the real engine, both are occurring simultaneously and cannot be separated in this manner.) Forcing the power piston down brings the displacer piston up, because of their

mechanical linkage. This upward motion displaces the remaining gas through the regenerator, where a quantity of heat is deposited. The regenerator consists of a packing of fine wires or ceramic shapes. After reaching its maximum extension, the power piston travels upward, forcing the now-cooled gas back through the regenerator where it absorbs the heat previously deposited there. The gas, at temperature T_b, is now back in the "hot space" with the displacer piston fully retracted, heating of the gas begins anew, and the other cycle starts.

Although actual operation of the Stirling engine does not correspond exactly to the idealized Stirling cycle, it can be made to operate very efficiently. There are other advantages: because the combustion of the fuel (in automobile applications) takes place continuously, not explosively, at atmospheric pressure and because the cylinders are dynamically balanced, the engine is exceptionally quiet. In fact, it can be operated without a muffler and, when installed on a bus, still passes the noise standards of those European countries that have set limits for buses.

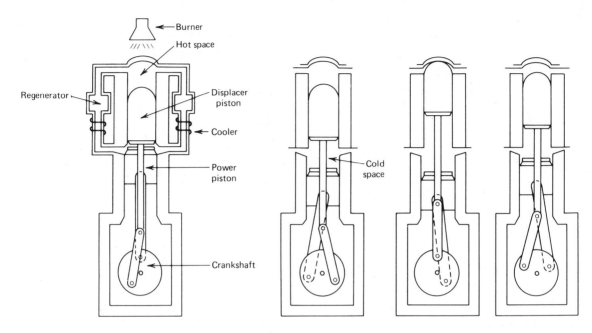

Figure 4.25 Simplified diagram of a Stirling cycle engine operation.

We mentioned the emissions advantage of the Stirling engine. Of course, any source of heat can be utilized: solar power, nuclear reactors, or even radioisotopes. It is unlikely that these last three heat sources could be used in an automobile propulsion system, however. The emissions aspect of the Stirling engine is so favorable that the Department of Energy produced a Stirling-powered research vehicle, the Genesis I, in 1979. Advanced Stirling

TABLE 4.5   Comparison of Stirling and Diesel Engines

Measurement	Stirling Genesis I	Diesel Vehicle	Advanced Stirling
Test weight	3376 lb	3418 lb	3100 lb
Horsepower	45	60	86
Fuel	Gasoline	Diesel	Wide-cut[a]
Miles per gallon			
Combined CVS cycle	19.4	24.8	44.2
Emissions (average)			
Grams per mile urban cycle			
CO	0.43	1.6	Approximately
HC	0.12	0.32	the same as
NO_x	0.37	1.48	Genesis I

Source; The Tenth Annual Report of the Council on Environmental Quality. Washington D.C.: U.S. Government Printing Office, 1979, p. 66.
[a]Wide-cut is a diesellike fuel.

engine designs are now being tested. Private industry is also interested, with the Ford Motor Company having made a licensing arrangement with a European Stirling engine producer. The DOE believes that if the research program continues to meet its goals the Stirling engine could reach the marketplace by the late 1980s. In Table 4.5 we compare some operating characteristics of the Genesis I Stirling, a diesel-powered vehicle, and an advanced Stirling power plant.

The major problem in the Stirling engine is the construction of the heated face of the cylinder. The face must be able to withstand continuous exposure to high temperatures over a long period of time without serious degradation. This is the reason that the Stirling engine is only now, after substantial developments in metallurgy, becoming a useful device.

4.4 REFRIGERATION

It is appropriate that we consider refrigeration in this chapter on heat engines, since a refrigerator is simply a heat engine traversing the thermodynamic cycle backward. In this device a quantity of heat is extracted from a low-temperature system, and a greater amount is rejected at a high-temperature reservoir, with the difference between the two amounts of heat being supplied as work from an external source.

The practical problem is finding a working substance that will absorb heat at low temperatures. For this to happen the temperature of the working substance should be below that of the low-temperature reservoir. Examine Figure 4.14; an isobaric process that works between two temperatures of reasonable extremes for a real substance must necessarily involve a change in phase of the substance. A refrigerator, then, involves a change of phase, either solid to liquid or liquid to gaseous, at or near the low-temperature reservoir and a reversal at the high-temperature reservoir.

What sort of substance do we need? It should be easily available and cheap. It should condense at the normal condenser temperature (ordinarily room temperature). Its boiling point should be sufficiently low so that it will not require vacuum operation, although this requirement may be waived in special circumstances. It should have a high latent heat of evaporation. It should not require an extensive capital investment in plumbing, pumps, and the like. Several refrigerants have been used; we discuss here some of those commonly employed.

Anhydrous ammonia is one of the oldest and most widely used refrigerants because of its high latent heat, moderate pressures, and small compressor capacity requirement. The evaporator pressure is above atmospheric in the usual installation where temperatures below $-33°C$ are not demanded. On the debit side, ammonia, while noncorrosive to the ferrous metals, is corrosive to brass and bronze; ammonia is toxic and also irritating to the eyes, nose, and throat.

The **halogenated hydrocarbons** (commonly called by one of their trade names, Freon) make up the most important group of refrigerants; they are hydrocarbons wherein chlorine and fluorine (halogens) or both have been substituted for part, or all, of the hydrogen. The Freons are nonflammable with low toxicity. Freon-12 is probably the most widely used of the Freons; although its latent heat is low, thus requiring a high mass-flow rate, the COP is essentially the same as for ammonia.

Carbon dioxide (CO_2) has been used quite extensively as a refrigerant on shipboard because of its nontoxic properties. However, the high pressures in the system and the low COP have decreased its popularity with the advent of the Freons. Its use today is primarily in the manufacture of dry ice.

Ammonia, having a high latent heat, is commonly used in large commercial refrigera-

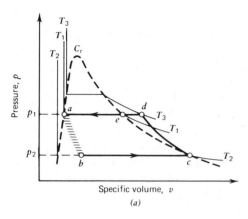
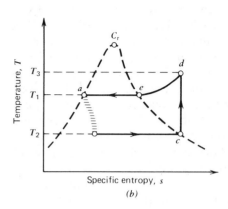

Figure 4.26 The vapor refrigeration cycle: (*a*) *p–V* diagram; (*b*) *T–S* diagram.

tion installations where very low temperatures, below 0°C, are required. For home refrigerators and air conditioners Freon is used. For large building air conditioners and some home refrigerators, a different system altogether, absorption refrigeration, is used. We shall examine the thermodynamics of these two refrigeration systems in some detail.

The Refrigeration Cycle

The ideal refrigeration cycle, indicated by the *p–v* and *T–s* diagrams of Figure 4.26, requires for its operation a *throttling process.* This process is nothing more than the throttle that we discussed earlier. In a throttling process the system proceeds irreversibly through a series of nonequilibrium states; it is not correct to say that the enthalpy is constant during this process. However, it is true that the initial and final enthalpies of the system are identical. The irreversible character of the throttling process is indicated by the dashed line in Figure 4.20.

It is the nature of a pure, real substance that, when a saturated liquid undergoes a throttling process, cooling and partial vaporization will take place. This is obvious from the diagrams. In the step b–c, heat is absorbed from the low-temperature system, completely vaporiz-

ing the working substance. The vapor is then compressed adiabatically from c to d and then condensed isobarically at the high temperature reservoir during step d–e–a, becoming again a saturated liquid.

The cycle as we have represented it is obviously an idealization. All real processes are irreversible, as we know. Nonetheless, as with all models, this one is useful for determining the limiting values of the performance of a given device.

Since work is not the output of a refrigerator, we cannot easily define an efficiency. We can define, however, a quantity that is a measure of merit of refrigerators. A high-performance refrigerator would be one that requires very little outside work to extract a quantity of heat from the low-temperature system. An appropriate measure of merit could be defined by

$$\omega = \frac{q_2}{W} = \frac{q_2}{(q_1 - q_2)}, \qquad (4.54)$$

where q_2 is the heat extracted from the low-temperature system, q_1 is the heat rejected at the high-temperature reservoir, and W is the external work supplied. The quantity ω is usually called the **coefficient of performance (COP)**. In a typical home refrigerator the COP may vary from 2 to 5.

For a Carnot refrigerator

$$\omega = \frac{T_2}{(T_1 - T_2)}. \qquad (4.55)$$

It can be shown using the definition of entropy that the work required by a Carnot refrigerator is the minimum work for any refrigerator operating between the same temperatures.

For a real refrigerator we can evaluate ω, the coefficient of performance, in terms of the enthalpies of the system at the states a, b, c, and d. Since b to c is an isobaric process,

$$q_2 = h_c - h_b, \qquad (4.56)$$

and similarly for d to c,

$$q_1 = h_d - h_a. \qquad (4.57)$$

But for a to b, the throttling process,

$$h_a = h_b.$$

Therefore,

$$\omega = \frac{(h_c - h_a)}{(h_d - h_c)}. \qquad (4.58)$$

We could now calculate the coefficient of performance using thermodynamic charts appropriate to the refrigerant in use.

Example 4.9 The temperature in the evaporator of an ammonia refrigeration plant (point c in Figure 4.26) is $-15°C$; and and the temperature at which ammonia condenses in the condenser (point a) is $30°C$. Find the coefficient of performance for this refrigerator, the vapor fraction at point b, and the compressor work.

From the tables for ammonia in Appendix B,

$h_a = 322.2$ kJ/kg (saturated liquid at $30°C$, 1166.6 kPa),

$h_c = 1422.8$ kJ/kg (saturated vapor at $-15°C$, 236.5 kPa),

$s_c = 5.534$ kJ/kg K.

From the tables, at $30°C$, $s = 5.534$ at greater than $200°$ superheat. We would need a Mollier chart to find the exact temperature—estimate at $100°C$. This would be point d. Estimate h_d:

$$h_d = 1.797 \text{ MJ/kg}.$$

And

$$\omega = \frac{1.423 - 0.322}{1.797 - 1.423} = 3.17 \text{ coefficient of performance.}$$

Find vapor fraction x by

$$h_a = x h_c + (1 - x) h_k; \qquad x = 16 \text{ percent},$$

$h_k = h$ of saturated liquid at $-15°C$.

The compressor work is given by

$$w = h_d - h_c,$$

since the process is isentropic, and

$$w = 1.797 - 1.423 = 374 \text{ kJ/kg}.$$

Note the significant amount of compressor work required in Example 4.9. This is a necessary consequence of the fact that a *vapor cycle* has been used. Naturally, it is advantageous to minimize this quantity for a given coefficient of performance.

Unfortunately, there are no government regulations concerning the efficacy of air conditioners or refrigerators, either the large commercial types or the varieties for home use. One can make that judgment for oneself, however. In the United States refrigerators or air conditioners are rated in British thermal units per hour, or tons. (One ton of refrigeration is historically the amount of energy required to melt one English ton of ice at $0°C$. This is 288,000 Btu. The commercial ton is really a rate: 288,000 Btu/24 hr.) One commercial ton of refrigeration is equivalent to 12,000 Btu/hr or 3.5 kW. This is the rate of extraction of q_2 in Eq. 4.56. The work in can be calculated from the amperage rating of the unit. Although manufacturers are not required to produce high COP units, they are required to tag each unit produced with an energy efficiency ratio (EER) which is the ratio of the cooling produced (in British thermal units per hour) to the electrical power input (in watts). The EER is related to the COP by unit changes:

$$EER = 3.414\,COP. \qquad (4.59)$$

It is surprising how much variation can be found between manufacturers (see Figure 1.10). Clearly, it pays to be careful in shopping for an air conditioner or a refrigerator, since you pay for the w—not the q_2!

As we have seen, one disadvantage to the vapor-cycle refrigerator is the large compressor energy requirement. If the working substance were a liquid throughout the cycle, this compression energy penalty would be substantially smaller. This can be the case with an **absorption system**. In this system the (gaseous) refrigerant is absorbed into a solvent; see Figure 4.27. Usually, this is ammonia dissolved in water. This strong ammonia solution is compressed with external work, ΔW; then ammonia is liberated by the generator with an external heat source, often low-quality spent steam from a power generator. The resulting ammonia vapor is condensed to a liquid with a value of ΔQ_R; cooling water is

required. This liquid is still under a high pressure, because of the compression of the earlier stage. The ammonia liquid undergoes a throttling process, thereby lowering its temperature considerably. Heat is extracted from the low-temperature system being cooled, absorbing ΔQ_A and vaporizing the ammonia. The vapor is then reabsorbed by the weak ammonia–water solution. This absorption being exothermic, some cooling is required. In actual practice, absorption systems tend to be much more complex in order to minimize the amount of water delivered to the condenser. Natural gas refrigerators for home use operate on essentially the same principle.

Because absorption refrigerants can use low-quality heat, they should in future find much wider application than in the past. Waste heat from power plants and even solar energy from rooftop collectors would certainly be appropriate heat sources. One drawback to the solar system is that a rather large amount of cooling water is required, which is usually in short supply in those areas of the country that are endowed with abundant solar resources.

An interesting refrigeration cycle based on **adsorption**, invented in Sweden and under development by the Institute of Gas Technology, is a good example of low-quality heat utilization. The physical operation of this device, the **MEC cycle** (for Munters Environmental Control), is illustrated in Figure 4.28, and a psychrometric chart for the cycle is shown in Figure 4.29. Warm moist room air at 27.7°C and 67 percent humidity passes through a heated molecular sieve, where moisture is absorbed. The heat of condensation raises the air temperature to 83.9°C and lowers the humidity to less thant 5 percent, point b. The air is cooled with a heat exchanger, exchanging heat with humidified outside air at point c. It is further cooled by humidification, and at point d it has a temperature of 13.6°C and a humidity of 53 percent, which should be quite comfortable.

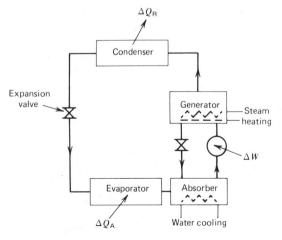

Figure 4.27 An absorption refrigeration system.

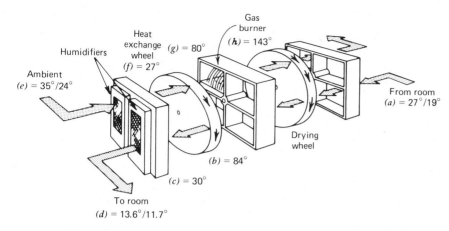

Figure 4.28 The MEC refrigeration cycle based on absorption. (Adapted from American Institute of Physics Conference Proceedings No. 25, with permission.)

Outside air is taken in at 35°C with 75 percent humidity, point e, and humidified so that at the point f it has a temperature of 26.7°C and a humidity of about 90 percent. Heat is absorbed from the rotating wheel, exchanged from household air being conditioned. At this point, g, the temperature of the air is 80°C. This is not high enough to dry the room air in the drying wheel, so that additional heat must be added by burning fuel, typically natural gas. At this point, h, the temperature is

143.3°C and the humidity less than 3 percent. This heated air is used in the drying wheel and then is rejected to the outside at a temperature of about 79.4°C. The performance claimed for this system is such that for a gas input rate of 46.9×10^6 J/hr, the unit delivers 2.7 tons of air conditioning at a COP of 0.73.

Since the entire process takes place at atmospheric pressure, this system has potentially large benefits. If more efficient heat exchangers could be developed so that the

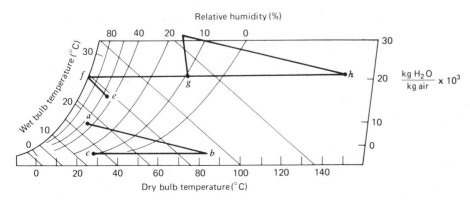

Figure 4.29 Psychrometric chart of the MEC cycle.

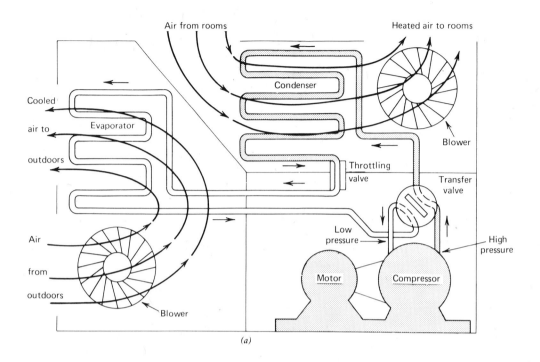

(a)

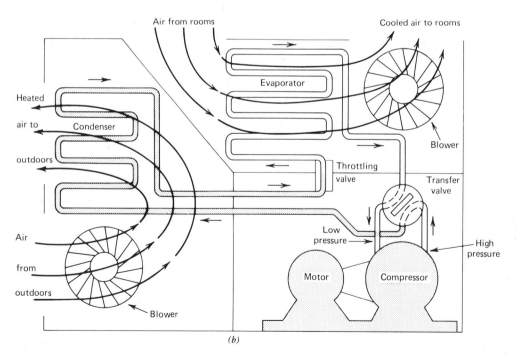

(b)

Figure 4.30 Schematic diagram of a heat pump: (a) winter operation; (b) summer operation.

added energy requirement were reduced or eliminated, the MEC system would be a clear winner over any refrigeration cycle device.

Heat Pumps

The fact that the coefficient of performance of a refrigerator exceeds unity suggests an intriguing possibility, that of using a refrigerator to provide heat. Let us assume $\omega = 4$. Then

$$\omega = \frac{q_2}{W} = 4, \qquad (4.60)$$

but

$$q_2 = q_1 - W, \qquad (4.61)$$

so

$$\frac{q_1 - W}{W} = 4 \qquad (4.62)$$

or

$$q_1 = 5W. \qquad (4.63)$$

This says that the heat rejected at the high-temperature reservoir is five times the work input! This work is typically provided by an electric motor driving a compressor. If W units of electricity were used in resistive heating, only W units of heat would be produced. The savings are obvious.

In fact, this use for refrigeration was first noted by Lord Kelvin in 1852! Devices to serve as refrigerators in the summer and heaters in the winter, called **heat pumps**, are available commercially. Figure 4.30 is a diagram of a possible such system.

Note the position of the transfer valve in the two diagrams. For winter heating, Figure 4.30a, high-pressure fluid is sent to the condenser. In Figure 4.26 this corresponds to path d–a. For summer cooling, low-pressure fluid is sent to the same device, which is now called the evaporator, path a–c. One drawback to this system is finding an appropriate low-temperature reservoir that can be cooled in the

winter. Some devices use outside air, others the earth, still others water from the mains. Care must be taken to prevent condensation or freezing at the cold outlet, the evaporator in Figure 4.30a; this would reduce the flow and subsequently the amount of heat available for extraction.

Real heat pumps, that is, those in actual use, never achieve the kind of performance suggested by the analysis above. In Figure 4.31 we see the discrepancy between the ideal and real worlds for heat pumps. In fact, below about $-2°C$, many commercial units switch automatically to resistive heating, because the heat-pump performance drops so low. Many improvements are being studied, so that future heat pumps may prove to be rather more effective both in cost and thermodynamic performance.

In this section we discussed a number of

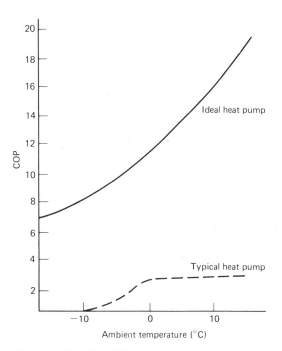

Figure 4.31 Coefficient of performance (COP) versus ambient temperature for an ideal and a typical heat pump.

devices called heat engines. We selected only a few for detailed analysis; there are many other kinds. We know that in the next half-century our reliance on fossil fuel reserves to supply heat energy will have to be curtailed or even abandoned; but this is not an imminent problem. Central-station electric power generation by means of turbines will very likely be the most efficient method to use for some time to come, expecially since nuclear reactors, converters, breeders, or fusion can be used as heat sources. The situation with automotive power is more critical; emission levels standards may become stricter as we learn more about the nature of photochemical smog, its formation and effects. Automobile manufacturers are turning to less common devices—rotary engines, Stirling engines, and turbines—in an effort to reduce undesirable emissions and yet keep the cost of automobiles within the reach of the average consumer while insuring profitable operation of their businesses.

SUMMARY

The *Internal Combustion Engine* is modeled thermodynamically by the *Otto cycle*. In this idealized model the efficiency of operation is related inversely to the *compression ratio*. In reality, efficiencies are about half the Otto cycle value.

Gasoline-powered engines emit several harmful products including carbon monoxide, oxides of nitrogen, and unburned hydrocarbons. These compounds, in the presence of sunlight, create harmful chemicals (**smog**).

Some improvement in emissions can be made by operating with a lean fuel to air mixture. Further improvement requires postcombustion processing, either with thermal reactors or *catalytic converters*.

Diesel engines rely on compression to generate the combustion—no spark. The efficiency of a model cycle also may be calculated in terms

of compression ratios. Diesel engines have emissions problems, but also have a higher fuel efficiency than a comparably sized gasoline-powered engine.

Rotary engines have not performed well in the past few years; there is probably little future for them.

A new thermodynamics variable, **enthalpy**, was introduced to simplify discussion involving fluid flow. In particular, in a *turbine* the work done is just the change in enthalpy of the working fluid.

Real substances can change phase; an ideal gas cannot. The p–V diagram of a real substance has a *vapor dome*; to the right of the dome the substance is a **gas**, to the left a **liquid**. Inside the dome the substance is a mixture of vapor and solid. Along the dome line the substance is **saturated**.

The model cycle for the steam turbine is the **Rankine cycle**. The efficiency of the Rankine cycle may be improved by *superheating* the steam. Other methods may also be used, including *reheat* and *regenerative heating*.

Gas turbines can have rather high efficiencies because of their high operating temperatures. The **Brayton cycle** is the model of the gas turbine.

The *Stirling engine* is an example of an external combustion engine. This engine has a theoretical efficiency equal to that of a Carnot engine operating between the same temperatures.

A *refrigerator* is also a heat engine, but one that traverses the thermodynamic cycle in the reverse direction to those previously discussed. The most commonly used refrigeration cycle is the *vapor cycle*, which relies upon the compression of a vapor and expansion of it through a throttle to extract heat from the surroundings.

Other types of refrigeration cycles include *absorption* and *adsorption* cycles.

A refrigerator may have a *coefficient of performance* greater than unity. In such a case it may be advantageous to use a refrigerator to supply heat in the winter, a *heat pump*. However, current models of heat pumps are not very efficient at temperatures below about $-2°C$.

REFERENCES

Internal Combustion Engines

Chinitz, W., "Rotary Engines," *Sci. Amer.* **220**(2), 90 (1969).

Cole, David E., "The Wankel Engine," *Sci. Amer.* **227**, 14 (August 1972).

Ebel, Robert H., "Catalytic Removal of Potential Air Pollutants from Auto Exhaust," in James A. Pitts and Robert L. Metcalf (Eds.), *Advances in Environmental Sciences*, Vol. I. New York: Wiley–Interscience, 1969.

Hurn, R. W., "Mobile Combustion Sources," in Arthur C. Stern (Ed.), *Air Pollution*, Vol. III. New York: Academic Press, 1968.

McEvoy, James E., *Catalysts for the Control of Automotive Pollutants*. Washington, D.C.: American Chemical Society, 1975.

Sittig, Marshall, *Automotive Pollution Control Catalysts and Devices*. Park Ridge: Noyes Data, 1977.

Taylor, C. F., and E. S. Taylor, *The Internal Combustion Engine*. New York: International Textbook, 1961.

Turbines

Glazebrook, Sir Richard, *A Dictionary of Applied Physics*, Vol. 1. New York: Peter Smith, 1950.

Hossli, Walter, "Steam Turbines," *Sci. Amer.* **220**(4), 101 (1969).

von Karman, Theodore (Chairman, Board of Editors), *High Speed Aerodynamics and Jet Propulsion*. Princeton: Princeton University Press, 1964.

Alternative Engines

Wilson, David G., "Alternative Automobile Engines." *Sci. Amer.* **234**(1), 39 (1978).

Group projects

Project 1. Study the relationship between engine size and performance. Develop data to display engine displacement versus gas mileage for a wide variety of cars. What factors influence the results? Should different model years be plotted separately? Do your data predict future trends?

Project 2. Visit the nearest steam-electric plant. Draw the appropriate $T–S$ diagram (similar to Figures 4.16, 4.17, 4.19, or 4.21); indicate the values of T and S. Calculate Q_{in}, Q_{out}, and W, and compare with the actual values. Discuss differences.

Project 3. Compare EER values for a number of air conditioning units. Develop a display similar to that in Figure 1.10. Evaluate operating costs for each unit over a 10-year period. How much improvement would be required to warrant replacing your existing unit?

EXERCISES

Exercise 1. Calculate the compression ratio for the engine described in Example 4.1.

Exercise 2. Automobile engines currently employ a compression ratio of about 9. A more typical compression ratio in 1920 was about 4.

How do the efficiencies compare for these two cases? What are the results of the improvements in the compression ratio? Are they all beneficial? Explain briefly.

Answer: Efficiency ratio = 1.38.

Exercise 3.   Use the p–V diagram for the air-standard Otto cycle to indicate the effect of the following two situations and discuss how they change the performance of the engine. (Treat them separately.)

(a) A supercharger for pressurizing the fuel–air mixture on intake.
(b) A plugged-up catalytic converter on the exhaust stroke.

Exercise 4.   What is the stoichiometric amount of O_2 (in moles) to burn completely 1 mol of propane (C_3H_8)? How much air would be required for complete combustion of this propane at STP?

Exercise 5.   About 2.4 kg of nitrogen oxides is produced for every 4.0 ℓ of gasoline burned in an internal combustion engine. An average car will use 2850 ℓ of gasoline each year, and there are about 100 million cars in the United States. What amount of nitrogen oxides is produced in a year? Assuming that approximately equal amounts of NO and NO_2 are produced, what is the volume of these products at STP?

Answer: 1.71×10^{11} kg/yr NO_x, 6.0×10^{13} ℓ/yr NO, 3.9×10^{13} ℓ/yr NO_2.

Exercise 6.   If pure propane or pure methane were burned in automobile engines, fewer undesirable emissions would result than if gasoline were burned. Why?

Exercise 7.   The NO_x emission from an ICE operated with a very lean air–fuel mixture is understandably small (see Figure 4.4). Why is it also small for very rich mixtures (i.e., for an air–fuel ratio that is less than about 13)?

Exercise 8.   Derive Eq. 4.20.

Exercise 9.   Compare W_{out} for a diesel engine having $r_C = 15$ and $r_E = 5$ and the Otto engine of Example 4.1. Assume they each have the same Q_{in}.

Exercise 10.   Show that the First Law of Thermodynamics can be written

$$dH = dQ + V\, dp.$$

Exercise 11.   The total work done by the fluid in a turbine is given by

$$W_{tot} = \int_{V_1}^{V_2} p\, dV.$$

Show that the shaft work W_s is given by

$$W_s = -\int_{P_1}^{P_2} V\, dp.$$

Exercise 12.   Apply Eq. 4.24 to a steady flow, adiabatic situation with an incompressible fluid and no shaft work and derive Bernouilli's equation:

$$mg\, \Delta y + \tfrac{1}{2} m\, \Delta v^2 + \Delta(pV) = 0.$$

Use this to show that in a horizontal pipe of variable cross section, it must be true at any place that

$$p + \frac{\rho}{2}\, v^2 = \text{const},$$

where ρ is the density of the fluid.

Exercise 13.   Use Eqs. 4.35 and 4.36 to find the critical temperature T_c in terms of a and b of a gas following the van der Waals equation:

$$\left(p + \frac{a}{V^2}\right)(V - b) = RT.$$

Answer: $T_c = (8/27)(a/bR)$.

Exercise 14.   If a steam turbine operates with a cycle as shown in Fig. 1.16 with $h_b = 69$ kJ/kg, $s_b = 0.232$ kJ/kg and $T_c = 200°C$, what should be the pressure at point a so that the efficiency of the cycle is 34.0 percent?

Exercise 15. A 38 percent efficient turbine system generates 1000 MW of electricity. What flow rate of water is required so that the temperature change of the water before and after the condenser is less than 8°C?

Exercise 16. For each of the three cycles A, B, and C shown in Figure 4.18 and the values of p, T, h, and s given there, calculate the heat and work. Calculate the overall efficiency of the total cycle. (Note the enthalpies and entropies shown are specific values.)

Exercise 17. For the Brayton gas cycle, verify that

$$\eta = \frac{T_c - T_d}{T_c}.$$

Exercise 18

(a) For the Stirling cycle, verify that

$$\eta = \frac{T_c - T_a}{T_c}.$$

(b) Obtain the form of the Stirling cycle on a T–S diagram and show that

$$W_{net} = (T_c - T_a)(S_d - S_c).$$

(c) Show that $Q_{in} = T_c(S_d - S_c)$, and hence verify the relation for η.

Exercise 19. The absorption refrigerator can be thought of in terms of the simplified processes shown in Figure 4.32.

(a) Show that a coefficient of performance in this case might be given by

$$\omega = \frac{Q_1' + Q_1''}{Q_3} - 1.$$

(b) Instead of a gas flame to provide Q_3, assume a solar collector system is used. If $\omega = 4$ and $Q_2 = 10$ MJ/hr, a typical figure for a house in the summer averaged over the day, what area solar collectors would be required?

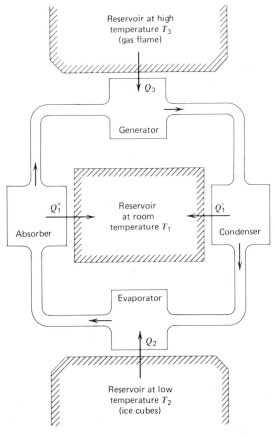

Figure 4.32 Simplified diagram of an absorption refrigerator.

Exercise 20. For a heat pump we have that

$$Q_2 = W \left(\frac{T_2}{T_1 - T_2} \right),$$

where the convention used is that the subscript 2 refers to the lower temperature. Show also that

$$Q_1 = W \left(\frac{T_1}{T_1 - T_2} \right).$$

Exercise 21. Consider an ideal refrigerator that removes heat Q_2 from the lower temperature T_2 and transfers heat Q_1 at the higher temperature T_1 by doing work W. Show that

$$Q_2 = \left(\frac{T_2}{T_1 - T_2}\right) W.$$

Exercise 22. The various methods of home heating can be compared in terms of cost as well as in terms of the amounts of energy used. Estimate the cost of providing 10^6 J each day by the following means.

1 An oil furnace with an efficiency of 60 percent.
2 A natural gas furnace with an efficiency of 60 percent.
3 Electrical heating with an efficiency of 100 percent.
4 An electric heat pump with one-third the theoretical maximum efficiency, operating between 4° and 20°C.

Use the average costs of 4.1 cent/kWh for electricity, \$2.50/(10^3 ft^3) for natural gas, and 75 cent/gal for fuel oil. (See if you can compare these average costs with the actual costs in your locality.)

Exercise 23. For ammonia refrigerator operating between $-15°$ and 30°C, calculate ω and the vapor fraction at the point b of the cycle, Figure 4.26. At d, $h = -854$ kJ/kg.

CHAPTER FIVE

DIRECT CONVERSION PROCESSES

The energy conversion systems discussed thus far have made use of an intermediate energy form, heat, in providing the usable end-product energy—usually electricity. Devices exist for converting energy in other forms to electrical energy directly, often bypassing the always inefficient heat-transformation stage. Systems using electrochemistry, photoelectricity, thermoelectricity, thermionics, magneto-hydrodynamics, and radioisotope power sources are but a few. With the clear need to reduce our reliance on fossil fuels (and the concomitant combustion stage) has come a renewed interest in several of these systems. To enhance our capability of making intelligent decisions of these various alternatives, we need to examine them in depth.

5.1 CHEMICAL CONVERSION

The first alternative we consider to the production of energy by the combustion of nonrenewable fuels—either chemical or nuclear—is the release of the potential energy locked into matter when electrons and nuclei combine to form atoms or atoms combine to form molecules. This energy, which is basically electrical to begin with, is generally called *chemical energy*, and a device for converting it directly into usable electrical energy is com-

monly—if generally erroneously—called a battery. Electrochemical conversion could become a substantial source of energy if some technical problems could be overcome. Many people are enthusiastic about the "electric auto" as a solution to the pollution problem caused by ICEs. We shall examine the various kinds of electrochemical devices in order to assess their values as energy converters.

Batteries

In an electrochemical **cell** a fuel material is *oxidized* (gives up electrons) at one electrode, called the **anode**, and an oxidizer material is *reduced* (accepts electrons) at another electrode, called the **cathode** (Figure 5.1). These two electrodes are separated by an **electrolyte** which is the means of conveyance of ions between the two; the electrons are transferred between the electrodes through the external circuit. A **battery** is a collection of two or more cells connected in series or parallel or some combination. Batteries are often classified as either primary or secondary, depending on whether or not the ionic reactions can be reversed, although there is no sharp dividing line between types. Batteries are convenient energy-storage devices, as they can supply relatively high currents at a reasonably con-

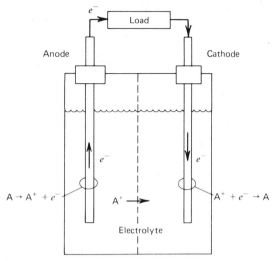

Figure 5.1 General diagram of an electrochemical cell.

Material	Kilogram / Ampere-Hours
Hydrogen	27,200
Beryllium	6,010
Lithium	3,920
Oxygen	3,400
Aluminum	3,030
Titanium	2,270
Magnesium	2,240
Fluorine	1,440
Iron^{3+}	1,465
Calcium	1,360
Sodium	1,185
Iron^{2+}	997
Zinc	834
Chlorine	770
Cupric oxide	682
Cadmium	485
Silver oxide (AgO)	440
Bromine	341
Lead^{2+}	262
Mercuric oxide	251
Mercuric chloride	200
Silver chloride	158

TABLE 5.1 Capacities of Various Materials

stant voltage for short periods of time; they are compact, require little maintenance, and pose very little environmental risk at present.

There are several types of batteries in general use today. What are the criteria we should use in selecting a particular type for a particular job?

■ Energy to weight ratio.
■ Power to weight ratio.
■ Voltage rating.
■ Availability of materials.

All these and others would be appropriate. We need to know some battery basics before proceeding.

Michael Faraday discovered in the nineteenth century that one gram molecular weight of material would be deposited from the anode on to the cathode of a cell for each 96,493 coulombs of electric charge passing through the cell (for monovalent atoms). This quantity of charge is called Faraday's constant (or faraday) F. The experimental results of Faraday can be summarized by

$$A = \frac{ItW}{Fn}, \qquad (5.1)$$

where A is the amount of material deposited in grams, I is the electric current in amperes, t is the time of deposition in seconds, W is the molecular weight of the material, F is 96,493 C, and n is the valence of the material. The results of Eq. 5.1 for various anode and cathode materials are summarized in Table 5.1. The table gives the number of ampere-hours required to deposit 1 kg of material; or, conversely, the number of ampere-hours of electricity available per kilogram of electrode material in a battery. The actual measured values approach those predicted by Eq. 5.1 very closely, indicating a high coulombic efficiency for electrochemical cells. The energy efficiency may not be as high, however.

Example 5.1 How much time is required to deposit a 0.01-mm layer of silver on a 10 × 10 cm plate with 100 A current?

The amount of material deposited is $10 \times 10 \times 10^{-3}$ cm^3 = 10 cm^3; silver has a density of 10.5 g/cm^3; therefore, $A = 105$ g. From Eq. 5.1,

$$t = \frac{FnA}{IW} = \frac{96{,}493 \times 1 \times 105}{100 \times 107.87}$$

$$\boxed{t = 939.26 \text{ s.}}$$

Another important factor in the choice of electrode material is the electrical potential difference realizable for a given configuration. This characteristic of a cell can be calculated by treating the cell as a thermodynamic system.

The only new feature of an electrical cell compared to the heat engines discussed in Chapter 4 is the cell's ability to do electrical work. The first law of thermodynamics must be modified to display this fact explicitly. When current is drawn reversibly from a cell, a certain amount of charge is transferred between the electrodes, say, dQ. But this charge may be written

$$dQ = F \, dN, \qquad (5.2)$$

where F is Faraday's constant of charge, and dN is the number of gram-equivalents of material transferred (volume × mass/molecular weight). Electrical work is charge times potential difference:

$$dW_e = \mathscr{E}F \, dN. \qquad (5.3)$$

The First Law applied to an electrical cell would then be

$$dQ = dU + p \, dV + \mathscr{E}F \, dN. \qquad (5.4)$$

If a means can be found to calculate $dQ - dU$ for a given system ($dV \approx 0$ for most cases), $\mathscr{E}$ can be determined.

We start by noting that most cell chemical reactions take place at constant temperature and pressure. Therefore, if we can find an expression containing dT and dp, simplification will result. For this purpose let us define a new thermodynamic variable by

$$G = H - TS. \qquad (5.5)$$

This is sometimes called the **Gibbs function** or, often, the free energy of the system. Combining this definition with that of enthalpy and Eq. 5.4, we obtain

$$dG = -S \, dT + V \, dp - \mathscr{E}F \, dN. \qquad (5.6)$$

For constant T and p this reduces to

$$\mathscr{E} = -\frac{\Delta G}{nF}, \qquad (5.7)$$

where nF represents the transferred charge, dN, and n is the valence of the ion.

The Gibbs function values relative to a standard value have been calculated along the same lines as entropy and enthalpy, so that, given the reacting species, the electrical potential $\mathscr{E}$ of the cell may be determined.

Example 5.2   In the Daniell cell of Figure 5.2, the overall reaction is

$$Zn + CaSO_4 \rightleftarrows Cu + ZnSO_4.$$

The valence is 2. Since ΔG for a pure metal is 0, ΔG for the above reaction is

$$\Delta G(ZnSO_4) - \Delta G(CaSO_4).$$

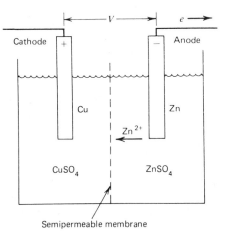

Figure 5.2   The Daniell cell.

From the *Handbook of Chemistry and Physics*,[1]

$$\Delta G = -208.31 + 158.2 = -50.11$$

and

$$\mathscr{E} = 1.109 \text{ V.}$$

This analysis has been developed for a *reversible* cell without load, that is, no current drain. In a real cell, the potential would be somewhat lower, and, of course, current drain also lowers a cell's potential because of internal resistance, the effect of poor electrical conductivity in the electrolyte.

The **electromotive force**, or emf (the electrical potential), available from several materials relative to a hydrogen electrode is given in Table 5.2. (A hydrogen electrode is a blackened platinum strip over which hydrogen at atmospheric pressure is bubbled.) The total open-circuit emf, $\mathscr{E}°$ of a cell is the algebraic difference between the emfs of the anode and cathode materials measured with respect to hydrogen.

$$\mathscr{E}° = \mathscr{E}_a{}^h - \mathscr{E}_c{}^h. \qquad (5.8)$$

The actual emf of a cell under load is less than this open-circuit value, primarily as a result of internal resistance. Other effects, called "polarization," also act to reduce the actual cell voltage. These emf reductions represent power losses in the cell. There are many sources of polarization losses; some are of chemical origin, some of physical origin, and some of electrochemical origin. As an example of one type of loss, the concentration of active reactant molecules in the immediate vicinity of an electrode is reduced during cell operation, resulting in a smaller charge transfer than in the open-circuit case. Many losses are rate dependent; that is, they are more severe at high current discharge than for low current

TABLE 5.2 Electromotive force at 25° C

Reaction	emf (V)
$Li \rightarrow Li^+ + e$	2.960
$K \rightarrow K^+ + e$	2.924
$Ca \rightarrow Ca^{2+} + 2e$	2.76
$Na \rightarrow Na^+ + e$	2.713
$Mg \rightarrow Mg^{2+} + 2e$	2.375
$Al \rightarrow Al^{3+} + 3e$	1.67
$Mn \rightarrow Mn^{2+} + 2e$	1.18
$H_2 + OH^- \rightarrow 2H_2O + 2e$	0.828
$Zn \rightarrow Zn^{2+} + 2e$	0.763
$Cr \rightarrow Cr^{2+} + 2e$	0.577
$Fe \rightarrow Fe^{2+} + 2e$	0.441
$Cd \rightarrow Cd^{2+} + 2e$	0.402
$Co \rightarrow Co^{2+} + 2e$	0.277
$Ni \rightarrow Ni^{2+} + 2e$	0.23
$Sn \rightarrow Sn^{2+} + 2e$	0.141
$Pb \rightarrow Pb^{2+} + 2e$	0.126
$H_2 \rightarrow 2H^+ + 2e$	0.000
$Cu \rightarrow Cu^{2+} + 2e$	−0.346
$2Ag + 2OH^- \rightarrow Ag_2O + H_2O + 2e$	−0.344
$Cu \rightarrow Cu^+ + e$	−0.522
$2I^- \rightarrow I_2 + 2e$	−0.536
$H_2O_2 \rightarrow O_2 + 2H^+ + 2e$	−0.582
$Fe^{2+} \rightarrow Fe^{3+} + e$	−0.770
$Ag \rightarrow Ag^+ + e$	−0.800
$Hg \rightarrow Hg^{2+} + 2e$	−0.852
$2Br^- \rightarrow Br_2 + 2e$	−1.065
$2H_2O \rightarrow O_2 + 4H^+ + 4e$	−1.229
$Sn^{2+} \rightarrow Sn^{3+} + e$	−1.256
$2Cl^- \rightarrow Cl_2 + 2e$	−1.358
$MnO_2 + 4H_2O \rightarrow MnO_4^- + 8H^+ + 5e$	−1.50
$Ag^+ \rightarrow Ag^{2+} + e$	−1.987
$2F^- \rightarrow F_2 + 2e$	−2.85
$2H_2O \rightarrow H_2O_2 + 2H^+ + 2e$	−1.77

drain. The evaluation and understanding of polarization losses have led to the development of improved cells in the past few years; see Tables 5.3 and 5.4 for listings of several types of batteries.

The battery most commonly in use today is the lead–acid storage battery, the type found in automobiles. The anode (electron donor) is made of *spongy lead*, while the cathode (electron acceptor) is made of a set of grills filled

---

[1]The free energy in this listing is labeled ΔF_f^o.

TABLE 5.3 Primary Cell Characteristics

| Type | Open-Circuit Voltage | Average Voltage | Capacity at Low Current | | Shelf Life (Years) | Operating Temperature Limits($°C$) |
			Wh/kg	Wh/m^3		
Air cell	1.5	1.25	315	8,537	5 ry)	4.4 to 60
Alkaline	1.52	1.25	213	14,634	2	−40 to 60
Indium	1.37	1.15	112	7,927	1 to 3	−29 to 88
Leclanché	1.5 to 1.65	1.25	210	13,415	1	−18 to 49
Magnesium-silver chloride	1.6	1.3	194	11,586		−2.2 to 29
Silver oxide-zinc	1.86	1.5	315	17,683	3 (dry)	−29 to 71
Solid electrolyte	0.69		7.3		20	−54 to 77
Thermal	3.0		19.4	366	10	−54 to 74

TABLE 5.4 Secondary Cell Characteristics

| Type | Average Voltage | Open-Circuit Voltage | Charge Loss per Month (percent) | Charge–Discharge Cycles | Specific Energy at Low Drain | |
					(Wh/kg)	(Wh/cm^3)
Nickel–iron	1.2	1.34	30	2000	24	0.06
Lead–acid (automotive)	2.0	2.14	25	300	33	0.08
Nickel–cadmium	1.2	1.34	2	2000	26	0.06
Silver oxide–cadmium	1.1	1.34	3·	2000	53	0.15
Silver oxide–zinc sealed	1.45	1.86	3	100	44–100	0.08–0.20
Silver oxide–zinc primary	1.45	1.86	—	—	121	0.23

with lead peroxide, PbO_2, called *brown lead*. The electrolyte is a dilute sulfuric acid solution. The reactions[2] at each electrode during discharge are

$$(Anode)\quad Pb + H_2SO_4 \rightarrow PbSO_4 + 2e^- + 2H^+ \tag{5.9}$$

and

$$(Cathode)\quad 2PbO_2 + 2H_2SO_4 + 2e^- \rightarrow 2PbSO_4 + 2H_2O + O_2^{2-}. \tag{5.10}$$

This cell will function until both sets of plates are covered with lead sulfate. Recovery is obtained by charging the cell, that is, by reversing the arrows in the above reactions.

[2]These reactions are simplified versions of the actual electrode reactions.

During charging, of course, the electrons are supplied by an external source. Unfortunately, the lead and lead peroxide are never replaced in the same physical condition after recharging, so that eventually the structure of the battery fails as a result of many discharge–charge cycles.

During discharge, some of the H^+ and O^- ions combine to form H_2 and O_2 at the appropriate electrode, so water must be continuously replaced. It has been found recently that a small percentage of calcium metal alloyed in the lead will scavenge the hydrogen and oxygen reversibly, eliminating the need for water replacement or for vents. This is the basis for the sealed batteries now on the market.

Lead–acid batteries are heavy and bulky,

have a low power-to-weight ratio (specific power), and a low energy-to-weight ratio (specific energy). In fact, they have only one advantage: they are cheap. Any hope of using batteries on a large scale for vehicular propulsion requires a breakthrough in battery technology.

Lead–acid storage batteries for automobiles are usually rated in terms of *cold cranking power*. This is the amount of current that can be delivered at 0° F with the battery potential being maintained above 7.2 V for a 12-V battery and 3.6 V for a 6-V battery. We can use this definition to estimate the quantity of material required for a given battery.

Example 5.3 A 500-cold-cranking-power battery has a mass of about 20 kg. How much material is tranferred between electrodes in 30 s of use at 500 A?

From Eq. 5.1

$$A = \frac{500 \times 30 \times 207.19}{96{,}493 \times 2}$$

$$\boxed{A = 6.2 \text{ g.}}$$

Since this is only a small fraction of the total mass of the battery, it is clear that considerable reserve *energy*, if not reserve *power*, exists in the battery.

Figure 5.3 indicates the tradeoff between specific power and specific energy for lead–acid batteries, as well as other propulsion systems. Example 5.3 illustrates a point on the upper left part of the lead–acid curve. Low-power discharge such as at highway cruising speeds would result in a higher energy output of the battery. For power-plant use, the operation would be more like that in Example 5.3.

High-Performance Batteries

New types of batteries are being developed in an attempt to satisfy demands for high

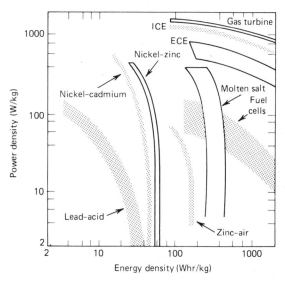

Figure 5.3 The trade-off between specific power and specific energy for various devices.

specific energy, high specific power, and compactness or low cost. For use in automobiles a battery should not only have a high specific power for providing adequate acceleration, but should also have a high specific energy so that a reasonable amount of operation between charges would be possible. Several industrial firms have been engaged in an intensive research effort over the past few years studying high-performance batteries. Because of the potentially very lucrative market for these devices, much of the research has been very successfully kept "under wraps." As a consequence, it is somewhat difficult to obtain detailed technical data about them. We shall present here only a cursory examination of the principles, but not the details, of high-performance battery operation.

There is an empirical rule concerning devices or processes under development: any device or process for which scientific or engineering details have been published doesn't work! This is probably the case with high-performance batteries, and is definitely the case with water-splitting cycles (as we shall see in the next chapter). However, the devices that have been publicized illustrate some of the characteristics and problems that these exotic batteries have.

In the last few years, electrochemical research has centered on batteries that have alkali metals and halides as electrodes (see Table 5.5). These materials have a much higher valence-charge-to-molecular-weight ratio than the materials of a typical storage battery; consequently, the electrical current per kilogram capability should be substantially higher. However, the effective use of these materials will require solutions to several problems. First, consider the electrolyte. It should have no effect other than that of providing a means of exchanging ions between the electrodes. Aqueous electrolytes, typical of most primary batteries, will react violently with the alkali metals but will dissolve the halogens only slightly. What is needed is a nonaqueous material that has a low resistance for the electrode ions but a high resistance for electronic conduction. Two approaches have been studied.

In one the electrolyte is the salt formed by combining the electrode ions. It must be kept at a high temperature in a molten state to improve the ionic conductivity. In the lithium–chlorine battery the anode reaction is

$$Li \rightarrow Li^+ + e \qquad (V_a = +3.05 \text{ V}); \quad (5.11)$$

the cathode reaction is

$$Cl_2 + 2e \rightarrow 2Cl^- \qquad (V_c = -1.36 \text{ V}). \quad (5.12)$$

The lithium and chloride ions combine in the electrolyte to form additional lithium chloride. To maintain the ionic mobility through the electrolyte it must be operated at a temperature in excess of 650°C. It would seem that the open circuit potential of this cell would be quite large; but the temperature dependence of the Gibbs function is such that, at the elevated temperature at which the cell must operate, the available potential difference is quite small—on the order of 0.5 V. The current density can be quite high, however, with values on the order of 10 A/m^2 not unreasonable.

The other approach to the electrolyte problem is the development of solid electrolytes having peculiar properties. The crystalline structure of certain oxides seems to be such that at elevated temperatures the resistivity for

TABLE 5.5 High-Temperature Batteries

System	Temperature (°C)	Materials Problems
Na–S	300	Compatibility of materials Electrode structures
Li–S	400	Cell containers Insulators
Li–Cl₂	650	Joining materials for seals Ion-conducting ceramics Purity of electrolyte and reactants Corrosion Solubility of reactants Long-term electrode performance Wetting characteristics of liquids

anion conduction is very small. This phenomenon has been known for some time, more than 80 years! A "magic" mixture of 85 percent ZrO_2 and 15 percent Y_2O_3 has been studied since about 1900. Until recently, the exact mechanism for the enhancement of anion conduction in this type of material was not understood. Then the crystal structure of **beta-alumina** was deduced.

Alumina is aluminum oxide: Al_2O_3. The Greek-letter prefix refers to a particular crystalline form. It was the complexity in the phase structure of alumina that made the determination of its properties so difficult. X-Ray crystallography has indicated that the atoms of beta-alumina are arranged in planes, with weak bonding between planes, similar to graphite. The spacing between planes is apparently such that alkali metal ions have little difficulty in traversing the crystal. This conductivity is enhanced when the alumina is compounded with an alkali metal, such as sodium. In sodium beta-alumina, the sodium atoms are located in the planar spacings between aluminum and oxygen atoms. Sodium ion conduction can then proceed by lattice vagrancy migration. Unfortunately, beta-alumina is difficult to fabricate, is fragile, and can undergo phase change readily.

The problems with electrolytes are only the beginning. Molten electrodes are very corrosive; battery seals and containers must be chosen with care. Recharging of these batteries does not proceed without problems. The metal is not reliably returned in the same state, but this is a problem common to all batteries. The potential for rupture of the battery must be considered, if the widespread application of high-performance batteries is to be contemplated.

Fuel Cells

Another electrochemical conversion device that has been extensively studied is called the fuel cell. The use of hydrogen–oxygen fuel cells on the Apollo moon flights, especially the unsuccessful Apollo XIII, has given this device a great deal of publicity and has stirred the imagination of some writers. A fuel cell is simply an electrochemical device in which the fuel is supplied externally; the electrodes are not consumed as they are in ordinary batteries. The fuel cell, illustrated schematically in Figure 5.4, operates just as a battery does: electrons are given up by the fuel at the anode and recombine with the oxidizer at the cathode, with ions transported across the electrolyte combining to form a stable product that must be removed from the cell.

The thermodynamics of fuel cell operation have already been developed; Eq. 5.7 applies equally well to any fuel system, internal or external, gaseous, aqueous, or solid. In fact, a variety of fuels have been examined for fuel cell use, most of these being gases because of their convenience in maintaining a constant supply of fuel material.

The H_2–O_2 cell has been studied extensively over the past 100 years. It has been used on space missions; a system used on Gemini flights is shown in Figure 5.5. The H_2–O_2 cell is representative of a general class of fuel cells that use an insoluble gas for the anode fuel. The anode reaction is

$$H_2 \rightarrow 2H^+ + 2e \qquad (V_a = 0.00); \qquad (5.13)$$

the cathode reaction is

$$O_2 + 2H^+ + 2e \rightarrow H_2O \qquad (V_c = -1.23 \text{ V}). \qquad (5.14)$$

The current density of this cell can be quite high, on the order of 10 A/m^2 at a potential of 0.7 V, depending on the design of the gas electrodes. In early designs the gases diffused through porous carbon plates to the electrolyte, usually a solution of potassium hydroxide. The electrolyte was held out of the carbon by a thin wax coating. The performance is this cell was limited by the poor

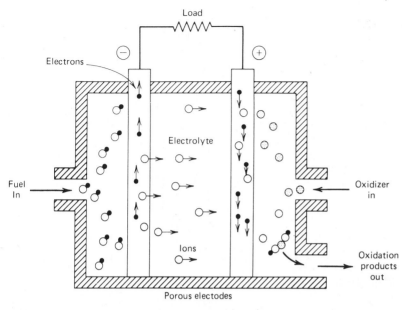

Figure 5.4 Typical fuel cell configuration.

solubility of the gases in the electrolyte and the fragility of the electrodes.

The H_2–O_2 cell has been improved by replacing the carbon electrodes with porous nickel on nickel supporting sheets, which act as a catalyst in ionizing the hydrogen. In addition, the gases are fed in at a high pressure, about 10 atm, and the cell is operated at a temperature of 400° C to increase the gas solubility and ionic conductivity. This improved H_2–O_2 cell (called the **Bacon cell**) has produced 90 A/m^2 at 0.6 V. The oxygen electrode is susceptible to corrosion, but chemical treatment of the nickel can help overcome this problem. It is estimated that a Bacon cell can store about five times more energy per kilogram than a conventional lead–acid storage battery.

The Bacon cell used two techniques to increase the reactivity of the fuel: elevated operating temperatures and catalysts. It is also possible to improve the cell performance by modifying the fuel. One cell has been developed that uses the entire methane series

of hydrocarbons. In this cell, which operates at about 1000° C, the fuel is converted to hydrogen ions:

$$CH_4 + 2H_2O \rightarrow CO_2 + 8H^+ + 8e. \quad (5.15)$$

The electrolyte is solid zirconia; this material has a high conductivity for hydrogen ions. At the cathode the reaction is

$$8e + 8H^+ + 2O_2 \rightarrow 4H_2O. \quad (5.16)$$

Current densities up to about 20 A/m^2 at 0.7 V have been produced. Other hydrocarbon cells have used kerosene, hydrazine, and formaldehyde with molten alkali carbonate electrolytes.

Hydrocarbon fuel cells have been studied for at least 100 years in an attempt to utilize fossil fuels directly, bypassing a conversion stage producing heat. The idea was that higher efficiencies could be realized in direct conversion, and therefore fossil resources could be made to last longer. This is especially important today. Although natural gas is probably much too valuable a resource to be used in any application, gas produced from coal could, in

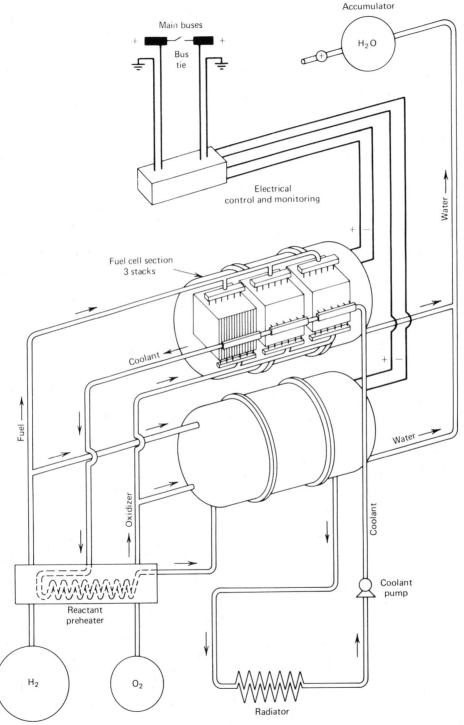

Figure 5.5 Schematic diagram of the H_2–O_2 fuel cell system used on Gemini spacecraft.

fuel cells, stretch our resources substantially.

There are a variety of technical problems yet to be solved before fuel cells can conveniently and economically displace other electrical generating divices. Electrocatalysts must be found that are effective, cheap, and do not enter into side reactions with the fuel or oxidizer. Electrolytes of high ionic conductivity that are also noncorrosive and can operate at reasonable temperatures must be developed. And, finally, schemes for developing inexpensive raw materials into fuel must be divised. Toward the latter end some work on using microorganisms and enzymes to break down garbage or sawdust biochemically to provide fuel seems at least interesting if not currently promising.

The extent to which any electrochemical device will be used in the future depends rather crucially—at the present time—on the cost of the electricity it can produce compared to that from other means. These costs are difficult to calculate, because, for one thing, the total environmental cost of electricity generation is never paid. For another, the costs of fossil-fueled generation are likely to increase substantially in the next few years. But clearly, the higher the efficiency of electrochemical conversion, the more likely it is to become economically competitive.

Efficiency of Electrochemical Conversion

Electrochemical devices, particularly fuel cells, have been hailed by some as the means of overcoming the Carnot limitation in efficiency. It has been stated that these devices can be 100 percent efficient. This is certainly not true. What is true is that there is a great deal of confusion concerning the definition of efficiency as applied to electrochemical devices.

There are in fact four different efficiencies that are useful to discuss. The first is a measure of the fraction of total energy available from the reactants that the device can possibly utilize. The Gibbs free energy ΔG in this case is the electrical work done by a cell. From Eq. 5.7,

$$\Delta G = - F\mathscr{E} \, dN. \tag{5.17}$$

This is the maximum amount of work that can be done by the cell, since we have assumed $p \, dV$ to be zero. Recall that the enthalpy ΔH is the energy available in the isobaric combination of reactants. Therefore, we can define an *ideal efficiency* by

$$\eta_i = \frac{\Delta G}{\Delta H} = 1 - \frac{T \, \Delta S}{\Delta H}. \tag{5.18}$$

The quantity $T \, \Delta S$ represents isothermal heat transfer—it may be *positive* or *negative*, depending on whether the heat is transferred *into* or *out of* the cell. Even when heat is rejected, the efficiency is less than unity, since both ΔG and ΔH are normally negative. This ideal efficiency is also the Second Law efficiency, which we defined in Chapter 3. Note that if $p \, dV$ is not zero, when gases are liberated, Eq. 5.18 will be somewhat different. See Chapter 6 for details.

The ideal efficiency of an H_2–O_2 cell at 298 K can be calculated from tables of ΔG and ΔH; it is 0.94. Both $\Delta G°$ and $\Delta H°$ are temperature-dependent. At 2000 K the ideal efficiency or the cell drops to 0.54; Table 5.6 compares ideal and Carnot efficiencies for the H_2–O_2 cell at various temperatures.

The highest thermodynamically *possible* efficiency from an electrochemical device is always less than unity. What is the highest *practical* efficiency? The actual efficiency is the electrical energy output divided by the chemical energy input:

$$\eta_{ac} = \frac{W}{\Delta H} = \frac{NFV_{ac}}{\Delta H}, \tag{5.19}$$

where N is the number of gram-equivalents of charge transferred and V_{ac} is the actual cell potential difference. Because of the various

TABLE 5.6 Relevant Thermodynamic Properties for the H_2–O_2 Fuel Cell

T (K)	$\Delta H°$ (kcal/mol)	$\Delta G°$ (kcal/mol)	η_i ($\Delta G°/\Delta H°$)	Carnot Efficiency[a] ($1 - T_C/T_H$)
298	−57.80	−54.64	0.94	0
400	−58.04	−53.52	0.92	0.26
500	−58.27	−52.36	0.90	0.40
1000	−59.21	−46.03	0.78	0.70
2000	−60.26	−32.31	0.54	0.85

[a]T_C = 298 K.

"polarization" losses in the cell the actual cell emf is always substantially less than the maximum open-circuit value calculated from Eq. 5.7. For the H_2–O_2 cell the actual voltage is around 0.7 V, depending of course on the current drawn from the cell and the operating temperature. This gives an actual efficiency of

$$\eta_{ac} = 0.56.$$

This is clearly less than the wild claim of 1.00 in the popular press, but it is, of course, much better than any room-temperature heat engine. (Especially so, if the rejection temperature is also room temperature!)

The other two efficiencies often used in the literature are the voltage and current efficiencies. The first is defined by

$$\eta_V = \frac{V_{ac}}{V_{rev}}, \qquad (5.20)$$

where V_{rev} is the reversible emf calculated from Eq. 5.7. The current or Faradaic efficiency is given by

$$\eta_F = \frac{I_{out}}{NFM}, \qquad (5.21)$$

where I_{out} is the output current, and M is the fuel reaction rate in moles per second. Voltage efficiencies are usually of the order of 50 to 60 percent, while current efficiencies may be much higher, depending on whether or not the fuel undergoes side reactions with the electrode of electrolyte that do not contribute electrons to the external circuit.

If fuel cells or other electrochemical devices can be made cheaply in large quantities to operate with fossil fuels at high efficiencies (60 percent), the fossil reserves can be stretched many hundreds of years into the future. Some of the problems of nuclear-power-plant siting can be overcome by transporting not the electricity a great distance from the plant but, rather, hydrogen. A nuclear plant on a floating platform could produce hydrogen from seawater by electrolysis. The hydrogen could be piped into residential fuel cells or central station generators much as natural gas is today. But before electrochemical devices can play these roles many problems associated with materials, electrode construction, electrolytes, and the like need to be solved.

5.2 PHOTOVOLTAIC CONVERSION

No device more typifies the concept of direct conversion than the "solar cell." The conversion of sunlight directly to electricity has been possible for several decades. But large, efficient, economic devices have yet to be produced. Nonetheless, with the pressures of environmental concern and traditional fuel scarcity, research in photovoltaic conversion continues, now at an accelerated pace. We should study this conversion system in some detail, to learn its advantages and disadvantages. Perhaps we will be able to assess the future impact of photovoltaic conversion on our energy systems.

Solar Radiation

In Chapter 6 we discuss solar radiation as a heat source in some detail. Here we are more interested in the devices that convert solar energy to electrical energy. For this purpose we need to study those aspects of solar radiation that are most relevant, primarily the wavelength distribution. Figure 5.6 is a plot of solar energy per unit time per unit wavelength versus wavelength. The distribution measured at the top of the atmosphere closely approximates the radiation spectrum from a 6000-K **blackbody**. A blackbody is any object that not only radiates at all frequencies but also absorbs *all* radiation of all frequencies incident upon it. There are no perfect blackbodies in nature, although they can be closely approximated. The blackbody is important in physics because, in correctly calculating the distribution of radiant energy from one as a function of energy, Max Planck was forced to use quantum principles for the first time. Real objects are neither perfect radiators nor perfect absorbers. On the surface of the earth the energy distribution of sunlight has large gaps caused by the absorption of specific wavelengths or bands of wavelengths by several gases in the atmosphere, namely, oxygen, ozone, carbon dioxide, and water vapor.

The energy-rate values of Figure 5.6 are for **normal solar incidence** (i. e., perpendicular), which does not occur anywhere in the United States. Actual solar energy calculations will have to be modified to include the effect of latitude. In addition, the condition of the atmosphere is very important; obviously, clouds affect the solar energy availability at the earth's surface. But so also do haze, smog, and high humidity. All these act as filters on the intensity and the wavelenths transmitted by the atmosphere.

Light, of course, is an electromagnetic wave, and so it carries energy. However, contrary to the belief widely held at the end of the nineteenth century, the energy content of an electromagnetic field is not spread over a wave front but, rather, is concentrated in packets or bundles that have come to be called **photons**. The photon concept was found by Albert Einstein to be necessary to explain the phenomenon of the **photoelectric effect**.

When light shines on certain metals, electrons are given off from the metal surface. The number of electrons is proportional to the intensity of the light, and the maximum energy of the electrons is proportional to the frequency of the light. This can be understood

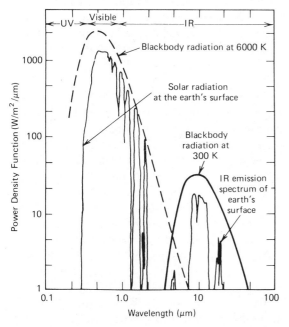

Figure 5.6 Electromagnetic spectra of solar and terrestrial radiation.

only if light is assumed to consist of photons of an energy proportional to the frequency of the light, making a collision with electrons and giving up all their energy to the electrons. The more photons the more electrons and, for different frequencies, different electron energies.

The relationship between frequency, or wavelength, and energy was found to be a simple one:

$$E = hf = \frac{hc}{\lambda}, \qquad (5.22)$$

where the constant of proportionality h—Planck's constant, 6.63×10^{-34} J s—comes from the blackbody-radiation formulation of Planck. Then, referring to Figure 5.6, the very-short-wavelength photons, shorter than about $0.4\,\mu$m, are high in energy content. This is the ultraviolet region of the spectrum. Ultraviolet light (UV) can be very damaging to tissue; it is responsible for sunburn and can cause skin cancer in large doses for very short wavelengths. Fortunately for us, the earth's atmosphere is relatively opaque to UV (but we may have problems; see Chapter 12) because of absorption by oxygen and ozone.

Example 5.4 What is the energy of a visible light photon?

From Figure 5.6, we see that the visible light region spans from about 0.4 to 0.7 μm. Using

$$E = \frac{hc}{\lambda},$$

we have

$$E = \frac{6.63 \times 10^{-34}\,\text{J s} \cdot 3 \times 10^8\,\text{m/s}}{0.4 \times 10^{-6}\,\text{m}}$$
$$= 4.97 \times 10^{-19}\,\text{J},$$

$$\boxed{E = 3.19\,\text{eV}.}$$

So, visible light photons range from about 1.7 to about 3.2 eV in energy.

At the other end of the spectrum are the very long wavelengths. Those greater than about $0.8\,\mu$m are known as infrared (IR). These are lower in energy content per photon, so not as damaging to tissue. The only known biological effect is simple heating. The far infrared wavelengths, of the order of $10\,\mu$m, are characteristic of the wavelengths emitted by the earth as it radiates away energy to outer space.

The major problem in photovoltaic conversion is finding an *inexpensive* device that will absorb as much of the total spectrum from the sun as possible, over this very broad range of wavelengths from the UV to the visible to the infrared. To select the right device we must understand something of the physics of their operation.

Some Solid-State Physics

Not all materials are photosensitive; only a few may be used as photovoltaic converters. There appear to be certain characteristics in the microscopic makeup of some materials that make them photosensitive. To determine these characteristics we need to look at some of the fundamentals of atomic physics as applied to solids.

There are obvious and not so obvious differences between solids. Chemical composition is one apparent difference, but this is really the same as saying there is a difference in the electronic structure of the atoms making up solids. The number and distribution of electrons in an atom determine its chemical identity, and when these atoms are packed together in a solid, these electrons determine not only whether the material will be a conductor or an insulator but also its heat conductivity, its elasticity, its transparency, and all the other physical properties of solids.

This relationship between the electronic structure of atoms and the physical properties of solids is rather well understood today. When we say "understood," we mean that mathematical models exist that enable us to calculate observed phenomena with precision in terms of certain fundamental constants. The mathematical procedures used to explain atomic phenomena are called **quantum mechanics**. This mathematics is rather complex, however, and we shall not dwell on it here. Instead we report only the results of calculations, which agree quite well with observations.

The energy available to an atomic system is **quantized**; that is, it can have only certain values, not a continuous range of values. Each atom has a large number of quantized energy states; the actual energy values are characteristic of that particular element. It is common to refer to *electron* energy levels, but the quantized levels are for the entire atom.

One of the consequences of quantum mechanics is that no two systems can be identical. The energy states of an atom can be identified by *quantum numbers* as well as by an energy value. These quantum numbers can be interpreted in terms of the physical degrees of freedom of the system: rotation, vibration, spin, and so on. No two systems can have identical quantum numbers. This means that only two electrons can occupy any given energy state (one spin "up," one spin "down").

When atoms are forced to be close together, as in a solid, the quantized levels split into many closely spaced levels, one for each atom. The spacing is so small that the many splittings form bands. In many solids the energy bands are separated by an energy difference called the **band gap**. The lowest energy states are occupied by electrons. The band that the outer or valence electrons occupy is called the **valence band**.

The electrical nature of a solid is determined by its band structure and the occupancy of the bands by electrons. In Figure 5.7 are shown four different kinds of band structures. In the first, the lower band is not filled; that is, the number of energy states in that band available for occupancy exceeds the number of electrons. It is possible, then, for electrons to move freely through the solid by a process of successive collisions, because allowed energy states are available. In the third type of structure, the first two bands overlap, so that even though the lower band may be full, there are nevertheless energy states available to electrons, and electrical conduction can occur. For this reason, the second band is called the **conduction band**.

In the second type of band structure the lower band is filled, and the next available energy state is separated from it by a large energy gap. Collisions that would give an electron, even one near the upper edge of the valence band, this much energy are unlikely, so conduction does not occur. A substance having this type of band structure is called an **insulator**.

The energy gap is very small in the fourth structure in Figure 5.7, so even at room temperature thermal agitation is sufficient to propel some electrons across the gap and into unfilled energy states. Materials of this type are called **semiconductors**.

The band gaps (E_g) for several types of

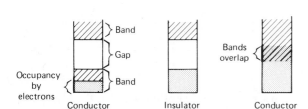
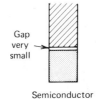

Figure 5.7 Four types of band structures.

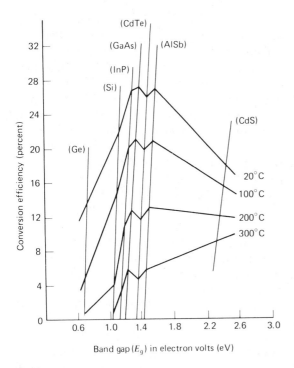

Figure 5.8 Band gaps and photoelectric conversion efficiencies for several semiconductors as a function of temperature.

photoelectric materials are shown in Figure 5.8. Note that E_g depends weakly on temperature, but that the photoelectric conversion efficiency is very strongly dependent on temperature.

We also see from this figure that all photoelectric materials have band gap energies in the visible range. Sunlight falling on them would free electrons to move about. But for each electron liberated a positively charged ion, called a **hole**, remains. These holes will recombine with free electrons unless a mechanism for separating electron–hole pairs can be provided. Recombination, of course, reduces the amount of photoelectrons available as current to an external circuit.

Even in the absence of sunlight there will be

a continuous production of electron–hole pairs. This is a consequence of the thermal excitations present for all materials at all temperatures above 0 K. The probability that a particular energy state will be occupied is given by the Fermi function

$$f(E) = \frac{1}{\exp\left[(E - E_F)/(kT)\right] + 1}, \quad (5.23)$$

where k is Boltzmann's constant, T is the absolute temperature, and E_F is the Fermi level. This function is plotted in Figure 5.9 for several temperatures. Note that at sufficiently high temperatures there is a reasonable probability that some higher energy levels will be occupied, conceivably in the conduction band.

The mechanism for separating electron–hole pairs is the **p–n junction**. Consider the specific case of silicon. This valence-4 atom forms covalent bonds with its neighbors, as in the top in Figure 5.10. If a small amount of energy is added by heating or by photon absorption, electrons can be liberated, leaving behind holes, as in the second drawing. If a few of the silicon atoms are replaced with valence-5 atoms, such as phosphorus or arsenic, one electron is left over per added atom after all

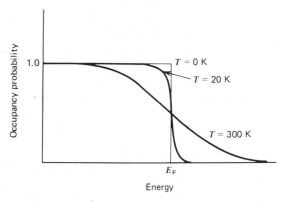

Figure 5.9 The Fermi function for three values of temperature.

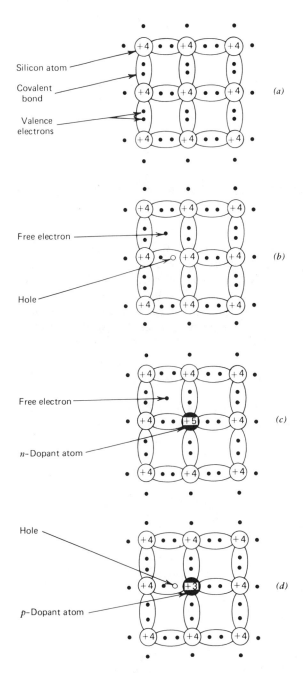

Silicon atom

Covalent bond

Valence electrons

(a)

Free electron

Hole

(b)

Free electron

n-Dopant atom

(c)

Hole

p-Dopant atom

(d)

the covalent bonds are formed. These electrons are loosely bound (0.03 eV) and in general move freely throughout the crystal. Since these valence-5 impurities (called **donors**) add electrons, this type of doped material is called **n-type silicon**.

If atoms of valence 3 are added, the reverse situation results. Too few electrons are available for form covalent bonds, and a hole is created that moves about the crystal. Dopants such as boron or gallium, called **acceptors**, are used to create this p-type silicon. Electrons in n-type and holes in p-type are called the **majority carriers**; holes in n-type and electrons in p-type are called the **minority carriers**.

If n-type and p-type silicon are brought into close contact, the free electrons and free holes in the junction region will diffuse toward the junction and combine to produce a region devoid of charge carriers, called the **depletion region**, as in Figure 5.11a. However, the impurity atoms in the junction region will now be ions, since their donated electrons or holes have disappeared. These donor or acceptor ions are fixed in the crystal and therefore produce an electric field. This field creates a barrier V_b to further block majority carrier flow, as illustrated in Figure 5.11b.

This figure is a plot of energy versus position along the p–n junction. After the two materials are placed in contact, the two Fermi levels

responding to the four valence electrons. (b) Silicon crystal lattice that contains a bond broken by thermal or light energy. When a valence electron breaks away from a covalent bond, it "leaves behind" a hole, which behaves like a positive charge carrier. (c) Lattice of a silicon crystal made into an n-type crystal (n-doped silicon). The *donor dopant* illustrated has five valence electrons, with five charges in its nucleus to balance these electrons. (d) Lattice of a p-type silicon crystal (p-doped silicon). The atom of an *acceptor dopant* with only three valance electrons has replaced a silicon atom, leaving a hole in the crystal lattice.

Figure 5.10 (a) Silicon crystal lattice at absolute zero. The "+4" in the center of each atom indicates the four positive charges in the atomic nuclei cor-

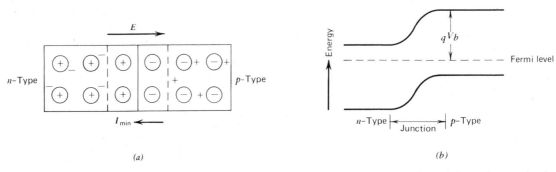

(a)　　　　　　　　　　　　　　　　　(b)

Figure 5.11 The p–n junction. (a) Creation of a barrier field by the depletion of free charges in the junction region; (b) the barrier as a displacement of energy levels.

must equalize. Minority current, from right to left in the drawing, sees no barrier, and so proceeds to its saturation value I_s. Majority current, from left to right, is impeded by the energy barrier qV_b. This barrier cannot be measured externally, because an equal but opposite barrier appears in the contacts made by the measuring device.

If a p–n junction is connected to a source of potential difference, as in Figure 5.12a, the electric field caused by the external potential difference, E_{ext}, is in the same direction as the

barrier field. There will be no change in the minority current, since it has reached saturation value; that is, all carriers being generated are swept across the junction. If the external potential difference were reversed, as in Figure 5.12b, the majority carrier flow would increase sharply:

$$I_{maj} = I_s \exp\left(\frac{qV}{kT}\right), \tag{5.24}$$

where V is the net potential across the junction, and I_s is the saturation current.

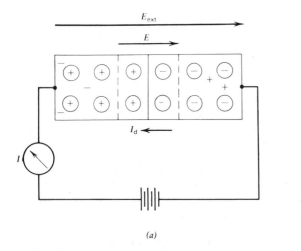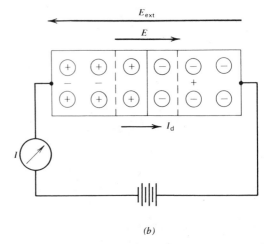

(a)　　　　　　　　　　　　　　　　　(b)

Figure 5.12 An external potential connected to a p–n junction (a) in the same sense as the barrier potential: no current; (b) opposite to the barrier: high current.

Expanded scales

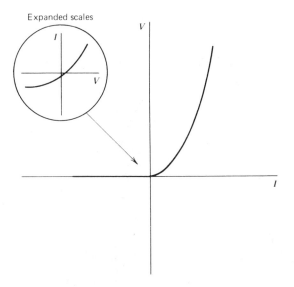

Figure 5.13 V–I curve for a p–n junction.

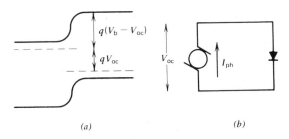

(a)   (b)

Figure 5.14 Effect of exposing a p–n junction to sunlight: (a) displacement of the Fermi levels; (b) equivalent circuit of the p–n diode.

The result can be summarized by the V–I curve of Figure 5.13. This may be recognized as the operating curve for a **diode**, which is exactly what a p–n junction is. In an ideal diode the total diode current is the difference between the majority and minority currents:

$$I_d = I_{maj} - I_{min}. \qquad (5.25)$$

Solar Cells

Exposure of an open-circuit p–n junction to sunlight causes a displacement of the Fermi levels, as in Figure 5.14. This change occurs because photons having an energy greater than the band gap E_g can create majority and minority carrier pairs in both the n- and p-type materials. These additional charge carriers elevate the Fermi level in the n-type material and depress it by electron–hole annihilation in the p-type material. This separation of the Fermi levels by qV_{oc} makes some majority carrier motion possible. Minority carriers will always be swept across the barrier. Under open-circuit conditions there will, of course, be

no net current; the minority and majority carrier motions will just balance out. The equivalent circuit for the open-circuit solar cell is shown in Figure 5.14b. An external device could measure this voltage.

The photoelectron current must be the difference between majority and saturation currents, since without light and load, these two are equal:

$$I_{ph} = I_s \left[\exp\left(\frac{qV_{oc}}{kT}\right) - 1 \right]. \qquad (5.26)$$

An external load connected to the junction, as in Figure 5.15, will be able to "see" a reduced barrier potential V_{ph}, so the device can act as a current source. With an external load, the only current available for the load is the photoelectron current minus the diode current:

$$I_{ext} = I_{ph} - I_s \left[\exp\left(\frac{qV_{ph}}{kT}\right) - 1 \right]. \qquad (5.27)$$

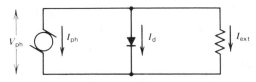

Figure 5.15 Equivalent circuit of a p–n junction connected to an external load and exposed to sunlight.

Example 5.5 Find the external current that maximizes the power output of a solar cell.

$$P = IV = \left\{ I_{ph} - I_s \left[\exp \left(\frac{qV_{ph}}{kT} \right) - 1 \right] \right\} V.$$

To maximize power, set $dP/dV = 0$.

$$\frac{dP}{dV} = I_{ph} - I_s \left[\exp \left(\frac{qV_{mp}}{kT} \right) - 1 \right]$$

$$+ V_{mp} \left(- I_s \frac{q}{kT} \exp \left[\frac{qV_{mp}}{kT} \right] \right) = 0,$$

where the subscript mp refers to maximum power values. This reduces to

$$\exp \left(\frac{qV_{mp}}{kT} \right) \left(1 + \frac{qV_{mp}}{kT} \right) = 1 + \frac{I_{ph}}{I_s}.$$

Equation 5.27 can be written

$$I_{mp} = I_s \left(1 + \frac{I_{ph}}{I_s} \right) - I_s \exp \left(\frac{qV_{mp}}{kT} \right).$$

In this expression, substitute for the exponential term from above, and

$$\boxed{I_{mp} = \frac{I_s(qV_{mp}/kT)}{1 + qV_{mp}/kT} \left(1 + \frac{I_{ph}}{I_s} \right).}$$

Using these definitions and other properties of the material, it is possible to calculate the current and voltage output of a particular cell over a wide range of temperature and illumination conditions. Such calculations for silicon are summarized in Figures 5.16 and 5.17 for 2-cm^2 cells. Figure 5.16 has been calculated for 16 mW/cm^2 illumination, Figure 5.17 for 0° C. The characteristics of this particular cell at other values can be extrapolated from these curves. We note a significant drop in performance with both increasing temperature and decreasing illumination.

These plots do not directly indicate efficiency. In fact, the maximum theoretical efficiency for a silicon cell is only about 45 percent. Since E_g for silicon is 1.08 eV, the

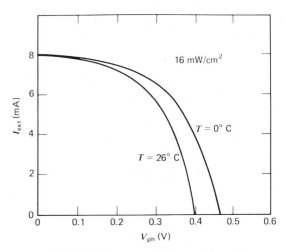

Figure 5.16 Performance of a solar cell as a function of temperature.

largest wavelength of light that will create an electron–hole pair is $\lambda = 1150$ nm. Since that part of the solar spectrum having $\lambda > 1150$ nm constitutes 22 percent of the total, all 22 percent is lost. For $575 \leqslant \lambda < 1150$, only 1.08 eV of each photon is used; the remainder is lost. This brings the total loss to about 45 percent. The maximum efficiency must be just the maximum deliverable power divided by the

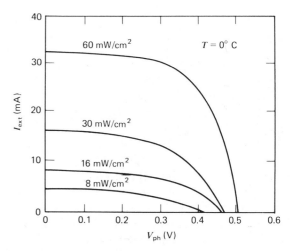

Figure 5.17 Performance of a solar cell as a function of illumination.

rate of energy incidence on the cell:

$$\eta_{max} = \frac{P_{max}}{N_{ph}E},\qquad(5.28)$$

where N_{ph} is the number of photons per unit time, and E is the average energy of those photons. Or, in terms of the cell parameters,

$$\eta_{max} \approx I_{ph}\frac{(q/kt)V_{max}}{1+(q/kT)V_{max}}\frac{V_{max}}{N_{ph}E},\qquad(5.29)$$

where V_{max} is the cell voltage at maximum power. Note that N_{ph} depends on E_g; as E_g increases, the number of photons having an energy greater than E_g decreases. However, as E_g increases, the saturation current also increases; therefore, the efficiency curve as a function of band gap should have a maximum, and indeed it does; see Figure 5.8.

Several factors influence the magnitude of the photocurrent. An important consideration is recombination of the charge carriers. Recombination removes charge from the photocurrent and is determined by (among other things) the thickness of the diode and its crystalline regularity. Imperfections such as grain boundaries, dislocations, and other effects causing broken bonds can serve as recombination sites.

There are also some practical considerations that must be taken into account. Two contacts on the photocell are required to make a complete electrical circuit. However, one of these contacts should be transparent! In the absence of a simple way to accomplish this, most cells are made with a gridded structure on the sunlight side (see Figure 5.18). Smooth silicon reflects up to 40 percent of the incident sunlight. Multilayer coatings or surface texturing or both can reduce this to less than 5 percent. Some photocurrent is lost in present designs because the cells are thicker than necessary. Charge carriers created near the cell surface may recombine at surface irregularities before being swept across the junction. Making the junction very near the surface should minimize

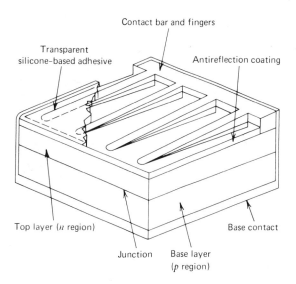

Figure 5.18 A typical commercial solar cell.

this effect. Several schemes have been suggested to increase the efficiency by using photons of all wavelengths more effectively. Two such ideas are illustrated in Figure 5.19. To date such schemes have not been used on a large scale, because the added expense is not paid for by the resulting increase in efficiency.

Because of these and other practical problems, the actual efficiences of silicon solar cells are in the 10- to 15-percent range. Future developments may increase this to 20 percent or so, but the largest single drawback to the widespread use of photovoltaic cells is their expense.

Silicon is one of the most abundant elements on earth, so you would think it would be relatively inexpensive. And it is, in a metallurgical grade. But the very high purity required for solar cells is very difficult to achieve. It is not simply a question of chemical purity, although that is important; structural purity is also very necessary. The solar cell should be made of a slab of silicon cut from a single crystal with as few imperfections as possible in structure.

Single crystal ingots of silicon are "grown"

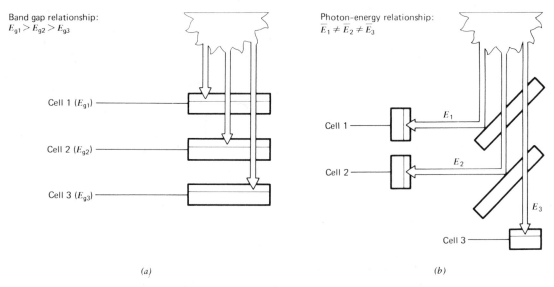

Band gap relationship:
$E_{g1} > E_{g2} > E_{g3}$

Cell 1 (E_{g1})

Cell 2 (E_{g2})

Cell 3 (E_{g3})

Photon-energy relationship:
$\bar{E}_1 \neq \bar{E}_2 \neq \bar{E}_3$

Cell 1

E_1

Cell 2

E_2

E_3

Cell 3

(a)

(b)

Figure 5.19 Two techniques for using a larger percentage of sunlight than is possible with a single silicon cell. (*a*) Tandem cell system; (*b*) filter–reflector system.

from a melt by slowly pulling a seed crystal from the surface (the Czochralski process). The ingot must be cut into thin wafers a few hundred micrometers thick (the saw blade is also this thick, so half the silicon crystal ends up as dust!). The wafers must be polished, where more material is removed, and then finally assembled.

The total cost of these various operations has decreased substantially over the years, but is still rather high. In 1980 the cost of a solar cell array was about $10/W (1975 dollars). The U.S. Department of Energy has set a number of goals for the next decade with respect to solar cell costs: in 1986 the costs should be no more than 50 cents/W (1975 dollars), and the efficiency should be 17 percent. If these goals can be met, photovoltaic conversion will indeed become competitive.

A variety of strategies are being pursued in an effort to meet or exceed these goals:

■ Production of high purity silicon in a continuous ribbon.

■ Production of polycrystalline silicon sheets yielding high efficiency.

■ Use of other materials: CuS/Cd, GaAs.

■ Use of several cells in tandem or with beam splitters to take advantage of the differing band-gap energies of different materials, as shown in Figure 5.15.

All of the above strategies have been studied in the laboratory; each has one or more technical or practical defects. The efficiencies of cells made from the continuous-ribbon silicon have not yet exceeded 12 percent. Polycrystalline sheets produce too many recombination sites, with a consequent very low efficiency. CuS and CdS cells have very low efficiencies; these cells must be very thin because they are opaque to sunlight. GaAs is a promising material because of its higher E_g and therefore higher efficiency. But gallium is not a very plentiful element, there being 10^4 times less gallium than silicon. It is not at all clear that sufficient quantities of gallium could be obtained at price level that would make the

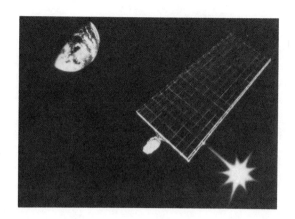

Figure 5.20 Artist's conception of a solar-satellite power station. (Illustration courtesy of NASA.)

GaAs devices competitive, even if high efficiencies could be obtained.

An interesting proposal for the use of solar cells in an orbiting satellite was made several years ago. This scheme, which received some initial funding from Congress in 1980, which would use a 65-km^2 solar collecting area, is illustrated in Figure 5.20. The DC electricity from the solar cells would be used to power microwave generators. These would direct a beam from a 2.6-km^2 antenna to a receiving station on earth. The microwave wavelength would be about 10 cm, taking advantage of the "window" in the long-wavelength absorption spectrum of the earth's atmosphere, which can be seen in Figure 5.6. A 93-km^2 receiving antenna system would be required on earth. Since the satellite would be in a geosynchronous orbit, only a single receiving system would be necessary, although two satellites would be required to avoid earth-eclipsing if an equatorial orbit were used, as is generally assumed.

It is estimated that, using the system described, 10,000 MW of electricity could be supplied; this is enough for all of New York City plus some of its environs. But, of course, what would it cost? When this proposal was first made in the early 1970s, it was estimated to cost about \$500/kW, somewhat more than coal-fired plants at the same time. But a number of somewhat unjustified assumptions were made: components an order of magnitude lighter than those available, the existence of the Space Shuttle, and high-efficiency solar cells.

Some concern has been expressed about the danger of an out-of-control microwave beam swinging about on the earth's surface. In fact, the power density in the beam described would be $(10^4 \text{ MW})/(2.6 \text{ km}^2) = 0.38 \text{ W/cm}^2$, which is about twice the power density of sunlight. However, the microwave portion of the 1.4 kW/m^2 of sunlight is rather small; so that 0.38 W/cm^2 of microwaves must be at least two orders of magnitude greater than the corresponding intensity in sunlight. This problem could be alleviated by increasing the beam size or decreasing the power level.

Aside from astronomically high costs, there is another factor to be considered. This system will add heat to the earth that would otherwise not be absorbed. This heat, concentrated in a metropolitan area, cannot be beneficial to the environment. This project is an example of a high-technology fix at high cost where lower-cost fixes would do the job—see Chapter 11.

The prospects for solar photovoltaic power production on a large scale are tantalizing. Many problems remain to be solved before it can become a reality. The need is clear, the problems are well defined, and pathways to the solutions exist. In the next decade or two, solar photovoltaic conversion may begin to contribute on a national scale. At the same time, other techniques for the direct conversion of radiant or thermal energy to electricity are being studied.

5.3 MAGNETOHYDRODYNAMICS

A number of devices have been built over the past 130 years or so that make use of the

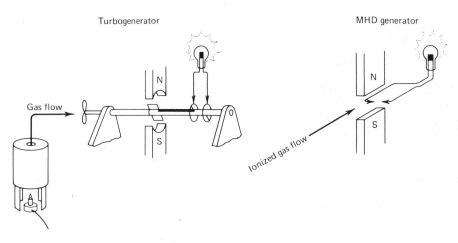

Figure 5.21 Comparison between a turbogenerator and an MHD generator.

interactions between electric and magnetic fields and the temperature or electron density gradients in solids. Most of these devices have limited potential to contribute to our energy conversion needs. There is one direct thermal conversion device that could have a very high efficiency and, therefore, should be very interesting to us as an alternative to steam-electric generation systems: the **magneto-hydrodynamic generator**. This device, usually abbreviated **MHD**, uses a hot, ionized gas directed through a magnetic field, as shown in Figure 5.21. This figure compares a turbogenerator and an MHD generator. In both cases, the motion of charged particles in a magnetic field is used to generate an emf. In the case of the turbogenerator, the charged particles are the free electrons in the rotating conducting wire; in the MHD generator the charged particles are the ions in the hot gas.

High efficiences are obtainable because, in theory, very-high-temperature gas can be used in the channel entrance. In practice, many difficulties need to be overcome before MHD will be practical. Let us spend some time looking at the basics for MHD so that we may appreciate the advantages and drawbacks of such a system.

A charged particle q moving with velocity $\mathbf{u}$ in a magnetic field $\mathbf{B}$ experiences a force given by

$$\mathbf{F} = q\mathbf{u} \times \mathbf{B}. \qquad (5.30)$$

This force can be thought of as the consequence of the interaction of the charged particles and an induced electric field:

$$\mathbf{E}_{ind} = \mathbf{u} \times \mathbf{B}. \qquad (5.31)$$

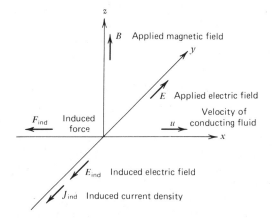

Figure 5.22 Vectors involved in charged-particle motion in magnetic fields.

The induced field gives rise to an induced current

$$\mathbf{J}_{ind} = \sigma \mathbf{E}_{ind}, \qquad (5.32)$$

which also interacts with the magnetic field to give a Lorentz force

$$\mathbf{F}_{ind} = \mathbf{J}_{ind} \times \mathbf{B}. \qquad (5.33)$$

Note carefully the directions of these various vectors (see Figure 5.22).

The induced current density $\mathbf{J}_{ind}$ is orthogonal to both the applied magnetic fields and the direction of the charged-particle motion. Collector plates oriented appropriately could be connected to a load to use this current, as in Figure 5.23. If life were this simple, MHD plants would be in use all over the world. Obviously, there is more to the analysis.

In fact, the electrons in the high-temperature, high-velocity gas cannot be directly deflected to the collecting plates—they are moving too rapidly. If they have a high enough mobility in the gas (i.e., the conductivity of the gas is high enough), they will be trapped into circular orbits in the magnetic field in the channel region. This dense screen of negative charge serves as a brake for the far more massive positive ions. A neutral plasma is then formed in the channel region, which serves to impede physically the flow of the neutral gas atoms by collisions. Electrons are thereby forced out of the screen and into the collector, through the external circuit. The net effect has been the extraction of energy from the gas and the delivering of it to a load.

The collector plates develop a net electric field $\mathbf{E}$ the applied electric field of Figure 5.22. The net induced current is then

$$\mathbf{J}_{ind} = \sigma(\mathbf{E}_{ind} - \mathbf{E}), \qquad (5.34)$$

or

$$\mathbf{J}_{ind} = \sigma(\mathbf{u} \times \mathbf{B} - \mathbf{E}). \qquad (5.35)$$

The MHD channel can be thought of as a current source, delivering specific energy at a rate given by

$$\dot{w} = EJ. \qquad (5.36)$$

Therefore, the electrical efficiency of this device would be the specific energy rate divided by the total specific energy rate for the channel:

$$\eta_e = \frac{\dot{w}}{\dot{w} + J^2/\sigma}, \qquad (5.37)$$

where J^2/σ is the Joule heating term. From Eq. 5.35 this is just

$$\eta_e = \frac{E}{uB}. \qquad (5.38)$$

The variation of specific power output and electrical efficiency is shown in Figure 5.24.

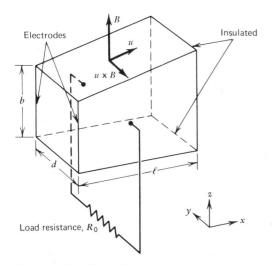

Figure 5.23 Simplified diagram of a continuous-electrode MHD channel.

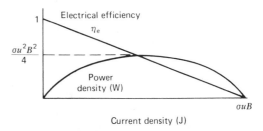

Figure 5.24 Electrical operating characteristics of an MHD generator.

Note the parabolic dependence of power output on current density.

We should also note that the not insignificant amount of power required to maintain the magnetic field and operate other devices in the system must be taken from $\dot{w}$, so that the practical efficiency will be rather considerably lower than that calculated from Eq. 5.38.

The calculation of the efficiency of conversion of thermal energy to electrical energy in an MHD channel is a rather more complicated process; we refer you to several textbooks cited in the bibliography for details and mention here only the ramifications. High efficiencies are achieved for the following MHD characteristics:

■ High gas σ.
■ Electrical efficiency of $\frac{1}{2}$.
■ High magnetic field.
■ High C_p and γ for the gas.
■ Input temperatures as high as possible and output temperatures as low as possible.
■ High gas velocity (high Mach number).

Practical considerations limit most of these characteristics, and some of them are mutually exclusive.

First of all, the electrical conductivity of a high-temperature gas is rather poor, unless extremely high temperatures are considered (see Figure 5.25). Temperatures in excess of 3000 K are really not usable, because of materials limitations. A technique for improving the conductivity of a lower-temperature gas is to "seed" the gas with an alkali metal in the form of a carbonate or chloride. This raises the conductivity of the gas considerably, especially at higher temperatures. While seeding increases the conductivity of the gas, it also increases the cost of the system, because the seed material will be too expensive to discard and must therefore be recovered and recycled.

High temperatures pose particular problems in the construction of the MHD channel. In-

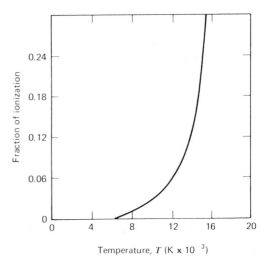

Figure 5.25 Fraction of air molecules thermally ionized as a function of temperature.

sulators should remain insulators at very high temperatures over long periods of time, and conductors likewise should retain their properties in this rather hostile environment. The problem is obvious: conductors become insulators and insulators become conductors at very high temperatures.

We should also note here that it is not possible to have the large inlet and outlet temperature difference required for high efficiency. At low temperatures, significantly lower than 2000 K, the gas conductivity, even with seeding, is too low. Consequently the outlet temperature will be high, in fact high enough to be the inlet temperature for a conventional steam cycle. In this sense, the MHD channel acts as a "topping cycle."

High magnetic fields can be obtained from conventional magnet designs at a high cost in power—Joule heating losses in the conductors. More efficient magnets can be made, using superconducting technology; higher fields are possible, and dramatically lower losses are incurred. Of course, power must be expended to operate the required cryogenic refrigerators.

All things considered, superconducting magnets are better in terms of operating costs than conventional magnets. Initial capital costs are about the same. Care must be taken to avoid having too high a Mach number for the gas. Flow velocities that are too large can cause flow choking or shock formation.

Any combustible material could be used to fuel the MHD plant, but for several reasons the most often projected fuel is coal. A more complete combustion of the coal could be obtained, including the char residue; SO_x would have to be removed, but this would be done before combustion for the MHD channel takes place, resulting in a clean effluent. The large amounts of NO_x produced by the very-high-temperature combusion in air could be profitably recovered in the form of commerical grade nitric acid. Figure 5.26 is a block diagram of a proposed combination MHD–steam power plant.

The MHD channel of this plant would be approximately 13 m long with a superconducting magnet supplying a 6-T field. The current density would average about 4000 A/m^2 along the channel at an average potential difference of 5400 V (DC). The net MHD output would be 800 MW. The gas flow would be 583 kg/s seeded with Cs_2CO_3 at 0.7 percent. The channel inlet temperature would be 2650 K, the outlet temperature about 2300 K. This plant is estimated to have an overall efficiency of 52 percent.

The increased overall efficiency of the combined plant over a fossil-fueled system of the same electrical output has two advantages: reduced heat rejection to the biosphere and extension of fossil fuel reserves.

A rather small research effort is presently underway in the United States and a few other countries, notably the Soviet Union. Several small (50 to 200 MW) units have been built and

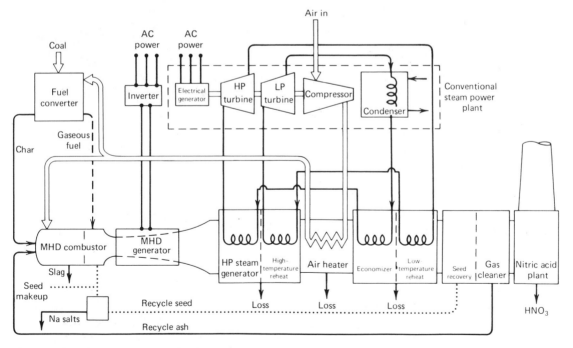

Figure 5.26   A proposed combination MHD–steam plant.

Figure 5.27 A portion of a 200-MW MHD generator with the magnet partially withdrawn. This device operated for 500 hr in 1978. (Photograph courtesy of Avco Everett Research Laboratory, Inc., Everett, Massachusetts.)

operated successfully for short periods of time. Figure 5.27 shows a portion of a 200-MW MHD generator that was operated in 1978. Materials problems are clearly the area where much original work is needed to make the entire system function. The high temperatures and corrosion of the combustion products and seed material have combined to make channel lifetimes short. Preheater and combustor designs also need upgrading before commercial operation can be contemplated.

Several different designs are under consideration, including closed-cycle and liquid-metal MHD systems. These have the advantage of producing no effluent, but they also must operate at a lower temperature, thereby lowering the overall plant efficiency.

In summary, MHD electrical generation has the potential to become not only competitive with conventional electricity-generating systems, but to surpass them in terms of effluent reduction, waste heat reduction, and fossil-fuel-reserves extension. MHD systems are not now commerical, but if sufficient research effort and money are expended in the next few years, they may well be.

5.4 WIND POWER

As every student of the history of technology knows, wind power has an old and honorable tradition. Wind provided the energy for ships and mills long before electricity was discovered in the mid-nineteenth century. When electricity began to dominate the energy mix in the twentieth century, a source of energy more reliable than wind was found in coal and oil. Wind, while "free" as a fuel, was variable, less concentrated and, as time went by, actually more expensive to use. Until recently, wind-powered electricity generators were found only in remote, hard-to-fuel locations such as fire lookouts and isolated farms. But the changes in the world oil picture have brought a renewed interest in wind generators as central stations, as well as for individual houses.

In this section we study the physics of energy removal from the wind and examine some of the existing wind machines—now called wind turbines. We also ask whether large-scale wind turbine usage will have any adverse environmental impact and, finally, we try to estimate the extent to which wind power can contribute to our national energy scheme.

Physics of Wind Machines

We can develop the physics of horizontal wind turbines by applying simple momentum theory to an idealized windstream. In Figure 5.28 wind with density ρ and velocity V impinges on a disk (windmill) of cross-sectional area A. The windmill extracts energy from the wind, so if the asymptotic velocity and pressure of the wind on the left are V and p_0, then the asymptotic values on the right are $(V - v_1)$ and p_o. The velocity of the wind as it ap-

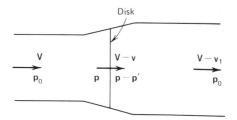

Figure 5.28 Fluid stream incident on a rotating disk (windmill).

proaches the disk is retarded to $(V - v)$, and this retarded axial velocity is continuous through the disk. The velocity-retarding values v and v_1 cannot be equal. Bernoulli's equation assures us that for *continuous flow*

$$p + \tfrac{1}{2}\rho V^2 = \text{const.} \qquad (5.39)$$

This relationship applies to the flow on the left side of the disk and on the right side separately. While the wind velocity is continuous through the disk, the pressure is not. The pressure just to the left of the disk is reduced to p, and the pressure just to the right is $p - p'$. The momentum extracted from the wind must be transferred to the disk. This momentum is just the mass transferred through the disk times the change in velocity. Therefore the transferred momentum per unit of time is given by

$$\frac{\text{Momentum}}{\text{s}} = (A\rho[V - v])v_1, \qquad (5.40)$$

where the mass flowing through the disk per unit of time is $A\rho(V - v)$. We use v_1 here, since v_1 is the net velocity change and $dp = m\,dv$.

The momentum change per unit time must be the drag force on the windmill. Of course, in a stationary windmill, this drag has no translational effect. It is this force, however, that provides the torque for the windmill and is therefore responsible for the development of mechanical power. The rate of energy extraction from the wind is then given by

$$\text{Power} = \text{drag} \times \text{velocity} = A\rho(V-v)v_l(V-v). \tag{5.41}$$

Here we must use $(V-v)$ as the velocity to determine power, since this is the actual wind velocity at the position of the disk.

There is a relationship between the asymptotic velocity reduction v_l and the reduction near the disk v. To find this relationship·we must again apply **Bernoulli's law**. We wish to obtain our results in terms of v, not v_l. The designer of a wind turbine can determine v by an appropriate physical design of the blades: number, size, shape, an so on. We equate the pressure head in the asymptotic region on the left with that on the left near the disk; and do the same thing for the right-hand side. The pressure head is the sum of actual pressure and $\frac{1}{2}\rho V^2$.

$$H_0 = p_0 + \tfrac{1}{2}\rho V^2 = p + \tfrac{1}{2}\rho(V-v)^2 \tag{5.42}$$

$$H_1 = p_0 + \tfrac{1}{2}\rho(V-v_1)^2 = p - p' + \tfrac{1}{2}\rho(V-v)^2. \tag{5.43}$$

Then,

$$p' = H_0 - H_1 = \rho(V - \tfrac{1}{2}v_1)v_1. \tag{5.44}$$

But p' is just the drag per unit area on the disk. Then comparing Eq. 5.44 with Eq. 5.40 it must be that $\frac{1}{2}v_1 = v$. So the power developed by the windmill is given by

$$\text{Power} = 2A\rho(V-v)^2 v. \tag{5.45}$$

The efficiency of converting windpower to mechanical power is just:

$$\eta = \frac{\text{output power}}{\text{input power}}, \tag{5.46}$$

$$\eta = \frac{2A\rho(V-v)^2 v}{(\tfrac{1}{2}A\rho V^3)}. \tag{5.47}$$

$$\eta = 4\frac{v(V-v)^2}{V^3}. \tag{5.48}$$

It may be easily demonstrated that this

efficiency is a maximum when $v = \frac{1}{3}V$ and that the maximum value is 16/27.

Of course, this does not indicate that all horizontal wind turbines will operate with this efficiency. Indeed, even well-designed units have difficulty achieving more than 70 percent or so of this value. By combining Eqs. 5.45 and 5.47 we can see that the maximum power obtainable from a horizontal wind turbine (operating at maximum efficiency) in terms of the diameter D of the disk and entering wind speed V is

$$P_{\max} = \frac{8\pi D^2 \rho V^3}{27}. \tag{5.49}$$

The problem confronting the wind-power engineer is the design of a system that is large and can operate at high wind speeds but at the same time maximizes the efficiency by making the velocity loss 1/3 the entering windstream velocity. This latter criterion depends on two basic factors: the physical configuration of the disk and the rotation speed of the blades, **tip speed**, as it is called.

Typically, a horizontal wind turbine usually has two or three propeller-type blades, depending on the operating conditions and the characteristics desired. In this configuration, the total power developed is the sum of the power developed along the blade:

$$\text{Power} = \int dP = \int \Omega\, d\tau, \tag{5.50}$$

where Ω is the blade angular velocity, and $d\tau$ is the element of applied torque at each point. Since the quantity Ω is related to the tangential velocity of the blade at each point, it is not unreasonable that the tip speed should be large so that the above integral over the blade should have the right magnitude.

Two-bladed propellers are more economical than three-bladed ones, but they sometimes experience vibrational problems that the latter do not. Centripetal force can be minimized by using lightweight blades. Wood, plastic, and especially reinforced fiberglass with a good strength-to-density ratio are suited to this purpose, whereas metal is not so adaptable to blade construction. Fiberglass in particular withstands storms and handles operating stresses well, in addition to possessing excellent fabrication properties. Windmills for pumping water have a greater number of blades so that they will run efficiently in low wind velocities.

Equation 5.48 appears to say that the maximum power extractable from the wind increases without bound as the wind velocity goes up. In theory, this is true; but in practice, of course, it is not. The reason is that the maximum power is extracted when the efficiency is at a maximum, which is true when $v = V/3$. For a given size and shape propeller, there is only one wind speed for which this condition can be met; there is then a maximum wind speed V_{max} at which the machine is designed to run. At wind speeds below V_{max} the output power will drop, and at speeds above V_{max} not all the available power will be used. This means that power output will be doubled with a 33 percent increase in wind velocity or cut in half with a 33 percent decrease in wind velocity. A more dramatic drop occurs if the wind velocity decreases 50 percent—only 12.5 percent of the original output power is then available.

There are both lower and upper wind-speed limits. A wind turbine is designed to start rotating at some minimum wind velocity, the cut-in speed; but the machine produces its full power output at V_{max}, the **rated speed**, which is 5 to 10 miles/hr faster than the average annual wind speed for the site. At wind speeds above this figure, the power output is controlled to remain at the maximum rated power capacity. To help maintain a constant level of power output, a practice called flat-rating is used, which keeps the wind generator at a constant output at all speeds above its designed rated speed. This is achieved by means of a

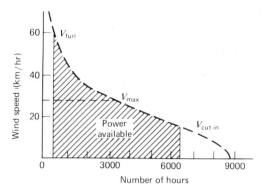

Figure 5.29 Power available at a "good site." The rated wind speed (v_{max}) is 27 miles/hr, the cut-in speed is 17 miles/hr, and the turbine speed is 60 miles/hr.

mechanical governor in most systems, or by changing blade pitch to make the momentum-transfer process less efficient. In either case, power is lost. Protection from high-speed damaging winds is provided by a furling mechanism that turns the rotor completely out of the wind at some appropriate wind speed. We see in Figure 5.29 that at a "good" wind power site, wind power is available more than 50 percent of the time.

We have so far dealt only with horizontal-axis wind turbines. This class of wind machines is more developed than the other: the vertical-axis wind turbine. Horizontal machines have one major drawback: they must be placed at the top of a tower for optimum operation. Not only is blade clearance needed, but also wind velocities usually increase with increasing height above ground. For very large machines this creates a problem: the large electrical generator must also be mounted on top of the tower. The tower then becomes a crucial economic element in determining the feasibility of wind power in any location. The vertical-axis wind turbine alleviates this problem, but of course it has problems of its own.

There are several different types of vertical

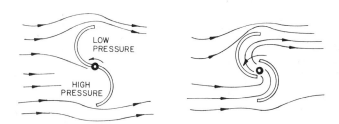

Figure 5.30 Schematic of a Savonius rotor, as seen from above.

machines. The **Savonius rotor** is diagrammed schematically in Figure 5.30. It usually consists of a cylindrical shell, split in half and mounted to rotate between top and bottom plates. The two halves are separated, as on the right, so that the passage between them can be opened and closed. With the passage closed, as on the left, the low-pressure region behind the top "wing" retards the motion and the system develops very little torque. Maximum power is extracted when the distance between the two semicircles is about half their diameter. This device cannot extract as much power as an efficiently operating horizontal propeller.

The other type of vertical axis machine that has been studied is the **Darrieus rotor**. This device has two or three thin blades, eggbeater style, which rotate quite rapidly (three or four times wind speed). The Darrieus rotor will not self-start, but must be brought up to operating speed by a motor.

All vertical machines have the advantage of being able to place the generator on the ground, thus simplifying the tower and associated support structure. Also, they automatically face the wind and so need no mechanism for rotating the device.

Wind Power Potential

The winds have been and are being used to generate electricity in relatively large scale at several places around the world—see Table 5.7. This table lists only the larger units; many

thousands of small (few-hundred-watt-sized) units were in use at various times, generally prior to the 1960s, to provide a small amount of electricity for rural users. The largest unit built in the United States to date was the Smith-Putnam generator on Grandpa's Knob in Vermont.

This horizontal-axis wind turbine system had a two-bladed propeller 53 m in diameter on top of a 33.5-m tower. Each blade weighed about 8 t and consisted of a stainless-steel sheet over stainless-steel ribs. The blade pitch could be adjusted, as could the opening angle of the two blades (normally fixed at 180°). The unit began operation on October 19, 1941, and ran more or less continuously for two and one-half years. It was then shut down for a main bearing replacement, which was not easy during the war years. It began operating again in March of 1945, but after three weeks one of the blade spars broke, hurling the blade over 228 m from the base of the tower. The unit was never rebuilt.

Why was this unit not rebuilt, and why do we not have large wind generators today? In a word, cost. The estimated cost of installed wind-power-generating capacity in the New England area in 1945 was $205/kW; the cost of oil- or coal-fired generating capacity at the same time was less than $100/kW. Again, we see that economics is the driving force, rather than technology.

But there are many forces at work. Since the cost of coal or nuclear installed electrical-

TABLE 5.7 Characteristics of Some Large Wind Generators

Designation	Rated at kW	Rated at km/h[a]	Swept Circle Diameter[b]	Blades	Rotation Speed (rpm)
Smith-Putnam, USA (Grandpa's Knob, Vt.), 1941–945	1250	51	53 m	2 D[c]	29
Aerowatt, France, 1957–1965	800	59	31 m	3 ?	47
NEYRPIC, France, 1963–1964	1000	59	35 m	3 ?	
NEYRPIC, France, 1960–1963	132	45	21 m	3 ?	56
Balaclava, USSR, 1930s	100	37	30 m	3 U[d]	30
Gedser, Denmark, 1957–1967; 1977–[e]	200	53	24 m	3 U	30
John Brown, UK (Orkney), 1952–?	100	56	15 m	3 D	130
Enfield-Andreau, UK, 1953–1956	100	51	24 m	2 D	95
Isle of Man, UK, 1958–?	100	66	15 m	3 U	75
Stotten-Hutter, Germany, 1957–1968	100	29	34 m	2 D	42
NASA-Lewis, USA, 1975–	100	29	38 m	2 D	40
NASA-USA (Clayton, N.M.), 1978–	200	35	38 m	2 D	40
DAF-Hydro, Quebec, Canada, 1977–	200	32	Darieus 38 m high	2 –	38
WTG-Cuttyhunk, USA, 1977–	200	45	24 m	3 U	30
Tvind, Denmark, 1977–	2000	53	54 m	3 D	Variable

Adapted, with permission, from Marshall F. Merriam, "Wind, Waves and Tides," *Annual Review of Energy* **3**, 29 (1978), © 1978 by Annual Reviews Inc.
[a]Windspeed measured at hub height.
[b]The DAF-Hydro Quebec unit is a vertical axis, Darrieus rotor. All other units in the table are horizontal axis, two-, or three-blade propellers.
[c]Blades were mounted downwind of the tower.
[d]Blades were mounted upwind of the tower.
[e]Gedser machine operated 1957–1967, stood idle 1976–1977, and began operating again in 1977.

generating capacity now is greater than $1000/kW (1980 dollars), it is not unreasonable to look again at wind power as an alternative source of electricity. The Department of Energy (DOE) in conjunction with the National Aeronautics and Space Administration (NASA) has built several pilot plants to test engineering processes. The 200-kW test turbine on Culebra Island, Puerto Rico, is shown in Figure 5.31. The characteristics of this device are given in Table 5.8.

So far these pilot plants have not demonstrated the commercial viability of wind power. Several factors must be clarified before they can do so:

■ Government policies.
■ Cost of conventional energy sources.
■ Cost of solar energy.

TABLE 5.8 Characteristics of the DOE/NASA 200-kW Experimental Wind Turbine on Culebra Island, Puerto Rico

Blade diameter	38.1 m
Blade material	Aluminum
Weight per blade	1045 kg
Cut-in speed	15.2 km/h
Rated speed	35.8 km/h
Furl speed	64.0 km/h

Source: NASA Lewis Research Center, Cleveland, OH.

■ Technological uncertainties of unconventional sources.

There is no orderly federal plan for the development of wind power and its assimilation into the national energy mix. Private util-

Figure 5.31 The DOE/NASA 200-kW test wind turbine on Culebra Island, Puerto Rico. (Photograph courtesy of the DOE.)

ity companies are unlikely to bear the investment burden by themselves, particularly with so many uncertainties in the energy picture. As coal and oil become more expensive, wind power costs become less a factor. However, new technology in the conventional resource recovery area—*in situ* gasification, oil shale distillation, and so on—may make any economic incentive smaller. Solar energy remains the sleeping giant. If solar energy can be used economically, it will put other forms of energy—both conventional and unconventional—out of business.

If we assume that some time in the near future, the economics of wind power relative

to other sources will be advantageous, we should then ask, "How much wind power is available, and are there any negative environmental aspects in the use of wind power?"

It is not relevant to ask the global wind circulation energy, since this total energy could not be used. To estimate wind power resources accurately would require a wind-speed measurement probram extending over several years for the entire country. No such comprehensive program exists in this country. (Great Britain has made such a study.) Data from 348 wind observing stations in the western United States were analyzed by one group.[4] Two percent of these stations have average wind speeds exceeding 7 m/s. At this speed, the average power density of the wind is just under 400 W/m^2, of which perhaps 175 W/m^2 could actually be converted to electricity. We should note that the measurements were made at 10 m elevation; wind turbines would have to be much higher than that, with consequently increased average wind speeds available.

This represents an apparently large source of electrical energy. A wind turbine with 50-m diameter blades could generate an average 3.5-MW power level at such sites. Of course, multiple machines could be placed at favorable sites. Is there any physical limitation to the density of wind generators?

Yes. This limitation results from two effects: wake interference and wind energy depletion. Wake interference is the turbulence caused by the passage of the wind through the blades. This turbulence reduces the net downwind wind speed. Studies with the DOE-NASA Sandusky machine indicate that wake interference can be minimized if the machines are spaced at least five rotor diameters apart and not aligned in straight lines.

[4]U. A. Coty, 1977 *Wind Energy Mission Analysis—Final Report.* Lockheed California Co. Rep. LR27611, Vols. 1, 2, 3. Springfield, Va.: National Technical Information Service, 1977.

The second problem, that of energy depletion, is not as well understood. The wind machine density is limited by the ability of the winds at large heights to couple energy into the layer of air moving next to the ground. It is estimated that this limitation would require that no more than 1500 m² of wind turbine area occupy each square kilometer of land. For machines with 60-m diameter rotors, this would limit the spacing to a minimum of about 35 rotor diameters.

There are also adverse environmental effects from wind generators:

- They are eyesores.
- They are noisy.
- The produce electromagnetic interference.

The first two items on this list are perhaps minor problems; after all, we have lived with such for many years—oil wells, factories, power plants . . . . The last one is not so easy to ignore, since much of the interference is in the television bands. There is not a very effective way to combat this problem, although experiments with fiberglass blades are encouraging.

What is the bottom line on wind power? As with most of the energy technologies we have studied, economics is the overriding conern. Wind power can contribute some—*not an extremely large amount*—to our overall energy supply. But only if the prices of conventional fuels increase relative to other prices.

SUMMARY

Energy in one form can be converted directly into electricity, bypassing a stage in which the energy appears as heat. This could result in a more efficient process.

An *electrochemical cell* consists of an **anode** and a **cathode** separated by a nonconducting *electrolyte*. Chemical reactions occur at these electrodes resulting in the liberation of electrons at the anode and the acceptance of elec-

trons at the cathode. A *battery* is a collection of cells.

Fuel cells are just electrochemical cells in which the electrodes are not consumed, but rather are supplied with fuel from an external source. Fuel cells have been used on spacecraft and could be used in homes and industry if they could be priced competitively.

Electrons can be liberated in certain types of *semiconductors* when they are struck by sunlight. This is the *photoelectric effect*; *photons* carry energy in proportion to their frequency.

Complete understanding of solid-state physics requires *quantum mechanics*. However, semiconductors can be thought of as materials in which the allowed energy states (*conduction band*) are separated from the next higher states (*valence band*) by a very small gap. Photons provide the energy to bridge this gap.

The *solar cell* acts as a diode in a circuit. There are two types of charge carriers: *majority* and *minority* carriers. The operation of a cell can be understood in terms of these quantities.

Silicon is usually used for solar cells, although other substances are being tried. None of these others is as abundant as silicon. The price of solar cells is still much higher than that required for competitiveness with fossil power.

Magnetohydrodynamics (**MHD**) is a way of using very high temperatures and consequent high efficiency to generate electricity. High-temperature ionized gas passes through a strong magnetic field. The charged particles of the gas are deflected to collecting plates, which serve as the "electrodes" of an equivalent battery. Pilot MHD plants have been built, but there are problems with maintaining the high magnetic fields, excessive wear on the collecting plates, and recovery of the material added to the gas to produce ionization.

Horizontal *wind turbines* can be understood in terms of momentum theory, in which the change in momentum of a volume of air is calculated as it passes through the disk of the wind machine. A maximum efficiency is found: 16/27.

Other kinds of wind turbines can be built, including vertical-axis types. These have not been studied mathematically. They have some advantages over horizontal machines, but they have not been demonstrated to be as usable in practice.

Wind is diffuse and irregular. It will be difficult to interface a wind generator into an electric grid system. There are also doubts about the economics of wind generators. They are not without environmental effects.

REFERENCES

Electrochemistry

Hagenmuller, Paul, and W. Van Gool (Eds.), *Solid Electrolytes.* New York: Academic Press, 1978.

Mahan, Gerald D., and Walter L. Roth, *Superionic Conductors.* New York: Plenum, 1976.

Fuel Cells

Austin, Leonard G., "Fuel Cells," *Sci. Amer.* **201**(4), 72 (1959).

Liebhafsky, H. A., and E. J. Cairns, *Fuel Cells and Fuel Batteries.* New York: John Wiley, 1968.

Solid-State Physics

Holden, Allen, *The Nature of Solids.* New York: Columbia University Press, 1965.

Kittel, Charles, *Introduction to Solid State Physics.* New York: John Wiley, 1960.

Solar Cells

Backus, Charles E. (Ed.), *Solar Cells.* New York: IEEE Press, 1976.

Direct Conversion

Angrist, Stanley W., *Direct Conversion of Energy.* Boston: Allyn and Bacon, 1971.

Corless, William R., *Direct Conversion of Energy*, AEC Understanding the Atom Series, Washington, D.C.: U.S. Government Printing Office, 1964.

Russel, C. R., *Elements of Energy Conversion.* Oxford: Pergamon Press, 1967.

Thermoelectricity

Joffee, Abram F., "The Revival of Thermoelectricity," *Sci. Amer.* **199**(5), 31 (1958).

MacDonald, D. K. C., *Thermoelectricity: An Introduction to the Principle.* New York: John Wiley, 1962.

Magnetohydrodynamics

Somers, Edward V., Daniel Berg, and Arnold P. Fickett, "Advanced Energy Conversion," *Ann. Rev. Energy* **1**, 345 (1976).

Wind Power

Merriam, Marshall F., "Wind Energy for Human Needs," *Tech. Rev.*, 29 (*January* 1977).

Metz, William D. "Wind Energy: Large and Small Systems Competing," *Science* **197**, 971 (1977).

Merriam, Marshall F., "Wind, Waves and Tides," *Ann. Rev. Energy* **3**, 29 (1978).

GROUP PROJECTS

Project 1. Make a wind power analysis of the nearest site where meteorological records are available. What is the average wind speed? How much average power could be generated? Extrapolate these results to your location. What do you conclude? (See *Popular Science*, July 1980, p. 100.)

Project 2. Estimate the total number of electrical kilowatt-hours you are personally responsible for in a week. Include appliance usage, classroom lighting and heating, movies, and so on. Estimate the number of automobile storage batteries necessary to store that amount of electrical energy.

EXERCISES

Exercise 1. Determine the maximum quantity of electricity (in ampere-hours) that can be produced by 1 kg of reactants (total) in a mercuric oxide–zinc battery.

Exercise 2. Lead–acid batteries are to be used in conjunction with a solar power plant to provide a peaking demand on cloudy days of 100 MW. What is the minimum number of batteries of the type in Example 5.3, if the load is continuous for 1 hr? What is the total mass and volume required?

Exercise 3. How much time is required to place a 25-μm chromium layer on a 4000-cm^2 car bumper with a 250-A current?

Answer: 17 min 50 s.

Exercise 4. Chrome plating at high current releases significant amounts of hydrogen gas, which traps chromic acid as it evolves. This chromic acid mist is very corrosive and toxic. The plating in Exercise 3 would release about 200 g of chromic acid per hour of operation. If a plating factory plates 5000 bumpers per day, how much chromic acid mist is emitted per day?

Exercise 5. Derive $dG = V\,dp - S\,dT$.

Exercise 6. In the circuit in Figure 5.32, show that the maximum power is delivered by the emf $\mathscr{E}$ to the load R when the cell's internal resistance r is given by $r = R$.

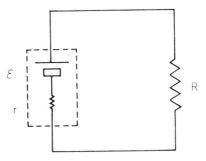

Figure 5.32 Diagram for Exercise 6.

Exercise 7. Show that

$$T\left(\frac{d\mathscr{E}}{dT}\right)_p = \frac{\Delta H}{nF} + \mathscr{E}.$$

Exercise 8. Determine the cost per kilowatt-hour of electricity from typical samples of

(a) 9-V transistor radio batteries.
(b) Alkaline 9-V transistor radio batteries.
(c) 1.5-V penlight cells (type AA).
(d) 1.5-V flashlight cells (type C).
(e) Mercury cells for watches or cameras.
(f) Lead–acid automobile batteries.

Compare with your current residential electricity rate.

Exercise 9. Calculate the actual efficiency of an H_2–O_2 fuel cell at 400 K, is the cell voltage is 0.71 V.

Answer: $\eta = 0.564$.

Exercise 10. About 10^{16} J/yr of energy is used annually in the form of natural gas burned in combustors to produce electricity at about 37 percent efficiency. If half of the electricity were produced instead by fuel cells using methane at 56 percent efficiency, how much natural gas would be saved?

Exercise 11. What is the energy of a 10-cm microwave photon? How many photons per second are contained in a 10,000-MW microwave beam?

Exercise 12. The internal resistance of a p–n junction is given by $R_0 = kT/qI_s$. Show that maximum power is transferred to a load when the load resistance is given by $R_L = R_0 \exp(-qV/kT)$.

Hint: Use Example 5.5 and Eq. 5.27.

Exercise 13. A 2-cm^2 solar cell operates at $0°$ C with 600 W/m^2 illumination and 6 percent efficiency. If the load resistance is $19\,\Omega$, at what voltage and current is the cell operating?

Answer: 0.369 V, 1.95×10^{-2} A.

Exercise 14. Using Si-solar cells at a temperature of $0°$ C with an illumination of 600 W/m^2, what total cell area would be required to supply 5 kW at 0.4 V to be inverted and stepped up for household use?

Exercise 15. What is the actual efficiency of the cell system in Exercise 14? What is the rate of heat production in the cells?

Exercise 16. Assume a 10-yr lifetime for solar cells. Estimate the kilowatt-hours available from a 500-W array, and express the cost of this array in terms of barrels of petroleum.

Exercise 17. For the 10,000-MW (delivered to the load) satellite system described in this chapter, assume an overall efficiency of power transfer from the solar cells to the load of 70 percent. If the illumination on the cells is 1.4 kW/m^2, what is the efficiency of the solar cells?

Answer: 16 percent.

Exercise 18. The radial distribution of power density in a microwave beam would be Gaussian:

$$P = P_0 \exp(-r^2/r_0^2),$$

where P_0 is the power density in W/cm^2 at the center and r_0 is a measure of the width of the

Gaussian and is 2×10^5 cm^2. What size circular antenna would be required for a total power of 10^7 W, if the power density is not to exceed the recommended value of 5 mW/cm^2?

Exercise 19. In one-dimensional MHD fluid flow the *Equation of Continuity* is given by

$$\frac{dp}{dt} + \frac{d(\rho v)}{dx} + \mathbf{J} \times \mathbf{B} = \text{constant}.$$

Use this to derive the equation of motion for the gas in an MHD channel:

$$\rho u \frac{du}{dx} = -\frac{dp}{dx} - J_{ind}B.$$

Exercise 20. What is the ratio of the waste heat produced by a 32-percent-efficient nuclear plant to that produced by a 51-percent-efficient combined MHD–steam plant of the same electrical output?

Exercise 21. Start from Eqs. 5.35 and 5.37 and derive Eq. 5.38.

Exercise 22. Assume there are 200 of the 800-MW MHD plants described in this chapter, each operating at 50 percent of capacity averaged over 1 yr and that the Cs_2CO_3 seed is recovered with 99.9 percent efficiency. How much cesium is required for makeup in a year? How does this amount compare with the annual U.S. production of cesium?

Answer: 1.05×10^7 kg Cs lost per year.

Exercise 23. Differentiate Eq. 5.46 with respect to v, and show that the maximum efficiency of a horizontal wind turbine is obtained when $v = \frac{1}{3}V$.

Exercise 24. Estimate the average peak power requirement of a typical residence. Assume a locality where the winds average 19 km/hr. What blade diameter for a horizontal wind turbine would be required?

Exercise 25. Compare the electrical power available from solar energy per unit area with that from wind power at the 348 stations discussed in the text.

Exercise 26. Show that $dP/P = 3(dV/V)$.

Exercise 27. Estimate the number of kilowatt-hours the Culebra Island unit could generate annually if the distribution of wind speeds were as shown in Figure 5.29.

PART THREE

HEAT GENERATION AND MANAGEMENT

CHAPTER SIX

HEAT SOURCES

There are two major sources of heat for heat engine operation: the combustion of flammable substances and the fission of certain fissile nuclei. In addition to these we could add solar radiation. Table 6.1 lists various heat sources, their energy content (or its equivalent), and the daily consumption of a 1000-MW electric generating plant using that source, along with the efficiency assumed in each case. We examine methods of calculating energy release for general chemical reactions in this chapter. These studies may help us determine the most appropriate heat source for a given application. But we must remember that decisions involving energy are usually made on purely economic rather than technical grounds.

Fossil fuels, primarily oil, and nuclear fuels account for the majority of the electrical power produced in the United States. Nuclear fuels currently produce electricity in most locations cheaper than coal; plutonium produced in breeder reactors is predicted by some to be an even cheaper resource, although many informed scientists disagree. We shall discuss nuclear power in detail in the next chapter.

All current heat sources are too cheap; their selling prices do not reflect all the costs incurred in producing them. In the case of fossil fuels the cost of repairing the land—restoring subsided areas after oil removal, returning to productivity strip mined regions—the human costs—uncompensated black-lung victims—and the incalculable costs of cleaning up an oil spill—are but a few examples. For nuclear energy the substantial costs of developing commercial power reactors were paid by the public out of taxes, not as part of the unit electrical energy cost; even now, the nuclear plant operators are able to obtain federally subsidized insurance. What is more, the long-range biological effects of the low-level radioactivity routinely released by such plants are not well understood; and of course, the solution to the problem of massive radioactive waste disposal has not been found and may turn out to be quite expensive.

The availability of inexpensive electric power has been very beneficial in the agricultural and commercial development of our country. At the same time, too cheap electric energy has undesirable effects. For one, it encourages an increase in consumption; but, more importantly, it encourages the substitution of the most convenient for the most energetically efficient. For example, many automobile engine blocks are now made of aluminum; they used to be made of steel. Blocks made of steel are just as effective as those made of aluminum, but they are heavier

TABLE 6.1 Heat Sources

Heat Source	Energy Release	Daily Consumption by 1000-MW Generating Plant
Methanol	$\sim 2 \times 10^7$ J/kg	1.8×10^6 kg ($\eta = 0.2$)
Coal	$\sim 3 \times 10^7$ J/kg	6750 metric tons, or 100 rail carloads ($\eta = 0.4$)
Oil	$\sim 4.3 \times 10^7$ J/kg	34,300 bbl or 100,000 gal ($\eta = 0.4$)
^{235}U fission	8.2×10^{13} J/kg	3 kg, or 430 kg natural uranium ($\eta = 0.3$)
^{2}D fusion	2.4×10^{14} J/kg	1 kg, or 30 m^3 of seawater ($\eta = 0.3$)
Sun	10^3 J/m^2 s	Collection area of 3×10^7 m^2 ($\eta = 0.1$) (assumes storage of heat for use during night)

and, therefore, not as convenient. But aluminum blocks require three times as much energy to manufacture as steel blocks! (This is a Catch-22 situation: steel blocks produce much lower overall efficiency in terms of fuel consumption!)

Another example of the substitution of convenience for efficiency brought about by the cheapness of power is electric space heating. (It is invariably more expensive to heat electrically, despite what your local electric company would like you to believe!) The conversion of electrical energy to heat is a very efficient process; for example, in a water heater it is 100 percent efficient. However, the conversion of fuels to electricity is rather inefficient (30 to 40 percent), so the overall efficiency of converting a fuel to space heat via electricity is low. If the fuel were used directly to heat the space at an efficiency of 60 to 70 percent, rather substantial savings in fuel and waste heat would be realized.

The use of electricity generated from fossil fuel burning for space or process heat is economically and energetically unsound and should not be condoned in an energy-conscious society. It should go without saying that advertising designed to increase the electrical consumption of the public is in direct conflict with the best interests of society, although understandable from the point of view of the utility. We should also note that rate schedules that penalize the frugal user and reward the profligate should be abolished.

The growing scarcity of fossil fuels and high-grade uranium ore along with stricter environmental control of resource acquisition and use will make all current fuels more expensive. If the increase in cost is sufficient, the historical trend of inefficient energy practices may be reversed. Pressure to develop new heat sources will increase; we should understand these various sources to know how to utilize best the resources available.

6.1 COMBUSTIBLE FUELS

We understand **combustion** to be a process of rapid oxidation with the release of heat. This energy is a result in part of the difference in potential energy of the reacting molecules before and after the reaction. If we could calculate these potential energies exactly, we could calculate the thermal release exactly for a given reaction. Unfortunately, atomic structure problems involving systems of many electrons can be, at best, only approximated. So rather than calculate heats of reaction, as they are called, we must measure them.

Physics of Combustion

If a chemical reaction occurs at atmospheric pressure, it is then an isobaric process, and

$$Q = \Delta H. \qquad (6.1)$$

A measurement in a calorimeter of the heat evolved (in an exothermic reaction) or the heat taken up (in an endothermic reaction) is a direct measurement of the enthalpy change ΔH. As a consequence, tables of ΔH have been compiled for various reactions; the pure state of a substance is assumed to have $\Delta H = 0$, and all enthalpy changes are recorded relative to the pure states of the reactants (often these enthalpy changes are written ΔH^0, the superscript zero indicating relative to the pure state).

To calculate the heat of reaction, ΔH_R^0, for a general chemical reaction at 25°C,

$$a\,A + b\,B \rightarrow c\,C + d\,D, \qquad (6.2)$$

where the small letters refer to quantity in moles, we have

$$\Delta H_R^0 + a\,\Delta H_{Af}^0 + b\,\Delta H_{Bf}^0 \rightarrow c\,\Delta H_{Cf}^0 + d\,\Delta H_{Df}^0, \qquad (6.3)$$

where $\Delta H_{Af}^0, \ldots$, refers to the heat of formation of that particular compound relative to the standard pure states obtained by reference to tables.[1] Most tables take the standard state to be 25°C and one atmosphere. If the reaction occurs at an elevated temperature, a slightly different procedure must be used.

Example 6.1 Calculate the heat of reaction of the combustion of methyl alcohol at 25°C

[1]See, for example, F. D. Rossini, D. D. Wagman, W. H. Evans, S. Levine, and I. Jaffe, National Bureau of Standards, *Selected Values of Chemical Thermodynamic Properties*, NBS Circular 500, Washington, D.C., 1952.

and 1 atm.

$$CH_3OH(l) + \frac{3}{2} O_2(g) = CO_2(g) + 2H_2O(l),$$

$$\Delta H_R^0 + (-57.02) + 0 = (-94.05) + 2(-68.32)$$
$$\text{from tables,}$$

$$\boxed{\Delta H_R^0 = -173.67 \text{ cal/mol.}}$$

At a temperature different from 25°C, the increase in enthalpy of the reactants resulting from this temperature difference must be included. Terms of the form

$$a\,\Delta H_A^0 = ac_p\,\Delta T. \qquad (6.4)$$

where c_p is the specific heat of A, should be added to Eq. 6.3.

Table 6.2 lists the heat of combustion of several, mostly organic, compounds. Note the very large figure for hydrogen. Hydrogen has been used to power an ICE, with obvious benefits from the emissions standpoint. And, of course, liquid hydrogen has been used as a rocket fuel for some time.

The available work from combustion is approximately equal to the enthalpy change. Combustion is usually adiabatic with no work done on the atmosphere. In this case the

TABLE 6.2 Selected Heats of Combustion

Substance	ΔH^0
Ammonia	2.4×10^7 J/kg
Alcohol:	
Methyl	2.5×10^7 J/kg
Ethyl	3.2×10^7 J/kg
Hydrazine	2.1×10^7 J/kg
Hydrogen	1.5×10^8 J/kg
Methane	3.7×10^7 J/m^3
Coal gas	$< 1.9 \times 10^7$ J/m^3
Power gas	$\sim 4.7 \times 10^6$ J/m^3
SNG	$\sim 3.7 \times 10^7$ J/m^3

available work (from Eq. 3.60) becomes

$$\Delta B = -T_0 \Delta S. \qquad (6.5)$$

But also in this case, from the First Law and the definition of enthalpy,

$$\Delta H = T \Delta S, \qquad (6.6)$$

so

$$\Delta B = -\Delta H. \qquad (6.7)$$

In general, only magnitudes are of interest so usually

$$\Delta B = |\Delta H|. \qquad (6.8)$$

ΔB is not exactly equal to this amount because of the contribution of diffusion. That is, the reaction products may diffuse into the atmosphere, increasing the entropy and thereby decreasing the available work.

This has been a very cursory introduction to chemical thermodynamics; the student inter-ested in obtaining more information should consult any of a number of good textbooks available on the subject, some of which are listed in the references for Chapter 3.

Synthetic Hydrocarbons

The most common heat source today is, of course, the combustion of hydrocarbon fuels, principally coal, oil, and natural gas. As we know, the easily obtainable reserves of the latter two will soon begin to dwindle, if they have not already done so. With our energy economy so heavily based on these fuels, it is desirable to ensure an adequate supply by substituting synthetic hydrocarbons for the difference between the supply and the demand. This is particularly true for natural gas, the supply of which has dropped steadily for the past few years, and has only begun to increase since prices were deregulated.

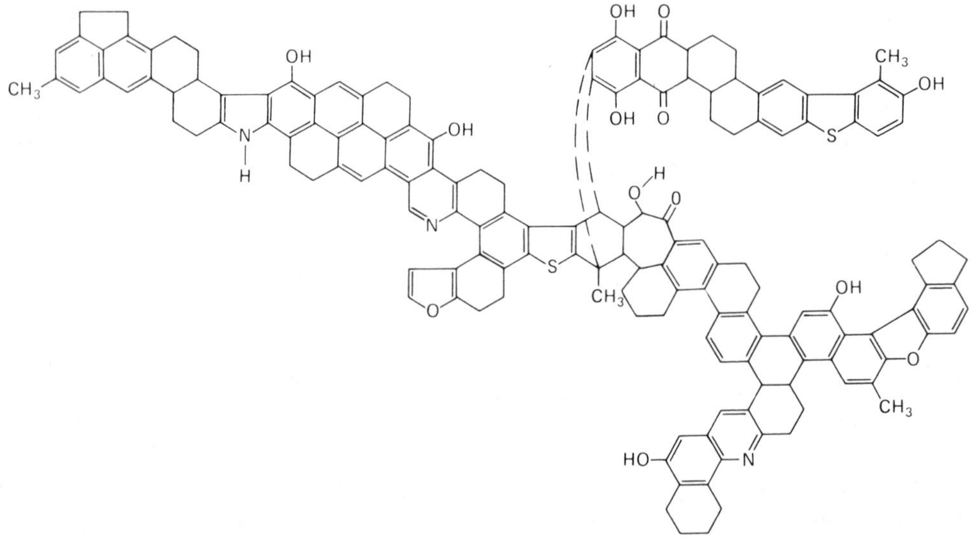

Figure 6.1 Possible molecular structure for coal. Note that since sulfur atoms are linked into the molecule, they cannot be removed by purely physical methods. (From T. E. Fisher, "Catalysis," in *AIP Conference Proceedings No. 19, Physics and the Energy Problem*, edited by M. D. Fiske and W. W. Havens, Jr. New York: The American Institute of Physics, 1974.)

Gasification Since natural gas is principally methane, CH_4, a cheap natural source of carbon is required, long with water to manufacture the synthetic substitute: SNG, as it is called. The ready availability of coal makes it a natural candidate, although naphtha—a petroleum product—has also been used. Coal gasification is not a new concept. Many cities manufactured combustible gas from coal from the early nineteenth century to just after World War II, when natural gas from large pipelines became available.

Coal is not simply carbon; it is made up of many complex, long-chain hydrocarbon molecules having a variety of trace constituents bound into the molecular structure. Figure 6.1 is one hypothetical model of such a molecule. Note that the sulfur is an integral part of the molecule and, as a consequence, cannot be removed from the coal by purely mechanical means. Other sulfur atoms are found in impurities in the coal and can be easily removed prior to combustion. Table 6.3 lists several constituents of a variety of oils and coals.

If we compare the carbon-to-hydrogen ratio for coal from Table 6.3 to that for the petroleum materials, we see the major problem in developing an economic coal gasification system: we must either add hydrogen or subtract carbon from the coal. There are also other problems: sulfur and nitrogen. These elements are quite common in coal, but are not as common in petroleum or petroleum products. Sulfur, as we know, has a deleterious effect on the catalysts used in petroleum refining.

Several processes have been used to produce a variety of combustible gases. The general idea of synthetic gas production is indicated in Figure 6.2. The feedstock, coal or naphtha, may require pretreatment for desulfurization or to prevent caking in the gasifier. Caking is a characteristic of bituminous coals. When heated, these coals become sticky and melt into a plastic blob, reducing the surface-to-volume ratio, providing less interaction surface for the gasification process. Such coals must undergo a pretreatment phase in which part of the surface is rapidly oxidized to protect the remainder of the coal. This pretreatment not only adds to the cost of the product, but also subtracts some otherwise usable carbon.

Early gasifiers simply distilled the coal without air, using the hydrogen content of the coal to produce some methane along with

TABLE 6.3 Typical Composition of Hydrocarbon Materials

	Net H/C Atomic Ratio[a]	S/C Atomic Ratio	N/C Atomic Ratio
Methane—CH_4	4	0	0
Gasoline—$(CH_2)_N$	2	0	0
Crude oil[b]	1.59–2.06	0.006–0.6	0.0014–0.15
Bituminous coal	0.6	0.016	0.018
Subbituminous coal	0.5	0.007	0.016
Lignite	0.25	0.005	0.015
Anthracite	0.05	0.004	0.001

From *Technology of Lignitic Coals*. Washington D.C.: U.S. Bureau of Mines, 1954.
[a]This assumes that all oxygen and sulfur present in the coal are in the form of H_2O and H_2S; all such hydrogen is excluded from the calculation.
[b]Crude oils vary significantly in composition; ranges are given.

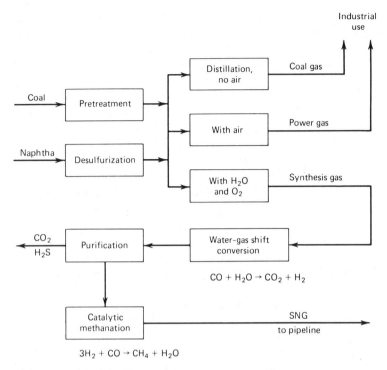

Figure 6.2 Steps in the coal (or naphtha) gasification process. Coal gas has a very low heating value, being about 50 percent $H_2 + CO$ and the remainder CO_2, CH_4, and others. Power gas has intermediate heating value useful for on-site power production but not for pipeline transport. It is about 14 percent CH_4, 11 percent CO, 16 percent H_2, 11 percent CO_2, 28 percent H_2O, and 30 percent N_2. Synthesis gas is high-quality gas useful for producing pipeline gas; it has about 10 percent CH_4, 21 percent CO, 40 percent H_2, 28 percent CO_2, and 1 percent other. SNG is usually 90 percent CH_4.

hydrogen and carbon monoxide. This gas was fairly high in heating value, about half that of natural gas; but this was not enough to warrant transportation over long distances. So this "coal gas" was restricted to residential and industrial use over short distances. This gasification scheme made very inefficient use of the coal, utilizing only about 30 percent. The remainder had to be sold or, more likely, discarded.

Current designs distill the coal or naphtha in the presence of either air or steam plus oxygen. The air process produces a rather

low-quality gas having a value only for on-site power production. (One disadvantage of using air for combustion is the presence of N_2 leading to the generation of large quantities of NO_x.) The $O_2 + H_2O$ process generates a somewhat higher-quality gas, which can be further treated to yield high-heating-value methane. This **synthesis gas**, also sometimes called *producer gas*, has a high CO and N_2 content. If the H_2-to-CO ratio of this synthetic gas is not close to the $3:1$ figure required for conversion to methane, further treatment is required. Some of the CO is converted into

CO_2 by reacting with H_2O in the "water–gas shift converter," thus liberating more hydrogen. The CO_2 and sulfur impurities are removed, and the remaining gas, consisting mostly of H_2, CO, CH_4, and H_2O, is passed through the catalytic methanation stage, where the CO and H_2 react to form CH_4. Water–gas shift and catalytic methanation are high exothermic reactions. Substantial and effective heat removal must be provided to maintain the reactants at the appropriate temperature. If the temperature becomes too high, this reaction dominates:

$$H_2O + CH_4 \rightarrow CO + 3H_2, \qquad (6.9)$$

and the process becomes ineffective.

By varying the catalysts a variety of end products can be produced:

$CO + 3H_2 \rightarrow CH_4 + H_2O$ (methane),

$6CO + 9H_2 \rightarrow C_6H_6 + 6H_2O$ (benzene),   (6.10)

$CO + 2H_2 \rightarrow CH_3OH$ (methanol).

Of course, the H_2-to-CO ratio would have to be adjusted appropriately.

The gas percentages indicated in Figure 6.2 are only approximate and vary substantially, depending on the exact type of feedstock, the particular process, the amount of air or steam plus oxygen relative to the feedstock, and the eventual use for the gas. Because the optimum process depends so critically on the feedstock, in particular on the hydrogen and sulfur content of the coal, full-scale production plants will be very expensive.

Several gasifying processes have been devised over the past 40 years; all have advantages and disadvantages. All also have one fact in common: the gas they produce is at least five times as expensive as natural gas. Gasification processes can be classified several ways: by the method of supplying heat for the gasification reaction (internal or external); by the method of achieving contact between the reactants (fixed bed, fluidized bed); by the flow

of the reactant (cocurrent or countercurrent); by the gasifying medium (hydrogen or steam plus oxygen, air or enriched oxygen); and by the condition of the residue removed (slag or ash). Nearly every combination of these various methods has been attempted in laboratory-scale projects. Only a few have reached the pilot-plant stage, and fewer yet have gone into commercial production. Several commercial Lurgi gasification plants are operating at various places around the world, although there are none in the United States. This process is rather expensive; many gasifiers are needed to produce a usable amount of gas. They require a great deal of maintenance and will not function on typical U.S. bituminous coal.

One technique for gasifying coal has been attempted that might eliminate the need for expensive gasification plants: *in situ* gasification. Coal seams underground are penetrated by drills, fractured (usually by water pressure) to provide for air flow, and then ignited. Air is pumped to the burning coal to support the combustion. The emerging gas, while low in heating value, could certainly be used for on-site power generation. If pure oxygen and steam could be pumped into the burning coal seam rather than air, a higher-quality gas, suitable for catalytic methanation, would result. Experiments so far with *in situ* gasification have been disappointing. The gas flow is not continuous, and the heating value is not constant. It is not clear what the coal utilization efficiency has been; problems of subsidence and water contamination need to be solved. These problems do not appear to be insolvable; *in situ* gasification may become a most desirable alternative to the very large capital investments required for above-ground gasification.

One major difficulty with *in situ* or mine-mouth gasification in the western United States is the lack of water. Some planners assume water could be pipe-lined from the Mississippi to the western coal fields, gas-produced there, and shipped back to the metropolitan areas of the Midwest economically.

The economics of gasification are frightening to contemplate. The U.S. consumption of natural gas was about 0.5×10^9 m^3 in 1978. The estimated cost (in 1975 dollars) for a 7.0×10^6 m^3 per day plant is $1 billion. To put this number into the appropriate context, note that the total capital investment in one U.S. gas-pipeline company that delivers about 6×10^7 m^3 per day is only about $250 million. On the same scale, the total U.S. investment in gas-transmission lines must be about $7.8 billion. The estimated difference between domestic supply of natural gas and the demand in 1980 will be about 2.8×10^6 m^3; if this were to be supplied by gasification plants, the cost of the plants would be $5 billion. In other words, U.S. industry would be required almost to double its capital investment just to increase its product by a factor of 1.05. Most utility owners see this as a poor investment.

Because of the high cost of the gasification plant and its rather low output, the cost of the product must be high. Estimating the costs of the product of nonexisting technology is a difficult—many would say foolhardy—task. Instead of attempting it, let us simply list the factors that must be accounted for in such a calculation.

Initial capital costs	($\sim$$1 billion for 7×10^6 m^3 per day)
Interest on loan	($\sim$15 percent??)
Stream factor (fraction of operating time)	(0.9)
Coal cost	($\sim$$1/10^9 J)
Operating costs	($\sim$$0.70/10^9 J)
By-product credit	($\sim$$0.30/10^9 J)
Shipping charges:	
coal	(16 cents/MT km)
gas	(2 cents/10^9 J)
water	(20 cents/10^9 J)
Environmental replacement costs	($0.20/10^9 J)

Depending on the location of the plant, interest rates, type of coal, and so on, product costs could ange between about $4/10^9 J and $10/10^9 J, rather considerably higher than the current controlled price of about 50 cents/10^9 J. Although complete deregulation of natural gas prices will force consumer prices higher, it should be noted that significant price increases in natural gas will also bring about significant shifts in primary energy sources. It appears that natural gas demand is rather more elastic than had been imagined. This elasticity will work against the construction of coal gasification plants. As in virtually all situations where we try to look into the future, we find it murky indeed and most difficult to predict.

Liquefaction   The same general procedure used to gasify coal can be used to liquefy it as well. After all, the long-chain hydrocarbon molecules of coal are very similar, if a little devoid of hydrogen and high in oxygen and nitrogen, to the complex molecules of crude oil. Two liquefaction processes have been used in the past, and several others are under development.

Germany, being devoid of native oil deposits, began to develop a coal liquefaction process in the late 1920s and early 1930s in preparation for future armed conflict. The United States also began similar studies, but the discovery of the vast west-Texas oil deposits just at the beginning of the Great Depression stifled further research. After World War II, there was a short revival of interest in coal liquefaction until the discovery of the Middle East oil fields. These ridiculously inexpensive deposits[2] effectively killed liquefaction research until the energy crunch of the winter of 1973–1974 showed us how vulnerable our energy economy has become.

[2]The actual production cost of a barrel of Saudi Arabian oil delivered at the dock as of mid-1981 is under 35 cents. The posted selling price of about $40 bears no relation to production cost or capital investment. It is just about what the market will bear—at least for now!

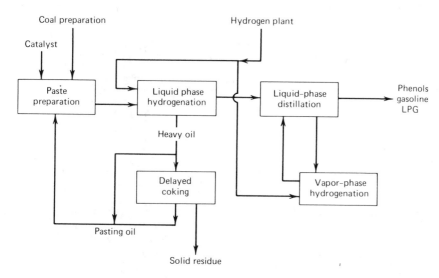

Coal preparation

Catalyst

Hydrogen plant

Paste preparation

Liquid phase hydrogenation

Heavy oil

Liquid-phase distillation

Phenols gasoline LPG

Delayed coking

Vapor-phase hydrogenation

Pasting oil

Solid residue

Figure 6.3 The Bergius process for liquefying coal. Several petroleum products from light oils and gases to heavy oils are produced. (From Harry Perry, "Coal Liquefaction," in *AIP Conference Proceedings No. 19, Physics and the Energy Problem*, ed. by M. D. Fiske and W. W. Havens, Jr. New York: The American Institute of Physics, 1974.)

Two commercial liquefaction processes have been used. The first of these, the Bergius process, is diagramed schematically in Figure 6.3. This process, not in use anywhere today, required a finely powdered coal slurried with an oil, a product of the process, and reacted with hydrogen, also derived from coal, at high pressure (700 atm) and high temperature (450°C). Several products are produced, including heavy oils, light oils, gasoline, and LPG. The process is rather expensive because of the extreme operating conditions of the reactor. Bureau of Mines estimates for gasoline produced by the Bergius process today place the cost at about 82 cents/gal, compared to about $1.50/gal for petroleum-refined gasoline. Of course, the refined gasoline price is artificial in the sense that it does not reflect the costs of production and could be reduced considerably at anytime by simple fiat. The Bergius cost, however, does represent actual production cost and cannot easily be reduced.

The second process, which is in use today in South Africa, is the Fisher–Tropsch process; it is diagramed in Figure 6.4. This process is, strictly speaking, a gasifier followed by appropriate catalytic conversion. Many types of hydrocarbon products can be obtained from the synthesis gas using an appropriate catalyst. For example, highly paraffinic oils can be obtained with a cobalt catalyst, whereas iron catalysts produce highly olefinic products; the hydrogenation process (Bergius) produces an aromatic oil. The F–T process is only about 38 percent efficient in terms of coal conversion, but it does not require the extreme operating conditions of the Bergius process. Nevertheless, estimates are that the Bergius process would produce fuels at a lower cost than the F–T process.

Several processes are under development currently, including the COED process (developed by FMC Corporation), Project Gasoline (developed gy the Consolidated Coal Corporation), the H-Coal process (developed by the Hydrocarbon Research Institute), and

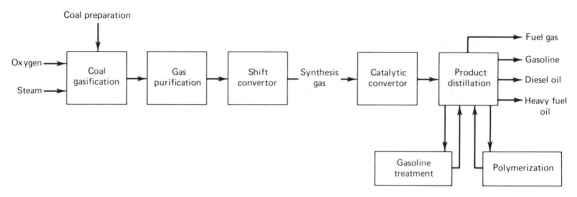

Figure 6.4 The Fisher–Tropsch coal liquefaction process. The end product is a synthesis gas that can be catalyzed to form a large variety of petroleum products. (From Harry Perry, "Coal Liquefaction," in *AIP Conference Proceedings No. 19, Physics and the Energy Problem*, edited by M. D. Fiske and W. W. Havens, Jr. New York: The American Institute of Physics, 1974.)

the Synthoil process (developed by the Bureau of Mines). All these processes have been tested with encouraging results in laboratory-scale models, processing up to a few hundred pounds per hour.

As with gasification, the capital investment required to operate an economically sized liquefaction unit is quite large, and the product costs are high. There has been an understandable hesitation on the part of utility owners to make such enormous investments when there is a continual threat of being undercut by the dumping of cheap Mideast oil on the market. The federal government finally seems to have realized this, and in 1979 established the Synthetic Fuels Corporation. This quasipublic corporation has begun to fund the construction of synthetic fuel demonstration plants. This has not happened without considerable opposition from within and outside government. In fact, the change of administration in 1980 may bring a return to the reliance on private industry and free-market forces for advances in synthetic fuels technology. However, the immediate past has shown this reliance to be misplaced.

Another question that should be asked is,

"Does the United States have the capability of significantly increasing coal production to supply gasifiers?" The answer seems to be no. To double the U.S. coal production in 10 years (to about 1.4 billion tonnes annually) would require the opening of 120 new underground mines (each producing 2×10^6 t/yr) and 120 new surface mines (each producing 7×10^6 t/yr). This averages to one underground and one surface mine per month. This also assumes that all current mines continue to produce at current rates! In fact, only 13 mines having a 2×10^6 t/yr capacity or greater were opened in the 10 years from 1960 to 1970. Mine companies usually take 18 months to open a new surface mine, and 5 years to open a new underground mine.

Finally, we must mention the points of interaction of synthetic hydrocarbon production with the environment. These points are indicated in Figure 6.5. Note that *aromatics* refer to hydrocarbon molecules of high C/H ratio, so that carbon atoms are forced to bond to other carbon atoms. These aromatics are toxic in general, and those derived from coal are also carcinogenic. Water represents a double problem: not only is there not enough

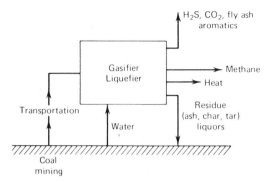

Figure 6.5 Energy and matter flow in coal gasification and liquefaction processes indicating points of interaction with the environment.

in the West, where the coal is located, but also the synthesis processes create a rather large amount of contaminated waste water, which if not treated would run off into the streams and rivers of the area.

Although hydrocarbon synthesis from coal is certainly a technical feasibility today, the fuels produced are so expensive compared to those naturally available that other alternatives must be found if we are to alleviate the current shortage and our dependence on overseas energy suppliers. There are environmental concerns, as well, in synthetic hydrocarbon production that must be addressed, if a significant program is embarked upon.

Hydrogen

There are really very few large energy sources: fossil fuels, nuclear energy, and solar

energy say it all. Our current crises are the result of an unfortunate distribution of reliances on these few sources. Clearly, any shifting of reliances from a scarce resource to a more abundant one will produce dividends. In particular, if the especially scarce fossil fuels could be saved for their petrochemical applications instead of being burned up for their heat value, we would be ahead of the game.

This is the basic idea behind the **hydrogen economy** concept (see Figure 6.6). In this proposal, hydrogen would not be a primary energy source, but would serve rather as an energy carrier or storage medium. Hydrogen, as it is currently produced, is too expensive to serve as a primary fuel; but it has some advantages that make it more economical than electrical energy transmission in many situations. It may even be the ideal motor vehicle fuel, thus liberating about 20 percent of our current petroleum usage for other purposes. In this section we look at some of these advantages as well as some of the methods of producing and utilizing hydrogen to see if the proponents of the "hydrogen economy" are correct in their assertion that hydrogen is the fuel of the future.

Table 6.4 gives a comparison of several properties of hydrogen and methane relevant to their energy-transmission capabilities. Note that, while hydrogen has about one third the energy per unit volume of methane, it also has about one third the viscosity, so current pipelines could be used to transmit the same total energy without modification. The pumps used

TABLE 6.4 Comparison of Hydrogen and Methane

	Hydrogen	Methane
Boiling point	$-259.28°C$	$-161.5°C$
Flammability limits (percent in air)	4.19–74.2%	5.3–14.0%
Density (vapor)	0.069	0.6
Autoignition temperature	585°C	538°C

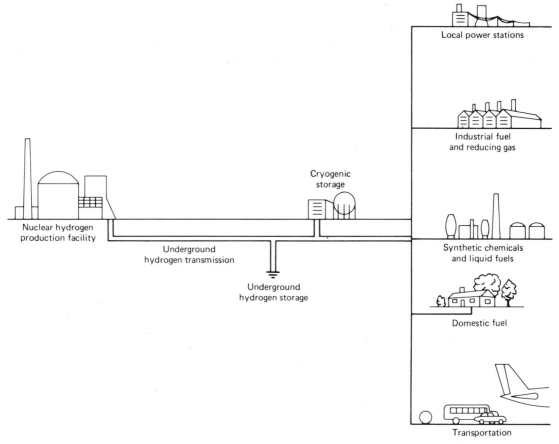

Figure 6.6 The *hydrogen economy* in which energy is transmitted and stored as hydrogen rather than electricity.

to compress the gas would have to be larger, thus transmission of energy via hydrogen would be somewhat more expensive than for natural gas. This is evident from Table 6.5,

TABLE 6.5 Energy Transportation Costs

Method	Cost ($/$10^3$ J)
Methane (pipeline)	0.03
Hydrogen (pipeline)	0.031
High-voltage transmission lines	0.20
Gasoline (tanker)	0.09

where energy transportation costs are compared. The figure for electricity assumes a transmission distance of more than about 100 km.

Other than cost, what are the advantages of energy transportation using hydrogen in existing pipelines? For one thing, pipeline owners will be able to use their very large capital investment, when the supply of natural gas becomes so small that its cost is prohibitively high. For another, the use of hydrogen enables the primary energy generator to be sited far away from the load at whatever location is most advantageous. For example, large clusters of nuclear plants could be located far away from metropolitan areas, near sufficient cooling water—on the coast or even offshore. Large solar energy plants could be located in the desert regions.

What about disadvantages? Since hydrogen diffuses more readily than methane, a larger amount would be lost in the pipeline through small leaks. But because of this diffusivity and the lightness of the hydrogen molecule, leaking hydrogen would not accumulate in basements as does methane. Hydrogen has very wide flammability limits compared to methane. This and the fact that hydrogen burns with a colorless flame might pose problems for household use. Additives can be used to provide the carbon necessary for color in the flame.

It should be noted that when it was first proposed to use methane a great controversy arose over the safety aspects of pumping a highly dangerous and flammable substance into the home. Of course, we think nothing of it now. It is true that there is a great fear of hydrogen, probably caused by the "Hindenburg syndrome," the fear of another great disaster like that which befell the dirigible Hindenburg at Lakehurst, New Jersey.

The Hindenburg was a rigid lighter-than-air ship (as opposed to a balloon; i.e., nonrigid) that was supported by hydrogen. The airship was docking on May 6, 1939, when an accident occurred, setting the hydrogen aflame. The airship fell to the earth, killing 36 people, most of whom died from the fall and not from the flames. Since hydrogen is lighter than air, it moves upward upon being released. So then the flames move upward, not downward. Dirigibles supported by helium have been proposed as low-cost alternatives for some types of transportation. We discuss this prospect in more detail in Chapter 11.

Another disadvantage to the use of hydrogen in pipelines stems from the ability of the hydrogen molecule to dissociate and diffuse readily into the interstitial spaces between atoms of pure metals and alloys. In steel, this causes the metal to become brittle and to fail. This "hydrogen embrittlement" is naturally more severe at higher pressure. A great deal of research is being done today to find resistant alloys.

The autoignition temperature of hydrogen is quite high, but the ignition *energy* is very low, only 0.02 mJ, less than 7 percent that of natural gas. Hydrogen can thus be used in catalytic heaters that burn at low temperatures ($\sim 200°C$), providing flameless, pollutant-free heat for cooking or space heating. With this system the formation of NO_x is eliminated; the only by-product is H_2O, which is useful for humidification. Using catalysts, very lean mixtures (<4 percent H_2) can be made to burn, ensuring complete combustion. With appropriate heat sinks, virtually 100 percent of the heat produced can be used, since no chimney vent would be required. Catalytic hydrogen burning could well revolutionize home energy production!

Another possible application for hydrogen in the home is the fuel cell, discussed in Chapter 5. If H_2–O_2 fuel cells can be perfected for home use, the flammability problem with hydrogen can be avoided. The cell can be used to convert the hydrogen energy to electricity, which can be used to power heat pumps for heating and air conditioning, as well as conventional electric stoves, in an economical manner. The only by-product of the fuel cell would be H_2O, which itself would be valuable to many households.

Industrial users of hydrogen would have little difficulty. Most boilers now using natural gas or oil can be converted easily and economically to burn hydrogen. The improvement in the air-pollution situation, at least along the East Coast, would be immediate and substantial.

The transmission and utilization of hydrogen present few difficulties; most scientists believe these can be overcome with present technology. The question then becomes one of whether or not hydrogen can be generated in sufficient quantity and at a sufficiently low cost.

Several techniques have been used or proposed for the commercial production of hydrogen. We have seen earlier in the section the *water–gas catalytic process*, in which $CO + H_2$ produced in a coal gasifier are reacted with steam over a catalytic bed to produce $CO_2 + H_2$. Another technique that has been used extensively but may not be available in the future is *steam reforming of methane* (natural gas). In this process we have

$$CH_4 + 2H_2O \rightarrow CO_2 + 4H_2. \qquad (6.11)$$

This process is widely used to produce hydrogen for ammonia synthesis. The final process, which is currently in use, is *electrolysis*. In the electrolytic decomposition of water,

$$H_2O \rightarrow H_2 + \frac{1}{2}O_2 - 242 \text{ kJ/mol.} \qquad (6.12)$$

This reaction is endoergic; it is economically feasible only when electricity is plentiful and cheap. Even though electrolysis of water is endoergic, it nevertheless requires much less energy per mole than other electrolytic reactions; for example, electrolysis of bauxite (Al_2O_3) requires 1.97 MJ/mol.

A 90-MW electrolysis plant is currently operating in British Columbia at the site of a hydroelectric installation. The 36 tons per day of hydrogen produced are used to make ammonia; the by-product oxygen is used in metallurgical processing. With present technology an electrolytic efficiency of about 60 percent is realized. Estimates are that 75 percent can be achieved and that even higher efficiencies are possible. In fact, the theoretical maximum efficiency is 120 percent (because some heat from the surroundings is utilized).

Several techniques have been proposed and studied as future hydrogen resources. These are **endothermic chemical decomposition, photolysis**, and **bioconversion**.

The first of these, it is hoped, will produce hydrogen from water, using several chemical steps and the relatively low temperature heat available from nuclear reactors or other sources, thereby avoiding the expense of the electrolytic process. One promising chemical sequence has been studied theoretically by scientists at the Euratom Research Center in Ispra, Italy. This process, called the Mark I, consists of the following steps:

1. $CaBr_2 + 2H_2O \rightarrow Ca(OH)_2 + 2HBr$ 730°C.
2. $Hg + 2HBr \rightarrow HgBr_2 + H_2$ 250°C.
3. $HgBr_2 + Ca(OH)_2 \rightarrow CaBr_2 + HgO$ $+ H_2O$ 100°C.
4. $HgO \rightarrow Hg + \frac{1}{2}O_2$ 100°C.

$$(6.13)$$

Recall here something mentioned in the last chapter—that any potentially vauable process that is common knowledge probably doesn't work! This is true of the Mark I; note the first step requires 730°C—much too high for a nuclear reactor (see Chapter 7). Also, the reaction kinetics of step 2 are also very slow. Another argument against Mark I is that it requires mercury, a rather expensive material. Of course, cycles can be found for which one or more of these arguments do not apply; but they all have drawbacks that have been severe enough that, as of today, there is no commercially feasible water-splitting thermochemical cycle.

The direct-photolysis-of-water proposal would use ultraviolet radiation from a fusion reactor. Such radiation would not normally be present in a D–T fusion process, but could be produced by injection of some heavy element into the fusion plasma. This proposal has not been pursued to the same extent as the thermochemical cycles. It appears there are a number of conceptual as well as engineering problems to be solved. For example, how do you keep the back reaction from depleting the H_2 and O_2 produced?

The other proposal referred to above is the direct production of hydrogen by biological processes. Certain bacteria or algae in sunlight have the capability to increase the oxidation potential of electrons from water to a level as much as 0.3 V more negative than the hydrogen electrode. Thus hydrogen ions could be converted to hydrogen gas more readily. A great deal of effort is being directed to this type of study, although so far the results indicate a rather low overall efficiency.

We have seen how hydrogen can be used in the home and examined several techniques for producing it. There is another way that hydrogen could contribute to our national energy picture: transportation.

Current internal combustion engines can easily be operated on hydrogen; minor adjustments in carburetion and spark timing must be made to account for the difference in the stoichiometric amount of air required and the difference in the flame-front propagation time.

As automotive fuel, hydrogen would be nearly perfect. The only by-products would be water vapor and NO_x; the NO_x could probably be controlled by catalytic converters. There would be no unburned hydrocarbons, no lead compounds, and, of course, no carbon monoxide. There is a drawback; otherwise, all cars would be hydrogen-powered today.

The problem is storage. A 76-ℓ (20-gal) gasoline tank has a mass of 53 kg when full. The energy equivalent of hydrogen would be only 19 kg, but how do you hold it? The large steel cylinders required to contain that much hydrogen would have masses of several thousand kilograms. The range of hydrogen-powered cars that have been built is typically under 100 km. It is possible to obtain a much higher hydrogen density by storing it as a liquid. But the low boiling point and severe flammability of liquid hydrogen make such storage impractical for automotive use, although not perhaps for aircraft.

It has been proposed that powering aircraft with liquid hydrogen produced at a central

TABLE 6.6 Comparison of Conventional Kerosene-Fueled versus Liquid Hydrogen-Fueled Subsonic Aircraft

	Subsonic Transport		
	Kerosene-Fueled	Hydrogen-Fueled	Change (percent)
Payload[a] (kg)	25,400	25,400	
Range (km)	6,000	6,100	
Cruise speed (mach)	0.820	0.820	
Gross weight (kg)	195,000	146,000	−26
Operating empty weight (kg)	108,700	97,880	−10
Fuel weight capacity (kg)	62,300	21,200	−65.9
Fuel volume (m³)	82.7	313	278
Wing area (m²)	322	263	−18.2
Wing span (m)	47	43	
Wing loading (kg/m²)	606	552	
Specific fuel consumption (kg/h/kg)	0.677	0.216	

Adapted, with permission, from D. P. Gregory and J. B. Pangborn, "Hydrogen Energy," *Annual Review of Energy* **1**, 270 (1976), copyright © 1976 by Annual Reviews Inc.
[a]Payload of 272 passengers plus 730 kg cargo.

location—the airport—would make air travel much more economical (see Table 6.6). There does not seem to be any movement toward this by the industry, however.

Another technique for storing hydrogen for automotive use that seems to be feasible is to combine the hydrogen with a metal as a metal hydride. Hydrogen, being a very small molecule, as we have noted, is able to penetrate between the atoms of regular crystalline arrays very readily. Hydride formation invariably results in embrittlement of the metal, causing loss of structural strength.

But as a method of carrying hydrogen for automobiles, the metal hydride technique seems very encouraging. The hydrogen density that can be obtained is actually higher in a metal hydride than in liquid hydrogen! The hydrogen can be released very easily by heating. But again there is a drawback. The metals that are most suited to this purpose are relatively rare and expensive (titanium) and are also rather heavy—the energy equivalent of the 76-ℓ gasoline tank would have a mass of about 628 kg (more about H_2 storage in Chapter 10).

We conclude that although hydrogen seems to be an ideal automobile fuel, the storage and distribution systems required for widespread usage are waiting for appropriate economic incentives in order to be developed. Such incentives do not appear to be in the immediate future. Whether the economics of the hydrogen economy can ever be sufficiently attractive is something we have not answered,

nor can we. But we should certainly keep in touch with technical developments that might change the situation as the need for fossil fuel conservation becomes more acute in the next decade.

Alcohols

The attractive properties of hydrogen make it valuable to study another class of fuels that are almost pure hydrogen: the alcohols. There is great interest today in the alcohol fuels primarily because they can be recovered from renewable resources, thus easing our dependence on foreign oil suppliers. But along with this interest has come a controversy: do alcohol fuels cost more in energy to produce than they release?

We shall try to answer this question and at the same time provide a reasonable overview of the entire alcohol fuels picture from manufacture through end use. The results may surprise you.

There are many alcohols, but only two of primary interest: methyl alcohol (methanol) and ethyl alcohol (ethanol)—the latter makes you drunk, and the former kills you. Methanol can be thought of as two hydrogen molecules made liquid by attachment to a carbon monoxide molecule; ethanol is not so easily visualized.

Methanol: CH_3OH,
Ethanol: CH_3CH_2O.

Some of the properties of these alcohols are compared to gasoline in Table 6.7.

TABLE 6.7 Comparison of Methanol, Ethanol, and Gasoline

Energy Content	Stoichiometric Air/Fuel	Boil	Freeze	Autoignition Temperature
Methanol 3.24×10^6 cal/ℓ	6.4	64.96°C	−93.9°C	467°C
Ethanol 4.33×10^6 cal/ℓ	9.0	78.5°C	−117.3°C	535°C
Gasoline 6.56×10^6 cal/ℓ	14.7	a	a	222°C

[a]Gasoline is a mixture of compounds that boil and freeze over a wide range of temperatures.

The late twentieth-century industrialized nations are only now discovering alcohol fuels, but we should point out that alcohol was extensively used for heating in Europe and North America in the early nineteenth century. Methanol was produced outside Paris by the destructive distillation of wood. It was used as a heating fuel and had several advantages—no ash to cart back out of town and considerably less pollution (which was, and still is, a serious problem). Methanol burns very cleanly with a faint blue flame. It was replaced by kerosene in lamps because the blue flame of methanol did no give off very much light. Kerosene burns much less completely, leaving a great deal of unburned carbon (soot), which gives off the characteristic yellow light in the flame. The widespread availability of petroleum products in the last half of the eighteenth century and the decline of forested regions available to the alcohol distillers effectively put an end to the use of alcohol as a fuel.

Ethanol and methanol appear to be superficially similar, but they are very different in many key properties important for their use as fuels. We list some of these in Table 6.8. Ethanol is produced primarily by the fermentation of fruit sugars (dextrose or levulose) or fructose. It may also be produced by converting several carbohydrates (sucrose, maltose, starch, or cellulose) into one of the above sugars by enzymatic action. This is the path

taken in the production of alcoholic beverages from grain. Methanol, on the other hand, is produced primarily by the destructive distillation of hardwood pulp; it may be obtained, in lower quantity per unit input, in a similar manner from waste materials having a high cellulose content (paper products). Both of these alcohols may be obtained from natural gas, petroleum, or coal by conversion to hydrogen in an appropriate manner and then catalytic reaction of the hydrogen and CO. Of course, obtaining alcohol in this manner does nothing to improve our petroleum resource picture!

It is very difficult to produce either ethanol or methanol as a 100 percent pure product. Higher-order alcohols—propanol, isobutanol, and so on—are made at the same time, and in the case of ethanol, the distilled product is at best a 96.54 percent water solution. For methanol, these added alcohols are actually a bonus, as we shall see. For use as a fuel, "methanol" is almost always a mixture of methanol and these higher alcohols and is usually referred to as **methyl fuel** (although not always; it is often simply called methanol).

As fuels, the alcohols could be burned in any situation where oil or gas is used. Of course, since the energy content of the alcohols per unit of volume is about half that of comparable petroleum products and since the cost of production is not that much different, it is not

TABLE 6.8 Comparison between Ethanol and Methanol

	Ethanol	Methanol
Production	Fermentation of carbohydrates, then distillation	Destructive distillation of hardwoods
Production cost (1980 $)	$1.50 to $2.00 per gallon	$0.50 to $0.75 per gallon
Water content percent first distillate	4.4 percent	0 percent
Maximum H_2O content for 10 percent alcohol–gasoline stability	0.46 g	0.1 g

obvious that there is any economic benefit to the use of alcohol fuels as replacements for petroleum products. But this is misleading; let us examine alcohol fuels for automobiles.

Any automobile today will operate satisfactorily on a blend of alcohol and gasoline. "Gasohol" is a blend of between 10 and 20 percent ethanol and gasoline. Many car owners report improved performance—mileage, acceleration, emissions—when gasohol is used. Others report cold-weather sluggishness, hard starts, poor mileage, and the like. Brazil has embarked on a national program of ethanol production and use as an automotive fuel—pure ethanol.

Pure alcohol has some significant advantages as a fuel. First, since the autoignition temperature of the alcohols is higher than that of gasoline, its octane rating is higher. Empirical studies have shown that, for every octane rating increase of one unit, the compression ratio of an engine may be increased by 4 percent and achieve the same knock-free performance. Any increase in compression ratio is an increase in efficiency of operation. In fact, for pure alcohol operation, the increase in performance cancels out the energy-per-unit-volume effect so the mileage obtained is approximately the same—perhaps slightly better or slightly worse, depending on operating conditions.

Example 6.2 Find the percent change in efficiency of an Otto engine for each percent change in R_c.

Start with Eq. 4.9:

$$\eta = 1 - \frac{1}{R_c^{\gamma-1}},$$

$$d\eta = \frac{(\gamma-1)}{R_c^{\gamma-2}} dR_c = \frac{(\gamma-1)R_c}{R_c^{\gamma-1}} dR_c,$$

$$\frac{d\eta}{\eta} = \frac{(\gamma-1)R_c}{R_c^{\gamma-1}} \left(\frac{R_c^{\gamma-1}}{R_c^{\gamma-1}-1} \right) dR_c,$$

$$\boxed{\frac{d\eta}{\eta} = \frac{(\gamma-1)R_c^2}{R_c^{\gamma-1}-1} \frac{dR_c}{R_c}.}$$

For $R_c = 8$ and $\gamma = 1.4$, $dR_c/R_c = 1$ percent produces $d\eta/\eta = 19.7$ percent, a significant improvement.

An engine burning pure alcohol produces far fewer emissions than an equivalently sized gasoline-powered engine. Alcohol vapor flame-front temperatures are relatively low, so about 50 percent as much NO_x is produced. The stoichiometric air-to-fuel ratio is lower for alcohol than for gasoline, and alcohol vapors burn much better in a lean mixture. Therefore, CO is also reduced to less than half the amount produced in a gasoline engine. Unburned hydrocarbons are essentially nonexistent because of the rather simple molecular structure of alcohol compared to gasoline. And, of course, lead compounds are not needed to raise the octane rating of alcohol, so the consequent lead pollution is also eliminated.

There are two minor problems and one major one in the use of pure alcohol fuels. The alcohols are solvents of most plastics and corrode many metals easily. The fuel system of a pure alcohol-fueled car would have to be built of resistant alloys with no plastics or rubber allowed. In fact, several changes would have to be incorporated in an alcohol car (see Figure 6.7). Since there is no national plan to use alcohol, such automobiles are not being built in this country. But many are being built by U.S. companies in Brazil, where such a plan does exist.

The other minor problem is connected with the vapor pressure of the alcohols. These fuels do not vaporize as readily as gasoline and consequently can make starting the engine difficult in cold weather. This problem has been successfully overcome by using a dual fuel system: starting on gasoline and switching to alcohol after the engine is warm.

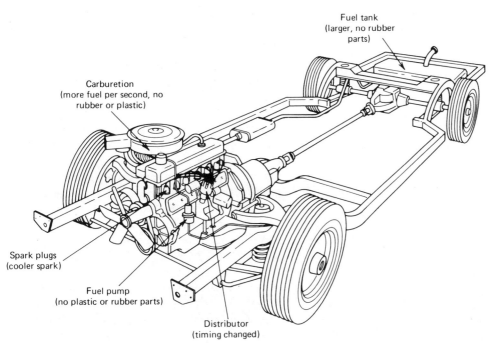

Fuel tank
(larger, no rubber
parts)

Carburetion
(more fuel per second, no
rubber or plastic)

Spark plugs
(cooler spark)

Fuel pump
(no plastic or rubber parts)

Distributor
(timing changed)

Figure 6.7 Changes required to operate an automobile on alcohol fuels. Most of these are necessitated by the corrosive nature of the alcohol and by the fact that the energy content per unit volume of alcohol is lower than that of gasoline.

The major problem in the use of alcohol fuels is the fact that we do not have a sufficient supply of either ethanol or methanol, we do not have an extensive distribution network for alcohol as we have for gasoline, and the automobiles for running on alcohol do not exist. A national commitment to alcohol fuels will involve a great deal of planning and a large expenditure. Of course, a switch to alcohol could be phased in over a period of many years, but who starts? Automobile manufacturers are not likely to produce cars for which there is no fuel supply; distillers are unlikely to produce a vast fuel supply for which there is no market. Perhaps an easing in by blending alcohol into gasoline is a good first step.

Alcohol–gasoline blends have been used in automobiles for more than 40 years with con-

siderable success. There are problems, of course, but with the price of gasoline heading ever upward, most of these problems seem small indeed. A few years ago, when the call for alcohol blending was first being (re)made, tests indicated a rather significant improvement in performance with a normal car—no alterations. Figure 6.8 indicates the improvement that was typically found. This was for a methanol–gasoline blend. The results for ethanol are comparable.

There is, however, a major problem: phase separation. The stability of an alcohol–gasoline blend is critically dependent on the amount of water present. As indicated in Table 6.8, less than 0.1 g of water per 100 g of methanol–gasoline is sufficient to bring the methanol out of solution and cause a phase separation. For

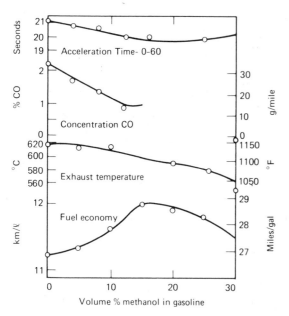

Figure 6.8 Performance of a 1969 Toyota Corona with methanol–gasoline blends. [From T. B. Reed and R. M. Lerner, "Methanol: A Versatile Fuel for Immediate Use," *Science* **182**, 1299 (1973). Copyright © 1973 by the American Association for the Advancement of Science.]

ethanol, the amount of water required is three times greater, and this is not so much a problem.

This difficulty comes about because both alcohol and water molecules are polar—gasoline molecules are not. The large polarizability of alcohol molecules accounts for the low vapor pressure for such a light molecular weight species: the molecules are strongly attracted to each other.

Adding 1 percent or more of the higher alcohols (butanol, etc.) seems to help prevent phase separation. However, this practice would significantly dilute, perhaps even remove, any possible economic advantage gained by the substitution of alcohol for part of the gasoline. Some reports have indicated that not even these additives have stopped the

problem entirely, especially at low temperatures. Note that this phase-separation difficulty is peculiar to methanol, not ethanol.

Another problem caused by the polar nature of the alcohol molecule is the enhanced vapor pressure of the alcohol–gasoline blend and subsequent difficulties with vapor lock. Even though pure alcohol has a low vapor pressure, when it is mixed with gasoline, the mixture has a much higher vapor pressure than gasoline alone. This comes about because the nonpolar gasoline molecules physically separate the alcohol molecules, destroying the polar bonds. The vapor pressure in this situation is then more like that to be expected from a substance with a relatively small molecular weight.

The vapor-lock problem can be dealt with either by removing some of the more volatile substances from the gasoline or by adjustment to the carburetion of the engine. Both of these solutions defeat the purpose of the idea. Removal of the butanes, pentanes, and the like, from the gasoline not only makes it more expensive, but also lowers the energy content per unit volume. Carburetor adjustment negates the advantage of using the blended fuel without modification.

It would seem that, from the standpoint of operation, an ethanol–gasoline blend is far superior to methanol–gasoline, and indeed it is. However, there is the question of economics. Methanol has been highly touted as a fuel because it can be readily obtained; ethanol is about three times as expensive. But, this is misleading. Ethanol is produced almost exclusively for the purpose of consumption in a beverage. As such, it must be rather high in purity and quality. Every one of the ethanol distilleries in the United States uses petroleum product fuels to provide heat. Yet distilleries could be built to produce burning-quality ethanol using plant materials as fuels.

In Brazil, the residue from the sugar cane (bagasse) that is crushed to provide the fermentable juice is burned to distill the ethanol.

It is not clear that a similar substance could be used on conjunction with U.S. grain fermentation plants. In Brazil the answer to the question, "Is alcohol usage energetically efficient?" is a resounding yes. In the United States, it is unfortunately no.

Using the current grain-fermentation–petroleum-product distillation technique requires on the average 130 MJ to produce 1 gal of ethanol, which has an energy content of 84 MJ. Clearly this is not an economically viable proposition. Or is it?

If the oil used to fire the distillery were replaced with coal, and if the petroleum-based chemicals used to make the fertilizers, insecticides, and herbicides used to grow the grain were replaced with nonpetroleum products, then even if the resulting ethanol were more expensive than gasoline, it would be worth it. Every barrel of oil replaced by alcohol is half a barrel not imported from a foreign supplier. Every additional gallon of alcohol supplied for fuel blending is another step toward a national alcohol economy.

We see again that there are no technological stumbling blocks in the way of converting to an alcohol-based automobile fuel economy. Economics and, to some extent, politics now do the driving. But there are hopeful signs: the introduction of gasohol is one. Brazil's program will be watched carefully by many nations. If it is successful in the next 5 to 10 years, many nations, perhaps even our own, will no doubt follow.

Fuel from Wastes

The United States produces wastes in great quantities: about 2 kg per person per day in 1980 and increasing at between 3 and 5 percent a year! In years past these wastes were generally regarded as nuisances. At best some agricultural wastes were used as fertilizers. Before the widespread introduction of "garbage disposals" in homes, municipal refuse

was generally a combination of:

Food residue	Glass
Paper	Wood
Textiles	Yard trimmings
Commercial swill	Metal (ferrous)
Rubbish	Metal (nonferrous)
Dirt	Rubber and leather
Plastics	Ashes
Street sweepings	Broken bricks
Dead animals	Concrete pieces

This often wet collection was usually buried in a nearby sanitary landfill. Eventually, simple disposal in landfills became less possible near large urban concentrations as land became harder to find and refuse from cities was burned. In the last few years we have begun to recognize society's discards as a potential resource. Methods of extracting energy or fuel from this resource have been studied extensively. Wastes can be divided into several classifications; the literature is full of detailed mechanisms for classifying wastes. For our purposes, we shall use the sources given in Table 6.9. Most of the ideas for producing fuel from wastes use municipal refuse as a starting point, although some use animal and crop wastes. We start with the former.

Studies by the EPA have shown that average municipal wastes consist of 60 percent combustile matter having a heat content of

TABLE 6.9 Solid Wastes in the United States

Source	Amount (Estimated for 1980)
Municipal	700[a]
Industrial	180
Mineral	1800
Animal	2300
Crop	900

[a]Figures are in 10^6 tonnes.

about 4170 kJ/kg and about 20 percent moisture. This compares quite favorably with low-rank lignite in heating value; the lignite also has a higher moisture content, as much as 50 percent. The sulfur content of municipal refuse is quite low compared to coal, being about 0.12 percent by weight. These features make it reasonable to consider straightforward combustion as one way to use the energy bound up in wastes.

We are not considering recycling in this section; we do that in Chapter 11. We should mention that combustion of refuse does not produce nearly as much energy as was required to manufacture the original products. Since most refuse is paper and since the average waste-paper product requires about 11 MJ/kg to manufacture, it is clear that only a small percentage of that energy can be recovered by combustion, even at high efficiency. On the other hand, waste sorting systems to separate and recover paper, metals, glass, and so on are also expensive. It is not clear at what point the tradeoff between combustion and recycling becomes viable.

Two types of combustion plants have been developed in this country to receive and burn municipal wastes directly: the water wall incinerator (WWI) and the modular combustion unit (MCU). The walls of the WWI are lined with water-carrying pipes. The water is heated to steam, which is then used as process heat to generate electricity or for space heating. In the MCU, generally smaller in size than the WWI, waste heat boilers are used to produce steam. Several units of both types are in operation today.[3]

Several systems have been designed to operate not with municipal waste directly, but with processed wastes called refuse-derived fuel (RDF). RDF is the lighter combustible fraction of municipal trash obtained by shredding wastes and separating heavier noncombustible materials. These separated materials are usually buried, as is the ash left after the combustion of the RDF. If all units currently under construction or under consideration were in operation, nearly 20 percent of this nation's solid municipal waste would be used to produce energy.

A substantial amount of the solid wastes produced in the United States is organic. About 15 percent of this could be treated to yield oil, methane, or other combustible hydrocarbons. Note that 170 million barrels of oil was about 3 percent of our requirement for 1971 and that 38 million cubic meters of methane was about 6 percent of our 1971 requirement of natural gas. Since 1971 our organic-waste production rate has increased about 3 percent per year and our demand rate about 4 percent per year (until 1978).

The reason for the large difference between the amount of wastes generated and the amount readily usable is that organic wastes are largely water. They must be dried to be convertible to combustible fuels. In addition, concentrations of wastes in quantities large enough to be usable are found only in the larger metropolitan areas. It is possible for an individual farmer to put togeter a manure digester to produce methane for his own use. He very likely could not produce enough to satisfy even his own requirements, but he certainly could reduce his dependence on commercial supplies.

For large-scale production of energy, sufficient solid waste must be accumulated in an economic manner to produce fuels that can compete with virgin fuels. With a current average municipal waste-generation rate of 2 kg per capita per day, a population of about 500,000 would be needed to provide enough refuse to supply one 1000-MT/day (TPD) facility. According to the 1970 U.S. Census, there are 26 cities in the country with populations in excess of one-half million. Of greater

[3]For details see *The Tenth Annual Report of The Council on Environmental Quality, 1979.* . Washington, D.C.: Council on Environmental Quality, 1978.

significance, there are 65 Standard Metropolitan Statistical Areas (SMSA) in the United States with populations in excess of 500,000. Though the collection-centralization problem still exists even for an SMSA, these areas, which represent about half of the nation's population, could easily support such a facility. The 1000-TPD size for such facilities does not need to be a requirement for economic running of a plant if other resources than energy are also extracted.

Of the 2.5 billion tonnes per year of waste generated by agriculture, only that involved in confined production, such as animal feedlots, tanneries, canneries, and the like, can be utilized on a large scale. About 23 percent of the 115 million cattle raised in the United States each year are bred and fattened on relatively confined feedlots. As an example, 10,000 cattle on one feedlot produce about 260 tonnes of manure per day. This represents an extremely

heavy environmental burden to the surface and ground waters of the surrounding area. The U.S. government recognizes this burden and cites energy extraction as one route to aid in processing the waste.

Three different processes have been studied for the conversion of wastes to fuel: hydrogenation, pyrolysis, and bioconversion. See Tables 6.10 and 6.11 for a comparison of the processes and their products. These alternatives to the incineration and landfill management methods, which account for 98 percent of the solid waste collected and disposed of in the United States, are in some respects a direct result of the Resource Recovery Act of 1970. This act put strong emphasis on recycling of wastes to recover products and energy and authorized funds for demonstration grants for recycling systems and for studies to encourage the recovery of resources.

In the hydrogenation process, developed at

TABLE 6.10 Comparison of Waste-Conversion Methods

Process Requirements	BuMines	Pyrolysis	Anaerobic Digestion
Form of feed	Aqueous slurry (15% solids)	Dried waste	Aqueous slurry (3–20% solids)
Temperature	320–359°C	500–900°C	20–50°C
Pressure	100–300 atm	1 atm	1 atm
Agitation	Vigorous	None	Slight
Other	Uses carbon monoxide	None	None
Form of product	Oil	Oil and char	Gas
Yield (percent of original material)	23%	40% oil; 20% char	26% maximum
Heating value (MJ/kg)	7.2	5.7 oil; 4.3 char[a]	11.4
Percent of original heat content recovered in product[b]	65%	82% (60% if char not included)	77% maximum

Source: NSF/NASA Solar Energy Panel.
[a]All of gas and half of char used to supply heat.
[b]Assumes heat content of dry waste is 3.8 MJ/kg.

TABLE 6.11 Comparison of Waste-Derived Oil and Number 6 Fuel Oil

Property	Number 6	Garrett	BuMines[a]
Elemental analysis, weight percent:			
Carbon	85.7	57.5	77.0
Hydrogen	10.5	7.6	10.7
Nitrogen		0.9	2.8
Oxygen	2.0	33.4	8.8
Sulfur	0.7–3.5	0.3	0.3
Energy value (MJ/kg)	8.7	5.0	7.2
Specific gravity	0.98	1.3	1.03
Density (kg/ℓ)	14.0	18.6	14.7
Energy content (MJ/ℓ)	41.50	31.78	35.91
Pumping temperature (°C)	46	70	60
Atomization temperature (°C)	104	116	ND

Source: Garrett Research and Development Co. and the U.S. Bureau of Mines.
[a]ND = not determined.

the BuMines Pittsburgh Research Center, the wastes and an alkaline catalyst such as sodium carbonate are reacted with carbon monoxide and steam at a high pressure (200 atm) and high temperature (300°C). Metals, siliceous materials, and moisture were removed prior to the reaction; the remainder was shredded. Studies showed that 85 to 90 percent of the organic material in refuse can be converted to bitumen, a water-soluble fraction, and a gas at 250°C and 217 kPa. At 380°C and 725 kPa, less of the water soluble material was produced. Complete conversion would give an oil yield of about 57 percent if the average carbon content of oil is assumed to be 78 percent. Since some carbon would still be converted to gaseous products, mostly CO_2, the actual oil yield would be smaller. Measured yields were 40 to 43 percent. Two barrels of oil per ton of dry waste was produced; some of this oil had to be used to provide the heat and CO required for the process. The oil produced (34 MJ/kg heating value) is very similar to Number 6 heating oil (41 MJ/kg heating value).

Pyrolysis is defined as the chemical change brought about by the action of heat. It can also be thought of as destructive distillation in the absence of oxidants, whereas incineration is combustion in direct flame in the presence of air or oxygen. Pyrolysis as a commercial process has been used for many years in the production of methanol, acetic acid, and turpentine from wood plus recovery of the residual charcoal. Waste pyrolysis arose out of work on conversion of coal to low-sulfur liquid fuels. The purpose of pyrolysis is the conversion of cellulosic materials to lower-weight organic molecules. The most important overall reaction involves the splitting out of oxygen to form compounds with high H/C ratios. Cellulose and other carbohydrates lose H_2O and CO_2 just on being heated. Hydrogenation, which is frequently a step in a pyrolytic process, consists of a feed heated under pressure in the presence of CO, steam, and a catalyst in a closed system. Oxygen can be removed by reaction with the added CO to form CO_2 by various disproportionation reactions, and by combinations of these reactions. The large number of possible reactions result in an oil made up of a complex mixture of different molecules.

Three chemical reactions take place simultaneously in pyrolysis:

$$H_2O + CO \rightarrow CO_2 + H_2 \quad \text{(water–gas shift)},$$
$$H_2O + C \rightarrow CO + H_2 \quad \text{(steam–carbon reaction)},$$
$$H_2O + CH_4 \rightarrow CO + 3H_2 \quad \text{(steam–hydrocarbon reaction)}.$$

CO minimizes the water–gas shift to CO_2, removes O_2 from the cellulose, and appears to inhibit dehydration of the cellulose into char. The oil from this process has a heating value equal to that of Number 6 heating oil.

One of the major difficulties in pyrolysis is the variation of effluent products from a change in operating conditions such as temperature, rate of temperature increase, gas flow, feed materials, and other parameters. Four broad classifications of products are obtained from pyrolysis reactions including tars, an aqueous mixture, an organic fraction, and a mixture of gases. Tars represent a relatively small proportion of the total products formed and decrease with increasing temperature. The aqueous portion is largely water plus water-soluble organics. The organic fraction contains a complex mixture including dissolved gases. The remaining gases are materials not condensed.

The organic liquid portion is the potential source of synthetic crude oil. It can be further separated into a water-soluble volatile portion representing about 10 percent of the organic material, which does not contain valuable petrochemicals. A water-insoluble, nonvolatile portion—a black tarry substance—foms an average of 85 percent of the organic liquid fraction. It consists of oxygenated compounds: esters and acids. The other major product is a high-ash char having a heat value of slightly more than 23 MJ/kg. These products—gases, liquids, and char—are separated and then upgraded to salable products. The gases may be sold or used as fuel in the operation of the system.

Bioconversion of methane involves the controlled anaerobic digestion of organic wastes by bacteria where organic matter decomposes in a regulated oxygen-deficient environment. This anaerobic digestion is frequently used in sewage-treatment plants as part of the secondary treatment of sludge. The process is not a new one. In 1895, street lighting in a section of Exeter, England, was fueled by gas from digesting wastewater. In the 1940s a dual-fuel engine was devised capable of utilizing fluctuating mixtures of sludge gas and other fuels. Digester gas is a relatively wet and dirty gas containing about 65 percent methane with an energy value of about 2 MJ/m^3, while natural gas has about 37 MJ/m^3. It is frequently contaminated with other gases from the sewage. However, it does have potential. In India and Africa it is a useful and economic supplement to natural gas for many small farms. Much research has been conducted on small-scale, low-cost plants and the various mixtures of organics used. In the United States, Los Angeles County sanitation districts produce enough gas to fulfill all on-site power requirements and have enough surplus to sell extra gas each year to a neighboring refinery.

Anaerobic digestion of complex organic wastes is a two-stage process. In the first stage, acid-forming bacteria act upon complex organics and change the form of complex fats, proteins, and carbohydrates to simple organic materials, commonly known as organic or volatile acids. The second stage involves the fermentation or gas-generation phase, which produces the desired methane gas. In this step, the methane-forming bacteria use the organic acids as substrate and produce the end products: CO_2, CH_4, and traces of H_2S, NH_4, mercaptanes, and amines. It is these traces that give the gas its objectionable odor. Basically this microbial fermentation is a mechanism that breaks C—C and C—H bonds to form CO_2 and H_2O, which is simply the reverse of photosynthesis. The quantities of off-gases vary, but the mixture consists of roughly 50 percent CH_4 and 50 percent CO_2. Varying the physical parameters influences the rate and quantity of CH_4 produced.

To attain continuous digestion, a proper balance between acid-forming and methane-forming bacteria is required. Optimum levels of five environmental parameters are essential to the establishment and maintenance of this balance. These parameters include *temperature*, which should vary only slightly over two levels: in the mesophilic range from 30° to 45°C, and in the thermophilic range from 45° to 60°C. The digester must be kept in an *anaerobic condition*, as even small amounts of oxygen can inhibit the bacteria. The *pH must be maintained* between 6.7 and 7.0. Below 6.2, waste stabilization ceases. The presence of *nutrients* in adequate quantities is needed by the bacteria. Specifically, a C/N ratio of 35:1 or lower must be maintained, which may require the addition of nitrogen. Finally, *toxic materials* in the form of inorganic salts must be controlled, either by chemical precipitator within the digester or by dilution below threshold levels by increasing the moisture content of the slurry. Once these five parameters have been established and maintained at optimum levels, production of gas should be spontaneous.

A 1000-t/d bioconversion facility could produce approximately $10^5 \, m^3$ of methane per day, based on a conservative value of $0.023 \, m^3$ of gas generated per kilogram of municipal solid waste and sewage sludge. Municipal solid waste is shredded to allow for efficient separation of the organic from inorganic matter (metal, cans, bottles, etc.) found in urban refuse and to reduce feed to a homogeneous size for easier digestion. Before the waste enters the digester, it must be mixed with nutrients and other chemicals (like, sodium bicarbonate, phosphorus) necessary for digester operation. Each digester (ten 17,000-m^3 digesters for a 1000-t/d plant) is maintained at

constant temperature and pressure while being continuously stirred. Stirring allows uniform digestion of the material to proceed in both stages. The methane produced is purified through a gas-cleaning process, and the residue is separated into a liquid and a solid fraction. The liquid could be returned to the digester for further treatment, while the solid has a heating value of 9.2 MJ/kg if dried to 25 percent solid and burned. This solid represents 20 percent of the incoming waste by volume.

In this section we have examined several methods of producing synthetic hydrocarbon fuels. We have seen that many methods exist and are technically feasible today. However, the costs of such fuels are higher than naturally available fuels such as oil, gas, and coal. There is a great risk for potential investors. There may be technical breakthroughs in the future that will dramatically lower the costs of synthetic fuels; the rapidly increasing scarcity of fossil fuels may make the high price of the synthetic fuels acceptable. Or perhaps we can find an entirely different technology on which to base our energy economy that would provide an abundant, cheap, clean source of energy.

6.2 GEOTHERMAL HEAT

Geothermal heat using current technology has the potential to supply a reasonable amount of our electrical energy requirements. Many people believe there are other applications of geothermal resources that could be even larger. In this section we examine several actual and proposed geothermal heat-utilization systems from several standpoints: the nature of the heat source, the mechanism for using the heat, and environmental concerns associated with that source. Some of the techniques for using geothermal heat are rather speculative; as a consequence our discussion of these must of necessity be less detailed, especially when dealing with the economics of that technique. We start by considering the nature of the various geothermal resources.

Geothermal Resources

Geothermal heat comes about because of the intrusion of hot magma from the interior close to the earth's surface. There are many locations where this has occurred, chiefly associated with tectonic plate boundaries and regions of geological activity. These areas, along with some producing geothermal sites, are indicated in Figure 6.9.

Geothermal reservoirs can be classified by several schemes. The most useful and most developed type of hydrothermal reservoir is one in which hot, dry steam is produced at the surface (this is also the least common type). Reservoir types are listed in Table 6.12. The most common reservoir now being used is the **liquid-dominated** type. Both liquid-dominated and dry-steam reservoirs have the same origin, as pictured in Figure 6.10.

Porous rock is heated by convection from the magma below. The porous rock is capped with a relatively impermeable rock layer that both insulates the heat and excludes water. Some surface water is able to find its way into the porous rock through fissures. Likewise, heated water can be forced up through other fissures. As the hot water nears the surface, its pressure drops, causing it to "flash" into steam. In some situations the beginning temperature is high enough and the pressure large enough so that all the hot water is converted to steam; this is the dry-steam reservoir. But in most locations the exiting fluid is a mixture of steam and hot water (really a *hot brine*, because of the large amount of dissolved

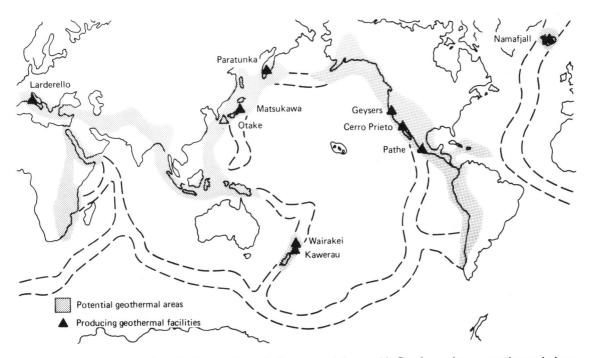

Figure 6.9 Potential and producing geothermal sites around the world. Geothermal areas are located along tectonic plate boundaries.

TABLE 6.12 Types of Geothermal Reservoirs

Reservoir Type	Status of Technology	Estimated U.S. Resource Base (quads)	Typical Sites or Regions
Hydrothermal (to 3 km depth) Dry-steam	Well-developed	100	The Geysers; Larderello, Italy
Liquid-dominated	Developed	3,000–10,000	Imperial Valley; Wairakei, New Zealand
Geopressurized (to 10 km depth)	Initial reservoir tests	44,000–132,000	U.S. Gulf Coast; Hungary
Dry, hot rock (to 10 km depth)	Extraction and exploration in R&D stage	48,000–150,000	Regions surrounding recent volcanism and those with high heat flow and shallow depth to basement
Magma (to 10 km depth)	Research topic	52,000–150,000	Near surface magma chambers, Hawaii

Source: U.S. Geological Survey Circular 726 (1975).

chemicals in the fluid). A producing steam well in the Cerro Prieto field in Mexico is shown in Figure 6.11. The steam is piped to a central generating station.

The U.S. Geological Survey lists 290 deposits of the hydrothermal type in the United States, about one fifth of which appear to have temperatures above 150°C. The amount of stored heat in these deposits is truly large: five regions in California and one in New Mexico each has more than 10^{19} J.

Example 6.3 Estimate the electrical power capacity of the six regions mentioned above.

We must make a number of assumptions: (1) ultimately 20 percent of the stored heat is recoverable, (2) conversion of heat to electricity takes place with 20 percent efficiency, and (3) heat is withdrawn uniformly, continuously over a period of 20 years. We then have

$$P = \frac{(6 \times 10^{19} \times 4.184 \times 0.2 \times 0.2)}{(20 \times 365.25 \times 24 \times 60 \times 60)},$$

$$\boxed{P = 1.6 \times 10^4 \text{ MW.}}$$

Although this stored heat energy represents a large potential resource, it is not clear that even the conservative assumptions in Example 6.3 are realistic. For a variety of reasons, which we discuss, none of the conditions made above can be easily achieved.

Geopressurized resources are currently being investigated but have not yet been commercially developed. These resources consist of water trapped in large, deep sedimentary basins, usually at temperatures below 200°C but at pressures from 500 to 900 kPa. The water contains sufficient quantities of dissolved methane that the gas would be almost equal in value to the electricity produced by the thermal system. These resources are found in an area of the United States extending from Texas to Louisiana both onshore and offshore. This area has been extensively drilled for oil and natural gas, although drillers avoid known geopressurized zones because of the difficulty in dealing with the high pressures. It has been estimated that the thermal energy of these regions alone would be capable of sustaining perhaps as much as 100,000 MW of electrical generation.

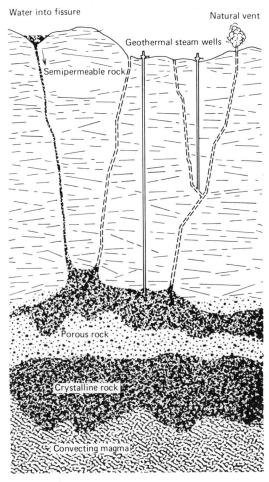

Figure 6.10 The origin of geothermal steam or brine.

Figure 6.11 One of the wellhead installations at the Cerro Prieto geothermal field in Baja California, Mexico. (Photograph courtesy of Lawrence Berkeley Laboratory.)

Figure 6.12 indicates a technique for recovering energy three ways from a geopressurized resource: (1) methane, (2) using the high pressure in a hydraulic turbine, and (3) using the heat to vaporize a working fluid such as isobutane. At the present time, the economies of recovering this energy militate against its use, although, as with so many of the energy resources we have studied, the future could alter this situation radically.

The third type of geothermal resource listed in Table 6.12 is **hot, dry rock**. This is simply large, impermeable rock formations that have been heated by magma bodies, conduction from the interior, or radioactivity of the crust. This resource is potentially very large. If we take the conventional figure of 22°C/km depth for the temperature gradient, we find that in the first 10 km of depth of the continental United States there is a total of about 13×10^{24} J of available thermal energy. Not all this energy is at a temperature high enough to be used for electricity generation. But even if we assume only 0.2 percent of this total amount to be ultimately recoverable, we find that this energy is equivalent to all the remaining coal in the United States!

This resource is not currently being

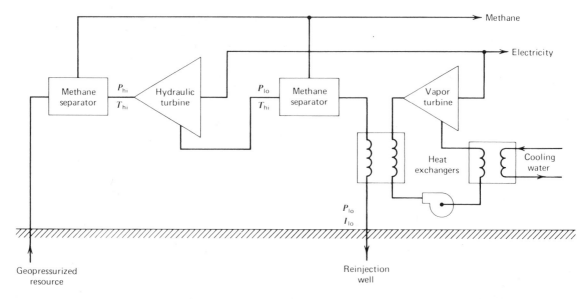

Figure 6.12 Three-way scheme for tapping geopressurized resource energy.

exploited, but because of its large potential many studies are underway. The difficulty with it is extracting the heat at a usable temperature. Because rock is not a good conductor of heat, the injection of water into a drill hole would soon cool off the wall of the hole. The resulting water temperature at equilibrium would be too low. To improve this situation a larger rock–water surface contact area needs to be provided.

One method that has been studied extensively recently is called **hydraulic fracturing**. This process, which is often used by oil companies, has proved to be safe and relatively inexpensive. The idea is to pump water down to the surface of the hot, dry rock, thereby fracturing the rock and sending the water into the crack. Increased pumping of water into the crack would result in the further fracturing of the rock, which would take place over a relatively large underlying area. The fracture would be in the form of a pancake-shaped wedge, up to thousands of meters in diameter. Once the fracture is formed and water is injected, another hole would be drilled into the

fracture area. Water reaching the surface would flow through a heat exchanger to produce steam and then be pumped into the fracture again to be heated.

The last potential geothermal resource in Table 6.12 is the **magma** itself. Estimates place the recoverable amount of energy from the magma to be in the range of 10^{23} J. However, utilization of magma heat is a great distance away, having many difficulties in its path (see, for example, the science fiction movie, *Crack in the World*). At this time magma studies are very preliminary and must be classified as very speculative.

Problems with Geothermal Energy

There are a number of difficulties associated with geothermal energy use. If it were not so, there would be much more use of it than there is! Principal among these are the economic aspects, which we shall mention later in this section. There are also a number of technical and scientific questions that must be addressed first.

TABLE 6.13 Chemical Discharges in the Waikato River at Wairakei Geothermal Area in New Zealand

Constituent	Increment[a] (ppm)
B	0.27
Li	0.13
Na	12.00
K	1.90
Rb	0.029
Cs	0.026
Mg	4.7×10^{-5}
Ca	0.17
F	0.077
Cl	20.8
Br	0.055
I	0.0047
NH_4^+	0.0014
SO_4^{2-}	0.24
As	0.039
Hg	1.5×10^{-6}
Silica	6.3

Source: R. C. Axtman, "Environmental Impact of a Geothermal Power Plant," *Science* **187** 795 (1975). Copyright © 1975 by the American Association for the Advancement of Science.
[a]Multiply by 4000 to obtain annual discharge in metric tons.

Since the temperature of the effluent (steam, water, or brine) from any geothermal reservoir is usually much lower than that of the steam from conventional boiler, special arrangements must be made to optimize the efficiency of energy conversion.

Not only are geothermal fluid effluents usually low in temperature, but also they usually have a rather high dissolved mineral content. Table 6.13 lists some of the principal chemicals discharged into a river near the geothermal area at Wairakei, New Zealand. These minerals are very corrosive and will damage turbine blades rather quickly if allowed to pass through the turbine along with the steam. One technique for removing these materials is to use a flash evaporator. In this device the brine is pumped into a region of lower pressure. Some of the water evaporates immediately to steam, leaving the minerals behind in a condensed brine (see Figure 6.13). The brine can be evaporated to produce salable quantities of the minerals; but, to be realistic, there is little market for most of these materials.

Because the Rankine cycle with steam is very inefficient at low temperatures, other materials have been examined as substitute

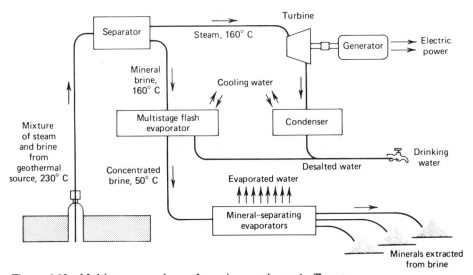

Figure 6.13 Multipurpose scheme for using geothermal effluents.

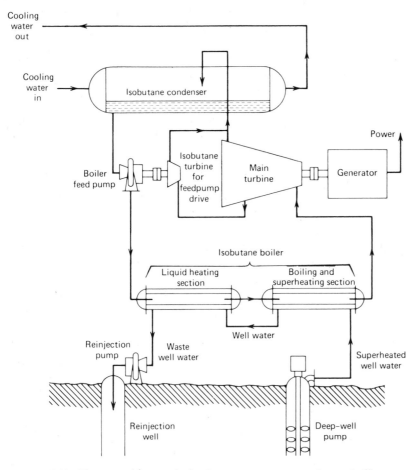

Figure 6.14 Vapor–turbine cycle for low-temperature geothermal effluents.

working fluids. Among these are ammonia, isobutane, and several fluoromethanes (freons). A typical vapor-turbine system is diagramed in Figure 6.14. In this system the geothermal fluid is used to heat and boil the working fluid, in this case isobutane. Cooling water is required to condense the working fluid prior to reheating. Note that the geothermal fluid is reinjected so there is no difficulty with releases of gases to the atmosphere or contamination of surface waters with the geothermal brine.

Nonaqueous working fluids such as those

listed above have larger low-temperature vapor densities than water and therefore would require smaller turbines for the same power output. However, these systems are more complex than the flashing cycle and are rather sensitive to exit temperature.

Example 6.4 What is the percent change in efficiency for each percent change in exit temperature as a function of input temperature for a vapor-turbine system?

Since we do not know the details of the system (fluid, temperature, etc.), we will assume Carnot efficiency and remember that the real-world case will be worse.

$$\eta = \frac{T_H - T_L}{T_H}, \quad d\eta = -\frac{dT_L}{T_H},$$

$$\frac{d\eta}{\eta} = -\frac{T_L}{T_H - T_L}\frac{dT_L}{T_L}.$$

For a system operating at $T_H = 250°C$, $T_L = 20°C$ and increase of T_L to $25°C$ (1.7 percent) produces a change of

$$-\frac{293}{230} \times 1.7 = -2.2 \text{ percent}$$

in efficiency of energy conversion.

There are several environmental problems that must be considered:

The possibility of earthquake triggering resulting from hydrofracture.
Land subsidence resulting from water extraction from support rocks.
Extensive noise generated by steam expanding in flash evaporators or heat exchangers.
Noxious emissions.
Handling the brine residue.

Earthquake triggering by hydrofracture (or brine reinjections) is a real possibility that must be seriously considered. Fluid-injection wells have long been known to be the sources of minor earth tremors. To date no adverse effects have been noted in the existing geothermal plants or research drill holes.

Subsidence as a consequence of water removal has been measured at the Wairakei plant. Both vertical and horizontal movements have been noted, with a total motion since 1956 in excess of 4 m. There does not seem to be a relationship between the underground water pressure and subsidence, as might have been expected. But there is evidence of a linear relationship between subsidence volume and fluid draw-off volume.[4] The Wairakei plant does not reinject the geothermal brine, so it may be possible to reduce the effects of subsidence in future installations by simply reinjecting the fluid back into the aquifer system from which it was withdrawn.

A great deal of noise is associated with the depressurization of the geothermal fluid at the earth's surface. This noise amounts to a deafening roar. This effect obviously would not concern those living out of range of the sound, but it is definitely of importance to people and animals living near such a facility.

There are several gaseous effluents from a geothermal plant. It is difficult to generalize about the quantities of each gas produced because the various geothermal systems worldwide are very different. For example, in the Monte Amiata field in Italy, the CO_2 discharge rate is 10 times the amount that would be generated by fossil-fueled plant of the same electrical output. However, in New Zealand, the CO_2 emissions are about a factor of 10 less than would be expected from a fossil-fuel-burning plant. The effect of CO_2 in the atmosphere is discussed in detail in Chapter 12.

Another commonly emitted gas is hydrogen sulfide, H_2S. This unpleasant-smelling gas (rotten eggs!) can be toxic at rather low concentrations. Detection level is only 0.002 ppm; eye irritation occurs at 10 ppm and lung damage at 20 ppm. Death occurs after 30 min at 600 ppm. Normally, H_2S is diluted to 5 ppm by carbon dioxide and air at the plant. But on foggy or windless days, the H_2S settles to the ground and can reach harmful levels.

Pollutants that do not go into the air sometimes enter surrounding waters. The presence of H_2S and CO_2, along with an increase in the temperature of the water, has led at Wairakei to the extra-rapid growth of a particular spe-

[4]R. C. Axtmann, *Science* **187**, 795 (1975).

cies of lake weed. The growth rate is so fast that it must be harvested every two weeks. In another lake located near the plant a species of plankton is host to a type of bacterium that can survive only if H_2S is present. The water in that lake is reported to smell like rotten eggs.

We should mention that the hot, dry rock geothermal resource would not have gaseous or mineral pollution problems. In fact, if it were desirable, chemicals could be placed in the cool water stream to leach out of the fracture zone any desired minerals, making the operation a combination energy plant and mine!

The major reason that geothermal energy has not been widely developed is economics. Geothermal resources may reach depths of up to 10 km, but it is usually only economical to drill no more than 2.5 km. The limits normally set in terms of drilling depths in relation to temperature of extracted water are:

1. For water temperatures less than 100°C, drill no more than about 2 km.
2. Temperatures must be more than 150°C to warrant drilling over 2 km.

For example, at the Geysers Plant, holes are drilled up to 2.4 km in depth, but the water released is in the form of dry steam, so it is worth the extra price to drill a little farther, considering the by-product of the operation.

An oil company drills a well, realizing that the process is very expensive and that the ratio of dry holes to producing wells is quite large (usually about seven to one). However, one successful oil well's profits are likely to pay off the money spent in the drilling of six unsuccessful wells, and then some. On the other hand, revenues from a successful geothermal well are not nearly as great (one-tenth to one-third those of a successful oil well), so that geothermal well drilling requires a higher success ratio. A moderately successful oil well (in 1978) producing 200 barrels of oil per day earned about $2,000 per day. A successful

geothermal well, producing 4000 liters of 150°C water per minute, earns $600 per day. There have not been many geothermal wells drilled in the United States—500 to date, versus 30,000 oil and gas wells drilled each year—but the success ratio of the 500 that have been drilled is more than 50 percent. Other encouraging facts are that, of six geothermal wells drilled in Idaho, three are successful, the other three are moderately successful, and the temperatures of the produced water are in accordance with those predicted from geophysical and geochemical data prior to drilling.

In addition to large initial costs, an entrepreneur is faced with a tangled array of legal problems. There is no clear legal definition of a geothermal resource. Some states treat geothermal resources as water resources, and others treat them as mineral resources. In each case there is a *different* set of complex laws and regulations concerning rights and regulation. Ownership rights are also difficult to resolve. The federal government, in giving away land to homesteaders, states, and railroads, in many cases retained mineral rights, but in some cases did not. Development of the resource also involves federal, state, regional, county, and municipal planning agencies, pollution agencies, power plant licensing agencies, and so forth.

It is not difficult to imagine that prospective developers are hesitant to commit large sums for geothermal exploitation while it is considered to be a high-risk venture and may involve long delays before income is produced.

We can conclude that, while geothermal energy does have a great potential in terms of the amount of the resource that may ultimately be recoverable, there are many problems involved in the recovery process. It is definitely not a totally clean, relatively cheap energy source, as we sometimes see in the popular press. Many technological, economic, and institutional problems need to be resolved before geothermal energy can contribute on a large scale toward solving our nation's energy problems.

6.3 SOLAR HEAT

In Chapter 5 we discussed the physics of the conversion of sunlight directly to electricity. Large-scale domestic application of such solar electric power is probably far in the future. A more immediate application for sunlight lies in its heating ability. Solar heat has been used for many years in a small way for domestic heating, desalinization, and other applications. The prospect of substantially increased costs for fossil fuels in the near future, coupled with an increasing level of environmental concern by the public, may bring research into solar heat systems finally into respectability if not prominence.

Insolation

The solar energy available at a given time at a given point on the earth's surface depends on four factors: the solar output at that time, the tilt of the earth with respect to the ecliptic plane (i.e., the season), the time of day, and the condition of the atmosphere above that point. The sun emits energy at the rate of 39×10^{26} J/s. At the position of the earth, 1.5×10^{11} m on the average, this corresponds to an energy rate per unit area of

$$S = \frac{3.9 \times 10^{26}}{4\pi(1.5 \times 10^{11})^2} = 1.38 \text{ kW/m}^2. \quad (6.14)$$

This quantity S is called the **solar constant**; it is measured in units of energy per unit area per unit time. The solar constant is not a constant, obviously, since it depends on the varying distance from earth to sun as well as upon fluctuations in the solar output, which may amount to as much as ± 1.5 percent.

The earth's tilt with respect to the elliptic plane varies seasonally. Figure 6.15 illustrates the daily variation in solar radiation *at the top of the atmosphere* as a function of latitude.

Note the high values for the polar regions near the summer and winter solstices. Since the poles are not the hottest regions on earth, the solar radiation present at the top of the atmosphere must not be absorbed easily by the earth's surface near the poles. This is the case, since snow and ice are good reflectors of radiation; that is, they have high albedo. Table 6.14 indicates the surface albedo as a percentage of the incident radiation for North America as a function of latitude and snow cover.

Absorption of sunlight by the atmosphere and by clouds also contributes substantially to reducing the energy available at the surface. A daily surface-solar-radiation measurement might have the appearance of Figure 6.16, where the large downward spikes correspond to rainy or cloudy days. Even on a cloudless day, about 14 percent of the incident radiation is absorbed by dust, water vapor, and other molecules in the atmosphere.

The radiation rate at a given point on the earth's surface at a given instant is given by

$$Q = S \left(\frac{\bar{d}}{d}\right)^2 f \cos Z, \quad (6.15)$$

where S is the solar constant, $\bar{d}$ and d are the average and instantaneous earth–sun distances, respectively, f is an atmospheric absorption factor, and Z is the sun zenith angle. This solar zenith angle, the angle between directly overhead and a line to the sun, can be expressed in terms of the sun's declination δ and the angle, h, to which the earth must be turned to bring the meridian of the observation point directly under the sun:

$$\cos Z = \sin \phi \sin \delta + \cos \phi \cos \delta \cos h, \quad (6.16)$$

where ϕ is the latitude of the observation point (refer to Figure 6.17).

Because of atmospheric absorption, latitude, solar declination, and cloudiness the exact value of solar energy available on the earth's surface is variable; but it is convenient for calculational purposes to assume an average

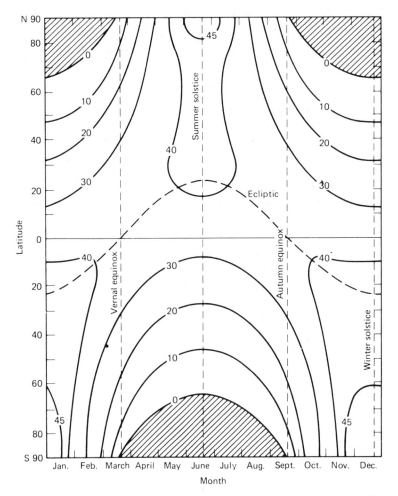

Figure 6.15 Energy available in megajoules per square meter at the top of the atmosphere as a function of latitude and day.

TABLE 6.14 Surface Albedo

Type of Surface or Vegetation	Albedo (percent)
Clean snow	85–95
Sand	35–43
Crops	10–25
Forests	10–18
Dark soil	8–15

figure of 1.0 kW/m^2. It should always be kept in mind, however, that this figure *is* an average; calculations for predicting energy generation or solar-heating capability and so forth must include the factors of Eq. 6.15.

Radiation

Solar energy has been used for a variety of applications, and recently several proposals to

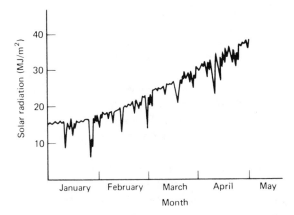

Figure 6.16 Typical result for the direct solar radiation on a horizontal surface for a several-month period.

use solar heat in large, central power-generating systems have been put forward. Most of these applications rely on the fact that emission and absorption of radiation occur simultaneously for all bodies, but the rates of the two processes are frequency dependent.

The understanding of the emission of radiation came in the nineteenth century with the development of the concept of a blackbody. This concept—none exists in reality—represents an object that absorbs all the radia-

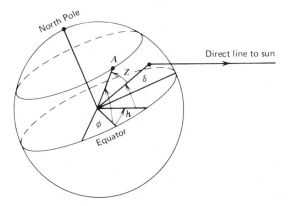

Figure 6.17 The relationship between solar zenith angle Z, latitude ϕ, declination δ, and hour angle h.

tion of all wavelengths incident upon it; it is a perfect absorber. Likewise, the emission spectrum of a blackbody can be calculated easily. Planck first developed a formalism for calculating the spectral distribution function $f(\lambda)$ for a blackbody in 1900 by assuming for the first time that oscillating electrons in atoms could have only certain allowed energies. He obtained

$$f(\lambda) = \frac{8\pi hc\lambda^{-5}}{\exp(hc/\lambda kT) - 1}, \qquad (6.17)$$

where h is Planck's constant (4.136×10^{-15} eV s), k is Boltzmann's constant (8.617×10^{-5} eV/K), and c is the speed of light. The calculated spectrum for a 6000-K blackbody is compared to the emission spectrum of the sun in Figure 5.6. The two curves agree quite well except at the shortest wavelengths, in the ultraviolet region.

The wavelength at which the spectral distribution function has a maximum λ_{max} shifts with frequency such that

$$\lambda_{max} T = \text{const.} \qquad (6.18)$$

In Figure 5.6, a second blackbody radiation curve corresponding to a temperature of 300 K is plotted. This curve estimates the emission from the earth; the maximum wavelength has shifted to about 9μm in the infrared region. A superconductor at 4 K would have its radiation maximum at about 0.06 cm wavelength in the super-high-frequency range used in some radars.

It is found that the energy flux rate from a blackbody is dependent solely upon temperature:

$$\dot{q} = \sigma T^4 \text{ energy/area-time,} \qquad (6.19)$$

where σ is the Stefan–Boltzmann constant 5.67×10^{-8} W/m² K⁴. Using the known energy emission rate of the sun and its apparent surface area in Eq. 6.19, an effective temperature of 5800 K for the sun is obtained. For any real surface the T^4 dependence still exists, but the

TABLE 6.15 Approximate Absorptivities of Selected Surfaces

Material	Temperature Range (°C)	Absorptivity
Polished metals		
Aluminum	250–600	0.039–0.057
Brass	250–400	0.033–0.037
Chromium	50–550	0.080–0.260
Copper	100	0.018
Iron	150–1000	0.050–0.370
Nickel	20–350	0.045–0.087
Zinc	250–350	0.045–0.053
Filaments		
Molybdenum	750–2600	0.096–0.290
Platinum	30–1200	0.036–0.190
Tantalum	1300–3000	0.190–0.310
Tungsten	30–3300	0.032–0.350
Other materials		
Asbestos	40–350	0.930–0.950
Ice (wet)	0	0.970
Lampblack	20–350	0.950
Rubber (gray)	25	0.860

flux rate depends on the nature of the material:

$$\dot{q} = \epsilon(T)\sigma T^4, \qquad (6.20)$$

where $\epsilon(T)$ is a temperature- and wavelength-dependent emissivity (or absorptivity) characteristic of that material. Table 6.15 gives the absorptivity range of several materials over various temperature ranges. Given the surface area, emissivity, and temperature of an object, one could calculate the energy emission rate from Eq. 6.19, but this does not give any information on the spectral distribution of radiation.

In fact, the spectral distribution, either emission or absorption, of a real body is highly frequency-dependent. Wavelengths that are not absorbed by a body must be reflected, unless the material is thin enough so that some of the nonabsorbed radiation may be transmitted through it. For example, a thin copper foil appears to be copper colored by virtue of light reflected from it, but when light transmitted through it is observed, it appears to be green. Figure 6.18 shows the variation of emissivity (or absorptivity) for conductors and insulators. Visible wavelengths range from about 0.4 to 0.8 μm. Metals such as aluminum and copper are typically poor radiators in the infrared region. This is the reason metals retain heat and cool slowly.

An object placed in sunlight absorbs and radiates energy simultaneously. The rate of energy absorption depends upon the solar constant, the area of the object perpendicular to the sun's rays (called the *normal* area), the object, and its absorptivity. The radiation rate depends on the emissivity (which is the same as the absorptivity) and the temperature. Eventually an equilibrium temperature is reached at which the rate of absorption equals the rate of emission.

This analysis is true, however, only for a blackbody, since only for a blackbody is energy emitted at all wavelengths. For real objects, we must take into account the wavelength dependence of the emissivity. At a given temperature, the wavelength of maximum emission (for a blackbody) is known—see Eq. 6.18. If a real object at that temperature has a very low emissivity, then the object would have to become hotter and hotter, assuming heat were not being conducted away, until the temperature were high enough to shift the maximum wavelengths into a region where the emissivity of the material is large. Do such materials exist?

Yes. Some very simple materials exhibit this effect, or a similar one, as do some very complex combinations of materials. Ordinary glass, for example, transmits most of the visible wavelengths with ease, but it is essentially opaque to the infrared and ultraviolet wavelengths—these are mostly absorbed (see Figure 6.18). Sunlight falling on an object in a glass case would cause that object to become

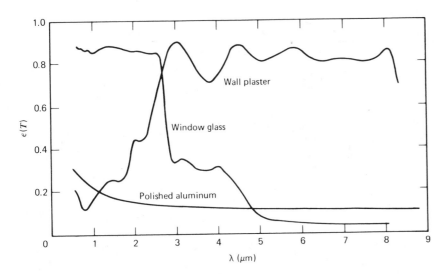

Figure 6.18 Variation of emissivity with wavelength for a typical metal and two insulators.

hotter than the same object in the open. Sunlight would be absorbed by the object, then reradiated. But the temperature of the object would be such that the wavelength corresponding to maximum energy emission would have to fall in the infrared. This radiant energy could not escape the glass case. That part of it absorbed by the glass would be reradiated isotropically. Half would then be radiated back into the interior, some of it causing additional heat (see Figure 6.19). This is the physics behind the operation of a botanical greenhouse, and for this reason this trapping of infrared radiation is often referred to as the **greenhouse effect**. There is speculation that this effect occurs in the earth's atmosphere: carbon dioxide and water vapor absorb a large part of the infrared radiation emitted by the earth. This trapping of this energy may be causing an increase in temperature of the earth; or it may be triggering other effects that could lead to a decrease in temperature. We examine it further in Chapter 12.

The other technique for suppressing infrared emission is to use frequency-selective surfaces. These surfaces are made up of layers of various materials that individually will not both pass visible and block IR, but taken together will.

Figure 6.20 shows a frequency-selective stack and its calculated spectral-absorption

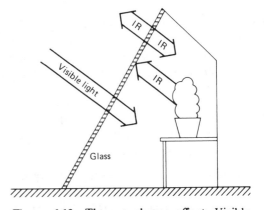

Figure 6.19 The greenhouse effect. Visible light is passed through the glass and absorbed in the interior. Long-wavelength (IR) radiation from the interior is absorbed by the glass; half of it is radiated back into the interior.

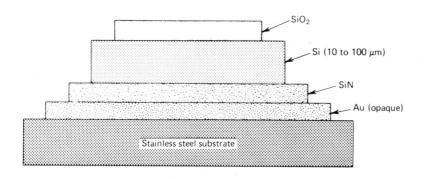

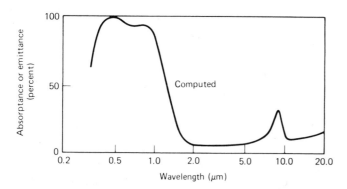

Figure 6.20 A possible frequency-selective solar absorption stack and its computed absorption curve. (From A. B. Meinel and M. P. Meinel, "Physics Looks at Solar Energy," *Physics Today*, February 1972.)

function; note that the horizontal scale is logarithmic. A stack of this composition is expected to have a ratio of absorptivity in the visible to emissivity in the infrared greater than 10.

We have seen so far that sunlight, while "free," is dilute and intermittent. It is not easily trapped without substantial losses, and therefore may be difficult to use.

Example 6.5 Assume 50 percent absorption efficiency. Calculate the surface area of collectors required for an average residence in winter to supply all space heating requirements.

Space heating requirements depend entirely on location. But the average Midwestern two-bedroom home requires about $420 \, \text{m}^3$ of natural gas for the worst month, $\approx 5 \times 10^9 \, \text{J/day}$. The sun shines about 6 hr with an average power density *on the collector* of $200 \, \text{W/m}^2$.

$$\text{Energy required} = (5 \times 10^9 \, \text{J/day}) \times 30 \, \text{days}$$
$$= 1.5 \times 10^{11} \, \text{J}.$$

$$\text{Area required} = \frac{(\text{energy required})}{(\text{energy density})} =$$
$$\frac{1.5 \times 10^{22} \, \text{J}}{200 \, \text{J/s m}^2 \times 30 \, \text{days} \times 24 \times 60 \times 60 \, \text{s/day}}.$$

$$\boxed{\text{Area} = 289 \, \text{m}^2.}$$

This is larger than can be reasonably accommodated on most houses (also beyond most budgets!).

The utility of solar heat depends a great deal on the temperature at which it is collected. For home applications the temperature rarely exceeds 100°C, a consequence of the fact that concentrating collectors that can track the sun are usually too expensive for domestic applications. To achieve temperatures in the intermediate range, from 100° to 500°C, tracking in at least one axis is required, as well as some kind of concentration. For example, long parabolic cylindrical mirrors can be used to direct sunlight to a collecting pipe filled with a heat-transfer medium (see Figure 6.21).

To achieve higher temperatures, two-axis tracking and more elaborate focusing schemes must be used. These added features naturally make the cost of the installation higher. A rather considerable debate has been in progress in the past few years on the advisability of pursuing "high-technology" solar heat applications at the expense of smaller, low-temperature technologies. In 1977, the Energy Research and Development Agency (ERDA) spent $60 million on large, central solar plant research and design and only $6 million on low- and intermediate-temperature applications. Since among high-technology advocates

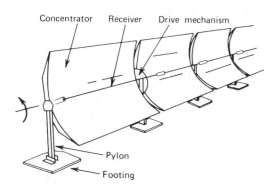

Figure 6.21 Single-axis tracking parabolic solar collectors.

there is a difference of opinion on the best system for transporting energy from collectors to a central converter, two systems have been proposed: energy transport by heat (i.e., pipes) and transport by light (i.e., mirrors). In fact, advocates of the latter system have won the argument, since it is that system that has received the lion's share of the funding from the federal government. The reasoning goes like this: the larger the collection array, the farther a heat transport piping system would have to carry the heat with heat losses by convection and radiation increasing. On the other hand, as a light-transport system's size increases, the cost of the reflecting surfaces should decrease because of the economies offered by the increased scale of operation.

Neither system has been built on a scale large enough for reliable projections to be made; no doubt the argument will continue. We look at both systems from a technical point of view and then discuss some of their relative merits, environmental impacts, and projections for the future. Then we look at small-scale applications of solar thermal energy and try to compare them with the larger projects.

The Power Tower

Federal funding for solar energy had been virtually insignificant until the energy crisis of 1973–1974 brought home in a very positive fashion the vulnerability of the U.S. energy system (see Figure 6.22). Since then, solar energy funding has increased dramatically, although early indications are that solar energy will receive rather considerably less attention from the Reagan administration. The bulk of the past 5 or 6 years' funding (60 to 70 percent) has gone to a series of projects that have come to be known as **power towers**. A 5-MWt test facility was constructed near Albuquerque, New Mexico, in the late 1970s, and a 10-MW pilot plant is currently under construction near Barstow, California.

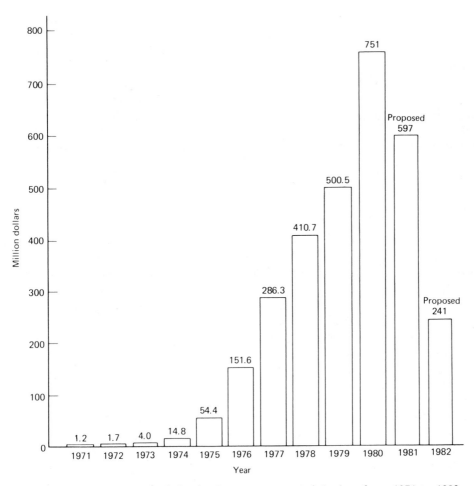

Figure 6.22 Growth of the federal solar energy research budget from 1971 to 1982. [From *Environmental Quality, the Ninth Annual Report of the Council on Environmental Quality.* Washington D.C.: U.S. Government Printing Office, 1978 and *Science* **211**, 1400 (1981).]

The desert regions are generally conceded to be the best for large, central solar plants because of the high annual solar insolation (see Figure 6.23). Figure 6.23 indicates the average annual solar radiation incident on a horizontal surface in watts per square meter (averaged over a full year and over all hours of the day and night). We should note that a rather considerable improvement can be made simply by tilting the collection surface at a fixed angle.

A substantial amount of solar energy is available in the southwestern United States. If only 10 percent conversion efficiency were possible, less than 15 percent of the land area of the state of Arizona would be required for collectors to supply the entire electricity requirements of the country. Of course, transporting the electricity from Arizona to the urban centers where it is required would not be a simple (or inexpensive) matter, and

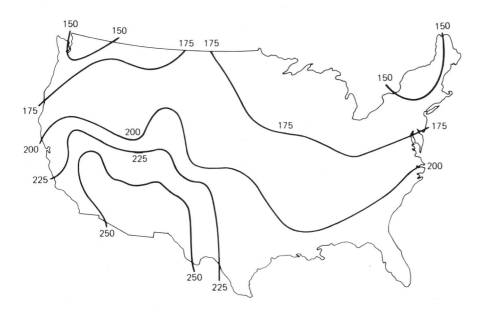

Figure 6.23 Average solar radiation in watts per square meter averaged over a full year and over all hours of the day and night.

Arizonans might object to this use of their state!

The operation of a power tower is very straightfoward, but there are some formidable problems to be solved before solar thermal electricity becomes a competitor to fossil-fueled power generation. Typically a tower containing the solar heat receiver is located near the south edge of a field of **heliostats**, two-axis sun-tracking reflecting mirrors. The mirrors direct the sunlight to the receiver in the tower. The solar heat is used to produce steam at a high temperature, which is then used in a Rankine-cycle turbine system. These steps are indicated schematically in Figure 6.24.

Example 6.6 At the Albuquerque power tower 333 heliostats having an area of about 37 m² each are used to reflect sunlight to the receiver. A peak flux of 2.5 MW/m² has been obtained. What is the receiver area? What are the receiver losses by radiation under these conditions?

We estimate the peak solar insolation at about 900 W/m². Therefore, 7.43 MW was the incident rate upon the heliostats. Heliostat reflectivity is typically 0.9 and the receiver absorptivity typically is 0.95 so

$$7.43 \times 0.9 \times 0.95 = 7.04 \text{ MW}$$

is the rate of absorption by the receiver. Therefore the receiver area must be

$$\frac{7.04}{2.5} = 3.81 \text{ m}^2.$$

We do now know of what materials the receiver is constructed. We must assume it to be a blackbody; then we know the losses we obtain will be *greater* than those of the actual receiver. The radiated energy flux is given by Eq. 6.15:

$$\dot{q} = \sigma T^4 \text{ W/m}^2.$$

The receiver temperature is estimated to be of the order of 600°C, so that

$$\dot{q} \approx (5.7 \times 10^{-8}) \times 873^4 \approx 3.3 \times 10^4 \text{ W/m}^2$$

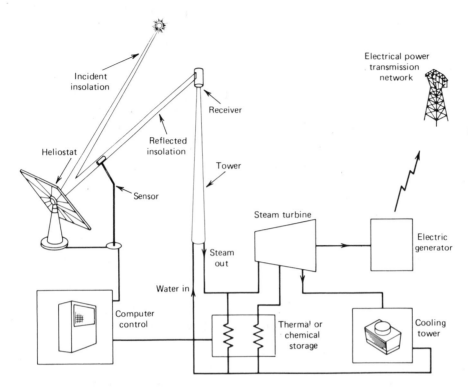

Figure 6.24 The power tower concept. Heliostats focus sunlight at a receiver on top of a tower. Conventional steam-turbine conversion to electricity follows. [From A. Hildebrandt and L. L. Vant-Hall, "Power with Heliostats," *Science* **197**, 1139 (1977). Copyright © 1977 by the American Association for the Advancement of Science.]

and the rate for the entire receiver is

$$(3.3 \times 10^4) \times (3.81 \text{ m}^2) = 1.26 \times 10^5 \text{ W}.$$

Convection losses would also occur; they would typically be half this amount.

The facility under construction at Barstow, California, is to be a 10-MW plant with a 152-m tower. This plant, along with the Albuquerque operation, will serve as a test bed for engineering studies and also as a means of determining eventual costs. A number of problem areas will be studied:

■ Heliostats.
■ Receivers.
■ Thermal storage.

The Albuquerque solar tower uses plane mirror heliostats, as in the lower left of Figure 6.25, although a variety of types have been tested. The Boeing design is novel in that the reflecting surface is aluminized mylar spread on a circular support inside a plastic canopy for protection. Heliostats will have to be built to very rigid design specifications: they must withstand high wind loads, they must be able to be stowed in several positions (vertical, to minimize hail damage; horizontal, to minimize wind damage; and inverted, to minimize sand storm damage), they must be aimed to within 0.1° and remain on track, they must be capable of being quickly "scrammed" in case of a loss of coolant in the receiver, and they should be

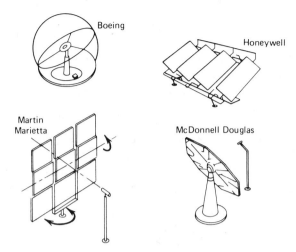

Figure 6.25 Several heliostat designs.

(relatively) inexpensive. The DOE estimates the heliostats should cost no more than about $80/m^2 in mass production quantities or economic operation. The heliostats produced to date have cost between $500 and $1000/m^2— clearly out of the "economic" range! (One critic of the program has said that the aerospace companies involved in the program are constructing aerospace heliostats at aerospace prices.)

The receiver must also perform a variety of tasks under adverse conditions: it must absorb most of the sunlight reflected to it by the heliostats, it must minimize radiation and conduction losses, it must withstand extreme and sometimes rapid temperature cycling, it should exchange heat efficiently to the cooling medium, it should be lightweight (it will be at the top of the tower), and it too should be inexpensive. Studies made for the DOE indicate that the receiver should contribute to less than about 15 percent of the total expense.

Thermal storage is required in any solar thermal electric operation. As we know, electrical energy demands are met by a combination of two types of generating plants: baseload and peakload. Solar power could be used for baseloading, but substantial thermal

or electrical storage would be required to do so. No one believes solar could be competitive as a baseload facility. The peak electrical load, above the base, follows the sun (a few hours behind it), so that solar power is ideal for peak loading, if a few hours of thermal or electrical storage are provided. A variety of techniques for providing this storage have been examined. We study them in detail in Chapter 10.

As of mid-1981 the Barstow Tower is well behind schedule and far over budget. There have been rumblings from Congress about future funding. There is a fear that this project might do for solar energy what Grandpa's Knob did for wind power, namely, set it back several decades. One of the by-products of the difficulties of bringing Barstow on line has been increased funding for domestic solar applications and renewed interest in large distributed collecting systems.

Solar Power Farms

There have been a variety of distributed collector proposals. The argument against them is that too much energy is lost in transporting the heat at a high temperature to the steam boiler. One proposal[5] would minimize these losses by making use of the selective-coated surfaces discussed earlier. In this proposal a steel pipe would be coated with the bulk absorber stack of Figure 6.20. The pipe would be enclosed in an evacuated reflecting cylinder to reduce heat loss by convection. A slot in the cylinder would allow entry of sunlight, which would be concentrated by a factor of 10 by glass Fresnel lenses.[6] Refer to Figure

[5]A. B. Meinel and A. P. Meinel, *Physics Today*, February (1972).
[6]A Fresnel lens is a device that makes use of Fresnel diffraction to form a focused image. In a circular geometry, it consists of rings of alternating dark and light circles whose nth radius is given by $R_n^2 = n\lambda r_0$, where r_0 is the focal length for light of wavelength λ. Additional information may be obtained from B. Rossi, *Optics*. Reading, Mass.: Addison-Wesley, 1959, p. 167.

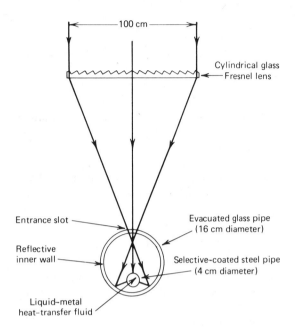

Figure 6.26 Proposed method of efficient heat transport in a distributed collector system. (From A. B. Meinel and M. P. Meinel, "Physics Looks at Solar Energy," *Physics Today*, February 1972.)

6.26 for a schematic illustration of these collectors. The heat accumulated by the steel pipe would be extracted by a cooling medium, probably a liquid metal such as sodium. This coolant could also be used as a thermal storage medium, since it has a rather high heat of fusion.

The Meinels estimate that a 30 percent conversion efficiency just over 8000 km^2 of surface area would be required to generate 10^6 MW of electrical power. To achieve this efficiency would require steam temperatures of the order of 500°C and more than 320 × 10^6 tonnes of sodium for thermal storage! To make the electricity competitive the cost of this system could not exceed $60/m^2, a very unlikely figure today.

There are scientific and engineering problems to be overcome. The frequency-selective absorbing stacks will deteriorate at unknown rates as a result of diffusion and UV exposure. Heat extraction and injection into the thermal storage medium need to be studied. Costs must be carefully analyzed. Comparisons between distributed collection (solar farms) and point processing (power towers) must be more accurately made with realistic data before serious conclusions can be drawn. But before all this is done, one other solar system should be considered for which the collector cost is zero.

Solar Sea Power

A ready-made solar collection system already exists: the oceans. A temperature gradient exists between the surface and bottom layers of the oceans because of the solar flux. Schemes to utilize this temperature difference in a heat engine date back to 1881.[7] An unsuccessful attempt was made in 1930 by Claude.[8] This lack of success has been attributed to the use of water as the working fluid for the turbines. A more modern proposal would use ammonia, freon, or other fluid commonly used in refrigeration systems. Another plant built in 1956 off the Ivory Coast of Africa successfully demonstrated the cold-water pipe technology, but was undercut in price by a new hydroelectric project and therefore was terminated.

Figure 6.27 illustrates schematically the operation of such a solar sea-power plant, usually called an **ocean thermal energy conversion**, or **OTEC**, plant. Cold water is taken into the system from lower levels and used to liquefy ammonia at 10°C. The liquefied ammonia is pressurized and then vaporized at 20°C by warm water from the surface. The exhausted water drops 2°C, from 25° to 23°C. The high-pressure ammonia vapor expands in the turbines, losing 10°C and substantial pressure. This ammonia vapor is condensed by colder water that gains 2°C in the process.

[7] J. d'Arsonval, *Revue Scientific* **17**, September (1881).
[8] G. Claude, *Mech. Eng.* **52**, 1039 (1930).

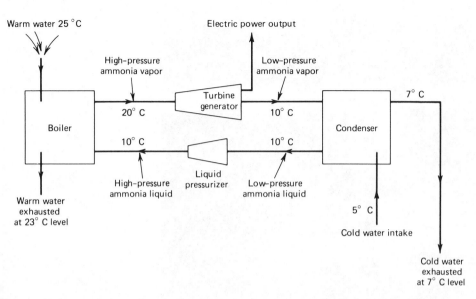

Figure 6.27 Schematic diagram of an OTEC electricity-generating system.

Example 6.7 What is the Carnot efficiency of the system in Figure 6.27?

$$\eta = 1 - \frac{T_c}{T_H} = 1 - \frac{273 + 10}{273 + 20},$$

$$\boxed{\eta = 3.4 \text{ percent.}}$$

The actual efficiency would be about half this.

The overriding concern is not the technical feasibility of the project, but the cost per unit of energy delivered. But the costs depend on the specifics of the design and the end product form of the energy.

Five research groups have produced detailed studies and designs for OTEC systems:

The University of Massachusetts. A submerged propane boiler about 20 km offshore of Miami to produce 400 MW, the electricity to be cabled onshore.

Carnegie–Mellon University. Submerged, possibly bottom-mounted in the Gulf of Mexico, ammonia boilers producing electricity to be used on-board, probably for hydrogen production.

Lockheed Corporation. A basically submerged, anchored spar producing 160 MW on-board (see Figure 6.28).

TRW Corporation. A surface vessel with power modules generating 100 MW on-board.

The Applied Physics Laboratory of Johns Hopkins University. A surface vessel generating 100 MW for on-board use in ammonia synthesis or aluminum reduction (see Figure 6.29).

Submerged plants are considerably less susceptible to hurricane damage than surface vessels. On the other hand, OTEC ships can "graze" to find the best ΔT. Onshore power connections—either electrical or hydrogen—eliminate on-board storage problems, but are expensive and fragile. Many technical problems need to be solved:

■ Large, efficient heat exchangers.
■ Stability of OTEC plant in currents.
■ Stresses on long cold-water pipes.
■ Biofouling of water pipes.

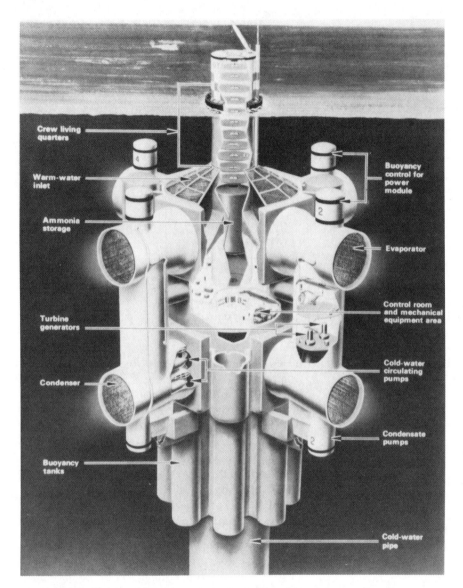

Figure 6.28 Artist's conception of a large OTEC plant. (Illustration courtesy of Lockheed Missiles and Space Co., Inc.)

These considerations make estimating costs difficult, but studies have been made. The cost of electricity from OTEC plants is estimated to be competitive with fossil or nuclear (refer to Table 6.16). Of course, a large number of assumptions have gone into the production of these numbers; it is by no means clear that all of them are justified. Some reviewers of these cost estimates have pointed out that the cold-water pipes in all designs would dwarf any marine structure ever attempted. It is also noted that marine construction is always many

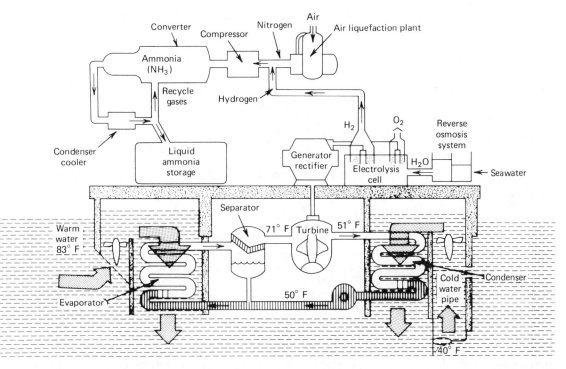

Figure 6.29 Schematic diagram of an OTEC/ammonia-plant ship. [From G. L. Dugger, E. J. Francis and W. H. Avery, "Technical and Economic Feasibility of Ocean Thermal Energy Conversion," *J. Solar Energy* **20**, 259 (1978).]

TABLE 6.16   Approximate Power Cost Comparisons (1975 Dollars)

| | Fossil Fuel | | Nuclear | Alowable OTEC |
	Oil	Coal	Low–High	Range
Investment ($/kWe)	465[a]	450[a]	500[a]–1000	1000–1900
Use factor[b]	0.75	0.75	0.6	0.9
Fixed charge rate	15%	15%	15%	13%
Costs (mills/kWh)				
Fixed charge	11	10	14–29	16–31
Operating cost	1	1	1	1
Fuel cost	20[b]	11–14[b]	3	0
Power cost	32	22–25	18–33	17–32

Adapted from G. L. Dugger, E. J. Francis and W. H. Avery, "Technical and Economic Feasibility of Ocean Thermal Energy Conversion," *J. Solar Energy* **20**, 259 (1978).
[a]Costs include $100/kWe for pollution and safety control costs, and costs for fossil fuel plants include 30-day fuel storage facilities.
[b]Oil at $11/bbl and eastern U.S. coal at $31 to $41/tonne. Based on heat rate of 1.05×10^7 J/kWh.

TABLE 6.17   Potential OTEC Capacity and Fuel Savings for the Year 2000

OTEC Plant Type/Product	Product Capacity	OTEC Power (GWe)	Onshore Plants Avoided (GWe)	Fuel Saved ($\times 10^9$ GJ/yr)
Tropical grazing plants				
Ammonia for fertilizer	45×10^6 t/yr	42	[a]	1.6
Aluminum refining	15×10^6 t/yr	23	26[b]	2.4
NH_3, H_2, electricity	36×10^6 t NH_3/yr; 17 GWe power inland U.S.	32	17	1.6
Subtotals, tropical plants		100		5.6
Offshore electric plants	18 GWe to utility grid	20	18	1.7
Totals		120	61	7.3

Adapted from G. L. Dugger, E. J. Francis and W. H. Avery, "Technical and Economic Feasibility of Ocean Thermal Energy Conversion," *J. Solar Energy* **20**, 259 (1978).
[a]Supplants 48 Mm³/yr natural gas consumption.
[b]Supplanted onshore capacity exceeds the OTEC busbar requirement because of improved plant integration and waste heat recovery.

times more expensive than the same system on land. Finally, we should mention that many experts believe these costs could be off by as much as a factor of 5.

The only way to know for certain is to forge ahead. Some OTEC proponents are quite optimistic about the future. If they are right, OTEC plants could have a significant impact on our energy picture by the year 2000 (see Table 6.17).

Because OTEC could be a large producer of electricity we should make certain there are no adverse environmental effects to be considered before a large and irreversible investment is made. The direct effect of large-scale OTEC will be the lowering of the temperature of the upper layer of warm water. The extent to which this could happen is not easily calculable. One estimate[9] is that an annual extraction of energy at the rate of 60×10^9 kW would

cause a temperature drop of only 1°C. Of course, 1°C could be a significant change that might have far-ranging effects, including advancement of Arctic glaciers and reduction of the summer growing season. Also we should keep in mind that the 1°C estimate is only an estimate based on a simple model; it could be quite wrong.

There are other effects, some of which may not be all bad. The cold water at the bottom of the ocean is rich in nutrients essential to the growth of marine life, but this cold water usually does not mix readily with the warm upper layers. At the poles the cold, dense water sinks to the bottom and descends southward. In some places natural upwellings bring the cold water to the surface. Off the coast of Peru the upwelling Humboldt Current accounts for about one fifth of the entire world's fish harvest. OTEC plants will produce a similar upwelling and so would be a natural breeding ground for fish and other seafood.

Several other schemes have been proposed

[9]C. Zener, "Solar Sea Power," *Physics Today*, January (1973), p. 48.

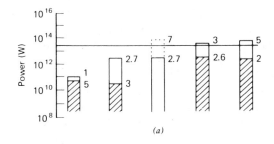

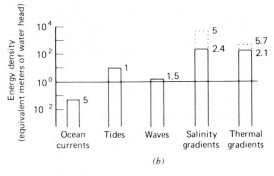

Figure 6.30 (a) Power or energy flux for ocean energy. On the "ocean currents" bar, the shading represents the power contained in concentrated currents. Estimated feasible tidal power is also shaded. The dotted extension on "waves" indicates that wind waves are regenerated as they are cropped. "Salinity gradients" includes all gradients in the ocean; the large ones at river mouths are shown by shading. On "thermal gradients," the shading indicates the unavoidable Carnot-cycle efficiency. The horizontal line at 30×10^{12} W is a projected global electricity consumption for the year 2000. (b) Intensity or concentration of energy expressed as equivalent head of water. "Ocean currents" shows the velocity head of major currents. For tides, the average head of favorable sites is given. For waves, the head represents a spatial and temporal average. The salinity-gradients head is for freshwater versus seawater, the dotted extension is for freshwater versus brine (concentrated solution). The thermal-gradients head is for 12°C; that for 20°C is dotted; both include the Carnot efficiency. [From J. D. Isaacs and W. K. Schmitt, "Ocean Energy: Forms and Prospects," *Science* 207, 265 (1980). Copyright © 1980 by the American Association for the Advancement of Science.]

to use the solar energy collected by the sea in the form of currents, waves, salinity gradients, and tides. Only the last of these has actually been used, as we mentioned in Chapter 2. Figure 6.30 compares these various sources. These are considered to be speculative, with the exception of tidal power: research in these areas is not heavily supported.

There is currently a vigorous research program in ocean thermal energy conversion. The next few years will provide many answers to some of the engineering problems and cost questions that we have discussed. Whether or not OTEC can become more than an interesting exercise in massive marine technology is one of these as yet unanswered questions.

Domestic Applications

Residential and commercial space and hot-water heating account for approximately 20 percent of the total U.S. energy consumption; residential space heating alone accounts for 12 percent. A considerable energy savings would result, then, from the application of solar technology to home heating alone. As short a time as 5 years ago, solar energy was considered "exotic" in most parts of the country. It is a very different situation today, since many people have discovered that solar energy can be utilized effectively and inexpensively now on a small scale in individual homes.

Two basic techniques are being used separately or together: active and passive technologies. We shall examine them both from several points of view. First, active systems.

Figure 6.31 illustrates the functional relationships and identifies the required conepts for the solar heating and cooling of a building. The major differences between this and a normal system are the solar collector and heat storage reservoir. A supplemental energy source is also required, since collection and storage facilities to meet peak load conditions

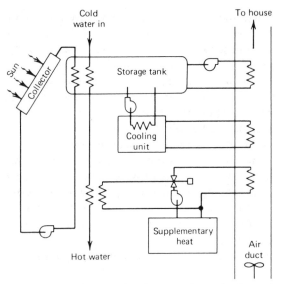

Figure 6.31 Schematic diagram of a solar-heated and -cooled building.

would be far too expensive in many localities to be practical. This supplemental source should be capable of supplying the full heating or cooling requirements of the building when demand is greatest.

In Table 6.18 the direct solar radiation on a horizontal surface (summed over 24 hr) as a function of month is shown for several U.S. cities. By properly tilting the collecting surface, substantially higher averages can be achieved. Solar collectors can contribute to the heating system in any city, but where the worst case average is over about 80 W/m^2 solar heating can handle all but a few days. The major question, of course, is, "Does solar heating cost more than fossil-fueled heating?" The answer, as usual, is "perhaps," or "maybe not," depending.

The key to an active solar heating (and/or

TABLE 6.18   Direct Monthly Solar Radiation on a Horizontal Surface[a]

Location	Jan.	Feb.	March	April	May	June	July	Aug.	Sep.	Oct.	Nov.	Dec.
Albuquerque, NM (35°3')	4.21	4.91	7.08	8.17	9.11	9.38	8.87	8.21	6.89	5.40	3.93	3.70
Astoria, OR (46°11')	1.27	2.12	3.78	4.96	6.39	6.17	6.94	5.82	4.23	2.52	1.24	1.02
Cleveland, OH (41°23')	1.72	2.34	4.49	5.07	6.68	7.40	7.26	6.27	4.56	3.27	1.60	1.48
El Paso, TX (31°48')	4.51	5.41	7.46	8.49	9.18	9.27	8.31	7.61	6.80	5.74	4.09	3.67
Fairbanks, AL (64°48')	0.063	1.18	3.49	5.53	6.24	6.70	5.74	4.07	2.11	9.36	0.23	0.046
Fresno, CA (36°46')	2.61	3.77	6.09	7.16	8.31	9.27	8.70	7.89	6.16	4.73	2.78	2.16
Indianapolis, IN (39°44')	1.93	2.71	4.37	5.19	6.30	7.18	7.06	6.25	4.89	3.61	2.02	1.69
Los Angeles, CA (33°55')	3.38	4.33	6.33	6.77	7.54	7.98	8.42	7.41	6.15	4.55	3.27	3.09
Miami, FL (25°47')	4.64	5.17	6.61	6.99	7.08	6.99	6.97	6.56	5.44	4.89	4.13	4.12
New York, NY (40°46')	1.98	2.71	4.37	4.99	6.01	7.02	6.46	5.47	4.39	3.22	1.86	1.66
Seattle, WA (47°36')	1.06	1.82	3.72	5.32	6.54	6.72	7.30	5.71	3.87	2.26	1.16	0.82

Source: T. Kusuda and K. Ishii, *Hourly Solar Radiation Data for Vertical and Horizontal Surfaces on Average Days in the United States and Canada.* Washington D.C.: U.S. Department of Commerce, 1977.
[a]In 10^8 J/m^2.

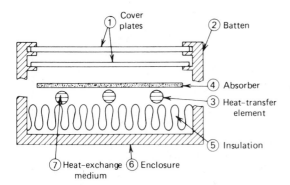

Figure 6.32 Expanded view of a flat-plate solar collector. (From D. McDaniel, *The Sun: Our Future Energy Source.* New York: John Wiley & Sons, 1979.)

cooling) system is the collector. It is usually a flat, metal absorbing plate with a heat exchanger and one or more devices for minimizing heat loss, as diagramed schematically in Figure 6.32. Some care must be exercised in the design and orientation of this plate to maximize solar collection.

Copper, steel, and aluminum plates have been used. In any case, the surface must be treated or coated to minimize reflection. At least 90 percent of the incident sunlight should be absorbed. Note that the absorptivity also depends on angle of incidence. Flat black paint or anodization, in the case of aluminum, should be used. Paint should be primed with a self-etching primer to prevent peeling. In cases where the heat-transfer medium is tap water

and this water comes into contact with the absorbing plate, neither aluminum nor carbon steel can be used, as these metals will be corroded by the minerals in water. Demineralized water could be used, but this makes the installation much more expensive.

The three metals have rather different thermal conductivity (see Table 6.19). The poor conductivity of steel means that many more meters of transfer tubing, or more rapid water flow, will be required to keep the collector plate from reaching excessively high temperatures (thereby promoting heat loss). Flow velocities should be minimized to reduce corrosion.

There has been some recent experimentation with black plastic films having air-filled pockets to provide insulation and water flowing through a central channel. These collectors are said to have the same collecting power as metal plates but cost rather considerably less. Actual manufacturing costs and long-range durability of these Tedlar–polyvinyl chloride–teflon films have yet to be determined.

One or more cover plates are required, usually of glass, to reduce radiation and convection loss from the collector plate. The total energy transferred from the collector can be written as

$$Q_a = \tau\alpha\Phi_0 - U\,\Delta T, \qquad (6.21)$$

where τ is the transmission factor through the cover plates, α is the absorptance of the col-

TABLE 6.19   Physical Characteristics of Several Flat Plate Materials

Item	Copper	Steel	Aluminum
Specific gravity (g/cm³)	8.92	3.04	1.09
Thermal conductivity (cal/s m °C)	93	11	50
Plate thickness for equal performance (mm)	0.4	1.6	0.8
Energy cost per kg (MJ)	110	27	271
Energy cost per square meter (MJ)	533	513	794

Adapted from D. McDaniel, *The Sun: Our Future Energy Source.* New York: John Wiley & Sons, 1979.

lector plate averaged over all angles of incidence and over an appropriate time period,
Φ_0 is the incident solar radiation, and $U \Delta T$ is
the sum of all heat losses. The factor U can be
thought of as an effective heat-transfer
coefficient, and ΔT is the temperature
difference between the plate and the outside
world.

The transmission factor τ depends on the
number of plates, their composition, and the
angle of incidence. Special low-iron-content
glass is desired to make the transmission
through each layer at least 90 percent. Glass is
almost completely opaque to the IR emissions
from the hot metal plate. Special multilayer
coatings, as discussed earlier in this chapter,
can be used to reduce reflection, but these are
expensive and little is known about their ability to withstand prolonged exposure. Mylar films
can also be used as covers. This material has
about the same transmission characteristics as
glass, but it is not as opaque to IR as is glass.
In addition, mylar degrades under UV
exposure and will tend to stretch and sag at
higher temperatures.

At least two cover plates should be used.
Figure 6.33 indicates the losses expected for
one, two, and three cover plates with varying
wind conditions. Clearly, convective losses
dominate, but these are reduced significantly
by using additional plates. Even greater loss
reduction can be achieved by evacuating the
space between plates.

Several different heat-transfer fluids can be
used. Among these are water, a mixture of
ethylene glycol (antifreeze) and water, and air.
Liquids are easy to handle and are more
efficient in transferring heat from the collector
plate. Air systems do not freeze, do not cause
corrosion, and are easier to use to extract
energy from the storage medium.

A complete home water- and space-heating
system is diagramed in Figure 6.34. In this case
water heated by the solar collectors is used to
heat water for household use directly in a

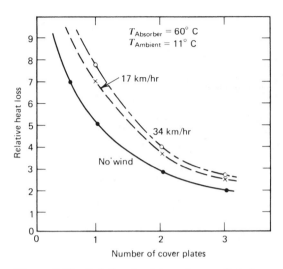

Figure 6.33 Relative heat loss from a flat-plate
collector as a function of number of cover plates
and wind speed. (From D. McDaniel, *The Sun:
Our Future Energy Source.* New York: John
Wiley & Sons, 1979.)

small 275-gal tank and then is stored in a larger
1600-gal tank. Heat is transferred to a surrounding stone bed by conduction. Air is
blown through the stone bed to be heated for
space-heating purposes. In the summer, hot air
from the house is blown through the stone bed,
thereby transferring heat to the water tank.
The water is pumped to an exposed surface on
the roof at night so that heat may be radiated
to the sky. External cooling may also be
needed as indicated. Different storage media
may also be used. Stone has the advantages of
being cheap and easy to install: no special
precautions are necessary. However, it does
not readily absorb or give up heat (see Table
6.20). Sodium sulfate decahydrate is an interesting substance that changes from a solid to a
liquid at 32.38°C. This phase change requires
about 80 cal/g. There are substances with
higher heats of fusion; these are more expensive and usually have phase-change temperatures that are somewhat high (see Chapter
10).

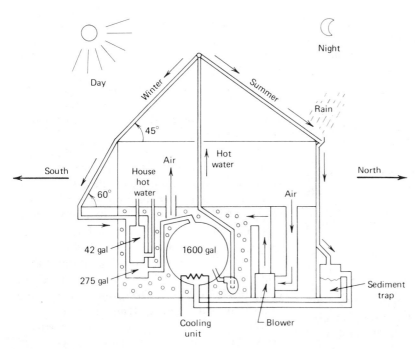

Figure 6.34 A possible active solar heating/cooling system for a home.

TABLE 6.20 Heat Storage Properties of a Few Materials

Medium	Specific Heat	
	cal/g °C	cal/m³ °C
Water	1.0	1000×10^3
Water and ethylene glycol (50/50 mix)		
Gravel	0.20	278×10^3
Sodium sulfate decahydrate	50	83×10^6

Example 6.8 How much energy can be stored in the tank in Figure 6.34? Assume a high temperature of 50°C and that energy can be withdrawn as long as the temperature exceeds 27°C.

$$Q = mc\,\Delta T = 1600 \text{ gal} \times 3.79 \; \ell/\text{gal} \times 10^3 \text{ g}/\ell$$
$$\times 1 \text{ cal/g°C} \times (50 - 27)°\text{C},$$

$$Q = 1.39 \times 10^8 \text{ cal} = 5.8 \times 10^8 \text{ J}.$$

This is only enough for about one day of heating in a mild winter climate! (This does not take into account heat stored in the stone.)

There are, of course, many variations possible on a solar heating/cooling system, and we cannot discuss them in detail. Many solar homes have been built in recent years, and a great deal of information is available in popular magazines and technical journals, which are cited in the bibliography.

So far we have discussed "active" solar systems. We now turn to "passive" systems and, first, inquire about the difference between the two. It is, in fact, not always easy to separate a solar energy feature of a house from a simple architectural feature (when is a window not a solar collector?). One simple definition of passive solar is:

Passive designs consist of architectural features, components, and assemblages thereof that make use of the natural transfer of solar-generated thermal energy (i.e., without the use of fans or pumps) for the purpose of water heating, space heating, or space cooling.[10]

Passive design is not limited to architectural features, however. Such considerations as placement of the building on the lot, distribution and type of trees, using natural contours to place part of the building underground, and the like must also be taken into account. The building should face south, at least the solar interceptors of the building should. This is not always possible in today's tract homes. Trees should not block the low-angle winter sun, if possible. Deciduous trees should be planted near the house to provide summer shade, but to let through winter sunshine. Partially underground houses can take advantage of the natural insulating ability of the soil, which is considerable. Totally underground homes in some of the very hot regions of central Australia need no air conditioning at all!

The significant feature of passive design is in letting the building act as its own solar collector and in allowing natural convection to provide space heating, and in some cases cooling. Five methods of passive solar collection are shown in Figure 6.35. These designs are reasonably self-explanatory. Note that many variations and combinations can be and have been used.

An important consideration for passive as well as active design is the construction of an energy-efficient building. Spacious, high ceilings waste heat; windows should be designed with care. Southern windows, of course, provide the solar collection. Northern windows should be minimized; in fact, the north wall could be backed by earth. All windows should be double-glazed; all exterior doors weather-

stripped. Insulation is essential, not only in the attic but also in the exterior walls.

Insulation is rated in "R-values." The R-value is the reciprocal of the heat-transfer coefficient, U in Eq. 6.18, when U is expressed in British thermal units per hour per foot per degree Fahrenheit. See Table 6.21 for some representative values.

Finally, we should mention the costs of solar systems. These designs are more expensive than you might think, even passive designs. The incremental cost for a heat-absorbing wall is about $33/m² (1977) with a delivered energy cost of between $15 and $24 per GJ, depending on geographic location. This should be compared with electricity costs of about $13/GJ and Number 2 fuel oil costs of about $7/GJ (all costs in 1977 dollars). The costs for roof ponds and thermosiphoning systems are higher.

Active systems are even more expensive. However, Congress recently passed the National Energy Act, which allows a residential tax credit of 30 percent on the first $1500 and 20 percent on the next $8500 of solar energy equipment, total credit not to exceed $2150. A study has been made with and without this credit for a variety of cities, comparing solar costs with electricity and fossil fuels. Certain assumptions have to be made in a study of this kind. It is assumed that most homeowners will expect only 3 years to elapse before the fuel cost savings become greater than the expenses of the solar system, after taxes; that only 5 years should elpase before the down payment for the solar system is recovered through fuel savings; and that only 10 years should elapse before the full cost of the system is recovered. The findings of the study are given in Table 6.22. We see that with these assumptions solar heating compares favorably only with electric heating in all cities, and then only with the tax credit. It does not compare favorably with fossil fuels in any city except Grand Junction, Colorado.

Many forces can change this picture,

[10]B. N. Anderson and C. J. Michal, *Ann. Rev. Energy* **3**, 57 (1978).

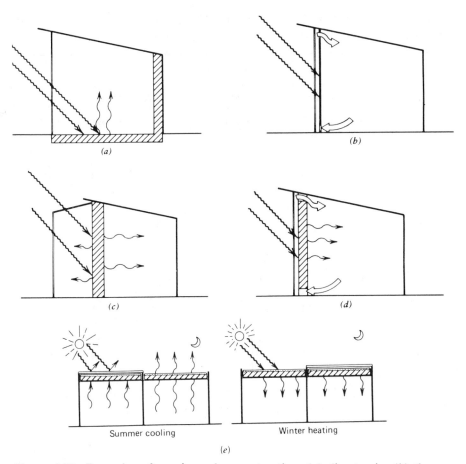

Figure 6.35 Examples of passive solar construction: (*a*) direct gain, (*b*) thermo-siphoning air collector, (*c*) solar greenhouse, (*d*) thermal storage wall, and (*e*) thermal storage roof. [Reproduced, with permission, from B. N. Anderson and C. J. Michal, "Passive Solar Design," *Annual Review of Energy* **3**, 57 (1978) © 1978 by Annual Reviews Inc.]

TABLE 6.21 Typical *R*-Values

Marble	0.05–0.07
Glass	0.15–0.2
Brick	0.15–0.3
Wet soil	0.5–1.5
Wet sand	1.0
Hardwood	2.0–3.0
Dry soil	5.0–10.0
5-cm polystyrene	8.0
Dry sand	10.0
9-cm fiberglass	11.0

however. Greater tax credits could be legislated, fossil fuel prices could go much higher (and no doubt will), and required payback periods could be lengthened. This final point should be mentioned: the assumed payback period is a flexible number. It depends on the perception of the homeowner on what is an economic proposition. Perceptions may well change in the next few years.

We cannot deal with the technical side of solar energy in a vacuum; it is unavoidably

TABLE 6.22 Comparison of Solar Hot-Water Systems to Three Others in Terms of Cost-Effectiveness of Payback Periods

City and Criterion	Alternative						Assumed Criterion
	Electricity		Fuel Oil		Natural Gas		
	Without	With	Without	With	Without	With	
Boston, MA							
Payback period, years	15	13	23	21	21	210	10
Years to recover down payment	8	1	18	12	18	14	5
Years to positive cash flow	1	1	8	8	9	9	3
Washington, DC							
Payback period, years	15	13	22	20	24	22	10
Years to recover down payment	8	1	16	1	21	18	5
Years to positive cash flow	1	1	7	7	12	12	3
Grand Junction, CO							
Payback period, years	11	9	14	12	23	21	10
Years to recover down payment	4	1	7	1	20	15	5
Years to positive cash flow	1	1	1	1	11	11	3
Los Angeles, CA							
Payback period, years	11	9	NU[a]	NU	24	23	10
Years to recover down payment	4	1	NU	NU	22	18	5
Years to positive cash flow	1	1	NU	NU	12	12	3

Source: R. Bedzek, A. Hirschberg and W. Babcock, "Economic Feasibility of Solar Water and Space Heating," *Science* **203**, 1214 (1979). Copyright © 1979 by the American Association for the Advancement of Science.
[a]NU, not utilized (fuel oil is not utilized for water or space heating in Los Angeles).

bound up with the politics of energy in general and nuclear energy in particular. Even though solar energy has been a subject of study for nearly a century, the development of its technology has taken place in an era of low fossil fuel costs. Thus, there has been little incentive for private interests to push solar energy. The burden has fallen on government and, until very recently, government has failed to rise to the challenge. A Presidential Commission pointed out in 1947 that we could have 3 million homes heated by solar energy by 1975. There were less than 100 at that time. But more than 50,000 solar homes were built in the last two years of the 1970s, with many more underway.

The reason solar research and development has never been a government priority is not difficult to understand. Despite warnings for more than two decades, many politicians would not admit to the obvious even during the height of the 1973–1974 energy crunch. Most of these who do recognize the very limited nature of fossil fuel resources now point proudly to their support of the Atomic Energy Commission in the past, and proclaim that breeder and fusion technology will solve everything.

In competition with nuclear energy (for R&D dollars) solar power has just crossed the starting line. If not for the space program of the 1960s, solar technology would not be nearly as advanced as it is today. It has only been in the past couple of years that solar energy has fully gotten its foot in the door.

A joint report by the National Science Foundation and the NASA Solar Energy Panel (*Solar Energy as a National Resource*, 1973) says that, by the year 2020, solar energy could provide 35 percent of heating and cooling in buildings, 30 percent of the nation's gaseous fuels, 10 percent of its liquid fuels, and 20 percent of its electrical requirements. The report proposed a 10-year R&D program costing $100 million to develop commercial availability of solar heating and cooling equipment. A $172 million R&D program was recommended to develop processes for producing artificial fuels from organic materials. In addition to wastes (farm and human), plants, algae, and trees might be grown specifically as renewable sources of fuel. The report suggests that generation of solar electricity will take longer to be competitive with nuclear, and will cost more in R&D. The total price tag placed on this proposal is $3.5 billion, $2 billion of which would go to costly demonstration plants.

In 1978 President Carter proposed a sweeping energy plan that would have given high priority to solar development. Unfortunately, Congress has not seen it necessary to concur. What is more, the Reagan administration has apparently decided to lower significantly government support of solar energy (see Figure 6.22). Besides the government itself, a number of other groups have a vested interest in energy policy, and must be considered in the policymaking process. By one count, there are at least eight such groups: (1) government-owned utility companies, (2) investor-owned utility companies, (3) petroleum product companies, (4) coal producers, (5) energy equipment suppliers, (6) owners of energy-producing land areas, (7) government conservation agencies and groups, and (8) private conservation groups.

Predicting the future is always a risky business at best. We shall not attempt it here. However, it should be pointed out that unforeseen events can have great impact on energy policy. It is not hard to imagine an immediate crash program to develop solar energy in the event of a catastrophic failure involving nuclear generation of electricity.

Whatever the prospects of a coherent national policy toward solar energy, skyrocketing fuel costs of the past few years ensure that it can never again be entirely pushed aside and forgotten. A final note: we quote without comment a statement by Sir George Porter of Great Britain, Nobel Prize winner in chemistry: "If sunbeams were weapons of war, we would have had solar energy centuries ago."

In this chapter we have reviewed most of the major sources of heat. Of all these potential heat sources only one, the combustion of fossil fuels, is currently economical enough to supply our demands without seriously disrupting our economy. It may be possible to develop some of these alternative sources, but the time scale is uncertain. In the meantime, a transitional heat source is required—one that is economical at the appropriate scale. Some believe our immediate—and even long-range—requirements can be satisfied by nuclear energy. Others do not. Clearly, nuclear power is here; we must look into it.

SUMMARY

The pricing of fossil fuels does not reflect their true costs, especially their environmental costs. However, combustile fuels will continue to be our most important energy source.

Since combustion usually occurs at a constant pressure, the heat liberated is just the change in *enthalpy*. Tables of enthalpy for pure substances have been tabulated, so enthalpy change calculations are straightforward.

Synthetic hydrocarbons can be produced from coal. The major problem is that the hydrogen-to-carbon ratio for coal is much smaller than that for any desirable hydrocarbon. Hydrogen must be added, usually in the form of water. Several large gasifiers have been built; these are rather inefficient and expensive.

Synthetic crude oil can also be produced, but the processes tried so far also suffer from efficiency and economic problems.

Synthetics production will tax the water supply of the West and produce several harmful by-products.

Hydrogen can be used as an energy-transport system. It has about one third the energy per unit volume as methane, but may be pumped at a higher pressure. Hydrogen has a higher range of flammability when mixed with air. Since it is lighter than air, it would rise if it leaked out of a home container. Hydrogen is produced commercially by *steam reforming* of methane or by *electrolysis*. *Chemical decomposition*, *photolysis*, and *bioconversion* could also be used. No commercially effective process for one of these has yet been discovered.

Alcohols, particularly *methanol* and *ethanol*, can be produced from forests or from agricultural by-products. Alcohols can be used in any situation where gasoline or methane is now used. Mixtures of methanol and gasoline, called *gasohol*, are now available but are more expensive than the gasoline by itself.

Alcohol-burning engines would produce fewer pollutants and would get better gas mileage than those burning gasoline. However, the technology does not today exist to make alcohol fuels economically competitive in this country.

In the United States wastes are produced at the rate of 2 kg per person per day. Much of this waste could be used directly or converted to a combustible fuel. Reasonably high-grade oil can be made from domestic wastes; agricultural wastes can be used to produce methane. These have been done on small scales, but no large plant has yet been able to operate economically.

Substantial reserves of heat exist near the earth's surface. These can be tapped in certain localities to provide steam that is hot enough to use in a turbine-electric generation system. Of the several types of *geothermal* reservoirs only the dry-steam and liquid-dominated types have been exploited to date.

There are environmental concerns with geothermal energy. The steam or brine is usually high in mineral content. Often noxious gases are emitted. The subsurface geology is usually not well known. Steam withdrawal or reinjection can cause instabilities.

Heat from the sun can and is being used. The sun is an intermittent source that varies in strength hourly, daily, and seasonally.

All objects above absolute zero will radiate energy according to the *Stefan–Boltzmann radiation law*. The emissivity (absorptivity) varies with substance and wavelength.

The *greenhouse effect* refers to the trapping of long-wavelength radiation by materials that are opaque at those wavelengths but transparent at short wavelengths.

The *power tower* is a method of concentrating sunlight in a collector to achieve high temperatures with relatively little losses. Mirrors are used to reflect the light to the top of a tower. The tower projects built so far have not been economic successes.

The opposite to the power tower is a distributed collecting system, called a *solar farm*. Heat is collected in each mirror and piped to a central steam generator. Heat-transport losses would be higher, but the capital costs would be lower. Such a system has not been built.

It is possible to use the ocean surface as a collector. This is the *ocean thermal energy conversion (OTEC)* plant proposal. This makes use of the temperature difference between the ocean surface and lower depths to operate a heat engine.

Small pilot plants have operated successfully, but it is not clear how to scale up to a very large plant. Thermal efficiencies are very low, and large quantities of water need to be processed.

Sunlight can also be used to heat homes. This can be a very economical process, depending on the location and construction of the home. Two systems are in use: *active solar* and *passive solar*. Passive homes are more likely to be economical than active ones at today's energy prices without substantial tax credits.

Flat-plate collectors are usually used in active systems. Special construction is needed to reduce heat losses.

REFERENCES

Chemical Thermodynamics

Hamill, William H., and Russell R. Williams, Jr., *Principals of Physical Chemistry*. Englewood Cliffs, N.J.: Prentice-Hall, 1959.

Synthetic Hydrocarbons

Hammond, A. L., W. D. Metz, and T. H. Maugh, *Energy and the Future*. Washington, D.C.: American Association for the Advancement of Science, 1973.

Linden, H. R., W. W. Bodle, B. S. Lee, and K. C. Vyas, "Production of High-Btu Gas from Coal," *Ann. Rev. Energy* 1, 65 (1976).

Morris, S. C., D. D. Moskowitz, W. A. Savian, S. Silberstein, and L. D. Hamilton, "Coal Conversion Technologies: Some Health and Environmental Effects," *Science* 206, 654 (1979).

Perry, Harry, "Coal Liquefaction," in *Physics and the Energy Problem—1974*, Fiske and Havens (Eds.). New York: American Institute of Physics, 1974.

Perry, Harry, "The Gasification of Coal," *Sci. Amer.* 230(3), 19 (1974).

Squires, Arthur M., "Chemicals from Coal," *Science* 184, 340 (1974).

Hydrogen

Dickson, Edward M., John W. Ryan, and Marilyn H. Smolyan, *The Hydrogen Economy*. New York: Praeger, 1977.

Gregory, Derek P., "The Hydrogen Economy," *Sci. Amer.* January (1973), p. 13.

Gregory, D. P., and J. B. Pangborn, "Hydrogen Energy," *Ann. Rev. Energy* 1, 279 (1976).

Jones, Lawrence W., "Liquid Hydrogen as a Fuel for the Future," *Science* 174, 367 (1971).

Wentorf, R. H., Jr., and R. E. Hanneman, "Thermochemical Hydrogen Generation," *Science* 185, 311 (1974).

Winsche, W. E., K. L. Hoffman, and F. J. Salzano, "Hydrogen: Its Future Role in the Nation's Energy Economy," *Science* 180, 1325 (1973).

Methanol

Chambers, R. S., R. A. Herendeen, J. J. Joyce, and P. S. Penner, "Gasohol: Does It or Doesn't It Produce Positive Net Energy?" *Science* 206, 780 (1979).

Hill, Ray, "Alcohol Fuels—Can They Replace Gasoline?" *Pop. Sci.*, March (1980), p. 25.

Reed, T. B., and R. M. Lerner, "Methanol: A Versatile Fuel for Immediate Use," *Science* 182, 1299 (1973).

Wigg, E. E., "Methanol as a Gasoline Extender: A Critique," *Science* 186, 785 (1974).

Fuel from Wastes

Blum, S. L., "Tapping Resources in Municipal Solid Waste," *Science* 191, 669 (1976).

Colueke, Clarence G., and P. H. McGanhey, "Waste Materials," *Ann. Rev. Energy* **1**, 257 (1976).

Kasper, William C., "Power from Trash," *Environment*, February (1974), p. 34.

Schneider, Jason, "Solid Fuel Alternatives," *Pop. Sci.*, January (1980), p. 120.

Solar Power

Anderson, Bruce N., and Charles J. Michal, "Passive Solar Design," *Ann. Rev. Energy* **3**, 57 (1978).

Dugger, Gordon L., Evans J. Francis, an Wiliam H. Avery, "Technical and Economic Feasibility of Ocean Thermal Energy Conversion," *J. Solar Energy* **20**, 259 (1978).

Hildebrandt, Alvin F., and Lorin L. Vant-Hull, "Power with Heliostats," *Science* **197**, 1139 (1977).

Kreith, Frank, *Principles of Heat Transfer.* Scranton: International Textbook Company, 1958.

McDaniels, David K., *The Sun: Our Future Energy Source.* New York: John Wiley, 1979.

Meinel, Adam Baker, and Marjorie Pettit Meinal, "Physics Looks at Solar Energy," *Phys. Today*, February (1972), p. 44.

Metz, William D., "Ocean Temperature Gradients: Power from the Sea," *Science* **180**, 1266 (1973).

Morse, Frederick H., and Melvin K. Simmons, "Solar Energy," *Ann. Rev. Energy* **1**, 131 (1976).

Sellers, William D., *Physical Climatology.* Chicago: University of Chicago Press, 1965.

Solar Energy, a monthly journal published by Pergamon Press.

Zener, Clarence, Solar Sea Power, *Phys. Today*, January (1973), p. 48.

Geothermal Energy

Armstead, H., and H. Christopher, *Geothermal Energy.* London: W. & F. N. Spon, 1978.

Cummings, Ronald G., Glen E. Morris, Jefferson W. Tester, and Robert L. Bivins, "Mining Earth's Heat: Hot Dry Rock Geothermal Energy," *Tech. Rev.*, February (1979), p. 58.

Kruger, Paul, "Geothermal Energy," *Ann. Rev. Energy* **1**, 159 (1976).

GROUP PROJECTS

Project 1. Interview as many gasohol users as possible. Compare their experiences with a comparable group of gasoline users. Are there any obvious advantages? disadvantages? Any messages for the rest of us?

Project 2. There are probably several active or passive solar homes in or near your community. Find out about them, their owners' experiences, economics, and so forth.

Project 3. For computer enthusiasts—calculate the available solar energy for your location. See George E. Mobus, "Harvesting the Sun's Energy," *Byte*, July (1981), p. 48.

EXERCISES

Exercise 1. Calculate the natural gas required (in cubic meters) and the heat wasted to raise the temperature of 189ℓ (50 gal) of water from 25° to 100°C each of two ways:

(a) Directly, as in a home water heater ($\eta = 0.62$).

(b) Electrically, with $\eta = 0.38$ for the conversion of heat to electricity at a central generating plant and $\eta = 1.0$ for the conversion of electricity to heat at the home.

Exercise 2. Calculate the heat of combustion (ΔH) of methane at 25°C; at 100°C.

Exercise 3. A particular coal is composed of 88 percent carbon, 6 percent hydrogen, 4 percent oxygen, 1 percent nitrogen, and 1 percent sulfur (by mass). Complete combustion is represented by:

$$C_w H_x O_y N_z S_u + e O_2$$
$$\rightarrow a CO_2 + b H_2O + c SO_2 + d NO_2,$$

where u, w, x, y, and z can be determined from the above information. Determine a, b, c, d, and e per tonne of coal burned.

Exercise 4. Calculate the heat of combustion of ethyl alcohol with oxygen at 25°C.

Exercise 5. A catalytic methanation chamber has a volume of $20 \, m^3$; a stoichiometric mixture of CO and H_2 is fed into the chamber at a total pressure of 100 atm and a temperature of 200°C. Assume the chamber is thermally insulated and that the reaction proceeds to completion with no back reaction. It is desired to maintain the temperature in the chamber below 300°C; how much heat must be removed by a cooling system?

Answer: 6.66×10^8 J per chamber filling.

Exercise 6. Use the data in Section 6.1 on costs for SNG to construct a model using the parameters given and any others you think relevant. Predict SNG costs for these situations:

(a) Minemouth coal in Wyoming; gas to Chicago (water transportation required).
(b) Wyoming coal in Chicago.
(c) Eastern coal in the East.

Exercise 7. The criteria for thermochemical water splitting cycles are

(a) The sum of ΔH^0 for the individual steps should equal (or exceed) 68 kcal/mol, ΔH for water.
(b) ΔG^0 for each step should be zero or a small negative number. (Why?)

Does the Mark I process fulfill these requirements?

Exercise 8. Using the Mark I process a plant processes $500{,}000 \, ft^3$ of H_2 per day. If 0.01 percent of the Hg is lost in the process, what is the total amount of Hg list per year in kilograms? in dollars?

Exercise 9. The autoignition temperature of gasoline is 247°C. Assume this to be the high temperature, and calculate the efficiency based on a Carnot cycle (use room temperature for the low temperature). Assume this value of efficiency to be equivalent to 100 octane. Repeat for hydrogen, and assume a linear relationship between efficiency and octane. What is the octane rating of hydrogen? Repeat for methanol.

Exercise 10. There are about $2 \times 10^{12} \, m^2$ of commercial forests in the United States intercepting annually $5.8 \times 10^9 \, J/m^2$ of solar energy. The conversion efficiency of solar-to-biomass energy is about 0.5 percent, from biomass-to-methanol about 10 percent, and from methanol to electricity 30 percent. If the forests have a replacement time of 50 years, what percentage of our electricity could be supplied by methanol-fueled generating plants?

Answer: 3.48×10^{16} J available; <0.1 percent electricity requirement.

Exercise 11. Assume a flow rate in a geopressurized resource of $3 \times 10^6 \, \ell/day$ at 26 MPa. How much power could be generated if all this fluid were used to turn a turbine and the exit pressure were 19 MPa?

Exercise 12. If the pressure of the fluid in a geothermal well changes from 6.2 to 2.8 MPa adiabatically and the fluid is initially a saturated liquid at 277.77°C, what is the character of the fluid at the top of the well?

Exercise 13. Figure 6.15 is asymmetric with respect to solar flux in the northern and southern hemispheres; in fact, the southern seems to have a higher total insolation. Why? It is a fact, though, that the southern hemisphere is on the average cooler than the northern. Again, why?

Exercise 14. One June 21 an observer at the North Pole will see the sun at 23.5° above the horizon *all day*. Use Eq. 6.16 to justify this statement.

Exercise 15. Using Eq. 6.15 and information from other references as required, compare the approximately annual solar insolation at Phoenix, Arizona, and Minneapolis, Minnesota. Relate sun availability to f in Eq. 6.15.

Exercise 16. Show that the amount of direct solar radiation received by a tilted surface is given in terms of the amount received by a horizontal surface by

$$I(\theta) = I_H \frac{\cos(\theta - \theta_T)}{\cos \theta},$$

where θ is the sun angle with respect to the zenith, and θ_T is the tilt angle. The zenith angle can be approximated with this equation:

$$\theta = \theta_L + 23.5 \times \cos \left[(N + 10) \frac{360}{365} \right],$$

where θ_L is the latitude, and N is the number of days elapsed since January 1. Use these results and average over a 24-hr period to find the value of I for a worst case, say January 31, at your latitude.

Exercise 17. Find the head required for a hydroelectric plant such that the energy given up by the water used to heat the working fluid in the OTEC system in Figure 6.27 is equivalent to the energy given up by the water in the hydroelectric plant. Compare the amount of electricity generated for a given quantity of water by each method.

Answer: 854 m; 8.37 kJ/kg hydroelectric, 0.28 J/kg OTEC.

Exercise 18. Calculate the heat absorption from a solar collector at your location in mid-winter for a system having two cover plates with $U = 0.008$ J/s m °C. Make the collector large enough to supply 10 MJ/day.

Exercise 19. A wall 4.5×2.4 m has three windows each 0.9×1.2 m, single glazed. The wall has 10 cm of polystyrene insulation. The windows are 0.5 cm thick. If the room is kept at 21°C and the outside temperature is −18°C, how much heat is lost through this wall?

Exercise 20. Design an active solar system for a 180-m^2 house at your location. Determine tilt angle, flow rates, storage medium, and quantity. Take into account heat losses through walls, floor, and ceiling.

CHAPTER SEVEN

NUCLEAR POWER

7.1 INTRODUCTION

Walk into any room today and say, "Nuclear Power!" and, unless the room is empty, you will probably start a vigorous discussion. The public perception of nuclear energy ranges from "the solution to the energy problem" through "just another power source" to "an evil foisted upon us by money-hungry utilities." The government's position has varied from "Atoms for Peace" in the 1950s, to "Project Independence" in the early 1970s in which nuclear power would more than double, to "proceed with caution" with worries about nuclear proliferation characteristic of the early 1980s. The stance of the electric utilities has gone from an enthusiastic embracing of the new technology in the early 1960s to the present rather cool attitude with worry about future fuel supplies.

Nuclear Power Policy

Nuclear technology has benefited for more than 30 years from federal policies designed to make it succeed. This has been a historical accident as much as a plan. The U.S. Atomic Energy Commission (AEC) was formed on December 31, 1946, to place under civilian management the nuclear weapons facilities that had been operated by the Army during World War II. The development of civilian power reactors was a natural offshoot of the nuclear submarine program. The AEC financed the bulk of the research required, much of it at the National Laboratories, which needed to diversify their programs since nuclear weaponry research had been scaled down considerably after the war.

The first commercial power reactor became operational at Shippingport, Pennsylvania, on December 2, 1957. In subsequent years many reactors were built, averaging three per year through 1979. Licensing of nuclear reactors was the responsibility of the AEC—indeed, any facility, process or operation using "special nuclear materials"[1] had to be licensed by the AEC. This dual aspect of AEC responsibility—funding research and licensing—produced an internal conflict of interest in the AEC that went unnoticed for some time. With the advent of large reactors (1100 MWe) in the mid-1960s, public uneasiness about nuclear energy, which had been diffuse and unorganized, became crystallized and vocal. Several organizational changes were made in the AEC in the early 1970s, but by the time of the OPEC oil embargo (fall 1973) it was clear that more serious "surgery" was required.

[1]"Special nuclear materials" include radioisotopes and fissile material.

The AEC was dismantled and replaced by the U.S. Energy Research and Development Agency (ERDA) and the Nuclear Regulatory Commission (NRC) in 1974. All responsibility for funding research was lodged in ERDA, while the NRC was designated the reactor-licensing authority. In the first ERDA budget (1975), nuclear power was given 56 percent of the total, which was $1.1 billion.

ERDA's lifetime was short. There was considerable public and congressional dissatisfaction with the rate of development of energy solutions. As a consequence, on October 1, 1977, the cabinet-level Department of Energy (DOE) came into being, replacing ERDA. The NRC, however, remained an independent commision. In the 1979 DOE budget, the amount devoted to nuclear energy was $885.5 million.

This short review of the history of federal nuclear policy indicates the extent of federal support for nuclear energy. This being the case, it is difficult to understand why the growth of nuclear power technology has slowed significantly. One factor is public opinion.

It comes as no real surprise that public perceptions differ so markedly. Nuclear power is a complicated subject; there *are* real benefits and real risks. But "experts" disagree on which outweigh the others. In this chapter we try to present the technical features of nuclear power operation and point out where the areas of concern lie.

Fission

We know from various experiments over the years that, while the diameter of the average atom is about 10^{-7} cm, the mass and positive electric charge of the atom are concentrated in the **nucleus**, a region about 10^{-12} cm in diameter. The normal atom is electrically neutral; for each positive electric charge carried by **protons** in the nucleus there is a cor-responding negative charge, an electron, in the extranuclear part of the atom. The chemical identity of an atom is determined by the number of electrons—and consequently the number of protons. In a chemical reaction the number of electrons associated with an atom will in general change; if the proton number ever changes (and it can!), then the identity does change. The proton number Z is most often referred to as the **atomic number**. Another physical characteristic of the nucleus is its mass. To measure mass let us choose a unit system in which the mass of the carbon atom is exactly 12 units. An **atomic mass unit**, abbreviated u, is defined as 1/12 the mass of the carbon atom; $1 u = 1.6598 \times 10^{-27}$ kg. In this system the mass of the hydrogen atom—that is, one proton and one electron—is very nearly $1 u$. Electrons have about 1/2000 the mass of the proton; they are assigned an atomic mass number of 0. Protons and electrons together cannot account for all the nuclear mass. A large part of it is carried by **neutrons**, electrically neutral particles having about the same mass as protons. Neutrons and protons are the only constituents of nuclei; they are given the collective name **nucleons**. The total number of nucleons in a nucleus is the **mass number** A.

The existence of neutrons makes it possible for the nuclear masses of two atoms to be different while their nuclear (and electronic) charges are the same. These two atoms would behave the same chemically; such atoms are called **isotopes**. All elements have isotopes; most of these are unstable; that is, they change identity by radioactive emission. Many elements have at least two stable isotopes, for example, ^{3}He and ^{4}He. The superscript is the atomic mass number–the number of protons plus neutrons. Tin has 10 stable isotopes. Some elements have only one stable isotope, like gold: ^{197}Au. Two elements, technetium and promethium, have no stable isotopes and are not found in nature. Naturally occurring uranium is a mixture of three isotopes: 0.006

percent ^{234}U, 0.711 percent ^{235}U, and 99.283 percent ^{238}U.

A given nucleus is unstable because the ratio of the number of protons to the number of neutrons in it is not correct for its total mass number. The appropriate ratio varies between about 1.0 for $1 \lesssim A \lesssim 40$ to about 0.6 at $A = 200$, where A is the mass number. Above about $A = 200$, all nuclei are unstable. Nuclear decay is an attempt to bring this ratio to its nearest best value for the given mass. If a nucleus has too much charge for a given mass, then a β^+ (positive electron, or **positron**) may be emitted. For example, the proton-to-neutron ratio for $^{11}_{6}$C is too high for stability (1.6 compared to 1.0). The subscript 6 here is the atomic number of carbon. In this case the nuclear decay would be in a manner to decrease this ratio; one of the protons bound in the nucleus would convert to a neutron:

$$p \rightarrow n + \beta^+ + \nu, \qquad (7.1)$$

where the ν is a neutrino.[2] The resulting decay would be

$$^{11}_{6}\text{C} \rightarrow {}^{11}_{5}\text{B} + \beta^+ + \nu. \qquad (7.2)$$

For completeness we should note that the half-life of ^{11}C is 20.5 min.

Half-life $(\tau_{1/2})$ is the time required for half the nuclei in a given sample to undergo radioactive change. Note that the time when any individual nucleus will decay cannot be predicted, since nuclear decay is a *random* process. Half-life is closely related to the idea of doubling time discussed in Section 1.1.

Example 7.1 An object had 100 μg of ^{14}C in the year 200 BC. How much did it have in 1980? ($\tau_{1/2}$ for ^{14}C is 5730 yr.)

[2]The neutrino and its companion antineutrino $\bar{\nu}$ are very light, probably massless, uncharged objects the existence of which, though long doubted by many, is now firmly established by experiments. Neutrinos were first postulated to account for the energy distribution of the electrons in beta decay.

Equation 1.5 is applicable. $C(t) = C(t = 0) e^{mt}$, in this case $m = -0.693/5730$. Note the negative sign.

$$C(1980) = 100 e^{-0.000121 \times 2180};$$

$$\boxed{C(1980) = 76.8 \ \mu\text{g}.}$$

A nucleus with the wrong charge-to-mass ratio for stability, for example $^{11}_{4}$Be with a proton-to-neutron ratio of 0.57, would decay by converting one of its neutrons:

$$n \rightarrow p + \beta^- + \bar{\nu}, \qquad (7.3)$$

where the β^- is simply an electron. For $^{11}_{4}$Be the resulting decay would be

$$^{11}_{4}\text{Be} \rightarrow {}^{11}_{5}\text{B} + \beta^- + \bar{\nu}. \qquad (7.4)$$

In the heavier mass region excess positive charge may also be reduced by the emission of an alpha particle (the ^{4}He nucleus):

$$^{203}_{83}\text{Be} \rightarrow {}^{199}_{81}\text{Tl} + \alpha. \qquad (7.5)$$

All nuclei beyond lead in the periodic table ($Z > 82$) are radioactive. For the nuclear equations 7.2, 7.4 and 7.5 to balance, two rules must be followed:

1. The total *mass number* before must equal the total *mass number* after.
2. The total *electric charge* before must equal the total *electric charge* after.

Note that the atomic number of the electron is -1.

In many cases of radioactive decay the emitted particles are accompanied by gamma rays. These are simply very-high-frequency photons that are used by the nucleus to rid itself of excess energy in much the same way in which atoms use visible-light photons in de-excitation processes. These three "rays," α, β, and γ rays, are collectively called **radioactivity**. Note that the "free" proton, that is, a proton not bound into a nucleus, is believed to be stable, but that the "free" neutron, such as is found in a reactor, is not. The neutron decays as in Eq. 7.3 with a half-life of about 780 s.

The $A = 235$ isotope of uranium has a property that is unique among naturally occurring isotopes: upon absorbing a low-energy neutron, the resulting nucleus ($A = 236$) splits or fissions. Symbolically, this process is written

$$^{235}U + n \rightarrow (^{236}U) \rightarrow {}^{A}_{Z}X + {}^{A'}_{Z'}Y + xn + E. \quad (7.6)$$

The letters X and Y represent product nuclei having charges Z and Z' and atomic mass numbers A and A'. Note that $Z + Z' = 92$. The sumbol xn indicates that x neutrons are given off during the fission process, so that $A + A' + x = 236$. The actual mass is not conserved; rather, part is converted to energy. The E indicates excess energy, chiefly in the form of kinetic energy of the fission products and neutrons; for ^{235}U, E has an average value of about 200 MeV (3.2×10^{-11} J). This energy is a result of the conversion of some nuclear mass to energy using the familiar equation from general relativity $E = mc^2$. The sum of the masses of the two product nuclei, ${}^{A}_{Z}X$ and ${}^{A'}_{Z'}Y$ is always less than the masses of the original ^{235}U and the neutron.

It is in general true that the mass of any stable nucleus is less than the sum of the masses of its constituents, as we see from the following example.

Example 7.2 Compare the mass of ^{4}He with the mass of its constituents.

$$2 \times m_{proton} = 2 \times 1.00782\ u = 2.01564\ u$$
$$2 \times m_{neutron} = 2 \times 1.00867\ u = \underline{2.01734\ u}$$
$$4.03298\ u$$
$$m_{4_{He}} = 4.00260\ u$$
$$\text{difference} = 0.03038\ u.$$

Since $1\ u = 9.31502 \times 10^8$ eV, this mass difference is equivalent to 28.3 MeV.

The difference between the actual mass and the constituent mass of an atom is called the **binding energy** (B). This is defined as the energy that would be released if the *atom* (not just the nucleus) were to be synthesized from the appropriate number of neutrons and hydrogen atoms. The binding energy may be calculated by

$$B = (M_A - ZM_H - [A - Z]M_n) \times 931.502\ \text{MeV}, \quad (7.7)$$

where M_A, M_H and M_n are the exact masses of the atom in question, the hydrogen atom and the neutron, respectively. A plot of binding energy per nucleon (B/A) versus mass number, Figure 7.1, gives some interesting insight.

We see that nuclei having mass numbers in the 50-to-60 range have the highest B/A. For both lighter and heavier nuclei the B/A is smaller. This means that very light and very heavy nuclei have a smaller difference between actual mass and constituent mass than do medium-mass nuclei. For example, if $A = 236$, the B/A is about 7.6 MeV or about 1794 MeV total. For masses in the region of $A \sim 115$, the B/A is about 8.5 MeV or about 977 MeV total for one nucleus and 1955 for two. Since the missing mass is smaller for heavy and light nuclei, energy must be released (more missing mass) when a heavy nucleus fissions or two light nuclei fuse together. Thus, 1 g of fissioning ^{235}U (2.56×10^{21} atoms) releases 8.19×10^{10} J of energy; this is equivalent to the heat of combusion of 2.7 t of coal or 13.7 barrels of crude oil.

The average number of neutrons given off in a single ^{235}U fission is 2.5. If these additional neutrons could be made to produce other fissions, a sustaining series of fission events could be produced. Each event would produce ~ 200 MeV of excess energy. This process is called a **chain reaction**. If it happens very quickly, say in a few microseconds, then a bomb results. If it happens in a controlled fashion with the chain reaction just barely

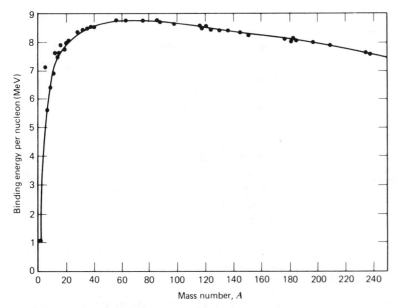

Figure 7.1 The binding energy per nucleon versus the mass number for several stable isotopes.

being maintained, then a nuclear reactor results.

The chain reaction will not continue if the energy of the neutrons being absorbed is above about 0.1 keV (1 keV = 1000 eV). Neutrons having energies below this value are called **thermal neutrons**. This designation indicates these neutrons have energies low enough to be comparable to the average vibrational energies of the molecules making up the interior of the reactor. These vibrations are what we usually associate with thermal energy. Reactors operating with neutrons in this energy range are called thermal reactors. The energies of neutrons emitted during fission are of the order of a few MeV. Some mechanism for slowing the neutrons down is required. In a typical reactor, material called a moderator is inserted between or interspersed among rods loaded with the fissionable fuel. The best moderator would be one that has only the effect of slowing the neutrons and does not capture any. Excessive capture would reduce the available

neutron flux and make it difficult to maintain the chain reaction. Graphite, light and heavy water,[3] and helium gas, among others, have been used as moderators. Of these, the most commonly used in power-generating reactors is light water.

CLASSIFICATION OF REACTORS

Several hundred nuclear reactors of various types have been built since the first one was completed under the bleachers at Stagg Field at the University of Chicago on December 2, 1942. Only a rather small percentage of these have been commercial power reactors. Table 7.1 lists the various properties by which reactors have been classified.

There are only two types of power reactors currently available commercially in the United States: the pressurized water reactor (PWR)

[3]Heavy water, D_2O, is water formed with the isotope of hydrogen having one proton and one neutron in the nucleus.

TABLE 7.1 Classification of Reactors

Energy of neutrons that produce fission
  Fast
  Intermediate (or epithermal)
  Thermal
Nuclear fuel
  Natural U (0.7 percent ^{235}U)
  Slightly enriched U (1 to 2 percent ^{235}U)
  Highly enriched U (>90 percent ^{235}U)
  ^{239}Pu
  ^{233}U
Method of heat removal, by circulation of
  Coolant only
  Fuel mixed with coolant
  Moderator–coolant
  Fuel, moderator, and coolant
Purpose
  Research
  Prototype
  Propulsion
  Heat source
  Electric power generation
  Isotope production
Arrangement of fuel and moderator
  Heterogeneous
  Homogeneous
Materials used in the following reactor components
  Moderator
  Coolant
  Structure
  Reflector
  Shield

and the boiling water reactor (BWR). A high-temperature gas reactor (HTGR) was developed, but has not been a commercial success in this country. The Canadian government has developed the Canadian Natural Uranium–Deuterium Reactor (CANDU), which has distinct advantages (and disadvantages) over the PWR and the BWR. The liquid-metal fast-breeder reactor (LMFBR) is under development; a small pilot plant was built in Michigan (the Enrico Fermi Plant, built in 1963), but it never operated entirely satisfactorily and has been shut down. The PWR and BWR use water for both coolant and moderator. They are often collectively called light water reactors (LWRs).

A variety of research reactors have been built in the past 40 years. These are usually small (<1 MW), water-cooled, high-enrichment reactors used for producing neutrons for nuclear studies or solid-state physics work.

Other countries have several different designs. Notable among these are the British magnox reactors, CO_2-cooled and graphite-moderated, and the French Phenix and Superphenix liquid-metal breeder reactors. Figure 7.2 gives some information about the distribution of reactors around the world. We shall examine the operation of all these designs, but we start with the U.S. thermal reactor, the LWR.

7.2 THERMAL FISSION REACTORS

On the average, a thermal neutron-induced fission of a ^{235}U nucleus produces 190 MeV of energy. That energy is distributed as follows:

167 MeV as kinetic energy of the fission fragments.
  5 MeV as kinetic energy of fast neutrons.
  7 MeV as prompt gamma rays.
  5 MeV from betas from decaying fission products.
  6 MeV from delayed gammas from fission products.
  11 MeV as kinetic energy of beta-decay neutrinos.

The 167 MeV and the 5 MeV from betas are absorbed almost entirely in the fuel elements; the fast neutrons lose most of their energy in the moderator. The energies of the gammas, both prompt and delayed, are distributed among the fuel, moderator, and reactor structure in a manner that is not straightforward to calculate, since gamma rays do not lose energy in the same manner as charged particles. The neutrino energy is lost because they escape the reactor (and probably everything else on

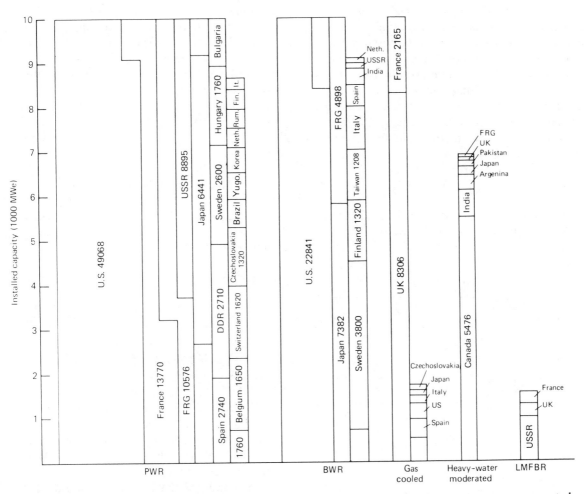

Installed capacity (1000 MWe)

10 — 9 — 8 — 7 — 6 — 5 — 4 — 3 — 2 — 1

PWR

- U.S. 49068
- France 13770
- FRG 10576
- USSR 8895
- Japan 6441
- Sweden 2600
- DDR 2710
- Spain 2740
- 1760
- Belgium 1650
- Switzerland 1620
- Czechoslovakia 1320
- Brazil
- Yugo.
- Korea
- Neth.
- Rum.
- Fin.
- It.
- Hungary 1760
- Bulgaria

BWR

- U.S. 22841
- Japan 7382
- Sweden 3800
- FRG 4898
- Finland 1320
- Taiwan 1208
- Italy
- Spain
- Neth.
- USSR
- India

Gas cooled

- UK 8306
- France 2165
- Czechoslovakia
- Japan
- Italy
- US
- Spain

Heavy-water moderated

- Canada 5476
- India
- FRG
- UK
- Pakistan
- Japan
- Argenina

LMFBR

- USSR
- France
- UK

Figure 7.2   1980 worldwide nuclear generating capacity, by reactor type. Only operating reactors counted.

earth!). Using the above figures it is not difficult to calculate that 3.3×10^{10} fissions per second are required to generate one thermal watt.

Example 7.3 In a 1000-MWe reactor operating at 32 percent efficiency, how many fissions occur in 1 yr of full power operation?

The required thermal power is $1000/0.32 = 3125$ MWt. Each year this corresponds to

$$(3125 \times 10^6 \text{ J/s})(3.16 \times 10^7 \text{ s/yr})$$
$$= 9.875 \times 10^{16} \text{ J/yr},$$

and since $1 \text{ MeV} = 1.6 \times 10^{-13} \text{ J}$ and 190 MeV/fission,

Number of fissions/yr

$$= \frac{9.875 \times 10^{16} \text{ J/yr}}{(190 \text{ MeV/fission})(1.6 \times 10^{-13} \text{ J/MeV})},$$

$$\boxed{= 3.25 \times 10^{27} \text{ fissions/yr.}}$$

The most probable neutron energy after ^{236}U fission is about 2.5 MeV. We have said that neutrons of this energy are not as likely to produce fissions as neutrons of much lower energies. We need a quantitative means of expressing this "fission likelihood." For this purpose we define the **microscopic fission cross section**. If I neutrons/s impinge on a target material t cm thick containing N nuclei/cm^3, and fission of the target nuclei occurs at a rate F fissions/s, then the microscopic fission cross section σ_f is

$$\sigma_f = \frac{F}{NtI} \text{ cm}^2/\text{nucleus}. \qquad (7.8)$$

Fission cross sections are usually in the range 10^{-22} through 10^{-26} cm^2; the value 10^{-24} cm^2 has been given the name **barn**.[4]

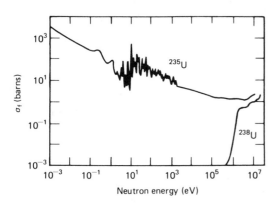

Figure 7.3 Fission cross section in ^{235}U and ^{238}U as a function of reactor energy. Averaged over many data points. [Data from *Neutron Cross Sections*, Vol. II (BNL 325), edited by D. I. Garber and R. R. Kinsey. Upton: Brookhaven National Laboratory, 1976.]

---

[4]The barn is not an accepted SI unit; neither is the term "fermis squared" (fm^2) (1 fm^2 = 10^{-26} cm^2). Both of these terms are used a great deal in nuclear physics and, even though there has been considerable impetus toward maintaining consistent use of SI units at least outside of the United States in other branches of physics, there has been no detectable movement toward adopting SI units for nuclear cross sections anywhere. Therefore, reluctantly, we use this nonstandard, but metric, unit, the barn.

Example 7.4 If a neutron flux of 10^{13} neutrons/cm^2 s impinges on a 1.0-mg/cm^2 foil of pure ^{235}U metal 10 cm^2 in area and the cross section for fission induced by these neutrons is known to be 100 barns, what is the fission rate?

From the definition above, the product Nt has the dimensions of nuclei/cm^2. The foil in question is characterized by an *areal density* ξ. We can convert areal density to nuclear density:

$$\text{Nuclei/cm}^2 = (\text{g/cm}^2) \times \frac{N_0}{A},$$

where N_0 = Avogadro's number (6.02 × 10^{23} nuclei/mol) and A is the atomic mass number of ^{235}U. Therefore

$$F = \frac{\sigma a \xi N_0 I}{A}$$

$$= \frac{100 \times 10^{-24} \times 10 \times 1 \times 10^{-3} \times 6.02 \times 10^{23} \times 10^{13}}{235},$$

$$\boxed{F = 2.56 \times 10^{12} \text{ fission/s.}}$$

In Figure 7.3 the microscopic fission cross section for ^{235}U and ^{238}U is shown as a function of neutron energy. We can see that neutrons of energy 0.03 eV are 1000 times more likely to induce fission than those of 3.0 MeV. This is clear indication of the need for moderation in LWRs.

Multiplication Factor

Life is unfortunately not so simple; there are many processes affecting the number of neutrons available for fission and their energies. The first of these is evident from Figure 7.3; there is a reasonable probability that the fast neutrons will induce fission in ^{238}U. But at the same time, additional neutrons are released in the fast fission, so the net result is an increase in fast neutrons. In order for the fission process to be self-sustaining, the neutrons

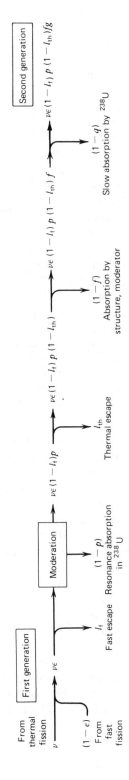

Figure 7.4 Schematic representation of chain reaction based on the fission of uranium nuclei by thermal neutrons.

released in one fission must produce at least one other fission. Let us examine all the factors that affect the neutrons in a reactor to see if we can relate the *second generation* of neutrons to the *first generation* in a simple fashion. A caution before we proceed: we are forced to discuss these effects serially (one after another). In fact these effects occur in parallel (Figure 7.4).

Let ν be the number of first-generation fast neutrons from thermal neutron-induced fission; this number is increased by the *fast fission factor* ϵ, about 1.03. That is, about 3 percent of the fission neutrons come from fast-neutron-induced fission. A fraction l_f, called the fast-neutron leakage factor, of the fast neutrons escapes the reactor without interacting in any way; of the original number, then, $\nu\epsilon(1 - l_f)$ are moderated. During the moderation process, as the neutron energies drop from the MeV range to the sub-eV range, another "neutron sink" is encountered. Figure 7.5 shows the total cross section for neutrons on ^{238}U as a function of neutron energy; there are several strong, narrow peaks. These are called resonant absorption peaks; the one at about 6.7 eV has a maximum cross section of about 8000 barns, 15 times the ^{235}U cross sections at the same neutron energy.

In physics a **resonance** is said to occur when a driving force and the driven system have the same frequencies. During resonance, there is a very large exchange of energy between the driver and the driven system. In nuclear physics a resonance is said to occur when there is an unusually large interaction between an incident particle flux and a target nucleus, usually apparent because of a large cross section for that interaction.

Let p, called the *resonance escape probability*, be the fraction of neutrons that are *not* absorbed by ^{238}U during moderation; then $p\nu\epsilon(1 - l_f)$ neutrons remain. Of these a fraction l_{th}, called the slow-neutron leakage factor, escape the reactor, and a fraction $(1 - f)$ are

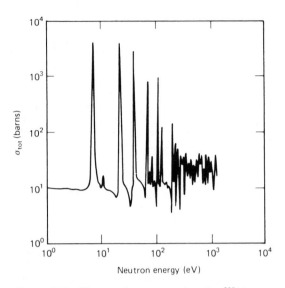

Figure 7.5 The total cross section for ^{238}U versus neutron energy. The cross section between resonances is mostly due to elastic scattering. Averaged over many data points. [Data from *Neutron Cross Sections*, Vol. II (BNL 325), edited by D. I. Garber and R. R. Kinsey. Upton: Brookhaven National Laboratory, 1976.]

absorbed by the moderator and structural parts of the reactor. Several absorption cross sections are listed in Table 7.2. This leaves $\nu\epsilon(1 - l_f)p(1 - l_{th})f$ neutrons; of these $(1 - g)$ are absorbed in ^{238}U, eventually producing plutonium, or in ^{235}U to produce ^{236}U. Hence, of the original ν first-generation neutrons, $\nu\epsilon(1 - l_f)p(1 - l_{th})fg$ second-generation neutrons are available to produce thermal-induced fissions in ^{235}U. The original number has been increased by a **multiplication factor** k_e, defined by

$$k_e = \epsilon(1 - l_f)p(1 - l_{th})f\eta, \qquad (7.9)$$

where $\eta = \nu g$. To have a self-sustaining chain-reacting system k_e must be at least equal to unity. If k_e is exactly unity, the reactor is termed **critical**.

The task before the nuclear reactor physicist in the 1950s and 1960s was to calculate in terms of the physical parameters of the reactor

TABLE 7.2 Thermal Neutron Absorption Cross Section for Various Substances

Substance	σ_a (barns)
Moderators	
H_2O	0.6640
D_2O	0.001
C	0.0045
Be	0.01
Controls	
B	775.0
Cd	2450.0
Fission products	
^{135}Xe	3.5×10^6
^{149}Sm	5.3×10^4
Fuel	
^{235}U	687.0
^{238}U	2.75
Structure	
Fe	2.53
Zr	0.18
Hf	105.0
(Contaminant in zirconium)	

the various terms that make up the multiplication factor for the types of reactor fuel-moderator combinations under consideration. The current reactor designs are the result of these years of intensive theoretical and practical labor. The terms of Eq. 7.8 can conveniently be divided into two categories: fuel factors (ϵ, p, and η) and geometrical factors (l_f, l_{th}, and f). Even in the case of the fuel factors, however, the geometrical arrangement of the fuel and moderator is important in determing the operating chearcteristics of the reactor. In either case, it is clear that the detailed nature of the neutron distribution inside the reactor must be known before the reactor problem can be solved.

Neutron Physics

The distribution of neutrons inside a reactor is a very complex function of space, energy, and time. Neutrons are produced only in the fuel and diffuse toward the edge of the reactor. Calculation of the neutron density near the reactor edge, far away from the fuel, is a very difficult problem, in which the exact shape of the reactor is crucial. For this reason many calculations are done for infinite reactors—that is, reactors that are assumed not to have boundaries. In such a reactor there can be no leakage, so the multiplication factor becomes

$$k_\infty = \epsilon p f \eta, \qquad (7.10)$$

often referred to as the **four-factor formula**.

The neutrons released during fission have energies greater than about 1 MeV; Figure 7.6 shows a plot of the number of neutrons per unit energy interval versus neutron energy for the thermal fission of ^{235}U or ^{239}Pu. To be effective in producing additional fissions, these neutrons must be reduced in average energy to the resonance energies near 10 eV, as shown in

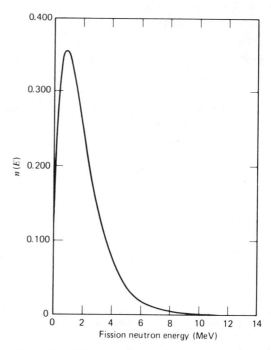

Figure 7.6 Fission neutron energy spectrum.

Figure 7.3. The moderator serves this purpose, of course, but this energy variation of the neutron flux makes analysis of the neutron distribution more difficult. For the first approximation it is usually assumed that the neutrons all have the same initial energy—the one-group approximation. Then refinements are made by assuming two, three, or more groups of neutrons having differing starting energies.

Time variations in the neutron distribution also occur, either from the gradual decrease of fissionable material over the operating life of the reactor or the sudden increase or decrease in reactivity caused by removal or insertion of control mechanisms. There is yet another time-dependent effect of importance in reactors—that of delayed neutrons. Some of the fission products are left with substantial excess energy so they can decay by neutron emission. This decay, however, does not occur immediately upon formation of these nuclides but, rather, occurs on the average after some mean lifetime. Table 7.3 lists the important delayed neutron groups that have been identified in the fission of the common fission-

able isotopes. Although these neutrons do not contribute substantially to the total number of fission neutrons, they are very important in the overall operating characteristics of power reactors.

The explosion risk—both nuclear and conventional—in a nuclear reactor is directly related to the problems of reactor control and the rate of power level change. The power level of a reactor is clearly dependent on the fission rate, which is in turn dependent on the thermal neutron density in the reactor. We need to understand the mechanisms that influence this neutron density to understand the potential for explosion.

The neutron density inside a reactor is a very complicated function of space, time, and neutron energy. In an actual reactor these parameters depend on the fuel, the moderator, their relative arrangement (homogeneous or heterogeneous) and the reactor size and geometry. We will not be able to derive results that are valid for all types of reactors because of this complex dependence. We shall, however, be able to draw some conclusions that will help us to understand some reactor

TABLE 7.3 Characteristics of Delayed Fission Neutrons in Thermal Fission

| Approximate Half-life (s) | Number of Fission Neutrons Delayed per Fission | | | Energy (MeV) |
	^{233}U ($\times 10^{-4}$)	^{235}U ($\times 10^{-4}$)	^{239}Pu ($\times 10^{-4}$)	
56.0	5.7	5.2	2.1	0.25
23.0	19.7	34.6	18.2	0.46
6.2	16.6	31.0	12.9	0.41
2.3	18.4	62.4	19.9	0.45
0.61	3.4	18.2	5.2	0.41
0.23	2.2	6.6	2.7	—
Total delayed	0.0666	0.0158	0.0061	
Total fission neutrons	2.50	2.43	2.90	
Fraction delayed	0.0026	0.0065	0.0020	

problems. The rate of change of thermal neutron density can be written as a sum of three terms: the rate of production of thermal neutrons, the rate of loss by absorption, and the rate of loss by leakage:

$$\frac{dn}{dt} = \text{production} - \text{absorption} - \text{leakage}. \quad (7.11)$$

The solution of this equation, while interesting, is rather complicated and would take us far off the track. We will, however, draw on some of the results. In particular, we note that the effective multiplication factor of Eq. 7.8 can be defined in terms of several reactor parameters. We denote it by k_{eff} to indicate it is not the same as the k in Eq. 7.8. Also, the generation time t_g, that is, the time required for a first-generation neutron to produce a second-generation neutron, is determined. Finally the time dependence of the neutron flux is given as

$$\phi(t) = \phi(0) \exp\left(\frac{\Delta kt}{t_g}\right), \quad (7.12)$$

where $\phi(0)$ is the flux at $t = 0$ and Δk is the **excess reactivity**, that is, the amount of reactivity in excess of that required for criticality. Reactivity is defined in terms of the effective multiplication factor:

$$\rho = \frac{k_{\text{eff}} - 1}{k_{\text{eff}}}. \quad (7.13)$$

For a ^{235}U reactor it should be 0.0065 for criticality, if only prompt neutrons are assumed to contribute to the reactivity.

Prompt-critical reactivity is used to define a unit of reactivity called the **dollar**. For a ^{235}U reactor, a reactivity of 0.0065 would be $1.00.

The effect of this exponential dependence of neutron flux (and, consequently, the power output) of the reactor on reactivity is illustrated in Figure 7.7. This figure has been drawn for ^{235}U assuming a generation time of 10^{-3} s. The excess reactivity has been set in a stepwise fashion to the reactivity shown for each curve. Note that when ρ is 0.001, it takes more

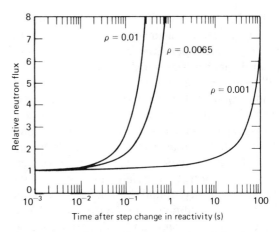

Figure 7.7 Neutron flux versus time following a reactivity increase.

than 100 s for the neutron flux to increase tenfold, but that when ρ is 0.0065, the same increase occurs in only about 1 s. For the higher value of reactivity the same increase would require only about 0.2 s. Clearly, if only prompt neutrons contributed to the reactivity, reactor control would be very difficult.

Reactor Control

Reactor control is made possible by the existence of the delayed neutrons mentioned earlier. These neutrons extend the generation time substantially and therefore flatten out the curves in Figure 7.7 for a given reactivity. In general, a reactor is operated so that the prompt neutrons alone provide a k_{eff} of about 0.9995 (less than required for criticality), and the remainder is provided by the delayed neutrons. In this fashion, sudden changes (control rod removal, cladding failure, etc.) are slower to affect the neutron density and can be counteracted before having an opportunity to cause permanent damage to the core.

A reactor consumes about half its fissionable material during a core lifetime. During this period it must, of course, remain critical. As a consequence, at the beginning it must contain rather considerably more fissionable material than is needed to sustain criticality. In order to ensure stable operation in this early period, control mechanisms must be introduced that will remove neutrons from the core without producing heat. Control rods of materials having high neutron-absorption cross sections are interspersed among the fuel elements and slowly withdrawn over the operating life of the core. Table 7.2 gives the absorption cross sections for control materials and other substances typically found in reactors. Several reactors, in additional to control rods, use soluble boron compounds directly in the moderating water for further control.

The control rods are operated in a fail-safe mode—that is, unless consciously withdrawn, they are normally in the core. And if neutron or heat sensors detect abnormal activity in the core, the rods are forcibly returned into the core in what is known as a **scram**. This action drops the neutron flux rapidly, so that thermal runaway may be avoided.

Another problem exists that requires that reactors start operation with substantial excess reactivity. This is the problem of fission products that also have high neutron-absorption cross sections—the so-called **reactor poisons**. Table 7.2 indicates xenon and samarium as the two major fission product neutron sinks; note that their absorption cross sections are indeed larger than that of the fuel. Xenon poisoning is easy enough to overcome in an operating reactor; the problem comes in restarting a reactor that, after operating for some time, has had to be shut down for a few hours.

A final mechanism affecting the reactivity of a reactor is the temperature. If the temperature of the core rises, and if as a consequence the reactivity increases, thermal runaway is inevitable, and the reactor will eventually destroy itself. If, on the other hand, as a consequence of a temperature increase, the reactivity decreases, the consequent reduction in

fission rate and heat-production rate will tend to lower the temperature and thereby act to stabilize the reactor. An important question in reactor design is, then, is the change in reactivity per degree temperature rise positive or negative? Or, in symbols, is $d\rho/dT > 0$ (undesirable), or is $d\rho/dT < 0$ (desirable)?

There are several effects that produce a temperature dependence: (1) the density of reactor materials changes with temperature; (2) the volume of the reactor changes; (3) the effective neutron temperature is correlated with the temperature of the moderator; and (4) the interactions between neutrons and nuclei vary with temperature.

If the temperature in a reactor core rises, this increase causes thermal expansion of the fuel elements as well as structural materials and moderator, the number of scattering nuclei per unit volume is decreased, the mean free path for neutrons is increased, and an increase in leakage, both l_f and l_{th}, results. Therefore, the effective multiplication factor is reduced along with the reactivity. There are other processes that make this effect more complicated than we have indicated. In general, in power reactors density changes caused by thermal effects are of secondary importance in determining the temperature dependence of the reactivity.

After a few collisions with the moderator nuclei, the average neutron energy is equal to the thermal vibration energy of the moderator atoms. The distribution of neutron energies is rather well predicted by the Maxwellian distribution. The elastic collision cross section for thermal neutrons is inversely proportional to their speed, so the effective cross section $\sigma(v)$ at a given velocity v is related to the cross section $\sigma(v_p)$ at the most probable velocity v_p by

$$\sigma(v) = \frac{\sigma(v_p)v_p}{v}. \qquad (7.14)$$

The velocity in a Maxwellian distribution is

proportional to the square root of the temperature. The fractional change in moderating cross section per unit temperature change is therefore given by

$$\frac{1}{\sigma(v)}\frac{d\sigma(v)}{dT} = -\frac{1}{2T}. \qquad (7.15)$$

This indicates that the moderating effect reduces as the temperature of the moderator increases; not only are the moderator nuclei farther apart (density effect), but they also appear smaller to the neutron (cross section effect).

The third temperature effect also produces a decrease in reactivity. This effect is the broadening of the resonance absorption peaks in ^{238}U (Figure 7.3). This effect comes about because of the change in relative motion between neutrons and nuclei when the effective temperature changes. It is often referred to by nuclear engineers as a **Doppler effect**, although there is no direct analogy between this effect and the familiar Doppler effect concerning the perceived frequency of waves. The result of this effect is an increase in the resonance absorption factors (η, p) and a consequent decrease in reactivity. This Doppler effect is an important temperature-dependent effect in today's power reactors.

In a thermal reactor the overall temperature coefficient of reactivity is negative; that is, an increase in temperature of the fuel or moderator produces a decrease in the fission reactivity in the core. In a typical water-moderated reactor in commercial service today $d\rho/dT = -3 \times 10^{-4}/°C$. The reactor is inherently self-regulating.

Because of the various control mechanisms, temperature effects, and so on that tend to lower reactivity, a reactor must normally be placed in service with a substantial excess reactivity to ensure criticality throughout the life of its core. Table 7.4 shows the excess reactivity in percent for various reactors with a breakdown of the several effects requiring that

TABLE 7.4 Reactivity Specifications for Several Power Reactors

Specification	Reactivity (or $\Delta k/k$) Requirements (percent)		
	Yankee[a]	Dresden[b]	Fermi[c]
Temperature	7.0	3	0.14
Fuel depletion	7.0	6	0.28
Xe and Sm	3.3	4	0.014
Operation, etc.	2.7	—	0.21
Total	20.0	13	0.644

[a]An early pressurized-water LWR.
[b]An early boiling-water LWR.
[c]The only U.S. commercial breeder reactor; sodium cooled. Now shut down.

excess. Commercial power reactors in the United States are either of the Yankee or Dresden type. This excess reactivity is essentially wasted during most of the life of the core. The extra neutrons are simply absorbed by extra control devices, rods, or boron solution. It also causes complications during start-up; control rods must be removed with great care.

If a reactor that has been running at a high power level for many days is shut down, how long does it remain hot in the core? This does not depend on the coolant, but on the fact that the radioactive fission fragments are undergoing decay and that this decay is depositing energy in the fuel elements. Studies have shown that, averaged over many nuclear species, the energy release per second per fission for a decay time t is given by

$$E = 2.66t^{-1.2}\,\text{MeV/s per fission}.$$

This figure includes energy released by both betas and gammas. If we assume the reactor is turned on for a time T_0 before being shut down and is operated at a power level P_0, then the energy release per second after a time t is given by

$$P = 6.1 \times 10^{-3} P_0 [t^{-0.2} - (t - T_0)^{-0.2}] \text{ W}. \quad (7.16)$$

Hence, a 1000-MW reactor having operated for 300 days will still have a core power output of 5.8 MW 100 days after shutdown. Obviously, cooling would have to be provided after shutdown to prevent overheating of the fuel.

The vast majority of the power reactors in use today use light water (as opposed to heavy water, D_2O) as a moderator–coolant. This dual use has advantages and disadvantages. Water, of course, has a high hydrogen content and, as a consequence, is a good moderator. It is readily available and presents no special problems in terms of pumps or piping. The fact that water is used as coolant means the temperature coefficient of reactivity is automatically negative; if the water temperature becomes too high, the reactivity of the fuel becomes negative; that is, the effective multiplication factor becomes less than unity, and the reactor is no longer critical.

There are disadvantages as well. First of all, the hydrogen in the water has a relatively high cross section for neutron absorption compared to other moderators. Table 7.2 gives a comparison of σ_a for several substances. The neutron absorption of D_2O is so much lower than that of H_2O that natural uranium (uranium not enriched in ^{235}U) may be used as a fuel. An-

other disadvantage is that the self-regulating temperature coefficient of reactivity limits the water temperature to relatively low values, compared with fossil-fueled steam plants. This means the overall efficiency of energy conversion from fuel to electricity is rather low for nuclear reactors, around 31 percent.

Commercial Power Reactors

There are two types of successful commercial power reactors in this country, as well as one so-far-unsuccessful type and one very successful one in Canada. Table 7.5 gives some of the operating details of the three successful types. The so-far-unsuccessful reactor is the high-temperature gas reactor (HTGR), which we discuss later. The characteristics of the PWR and BWR are very similar, but these are both quite different from the Canadian CANDU system. The most notable difference is the fact that the CANDU reactor uses naturally occuring uranium, instead of isotopically enriched fuel, and it therefore requires heavy water as a moderator and coolant. All three types use the oxide fuel pellet.

The fuel is in the form of small UO_2 cylindrical pellets ~ 0.89 cm in diameter and 1.5 to 2.3 cm long. The UO_2 powder from the UF_6 conversion plant (for BWRs and PWRs) is

TABLE 7.5 Comparison of Thermal Reactor Characteristics

	PWR	BWR	CANDU
Primary coolant	Light water	Light water	Heavy water
	15.5 MPa	6.65 MPa	10.00 MPa
	320° C	280° C	310° C
Steam conditions	7.88 MPa	6.65 MPa	3.93 MPa
	285° C	280° C	250° C
Core power density	~ 35 MW/m^3	~ 22 MW/m^3	~ 15 MW/m^3
Fuel	UO_2	UO_2	UO_2
	1.2 to 4%	1.1%	Natural U
Burnup (MW days/t)	$\sim 35,000$	$\sim 35,000$	$\sim 10,000$

Figure 7.8 Nuclear power reactor under construction at the Tennessee Valley Authority's Sequoyah nuclear station. The reactor will be built in the center cavity; the four domed cylinders are steam generators. (Photograph courtesy of Westinghouse Electric Corporation.)

pressed into this form, then sintered and loaded into zircalloy[5] tubes 4.0 to 4.5 m long. A typical 1000-MWe LWR will contain about 35,000 such tubes, each holding about 150 pellets. The density of UO_2 is about 11 g/cm^3; it has a melting point of 2500° C. For pellets of the average size described above, a 1000-MW reactor would contain about 8.4×10^4 kg of uranium. Today's power reactors are very large, as might be expected from their heat-generating capacity, 3800 MWt. Figure 7.8 shows a large, 1148-MWe reactor under construction.

[5]Zircalloy is an alloy containing zirconium and about 1.59 percent Sn, 0.1 percent Fe, 0.1 percent Cr, and 0.05 percent Ni.

The individual fuel pellets have to be isolated from the water coolant. Reactivity and thermal stresses cause substantial deformations and density changes in the fuel. Fission by-products must also be contained. The tubes used for this purpose are usually fabricated of zircalloy, although stainless steel has also been used. Zircalloy has mechanical strength and resistance to corrosion as well as superior nuclear properties; this is, its overall neutron absorption is less than that of stainless steel.

With the high cost of nuclear fuel and the high cost of downtime, it is important that fuel reloads be as infrequent as possible. Therefore, it is desirable to use as high a percentage

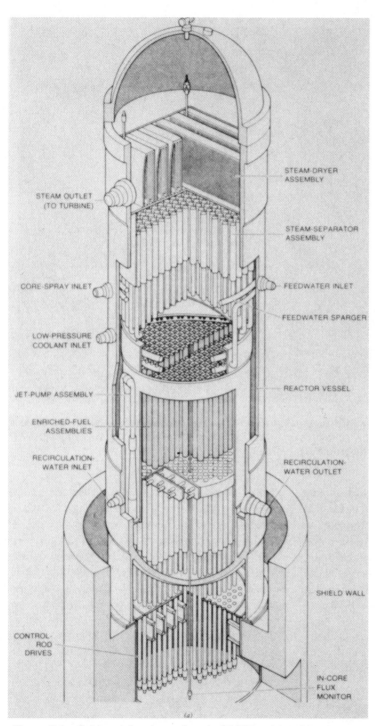

STEAM-DRYER
ASSEMBLY

STEAM OUTLET
(TO TURBINE)

STEAM-SEPARATOR
ASSEMBLY

CORE-SPRAY INLET

FEEDWATER INLET

FEEDWATER SPARGER

LOW-PRESSURE
COOLANT INLET

JET-PUMP ASSEMBLY

REACTOR VESSEL

ENRICHED-FUEL
ASSEMBLIES

RECIRCULATION-
WATER INLET

RECIRCULATION-
WATER OUTLET

SHIELD WALL

CONTROL-
ROD
DRIVES

IN-CORE
FLUX
MONITOR

(a)

Figure 7.9   Cutaway drawings of typical LWRs. (*a*) PWR.

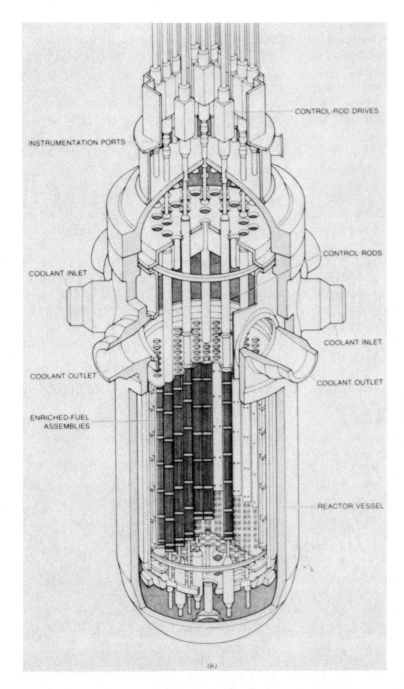

CONTROL-ROD DRIVES

INSTRUMENTATION PORTS

CONTROL RODS

COOLANT INLET

COOLANT INLET

COOLANT OUTLET

COOLANT OUTLET

ENRICHED-FUEL
ASSEMBLIES

REACTOR VESSEL

(b)

Figure 7.9 (*b*) BWR. (From Hugh L. McIntyre, "Natural Uranium Heavy-Water Reactors," *Scientific American* **233**, No. 4, p. 17, copyright © 1975 by Scientific American, Inc. All rights reserved.)

of fissile material in the fuel as possible. This utilization is expressed in the industry in terms of megawatt-days of operation per tonne of ^{235}U in the core. Early LWRs achieved about 15,000 MW days/t uranium; current reactors achieve about three times this figure. Fuel replacement is ordinarily dictated by the dropping of the power level of the reactor or an indication of cladding failure. The typical ^{235}U burnup before replacement is about 60 percent.

The concern over refueling downtime is specific to LWRs. A significant difference in construction between the CANDU and LWR minimizes the problem in the Canadian design. In LWRs the entire core is under high pressure (see Table 7.5 and Figure 7.9). In the CANDU, reactor coolant is pumped through many small, pressurized pipes containing zircalloy-clad fuel elements (see Figure 7.10). There is an interlock system that allows fuel elements to be inserted without necessitating a shutdown of the reactor. This enables operation over a significantly longer time span and much more uniform fuel burnup.

Of the currently active LWR power reactors in the United States, about 60 percent are PWRs and 40 percent are BWRs. Both have negative coefficients of reactivity. BWRs have a more difficult steam-handling problem (moisture removal, radioactivity). PWRs require a more complex and expensive reactor containment vessel, but have a higher thermal power density and consequently better fuel utilization.

Several countries use gas-cooled reactors. A gas-cooled technology was developed in this country and used in a small plant, the 40-MWe Peach Bottom Unit Number 1. This reactor has been shut down after about eight years' successful operation. A larger facility (330 MWe) has been constructed at Fort St. Vrain, Colorado, by the General Atomics Company. It became critical in July 1976, but did not begin commercial operation until 1979.

It has been plagued with a variety of problems that have caused the NRC to derate the unit to 220 MWe. These problems include rapid temperature excursions that are not understood. As a result of the difficulties of bringing this unit to full power, HTGR orders placed by other utilities have been canceled, and General Atomics now finds itself in the awkward position of having a large investment in a potentially superior technology that it cannot sell.

Coolant gases in common use today are helium and CO_2; the choice is dictated by cost and moderator. The coolant gas, whatever the choice, has little moderating effect, so additional moderating material in the reactor core is required. In the United States, graphite has been used exclusively in HTGRs' this choice of moderator precludes the use of CO_2, since CO_2 reacts with carbon at high temperatures to form CO. This can result in a net loss of moderator. Outside the United States, CO_2 is usually used in HTGRs, since almost all the commercially available helium in the world comes from the United States (see Chapter 10) and is quite expensive to export.

The CO_2–graphite reactors also have serious explosion risks. The reaction

$$CO_2 + C \rightarrow 2CO \qquad (7.17)$$

that extracts carbon from the graphite also causes a structural realignment of the carbon atoms. Periodically, it is necessary to shut down the reactor and "anneal" the graphite. This atomic rearrangement is endothermic and must be done with great care. One of the first reactor accidents was caused by this effect in Britain, the "Windscale incident."

The Fort St. Vrain plant has the following operational characteristics: helium pressure 47 atm, maximum helium temperature 538° C, and thermal efficiency about 39 percent. The expected fuel burnup was to be in the neighborhood of 100,000 MW days/t uranium, making this reactor extremely efficient in terms of fuel utilization.

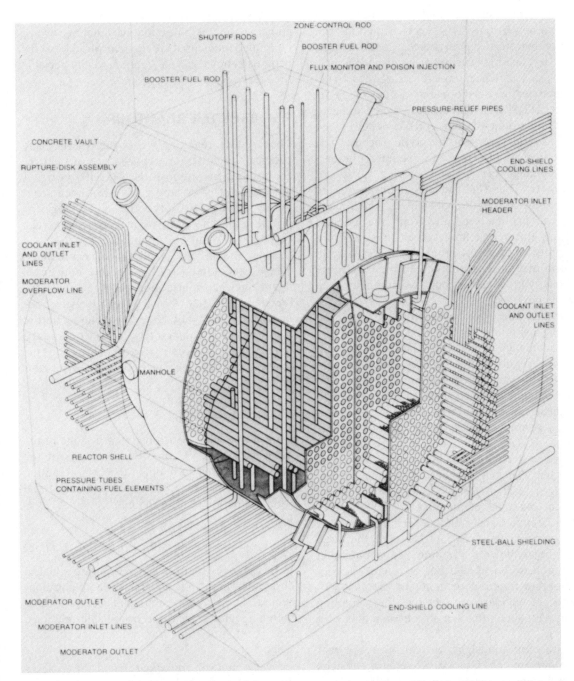

Figure 7.10 Cutaway drawing of a CANDU heavy-water reactor. (From Hugh L. McIntyre, "Natural Uranium Heavy-Water Reactors," *Scientific American* **233**, No. 4, p. 17, copyright © 1975 by Scientific American, Inc. All rights reserved.)

To achieve such a high burnup, a high thermal power density is required; in addition, high-temperature performance, good neutron economy, and high fuel integrity are necessary. As a consequence, much effort has been devoted to the development of fuel elements for the HTGR. Those currently in use utilize a uranium carbide–thorium carbide pellet coated with several layers of pyrolytic carbon. The pellets are about $100 \mu m$ in diameter; the carbon coating serves as a barrier to fission products. The fuel pellets themselves are loaded into a graphite-core block, which has a flow of helium separate from the coolant stream; this purge flow serves to sweep out fission-product gases, providing a very careful control over fission products. In the HTGR the radioactivity release should be minimal; even zero is possible.

Besides the low radioactivity release in normal operation, the HTGR has another distinct advantage over LWRs: that is, its thermal efficiency is about the same as a fossil-fueled plant of the same electrical output. Hence, the same amount of waste heat is generated, and this heat can be dispersed by cooling towers, not by heating a nearby river or lake. In addition, HTGR techniques carry over in a natural way to the next generation of nuclear reactors, the fast breeder. There are disadvantages to the HTGR as well. The helium coolant must be pumped at very high velocities, around 60 m/s, and rather large volumes are required, about 28.5 m^3/MWe. If the large centrifugal blowers required for maintaining this coolant flow should fail, adequate back-up pumps are required to maintain the fuel elements at temperatures below their melting point. Of course, this same problem exists with LWRs.

With clear hindsight, we can now see that development of alternative reactor designs should have received more attention in the early days. It may be too late now. The entire industry is in a very uncertain position, which may take some time to clarify. One factor that could swing the direction of opinion toward continued nuclear development would be the demonstration of a commercially viable breeder reactor.

7.3 BREEDER REACTORS

The breeder reactor, if it is successful, may become the most important source of heat for electricity generation in the next 100 years. A great deal of effort and money has been spent in an attempt to solve the many lingering problems in the path of making commercial breeders a reality. The breeder's future—in this country, if not in others—is very cloudy. The Carter administration became concerned about nuclear proliferation and the role of the breeder in that area. The Reagan administration may not share these views. Before we discuss these difficulties, let us examine the operation of breeder reactors based on the reactor physics we have developed to this point.

Physics of Breeding

In the breeder, neutrons from fissions in the fissile material, ^{235}U or ^{239}Pu, in excess of the number required to mainain the chain reaction are used to produce additional fissile material from fertile material in the core, ^{238}U:

$$^{238}U + n \rightarrow {}^{239}U \rightarrow {}^{239}Np + \beta^- + \bar{\nu}$$
$$\hookrightarrow {}^{239}Pu + \beta^- + \bar{\nu}. \qquad (7.18)$$

In this manner fuel is created at a rate equal to or greater than that at which it is being consumed. The far more abundant isotope ^{238}U is thereby converted to a fissionable nucleus, thus stretching our fuel resources substantially.

Every neutron removed from the reactor flux reduces the ability of the reactor to sustain the chain reaction. In order for breeding to occur, a sufficient number of neutrons must be

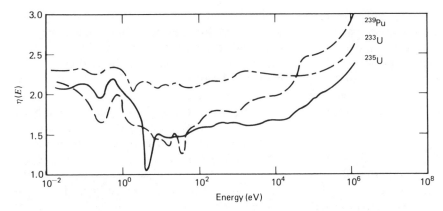

Figure 7.11 Energy variation of η for ^{233}U, ^{235}U, and ^{239}Pu. (From *Evaluated Nuclear Data File*, National Neutron Cross Section Center, Brookhaven National Laboratory, ENDF/B-II.)

emitted per fission to provide for (1) producing a further fission, (2) providing for capture in ^{238}U, and (3) providing for leakage and capture in structural and contaminant materials. We can define a **conversion ratio**, CR, which will be a measure of the breeding effectiveness of a given reactor. This ratio is given by

$$CR = \epsilon\eta - 1 - L, \qquad (7.19)$$

where ϵ and η are as defined in Section 7.1, and L is a leakage factor. If CR = 1, the reactor just maintains its own fuel supply; for CR > 1, the reactor can actually increase its inventory of fuel while producing heat for use in generating electricity.

The extent to which CR exceeds unity is determined by the factors ϵ, η, and L. Both ϵ and L depend on the specific design details of a given reactor. The fast fission factor ϵ has been measured for a variety of fuel–moderator geometries for thermal reactors and is found to vary between about 1.00 and 1.05. The leakage factor can be made small by appropriate design. Then to a first approximation it is η, the number of fission neutrons released per absorption, which is important in determining the breeding effectiveness of a reactor; and this quantity depends essentially on the fuel

only:

$$\eta = \nu\,\frac{\sigma_f}{\sigma_a}, \qquad (7.20)$$

where ν is the number of first generation neutrons, σ_f is the fission cross section of the core (fissile plus fertile), and σ_a is the total absorption cross section of the core. For successful breeding it is clear that η should be somewhat greater than 2. But fission and absorption cross sections are functions of neutron energies. Figure 7.11 shows the energy variation of η.

Since thermal reactors operate in the resonance energy regime, $0.01 < E_n < 100$ eV, it is clear from Figure 7.11 that, with the exception of ^{233}U-fueled reactors to be discussed later, breeding is not possible with this type of operation. It is also clear from this figure that if the energy spectrum of the neutrons is shifted so that the peak energy is in the neighborhood of 1 MeV, breeding becomes a real possibility. Added to this is the bonus that, at the higher neutron energies, the fast fission factor may be as large as 1.20; that is, 20 percent of the fission neutrons come from fissions occurring in the fertile material, rather than the fissile core.

A reexamination of Figure 7.5 will convince you that the fission neutron energy spectrum without moderation would be the most desirable distribution for breeder operation. Neutrons in this energy range are termed **fast**, and so a reactor operating in this region would be called a **fast reactor**, or fast breeder reactor.

Fast Reactor Parameters

A fast reactor must be constructed very differently from a thermal reactor. Fission cross sections for fast neutrons are two orders of magnitude smaller in a fast reactor (refer to Table 7.6). As a consequence, high fuel concentrations (~50 percent) are necessary to achieve criticality and to ensure moderation of the neutron spectrum. This concentration of the fuel leads to very high thermal power densities, which in turn lead to difficulties in extracting the heat.

Early breeder reactor designs, for weapons-grade plutonium production, used pure uranium metal, 25 to 50 percent enriched, in ^{235}U fuel pins clad in stainless steel grouped together in a region of only a few cubic meters (or even less than 1 m^3). Liquid sodium metal was chosen as the coolant for several reasons: it has high

thermal conductivity and high boiling point, it has a fairly low neutron absorption cross section, and it is not very effective in moderating fast neutrons. Surrounding the fuel core was a blanket of fertile ^{238}U, in which most of the breeding occurred.

In the early breeder designs the emphasis was placed on maintaining the "hard" neutron spectrum, that is, on minimizing moderation. Little attention was paid to fuel cycle economy. But in subsequent years, as the development of commercial breeders was started, it became clear that fuel cycle costs had to be held to a minimum. In operation terms, this meant that the reactor should operate as long as possible with a given fuel load, "burning up" as high a percentage as possible of the fissile nuclei. This would reduce the plant downtime, and also reduce the number of passes a nucleus would make through the reactor–reprocessor system before being used.

Current LWRs achieve a fuel utilization of about 35,000 MW days/t uranium. The value predicted to be required for economical breeder operation is about 100,000 MW days/t uranium. The major deterrent to higher burn-ups in LWRs is the drop in reactivity of the

TABLE 7.6 Nuclear parameters for Fast[a] and Thermal[b] Reactors

	^{235}U		^{238}U		^{239}Pu	
	Fast	Thermal	Fast	Thermal	Fast	Thermal
σ_f (barns)	1.44	582.0	2.20	527.0	1.78	746.0
ν	2.52	2.47	2.59	2.51	2.98	2.90
σ/σ_f	0.152	0.19	0.068	0.102	0.086	0.38
$\eta - 1$	1.18	1.07	1.42	1.28	1.74	1.10

Adapted, with permission, from L. J. Koch and H. C. Paxton, "Fast Reactors," *Annual Review of Nuclear Science* **9**, 437 (1959), copyright © 1959 by Annual Reviews Inc.
[a]Fast neutron parameters are averaged over the typical reactor neutron system.
[b]Thermal neutron parameters are appropriate to 0.025 eV.

reactor as fissile nuclei are used up; in the breeder it is likely to be radiation damage to the fuel elements.

Metallic fuel elements were found to swell, both radially and axially, during operation. The decreased nuclear density resulting from the swelling had a negative effect on the reactivity, and the radial swelling caused ruptures of the steel cladding. The radiation damage was even more pronounced when a small amount of plutonium was mixed into the fuel. Serious swelling was found to occur at burnups as low as 1 percent (atomic) and temperatures as low as 400° C. It was evident that the more hostile environment of the breeder core—higher neutron flux and higher temperature—meant that a

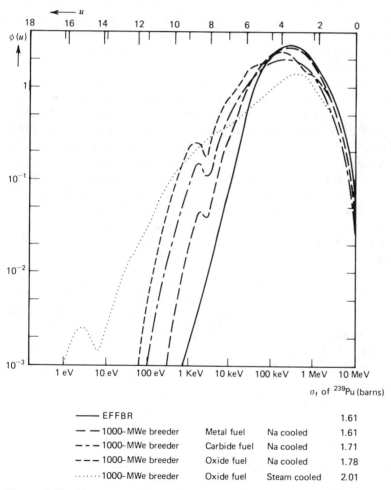

			σ_f of ^{239}Pu (barns)
—— EFFBR			1.61
— — 1000–MWe breeder	Metal fuel	Na cooled	1.61
– – – 1000–MWe breeder	Carbide fuel	Na cooled	1.71
– – – 1000–MWe breeder	Oxide fuel	Na cooled	1.78
·······1000–MWe breeder	Oxide fuel	Steam cooled	2.01

Figure 7.12 Neutron spectra for various fuel–moderator combinations. (Reproduced, with permission, from N. Häfele, D. T. Raude, E. A. Fischer and H. J. Laue, "Fast Breeder Reactors," *Annual Review of Nuclear Science* **20**, 393 (1970), copyright © 1970 by Annual Reviews Inc.)

different fuel material would be required to realize the high burnups demanded by fuel economics.

A shift was made to oxide fuels for breeders. Many years of operating experience with oxide fuels for LWRs indicated that, while they were not without problems, they did not seem to suffer from radial swelling and they have a higher melting point, which somewhat offsets their lower heat conductivity. But the oxygen in the fuel serves as a moderator and thereby "softens" the neutron spectrum (see Figure 7.12). As a consequence, the breeding performance of oxide-fueled breeders is expected to be inferior to that for the metallic-fueled varieties. Table 7.7 gives a comparison for different fuels. These values are based on an idealized core containing no fission products; equilibrium values of CR in a real reactor would be substantially lower. Carbide fuels, such as could be used in HTGRs, are also being studied for possible breeder application, but a backlog of experience does not exist for them as it does for the oxide fuels. There are reasons to believe, however, that carbide fuels would be superior to oxides. They have higher heat conductivity and melting points and increased resistance to radiation damage.

TABLE 7.7 Study of 800-ℓ Spherical Reactors Fueled with Metal, Oxide, and Carbide Plutonium Fuels[a]

	Metal	Oxide	Carbide
Density (g/cm³)	19.0	8.4	11.4
^{239}Pu + ^{241}Pu (kg)	533.0	294.0	312.0
Breeding ratio			
Core only	0.86	0.37	0.55
Total	2.12	1.81	1.90

Adapted, with permission, from L. J. Alexander, "Breeder Reactors," *Annual Review of Nuclear Science* **14**, 287 (1964), copyright © 1964 by Annual Reviews Inc.
[a]Fuel: 40 percent ^{239}Pu, 10 percent ^{240}Pu, 25 percent ^{242}Pu, 25 percent ^{238}U.

One major benefit from the use of ceramic (oxide and carbide) fuels is that their densities are about half that of the metals. This, coupled with the fact that fuel enrichments for ceramic-fueled breeders will be about 12 percent compared to 50 percent for metallic-fueled breeders (for reasons to be discussed shortly), lowers the thermal power density a factor of 4, greatly alleviating heat-transfer problems. In fact, the most important reason for using liquid sodium as a coolant in the first place in metal-fueled breeders is no longer persuasive in the case of ceramic-fueled reactors.

A very important consideration in a breeder reactor discussion is the doubling time, that is, the time required for a reactor to produce twice as much fuel as it has used. Clearly, this time period should be less than the estimated useful life of the facility, preferably much less. If a breeder having a total fuel inventory of M_c kg operating at a power level of P MW produces fuel at the rate $\dot{m}$ kg/MW day, then the doubling time t_2 is such that

$$t_2 \dot{m} P = 2M_c. \qquad (7.21)$$

The total fuel inventory is the sum of fuel in the reactor, M_r, and fuel for and from the reactor in the fabrication and reprocessing steps of the fuel cycle. Note that

$$M_r = \frac{P}{p_s} \qquad (7.22)$$

and

$$M_c = \left(\frac{P}{p_s}\right)\left(1 + \frac{t_c}{t_r}\right), \qquad (7.23)$$

where p_s is the specific power (in megawatts per kilogram) of the reactor and t_c and t_r are the fuel cycle time and the reactor residence time of the fuel, respectively.

We can write the fuel production rate $\dot{m}$ in terms of a parameter g_B, called the breeding gain. And the net fissile material destruction rate $\dot{m}_d$ as

$$\dot{m} = (1 + g_B)\dot{m}_d. \qquad (7.24)$$

It is, in theory, possible to calculate g_B from fuel and geometry parameters, but this calculation is not simple. For fast reactors this parameter is about 0.33. The factor $\dot{m}_d$ can be calculated from the thermal power of the reactor and the fact that in a fast fission 205 MeV of energy is released on the average. Most fast reactors have predicted doubling times of about 20 years, although many yet-to-be-determined factors can influence this figure significantly—in particular, reprocessing time.

Fast Reactor Control

Of primary concern in fast reactor design, as it is in LWRs, is the temperature dependence of the reactivity. Clearly, for safe operation $d\rho/dT$ should be negative. The problem has been studied extensively only for sodium-cooled reactors LMFBRs. In reactors of this type, the major concern is the effect of voids in the sodium from boiling or even the total loss of sodium resulting from an accident. Sodium voiding has several effects. First, the neutron spectrum is hardened as a result of the reduced moderating effect of the voids compared to the sodium. Second, leakage from the reactor core increases because of the reduced scattering effect of the voids.

For a small LMFBR where leakage is the dominant factor, sodium voiding yields $d\rho/dT < 0$. But for a large power reactor of the size contemplated for commercial usage, the increased reactivity resulting from neutron spectral hardening dominates, and $d\rho/dT > 0$. This is evident from the plot of η versus neutron energy, as shown in Figure 7.11. When this was discovered, reactor designers were stymied until another important temperature effect—discovered earlier for thermal reactors—was found to play an imporant role in fast reactors: the **Doppler effect.**

This effect—the broadening of absorption resonances to span a greater energy range—is a profound determining factor for the magnitude of the neutron flux in a breeder. Since these are nonfission resonances in fertile material, the net result of Doppler broadening is a substantial reduction in neutrons and consequent loss of reactivity. This reactivity loss can more than offset the reactivity increase by spectral hardening, if the fertile content of the core is high enough compared to the fissile content. As a consequence LMFBR fuels will be limited to enrichments between 12 and 25 percent. Even with this precaution, fast-reactor temperature coefficients, while negative, will be quite small, about 2×10^{-5}.

The small temperature coefficient means that control of fast reactors will be more difficult. The fact that the overall temperature coefficient is a sum of positive and negative effects means that engineering safeguards must be built into the reactor system. Even with these, there is little doubt that fast reactor operation will be more difficult and more dangerous. While the control problems discussed above apply specifically to the liquid-metal-cooled fast breeder, there are similar or other problems with the other types of breeders receiving attention in the United States today.

Current Breeder Reactor Designs

There are three breeder reactor designs currently under study: the liquid-metal fast breeder reactor (LMFBR), the gas-cooled fast breeder reactor (GCFBR), and the molten salt breeder reactor (MSBR). Only one of these the LMFBR, is being developed in earnest (although not without problems; see the next section). The other two designs have rather substantial advantages over the LFMBR as well as some disadvantages. We examine all three types.

LMFBR We have already discussed some of the physical aspects of the LMFBR. Figures

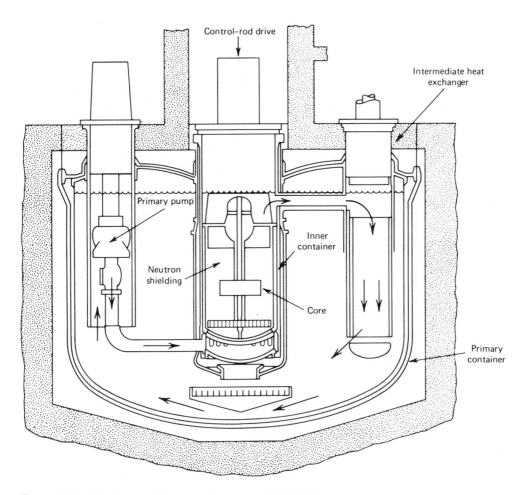

Figure 7.13 Liquid-metal fast breeder reactor (LMFBR), pot system.

7.13 and 7.14 illustrate the two basic LMFBR designs. In the pot system the reactor core assembly, sodium pumps, and intermediate heat exchangers are all immersed in a large tank of sodium. This arrangement will prevent the core from being uncovered in the event of a failure of the primary cooling system. This pot system requires a large amount of sodium and limits access to the reactor and peripheral equipment.

In the pipe system only the reactor core vessel is filled with sodium. The primary coolant loop—including pumps and heat exchangers—is located outside the reactor assembly. By proper engineering of the primary cooling system components, core submergence can be ensured with the same degree of reliability as with the pot system. The pipe system requires less sodium and greatly simplifies maintenance.

We have discussed earlier the historical reasons for the choice of sodium as a coolant—primarily its excellent heat-transfer characteristics. In addition, we should mention that, because the sodium coolant loop operates at low pressure, it is safer than a pressurized

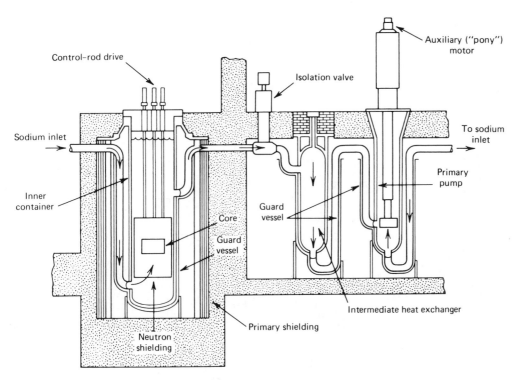

Figure 7.14 Liquid-metal fast breeder reactor (LMFBR), loop type.

system. Likewise, natural circulation of the sodium would help to cool the reactor in the event of an emergency shutdown.

Offsetting some of the advantages of sodium cooling are a number of serious drawbacks. Sodium becomes intensely radioactive in the neutron environment of the core:

$$^{23}\text{Na} + \text{n} \rightarrow {}^{24}\text{Na} \rightarrow {}^{24}\text{Mg} + \beta^- + \bar{\nu}. \quad (7.25)$$

This reaction has a neutron absorption cross section of about 0.001 to 0.002 that of the fission cross section of ^{235}U, the smaller number being for thermal neutrons and the larger for fast neutrons. The half-life of ^{24}Na is 15.0 hr. Because of this radioactivity it is necessary to employ an intermediate coolant loop using sodium, which does not become radioactive and which serves to transfer heat from the primary loop to the steam generator and superheater (see Figure 7.15). Since sodium reacts violently with water, the steam generator and superheater design must eliminate the possibility of leaks between the two fluids. Both the British and French prototype reactors (the Dounreay Fast Reactor and the Phenix) have experienced difficulties with leaking welds in the steam generators. The Soviet breeder, the BN-350, experienced a spectacular sodium ejection when a steam generator failed. This event was so striking it was observed by U.S. reconnaissance satellites! And, finally, since sodium is opaque, maintenance and refueling of the core must be carried out without visual assistance.

Figure 7.16 shows the Experimental Breeder Reactor II (EBR-II) located at the Idaho National Engineering Laboratory and operated by Argonne National Laboratory for the DOE. This LMFBR came on line in 1964 and has served as a test facility for fuels, structural

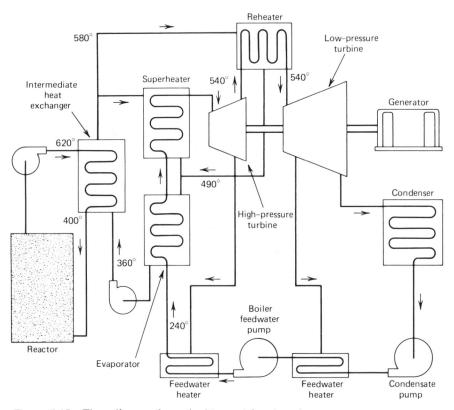

Figure 7.15 Flow diagram for a liquid-metal fast breeder reactor.

material, absorbers, and sensors. A 325,000-ℓ sodium tank encloses the core; the sodium is at a temperature of 370° C. An electrical generator capable of 20-MW operating has been used with the EBR-II to produce more than 10^9 kWh of electricity. In 1976 the reactor operated 76.9 percent of the time available.

GCFBR One alternative to the LMFBR is the gas-cooled fast breeder reactor. With the poor performance of the gas-cooled thermal reactor, it is not clear what future the GCFBR has. But since there are some theoretical advantages to it, we should look at it carefully. This type of reactor incorporates existing cooling-system technology; the fuel pins will be the stainless-steel-clad mixed oxide fuels being developed for the LMFBR. The overall

height of a 300-MWe reactor would be about 3 m, with the core itself about 1 m high. The coolant gas is helium at 70 to 100 atm pressure. The entire reactor is contained with a prestressed concrete vessel about 25 m in diameter and 20 m high. This vessel is designed to contain the helium in the event of a pumping system failure.

Helium as a coolant has several advantages. Since it does not interact with the fast neutrons in the core, it does not enter into the reactivity temperature coefficient of the reactor, thereby simplifying control problems. There is not a problem in the helium-cooled reactor equivalent to the sodium void difficulty with the LMFBR. Furthermore, the absence of a moderating effect maintains the hard neutron spectrum and enhances breeding performance.

Figure 7.16 The Experimental Breeder Reactor II (EBR-II) located at Idaho National Engineering Laboratory and operated by Argonne National Laboratory for the U.S. Department of Energy. (Photograph courtesy of Argonne National Laboratory.)

Helium is noncorrosive, and its transparency permits direct viewing of the refueling and maintenance operations. Since helium does not become radioactive, hot helium can go directly to the superheater and steam generator without passing through an intermediate heat-transfer loop, as is normally required with sodium. One can even conceive of the possibility of a direct power cycle, where the helium coolant leaves the reactor and flows directly to a gas turbine that drives a generator.

The many advantages of this reactor over the LMFBR would seem to make its development a natural thing; but, as we know,

scientific judgments alone are not sufficient to form major policy decisions.

MSBR The final type of breeder reactor that we shall discuss is the molten salt breeder reactor (MSBR). This reactor had been funded at a minimal level, just to keep it alive as a technology to fall back on in case some major problem developed with the LMFBR. But, in 1972, a decision to reduce the funding further was apparently made by the AEC, which had the effect of turning off further research in this area. Perhaps the recent concern about energy will bring about a revival of MSBR research.

The MSBR is a thermal breeder—it does not use the fast-neutron spectrum required by the LMFBR and GCFBR. In order to obtain the $\eta = 2$ requirement for breeding, ^{233}U is used as the fissile material. The fertile material in this reactor is ^{232}Th; the breeding reaction is as follows:

$$^{232}Th + n \rightarrow {}^{233}Th \quad (\tau_{1/2} \times 22 \text{ min})$$
$$\hookrightarrow {}^{233}Pa + \beta^- + \bar{\nu} \quad (\tau_{1/2} = 27.4 \text{ days})$$
$$\hookrightarrow {}^{233}U + \beta^- + \bar{\nu}. \quad (7.26)$$

Thorium is a rather abundant heavy element; it is common in granites and shales. The mass-232 isotope is the only one that occurs naturally. The MSBR employs a fuel composed of ^{232}Th and ^{233}U dissolved in molten lithium and beryllium fluorides. This viscous liquid passes through a moderating graphite matrix within which most of the fissions occur. It also circulates around the outside of the core region, forming a blanket within which most of the breeding occurs.

The major advantage of the MSBR is that it eliminates fuel pins entirely and makes on-site fuel reprocessing practical. The first advantage may become a crucial point in the development of commercial breeder capability because of the great difficulty that has been experienced in testing current fuel-pin assemblies under conditions similar to those of the LMFBR or GCFBR neutron and heat environment. The on-site reprocessing is also an extremely desirable feature, although adequate safety

TABLE 7.8 Characteristics of a 1000-MWe MSBR

Reactor thermal power (MWt)	2250
Overall plant thermal efficiency (percent)	44
Fuel salt inlet and outlet temperatures (°C)	566, 704
Coolant salt inlet and outlet temperatures (°C)	454, 621
Throttle steam conditions	240 atm, 538° C
Core height/diameter (m)	4.0/4.3
Radial blanket thickness (m)	0.5
Graphite reflector thickness (m)	0.8
Number of core elements	1412
Size of core elements (cm)	$10.2 \times 10.2 \times 396$
Salt volume fraction in core (percent)	13
Salt volume fraction in undermoderated zones (percent)	37 and 100
Salt volume fraction in reflector (percent)	< 1
Average core power density (W/cm³)	22
Maximum thermal neutron flux (neutrons/cm² s)	8.3×10^{14}
Graphite damage flux (> 50 keV) at point of maximum damage (neutrons/cm² s) (neutrons/cm² s)	3.3×10^{14}
Estimated graphite life (yr[a])	4
Total salt volume in primary system (ℓ)	48,700
Thorium inventory (kg)	68,000
Fissile fuel inventory of reactor system and processing plant (kg)	1470
Breeding ratio	1.07
Fissile fuel yield (percent/yr)	3.6
Fuel doubling time (exponential) (yr)	19

Reprinted, with permission, from A. Perry and A. Weinberg *Annual Review of Nuclear Science* **22**, 318 (1972), copyright © 1972 by Annual Reviews Inc.

[a]Based on 80 percent plant factor and a fluence of 3×10^{22} neutrons/cm² (> 50 keV).

standards might be difficult to achieve—it would be necessary to pump highly radioactive materials throughout the system.

Another important advantage of the MSBR is that uranium and not plutonium is the end product. Uranium—although radioactive—is not nearly so toxic and dangerous as plutonium. Many critics of the LMFBR program have cited this one fact alone as reason enough to switch to the MSBR.

A major drawback to the MSBR is its rather poor breeding performance (see Table 7.8). The neutron flux must be very carefully managed to provide the excess required for breeding. In addition to the leakage and fission-product poisoning common to all reactors, the MSBR has an additional problem: neutron absorption by an intermediate stage nuclide in the breeding chain, ^{233}Pa (protactinium). This nucleus has a neutron absorption cross section—integrated over the resonance energy region—of about 850 barns, compared to the ^{233}U cross section of about 525 barns. It would seem that breeding would be an impossibility as long as the ^{233}Pa were kept in the neutron environment.

The solution is to circulate the molten fuel–fission-product mixture; on-line separation of the protactinium and fission-product poisons would then be possible, so the recirculated material would contain only a mixture of fissile and fertile nuclei. On-site circulation of the fuel mixture to remove contaminants thereby alleviates one of the major problems with LWRs today, that of transporting spent—and high radioactive—fuel elements for reprocessing.

The are many engineering difficulties associated with practical operation of MSBRs,[6] just as there are with all types of breeder reactors. Whether or not additional research on these difficulties will be done remains to be seen.

7.4 REACTOR SAFETY

How safe is nuclear power? This seems to be a very straightforward question. We would like to think—since there are many nuclear power plants in operation around the world—that nuclear power is as safe as any other power source. If it is, is this enough?

Nuclear power plants release small amounts of radioactivity in routine operation. Although these releases are far below background levels,[7] they can lead to a slight statistical excess of cancers in the exposed population. A large nuclear reactor that has been in operation for more than a few months will have an inventory of radioactivity in the core exceeding 10^{20} Bq. If that radioactivity were released suddenly into the atmosphere, a very large number of immediate deaths would result as well as an equally large number of delayed deaths by induced cancers. Then perhaps we should insist that nuclear power plants be safer than conventional plants. Again, how safe?

This question has been with us as long as nuclear power has; it forms the root of the nuclear dissension movement. It has spurred the spending of millions of dollars on studies, experiments, and tests over the past 20 years or so. Yet, the question remains. In this section we look at reactor safety in detail from several points of view: the physics and engineering of safety, past history of safety-related incidents at nuclear plants, and the assessment of risk and assignment of benefits and penalties. We may not answer the question, "How safe is nuclear power?" to everyone's satisfaction, but then that is part of the problem. The public perception of the hazards from nuclear power is very different from those of other power sources, even though these other sources may be *in the long run* more injurious to the public health.

---

[6]See, for example, A. M. Perry and A. Weinberg, *Ann. Rev. Nucl. Sci.* **22**, 317 (1972).

[7]See Chapter 14 for a discussion of radiation units, background levels, and radiation effects.

Safety Assurance

At the beginning of the civilian reactor program the AEC established the following aspects of reactor safeguards:

- Site selection.
- Design criteria.
- Engineered safeguards.
- Operations planning.
- Accident containment.

By examining these aspects and following the criteria thereby developed, it was hoped that accidents could be avoided entirely; but, if not, that the consequences of accidents could be minimized.

The concern over *site selection* was to ensure that reactors built in areas prone to seismic activity or adverse weather phenomonena (tornadoes) be specially constructed to withstand any adversity. There is little evidence that serious consideration has been given to siting reactors *away* from such sites. One plant, the Rancho Seco unit in California, is built virtually on top of the San Andreas fault! Nuclear plants are routinely sited near large metropoliton complexes; this is only reasonable, since it is these large areas that have the greatest need for electric power. It would seem in retrospect that the site allocation criterion has not added much to the overall safety picture for nuclear power.

The *design criteria* chosen to maximize the safety of operation cover a wide range of operating parameters: maximum water temperature and pressure, maximum fuel cladding temperature, maximum fuel enrichment, and maximum thermal power generation, among others. These specifications ensure that the reactor will be inherently self-regulating with respect to fissions at all stages of its operation. This includes start-up, where a rather large amount of excess reactivity must be included, through full-power operation, where thermal transients caused by nonuniform fissioning in the rather large core must be dealt with, to shutdown for refueling, where the fission rate must be reduced slowly and uniformly throughout the core.

The maximum fuel cladding temperature allowed by the AEC design criteria during a loss of coolant accident is 1204° C. During normal operation this cladding is at approximately the coolant temperature, about 315° C, and the fuel itself is estimated to have a temperature of 2300° to 2500° C at the center. This should be compared to the melting point of UO_2 given in Table 7.9.

There is some disagreement over whether or not these design criteria are conservative enough. The maximum allowed thermal power has risen since 1957 to the current value of 3800 MW for reactors now under construction. Very little operating experience has been logged with such large reactors; it seems to be the case that, as the thermal power of reactors has increased, their operating reliability has decreased (although this is disputed by many

TABLE 7.9 Characteristics of Uranium Dioxide

Melting point	2800° C
Crystal structure	Face-centered cubic
Lattice parameter (A)	5.468
Theoretical density (g/cm^2)	10.96
Thermal conductivity (cal/s cm^2 ° C)	0.02 (at 20° C)
Thermal expansion coefficient (per ° C)	$\sim 1 \times 10^{-5}$ (0 to 1000° C)
Fracture strength (MPa)	60
Modulus of elasticity (GPa)	150

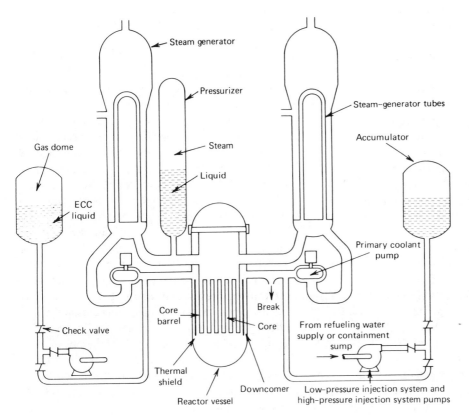

Figure 7.17 Simplified PWR system showing the emergency core cooling system (ECCS).

utility companies). Not all failure modes of smaller reactors have been found; surely, large reactors will be even more complex. Arguments along this line would lead to a reduction in reactor thermal power, rather than an increase. We should note that larger reactors exist or are under construction in other countries.

While design criteria were developed to ensure safe, normal operation, the *engineered safeguards* were promulgated to maintain safety during abnormal operations or accidents. There are many such devices or systems, but the most important are the emergency core cooling system (**ECCS**), the backup system, the reactor scram, and emergency diesel power.

The cooling systems for a typical PWR are shown in Figure 7.17. This drawing shows only two steam generators and primary coolant pumps; in an operating system there may be as many as four. In the event of a cooling line break, a "loss of coolant accident," or **LOCA**, as shown, high-pressure water would be expelled from the vessel, called "blowdown." The core temperature would begin to rise; a mixture of steam and water would continue to be expelled until the containment pressure were reached. Just before this point, the ECCS injection pumps would automatically begin to operate to flood the core with water. There is also a low-pressure injection system that is used to flood the core from the bottom up.

There has never been a full-scale test of the

ECCS in an operating power reactor. This fact is a significant source of concern among many people who are uneasy about nuclear power. All of the information on the operation of the ECCS in a LOCA comes from computer simulation based on mathematical models and extrapolation from existing technology and a few semiscale tests.

Therefore, even though the NRC regulations state that an applicant must demonstrate that the ECCS of the proposed reactor meets the design criteria, none has in fact done so—except through the simulation and extrapolation mentioned above.

The ECCS debate simmered (and occasionally boiled) throughout the 1960s. Two semiscale tests in 1971 provided the trigger for a detailed review of the ECCS design criteria by the AEC. In the first experiment at Oak Ridge National Laboratory, the conditions of a blowdown were simulated in a small test reactor. The rods were exposed to steam; residual fission-product heating, characteristic of a real accident, was provided by in-plant fission heat generation—but only until a predetermined temperature was reached. This maximum temperature of 966° C was maintained for 2 s. Extensive fuel rod rupture occurred, with all fuel elements swollen and bowed. One rod failed dramatically and appeared to start an unexpected, self-propagating fuel-element failure mode. Hydrogen was liberally generated by metal–steam reactions. There was a 48 percent blockage of coolant channels.

In the other experiment by the Idaho Nuclear Corporation, 12 bundles of fuel rods filled with AlO_2 rather than UO_2 simulated the reactor core. The rods were heated electrically to a temperature of 1150° C. Upon exposure to steam, there were sharp temperature spikes up to 1616° C, severe oxidation of the zirconium, and considerable fragmentation and fusing together of the cladding. Several rods were punctured, allowing alumina–steam reactions,

suggesting reactions of the sort

$$3UO_2 + 2H_2O \rightarrow U_3O_8 + 2H_2 \qquad (7.27)$$

in an operating reactor.

From these experiments it was concluded that cladding temperatures above about 980° C during blowdown may result in extensive fuel-rod swelling and constriction of coolant flow channels. At higher temperatures, unexpected fuel cladding damage and fuel reactions with steam will lead to sharp temperature spikes, which in turn will enhance melting, fuel rod ruptures, and general core damage. Thus, if a LOCA occurred in a reactor operating at the maximum allowable cladding temperature of 1200° C, it is quite likely that the ECCS would not only not be able to prevent severe core meltdown, pressure buildup, and radioactivity release, but might in fact precipitate these effects by causing "waterhammer."

It was also quite clear that the method of predicting the behavior of a reactor system was not satisfactory, since this method, computer simulation, had failed to predict the results of these two tests.

By this time, the concern over the effectiveness of the ECCS was being expressed more vocally and more urgently. The AEC ordered a series of hearings on the ECCS, which lasted from January 29, 1972 until July 25, 1973 (intermittently). In December of 1973, the AEC issued a set of revised criteria for the ECCS that addressed a number of the shortcomings of the earlier criteria. In particular, the computer codes used to predict the course of a LOCA had to be rather more sophisticated than before, and had to take into account the following:

The dynamic conditions of coolant flow throughout the primary cooling loop, particularly around hot spots.
Cooling provided by the exiting primary coolant.

Heat transfer at various stages of the accident.

Core geometry changes, flow blockages, core fragmentation, propapagating fuel element failures, steam expansion, and chemical reactions between the fuel and cladding. Zirconium—water/steam reaction.

Time required to bring pumps to speed and effect of reactor vessel pressure on pump performance.

But even these newer codes have not been tested in an operating reactor. As long ago as 1963, the AEC began the construction of a 50-MW test reactor to be used in a loss-of-fluid test (the LOFT reactor). In 1969, while the reactor was still under construction, the mission of the program was changed from a core-meltdown test to a test of the ECCS. Then later, somehow, the program seemed to shift to quality-control tests. Needless to say, the entire program was slowed almost to a standstill. The first *actual* tests were reported in 1979, and although they have been hailed by the nuclear industry, little information has been made public.

The LOFT tests have also been criticized because of scaling difficulties. Since the LOFT reactor is small compared with current LWR designs, it is not clear whether a meaningful comparison can be made between them. Because of these differences, a study by the American Physical Society[8] suggests that LOFT cannot be used to validate the computer codes used in ECRS acceptance but, rather, it can be used as a test bed for the rather large number of separate effects that interact during a LOCA. We should also note that the LOFT reactor simulates only a PWR, and that a similar test for BWRs is not now being proposed.

[8]Report of the American Physical Society by the Study Group on Light-Water Reactor Safety, *Rev. Mod. Phys.* **47**, S1 (1975).

In summary, we can say that no full-scale test, which would be unambiguous in its interpretation, has ever been made of the emergency core cooling system of a light water reactor. Several small-scale tests cast doubt on the utility of the computer codes that are used to certify the ECCS, even though these codes are much improved now over earlier versions. A substantial amount of experimental verification of these computer codes as well as increased research on the ECCS systems is required. In light of recent occurrences (see Three Mile Island, below), perhaps this research will now be done.

The final engineered safeguard is that of accident containment. The AEC evolved a triple-barrier approach to the problem of radioactivity release from reactors in an accident. The first barrier is the fuel cladding itself. This material, usually zircalloy or stainless steel, under normal operating conditions resists the high-temperature, high-neutron flux environment. Most of the radioactivity in a reactor results from the fission products that remain inside the fuel pellet, as long as the cladding integrity is maintained. It is not uncommon for some fuel elements to rupture during *normal* operation or to develop pinholes, although the number doing so has been relatively small.

In the event of radioactivity release from the fuel elements, the second barrier is designed to keep the material from spreading to the outside world. This barrier is the reactor pressure vessel. Typical BWR pressure vessels are designed to withstand a pressure of about 85 atm with a normal operating pressure of 68 atm. In a PWR the equivalent figures are 170 atm, design and 153 atm, operating. As these figures indicate, the pressure vessels are designed to withstand minor excursions from normal operating conditions. They would, for example, contain radioactivity spilled from a ruptured fuel element in the cooling system;

but a major accident could cause a rupture of this barrier.

The final barrier is the reactor building, called the containment structure. These buildings, having characteristic spherical or cylindrical shapes, are the signatures of nuclear power plants in this country. They are designed to withstand an overpressure of from about 3 to 5 atm. These numbers are calculated by computer codes, with assumptions being made about the most likely kinds of chemical reactions, nuclear reactions, and so forth, to occur for a particular kind of accident, the so-called maximum credible accident.

The most disastrous accident that could happen to an operating nuclear reactor would be for *all* the pipes carrying cooling water to the core to break simultaneously. This is not considered to be likely enough by the NRC to warrant consideration. This agency has determined the maximum credible accident to be one in which *one* of the coolant pipes breaks, although it has always been maintained that the probability of such an accident's happening is vanishingly small.

It is not at all clear that the multiple-barrier approach could contain a steam or hydrogen explosion, which could result from a partial core meltdown. A molten uranium dioxide core would be so hot that it would melt through the bottom of the reactor vessel, through the concrete of the containment building, and into the earth (hence, the name "China Syndrome" for such an event). This molten mass would explode violently upon encountering significant groundwater, sending the radioactive debris back up into the reactor building. Very little research has been done on this area of reactor safety, and such events are not included in the computer codes.

There is little agreement among critics and proponents on the efficacy of reactor-accident containment procedures or designs. Indeed, there have been radioactivity releases—see

below. In all cases it is believed that the amount of release was small enough that it did not constitute a hazard, but there is also disagreement about that. It is fair to say that the triple-barrier containment approach, *as we now have it*, is probably not adequate for all possible failure modes.

Safety History

On the face of it, and in comparison to other technologies, nuclear power seems remarkably safe. There have been only two accidents at a commercial reactor with significant radiation release to the public. There have been accidents at other reactors; there have been near misses at commercial plants, and there have been literally hundreds of minor accidents that could have been significant except for the safety systems described earlier.

The nuclear industry is required to report all safety-related incidents to the Nuclear Regulatory Commission, which publishes them in a publication, *Nuclear Safety*, every two months. This journal has, in addition, articles related to safety and is a valuable source of information.[9] The large number of safety-related incidents can be sorted into three major categories:

■ Operator error.
■ Equipment failure.
■ Poor design.

All of the major accidents at nuclear plants can be attributed to one of these three categories, principally the first.

There have been five major accidents at nuclear reactors worldwide. These are summarized in Table 7.10. Three of these accidents

[9] A number of these incidents have been discussed in detail from the viewpoint of their serious potential consequences in the book *The Nugget File*, edited by Robert D. Pollard and published by the Union of Concerned Scientists, Cambridge, Massachusetts, 1979.

TABLE 7.10 Major Nuclear Reactor Accidents

Accident	Year	Reactor Type	Commercial	Radiation Release	Loss of Life
Chalk River, Chalk River, Canada	1952	Research; D_2O, H_2O		Little	
Windscale, Windscale, England	1957	Power; graphite, CO_2	X	Significant	
Idaho Reactor Test Station, Idaho	1961	Experimental		Little	2
Enrico Fermi, Lagoona Beach, Michigan	1966	Power LMFBR	X	None	
Three Mile Island, Harrisburg, Pennsylvania	1979	Power PWR	X	Significant	

were at commercial plants, as indicated. Only one of these, the Three Mile Island accident, was at a reactor that is of a design currently in use in the United States.

The Chalk River accident was caused by two operator errors, the first of which could only have occurred in a D_2O-moderated, H_2O-cooled reactor. A technician closed the wrong valve, reducing the H_2O coolant supply, but not the D_2O moderator supply. The chief operator came down from the control console to the basement, discovered the error, and telephoned back to the control desk to have the control rods inserted. The operator there, however, pushed the wrong button; the control rods were not inserted. The core eventually overheated, melted, and caused the formation of hydrogen, which exploded, destroying the core. There was no significant release of radiation to surrounding buildings, and no one was injured.

The Windscale accident was also caused by operator error, but in a reactor of a type for which this kind of error in unique. This type of graphite-moderated, gas-cooled reactor needs to be shut down periodically to anneal the graphite, as discussed earlier in this chapter. This annealing process must be done slowly and carefully, or the graphite can overheat and burn. This is exactly what happened. Un-

fortunately, sensors inside the core and in the stack did not detect the fire and consequent radiation loss for several days.

The third accident was at an experimental reactor being operated by the U.S. Army in Idaho. Two technicians were attempting to remove a control rod by hand, which was apparently sticking, when it suddenly gave way, releasing a burst of neutrons. These neutrons apparently killed the men instantly and activated the area. The core remained intact, and no radiation was released to the environs.

The accident at the Enrico Fermi LMFBR has already been discussed. This was a case of design error or faulty construction or both. The accident did result in a partial core meltdown, but no adverse effects followed, and no radiation was released.

The Three Mile Island accident is the most serious accident yet to occur in the commercial nuclear power field. Substantial amounts of radiation were released, although the immediate effects on the population are believed to have been small. The accident was caused by a series of equipment failures, design deficiencies, and operator errors that would be very difficult to anticipate in any model, but that are obviously not of trivially small probability. Three Mile Island will no doubt be the most investigated accident in nuclear power

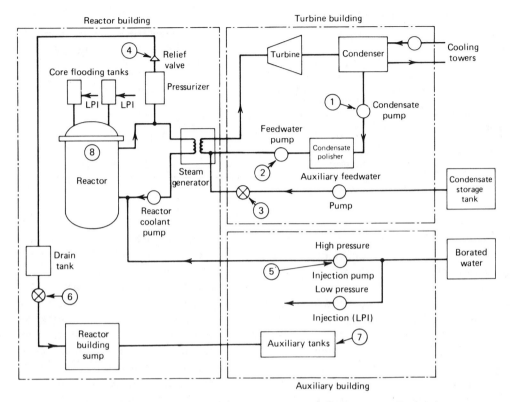

Figure 7.18 Simplified diagram of the Three Mile Island Unit 2 reactor. Redundant systems are not shown (i.e., one steam generator instead of two, etc.).

history (to date), with no fewer than 14 committees examining it.[10]

Most of the salient facts concerning the accident are known, and we shall reconstruct it, referring to Figure 7.18.

Three Mile Island Nuclear Power Station

[10]Presidential Commission, Nuclear Regulatory Commission, Electric Power Research Institute, Advisory Committee on Reactor Safeguards, Metropolitan Edison Company, Babcock and Wilcox Company, Energy Subcommittee of the House Interior Committee, Health Committee of the Senate Human Resources Committee, Energy Subcommittee of the Joint Economic Committee, Environmental Protection Agency, Energy and Nuclear Proliferation Subcommittee of the Senate Governmental Affairs Committee, Energy Subcommittee of the House Government Operations Committee, Energy and Power Subcommittee of the House Commerce Committee, and the General Accounting Office.

contains two Babcock and Wilcox 907-MWe (2772-MWt) PWRs. On March 28, 1979, Unit 1 was shut down for refueling, and Unit 2 was operating at 98 percent power. At about 4:00 A.M., the accident occurred.

1. Condensate pump shuts off (No. 1 in Figure 7.18).
2. Loss of suction causes feedwater pumps to trip off (No. 2); turbine trips off.
3. Within 2 s, reactor "scrams," auxiliary feedwater pumps start.
4. Within 6 s, pressure builds in steam generator (2255 psi) and causes relief valve in pressurizer to open (No. 4).
5. Within 12 s, pressure inside reactor reaches 2355 psi and causes Reactor Cooling System to start.

6. Auxiliary feedwater pumps are running, but no flow because valves were not opened after maintenance a few days earlier (No. 3).

7. Reactor vessel pressure drops to 2205 psi, relief valve should have closed, but was stuck open.

8. After 1 min, pressurizer level indicator rises rapidly, steam generators drying out.

9. After 2 min the ECCS automatically activates at 1600 psi (No. 5).

10. After 4 min, 30 s, operator turns off one high-pressure injection (HPI) pump because pressurizer level indicator erroneously indicates level high.

11. After 8 min, auxiliary feedwater flow is initiated by the opening of the closed valves (No. 3).

12. After 10 min, 30 s, the second HPI pump is manually turned off.

13. After 15 min, the drain tank rupture disk (No. 6) blows at 190 psi (designed to blow at 200 psi) because relief valve (No. 4) had still not closed.

14. Sump pump sends radioactive water to auxiliary tanks (No. 7).

15. Between 20 and 75 min after initiation, system parameters stabilized: 1015 psi and 550° F, relief valve open, auxiliary feedwater pumps on, ECCS on, sump pump on.

16. Between 1 hr, 15 min and 1 hr, 40 min after initiation, the operator turns off both reactor coolant pumps because of excessive vibration.

17. Core temperature begins to rise; off scale within 14 min of HPI trip. Natural circulation of coolant should have been achieved, but was not; partial core blockage or voiding suspected; that natural circulation was not achieved was *not* recognized.

18. At 2.3 hr, relief valve closed by operator (No. 4).

19. After 3 hr, reactor vessel pressure increases to 2150 psi and opens relief valve.

20. Between 3.8 and 10 hr after initiation, several pressure spikes noted; probably small hydrogen explosions. Reactor pressure decreases to about 500 psi; core was probably partially uncovered during this period, causing some meltdown and release of radioactive fission by-products to coolant.

At this point, the presence of the hydrogen "bubble" inside the reactor was noted; it continued for several days. The hydrogen concentration in the containment building was measured to be 1.9 percent (compared with a 4 percent flammability limit and a 6 to 8 percent explosive limit). The hydrogen was extracted from the reactor by "spraying" it out through the relief valve with the Low Pressure Injection system. The gases in the containment building were cycled through a hydrogen recombiner, which was installed in the auxiliary building.

The hydrogen-recombining system was installed on March 31 and continued in operation through April 13, until it burned out a heater. During this period, the reactor temperature was brought from about 290° to about 120° C with one reactor coolant pump on. Radioactivity was released from the auxiliary building because the holding tanks there overflowed and xenon and iodine outgassed.

What have we learned from Three Mile Island? On April 5, the Nuclear Regulatory Commission issued a bulletin in which it identified six potential human, design, and mechanical failures that contributed to the accident.

1. Auxiliary feedwater systems valves out of service.

2. Pressurizer relief valve failed to close.

3. Pressurizer level indicator gave erroneous readings.

4. Reactor vessel not isolated when HPI initiated, leading to transfer of radioactive water to auxiliary building.
5. HPI system operated intermittently, leading to reduction in primary coolant.
6. Turning off reactor coolant pumps to avoid vibration damage led to uncovering of core and subsequent damage.

This accident made it very clear that a series of *individually* small events could lead to consequences as serious as those postulated for the worst-case design base accident, a break in the primary coolant system. This possibility had never been considered credible, and must force some serious rethinking about accident prediction and prevention.

Only a few reports have been filed; the extent of the damage to the reactor core is believed to be extensive. The damage to the nuclear power industry also seems substantial. Many orders have been canceled, and several completed plants sit idle for lack of licenses. (Since 1978 not a single nuclear power reactor has been licensed for operation.) The credibility of the nuclear power industry with the general public is very low, but at the same time, external forces are making conventional options (in particular, oil) quite undesirable.

The alternatives to nuclear power are few and expensive. Because of the great advantage of low fuel costs enjoyed by nuclear plants, the industry may yet survive, grow, and become healthy again. And as a result of Three Mile Island, it may be even safer than before; we should watch carefully the developments of the next few years.

7.5 THE NUCLEAR FUEL CYCLE

There can be little doubt that nuclear fuels are cheaper for the generation of electricity than fossil fuels (with the exception of natural gas). It is not sufficient to consider simply the cost of obtaining nuclear fuel. Because of the nature of the fission-derived energy-generation

process, the entire cycle of ore–fuel–waste–fuel recycle must be considered (see Figure 7.19). Table 7.11 shows the costs associated with the nuclear fuel cycle in years past and projections for the future compared with the costs for coal (1 mill = 0.1 cent). Studies of this type can be faulted on several grounds: the fuel cycle has never been closed; that is, reprocessing and waste disposal have not yet been achieved successfully commercially, so their costs are not known. Ore costs have been assumed to increase slowly; in fact, the cost in 1979 is in excess of $80/kg (the study in Table 7.11 was published in 1976).

In addition, several of the environmental costs associated with the nuclear fuel cycle have not been adequately assessed. In particular, the excess cancer hazards associated with mining and milling and the as-yet-unknown hazards involved in reprocessing have not been included. Finally, the threat of diversion of weapons-grade nuclear material at one or more points in the cycle will have to make the cycle more expensive than indicated because of higher security requirements.

In this section we examine the entire nuclear fuel cycle from several points of view: the science involved, the environmental impact of each step, and the relationship of each step to the whole. Finally, we try to assess the present state of nuclear power with respect to the present state of the nuclear fuel cycle.

Mining and Milling

The end product of these operations is relatively pure U_3O_8, *yellowcake*. At present there are eight uranium mills in operation in Wyoming and two still functioning in Utah. There are, of course, similar facilities in many other countries, including Canada, Australia, the Soviet Union, Zaire, Gabon, and others. The processes used to obtain the yellowcake are different in the various facilities because of the wide range of chemical constituency of the

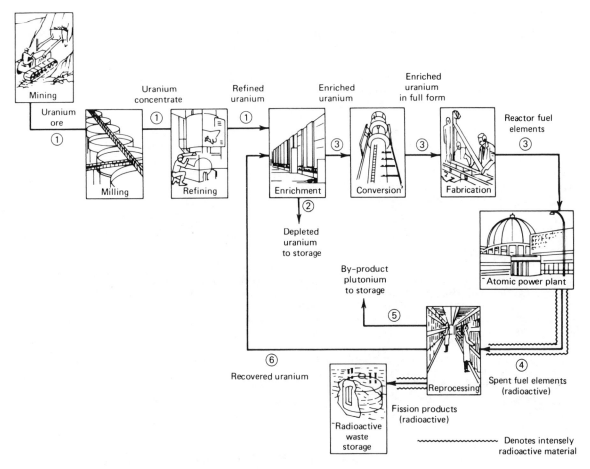

Figure 7.19 The nuclear fuel cycle, as originally envisioned.

Fuel Flow: Annual Flow of Fuel Material in 300-MW BWR[a]

Ray	Uranium Content (tons)	^{235}U (percent)	^{235}U (lb)	Plutonium (lb)
1	50[b]	0.7[c]	700	—
2	49.6	0.2	200	—
3	20[d]	2.0	800[d]	—
4	19.6	0.8	300	250
5	—	—	—	250
6	19.6	0.8	300	—

[a]Equilibrium conditions.
[b]Corresponds to about 25,000 tons of uranium ore.
[c]Natural concentration of ^{235}U isotope in the uranium element.
[d]Corresponds to one fourth of reactor core.

TABLE 7.11 Nuclear Fuel Costs 1965 to 1975

Factors (per Unit Uranium)	Range (mills/kWh)[a]				
	1965–1973	1974–1975	1984–2085 (Estimated)	1975	1985
Ore cost ($/lb)	6–8	8–15[b]	15–25[a]	0.94	1.43
Conversion ($/lb)	1.0–1.25	1.25–1.40	1.40–1.80	0.09	0.10
Separative work ($/SWU)[c]	28–38	38–43	50–100	0.58	1.24
Fuel fabrication ($/lb)	28–41	35–41	41–53	0.33	0.40
Shipping, reprocessing, and waste disposal ($/kg)[d]	20–40	40–80	100–150	0.15	0.28
Interest charges	12–15	12–15	15–18	0.61	1.23
Total				2.70	4.68
Coal (cent/10⁶ Btu)[e]	30–45	45–150	150–220	14.10	20.70
Incremental cost advantage, nuclear over coal				11.40	16.00

Adapted, with permission, from E. Zebraski and M. Levenson, "The Nuclear Fuel Cycle," *Annual Review of Energy* 1, 101 (1976), copyright © 1976 by Annual Reviews Inc.
[a]Using highest value in range for each time period; all figures in 1975 dollars.
[b]Spot purchases over $35/lb have been recorded, but on a very thin market.
[c]Unit of separative work.
[d]Or cost of long-term fuel storage and safeguards without reprocessing and neglecting present worth of uranium and plutonium credits.
[e]Assuming a heat rate of 9000 Btu/kWh for coal versus 10,000 Btu/kWh for nuclear. Considerably higher spot-purchase prices have been recorded for both uranium and coal, but on thin markets. Oil prices are 30 to 50 percent higher than coal.

uranium. These processes are also different from those normally used to extract metals from ores.

The rather large abundance of uranium on the earth's surface is a consequence of the large ionic radius of quadruply ionized uranium. Because of this large radius, uranium is fractionated out of partial melts in the mantle upward to the surface. At the surface, the existence of several ionic states allows uranium to form a wide range of complexes, many of which are water-soluble. In order for uranium to concentrate in ores, several circumstances must coincide:

■ Source of the element.
■ Water to transport it.

■ Suitable subsurface conduits.
■ Complexing agents.
■ Precipitating elements.

The probability that all these circumstances will occur together very often is small, so ore bodies of high concentration are rare (and, indeed, have already been mined!).[11]

The uranium being commercially exploited today comes primarily from fossil placers and sandstone, the ore having a uranium content in the 1 to 10 parts per 1000 range. Significant

[11]These ores were in the form of *pitchblende* or *uranite* of from 1 to 4 percent grade found only in Canada and Zaire (known previously as the Belgian Congo). The chemical composition is of the form $xUO_2 \cdot yUO_3$, with $0 \leq y/x \leq 2$.

quantities of uranium are believed to be available at lower grade.

To produce concentrated yellowcake, the crushed ore is first treated with a leaching solution (either acidic or basic, depending on the ore). Then the uranium is removed from the leach liquor by either solvent extraction or ion exchange. The final product contains 70 to 80 percent U_3O_8.

The production of yellowcake is not without environmental hazards. Uranium mining is uniquely hazardous because of the radium, thorium, and daughter nuclei concentrations found along with the uranium. All these nuclear species decay through the daughter, *radon*, a shorter-lived radioactive gas. The radon daughters (also radioactive) are quickly absorbed on dust particles that may be inhaled and deposited on inner lung surfaces of miners.

We discuss the biological effects of nuclear radiation in detail in Chapter 14. For the time being, we note that the excess cancer risk for uranium miners has been well established by epidemiological studies.[12] Because of the large number of physical and physiological variables involved in estimates of radiation dose for miners, a unit based on the concentration of radon daughters has been developed. Any concentration of radon daughters in $1\ \ell$ of air that will result in the *ultimate emission* of a total of 1.3×10^5 MeV of alpha-particle energy is called one working level (WL).

Occupational exposure to 1 WL for a period of 1 month is called a working level month, or 1 WLM. Since July 1, 1971, occupational exposures have been limited to 4 WLM per year. The epidemiological study referred to earlier estimates that one man working 1 yr at 4 WL exposure incurs an additional cancer risk of 0.00053. The extent to which this represents an acceptable risk can be debated. It is difficult, but certainly possible, to place numerical values on such risks, and we discuss this process, "radiation axiology," in Chapter 14.

An additional environmental hazard associated with nuclear fuel procurement is present by the uranium mill **tailings**, that is, the residue from the ore after the uranium has been removed. These tailings typically have a high thorium concentration, and therefore high radon emissions.

These emissions pose serious local problems if the tailings are not appropriately stored—and sealed. In Grand Junction, Colorado, tailings were used in home and school construction, resulting in radiation levels significantly higher than background at those sites. (The houses and schools were subsequently destroyed, and the homeowners were compensated at federal expense.) The seriousness of the radon emissions from tailings has been estimated by several authors, who do not agree. Several factors must be taken into account:

- Quantity of tailings produced.
- Amount of radon escaping tailings.
- Health effects of radon.
- Length of time.

Since the thorium isotope in question (^{230}Th) has a very long half-life (800,000 yr), the potential for harm is great. One estimate[13] places the number of deaths attributable to uranium mill tailings over an 80,000-yr period at about 5.7×10^6. Another estimate[14] is 0.2 to 0.25 deaths per year. This sort of discrepancy underlines the difficulties in making such estimates as well as

[12]F. E. Lundin, Jr., J. K. Wagoner, and V. E. Archer, *Radon Daughter Exposure and Respiratory Cancer, Qualitative and Temporal Aspects*, National Institute for Occupational Safety and Health and National Institute of Environmental Health Sciences *Joint Monograph No. 1*. Springfield: NTIS, 1970.

[13]D. D. Comey, "The Legacy of Uranium Tailings," *Bull. Atomic Sci.* September (1975), p. 43.

[14]Bernard L. Cohen, "The Impacts of the Nuclear Energy Industry on Human Health and Safety," *Amer. Sci.* **64**, 550 (1976).

the need for better estimates. Again, the death level that could be considered acceptable can and is being debated, but a serious debate cannot proceed without accurate, reliable information on effects, costs, and benefits.

Enrichment

Light water reactors of the type built in the United States require enriched fuel—of the order of 3 percent ^{235}U compared to the natural abundance of 0.71 percent. Chemical reactions have very little sensitivity to the atomic mass of the reacting species; consequently, they are of no use in providing a means for concentrating the ^{235}U isotope.

Many physical processes are dependent on atomic masses; two have been used to provide enriched uranium, and a third is in the development stage. Both gaseous diffusion and centrifuges are currently in use for enrichment. Laser separation has worked in the laboratory and is being scaled up for commercial application. These three processes can be compared from several points of view (see Table 7.12). Although we haven't defined all the terms in this table yet, it is clear there is an economic advantage to the laser process. This, coupled with rather-low first plant costs, means that many nations now without uranium enrichment capabilities (and consequent nuclear-weapons-making ability) may soon have them.

Enrichment processes are characterized by a **separation factor**, that is, the ratio of the percentage of desired isotope after the enrichment stage to that before. When the separation factor is small, as it is in the gaseous-diffusion and centrifuge processes, many stages must be used to enhance the desired isotopic content.

In the gaseous-diffusion process, a set of barriers as in Figure 7.20 is used. The barriers contain hundreds of millions of pores per square inch, the average diameter being about 5×10^{-6} cm.

The diffusion rate of a gas is inversely proportional to its molecular weight; therefore, the theoretical separation factor for gases of two different weights is given by

$$\alpha = \sqrt{\frac{M_h}{M_l}}, \qquad (7.28)$$

where M_h and M_l are the molecular weights of the heavier and the lighter species, respectively.

The gas used in ^{235}U separation is UF$_6$. This substance is a solid at STP, but sublimes to a gas at 56.4° C. This compound has two advantages for use in the diffusion process: fluorine has a relatively low molecular weight, and it exists in nature in only one isotopic form. Unfortunately, UF$_6$ is very corrosive to

TABLE 7.12 Comparison of Three Types of ^{235}U Enrichment Processes

	Laser	Centrifuge	Diffusion
Separation factor	10–1000?	2–10	1.0043
Energy requirement (kW h/SWU)	170	210	2100
Capital cost ($/SWU)	195	233	388
Economic size (metric tons)	3000	3000	9000
Process area (ha)	7.2	8	2.4

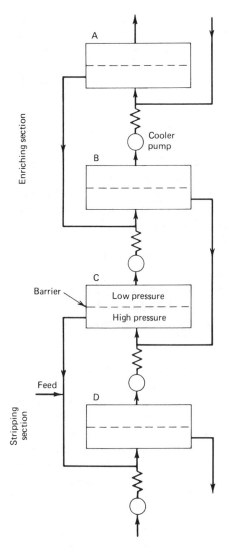

Figure 7.20 A gaseous-diffusion cascade.

many metals and reacts readily with moisture to form a solid (UO_2F_2). For UF_6, $\alpha = 1.0043$.

There are three gaseous-diffusion plants in the United States operated by the Department of Energy at Oak Ridge, Tennessee; Paducah, Kentucky; and Portsmouth Ohio. Together, these plants have the capacity to produce about 17×10^6 SWUs per year. An **SWU** is a

unit of separative work. It is not easy to define an SWU, since the amount of enriched uranium produced per SWU is not a constant but depends on the total amount of natural uranium being processed and the concentration of ^{235}U in the residual materials, or **tails**. For a tails concentration of 0.2 percent ^{235}U, a gaseous-diffusion plant produces about 200 g of 3.2 percent enriched ^{235}U per SWU. At a tails concentration of 0.3 percent, 300 g/SWU is produced. But operating at the higher tails assay requires more feed materials.

The separative work *per stage*, SW, may be defined. It is simply

$$SW = L\theta(1 - \theta)(\alpha - 1)^2, \qquad (7.29)$$

where L is the throughput in moles, θ is the mole fraction of material in one exit stream and $(1 - \theta)$ is the mole fraction in the other.

If we denote the molecular fractions of ^{235}U in the feed, product, and tails streams by E_0, E_p, and E_t, respectively, then the number of moles of feed and the number of moles of tails required to provide 1 mol of product is given by

$$\text{Feed} = \frac{E_p - E_t}{E_0 - E_t},$$
$$\text{Tails} = \frac{E_p - E_0}{E_0 - E_t}. \qquad (7.30)$$

Example 7.5 How many moles of feed are required and how many moles of tails are produced in providing 1 mol of 3 percent enriched ^{235}U, starting with natural uranium? (Assume 0.2 percent tails.)

$$\text{Feed} = \frac{0.03 - 0.002}{0.0071 - 0.002} = 5.49 \text{ mol},$$
$$\text{Tails} = \frac{0.03 - 0.0071}{0.0071 - 0.002} = 4.49 \text{ mol}.$$

Because the separation factor for an individual gaseous-diffusion barrier is too small, the barriers must be arranged in a cascade of many stages as in Figure 7.20. Note that the enriched gas is always moving "up" in this drawing, and the stripped gas is moving "down." Nearly 2000 such stages are required to produce 90 percent enriched ^{235}U with tails between 0.5 and 0.1 percent (a smaller tails assay would require even more stages). This requirement of a large number of stages accounts for the large capital investment, plant area, and high energy consumption of gaseous-diffusion plants. In 1970, the three U.S. plants used 4 percent of all the electricity produced in the United States.

It is for these reasons, plus one other, that private industry has not been induced to get into the enrichment business, even though overtures were made by the federal government in the early 1970s. The other factor has to do with other enrichment technologies. Businessmen do not want to invest multibillions of dollars in a technology that seems to be on the verge of being made antiquated by newer processes.

The centrifuge was studied for uranium enrichment back in the early days of the **Manhattan Project**,[15] but was discarded because the high-speed rotor technology required was not reliable enough. In recent years, a new design has been developed, and that, along with advances in materials science, has led to the development of reliable, advanced centrifuge systems, which are in use in the United States today and are being planned by a European consortium of commercial firms.

In the centrifuge the separation is proportional to the mass difference, not the ratio, so that a much higher separation factor can be achieved. Just how high is classified information in the United States; factors between 2 and 10 are believed to be possible, and even much higher factors have been hinted at.

A cutaway drawing of a centrifuge is shown in Figure 7.21. The thin-walled rotor is driven

by an electromagnetic motor, shown at the bottom of the casing. Gas (UF$_6$) is fed into and withdrawn from the rotor through the stationary center post. The bottom scoop, also stationary, serves as a means of removing the depleted gas, and also as the driver for a vertical gas flow. It is the vertical flow, more than the rotary action, that provides an *axial* isotopic separation. The motion of the gas can be analyzed by solving the rather complex equations resulting from the application of conservation of energy, momentum, and mass. Since the motion is turbulent and very complex, straightforward, analytic solutions are not possible. Indeed, high-speed computers fast enough and large enough to do the job have become available only in recent years.

While the separation factor for a centrifuge may be large, the throughput (i.e., the amount of material processed by an individual unit) is small. This is a consequence of the fact that it is difficult to make large rotors that will spin at the high speeds required (400 m/s) and maintain physical integrity. Also, the gas must be kept at a rather low pressure to prevent the deposition of solid UF$_6$ in the interior of the centrifuge. With a separation factor of 2, a gas-centrifuge plant having the capacity of 9×10^6 SWU/yr would require 500,000 centrifuges. Clearly, this enrichment technology is not "cheap," even though the operating expense of this system compared to a similar-sized diffusion plant would be smaller. The capital outlay required is still large enough to discourage all but the largest governments and the one European consortium, which is beginning to rethink its position in light of the recent reduced interest by industry in nuclear technology.

There is another technology that may eventually be even cheaper and therefore more accessible to smaller countries, leading to possible nuclear weapons proliferation: laser isotope separation.

As we noted in Chapter 5, the amount of

---

[15]*Manhattan Project* is the code name given to the atomic bomb development project during World War II.

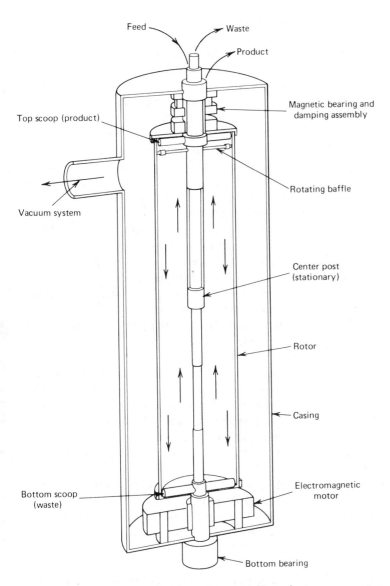

Feed

Waste

Product

Magnetic bearing and
damping assembly

Top scoop (product)

Rotating baffle

Vacuum system

Center post
(stationary)

Rotor

Casing

Bottom scoop
(waste)

Electromagnetic
motor

Bottom bearing

Figure 7.21 Cutaway drawing of a centrifuge isotope separator
stage.

energy an atomic system (atom, molecule, nucleus) may have is quantized, that is, limited to selected values. Therefore, atomic systems absorb and emit energy in quantized amounts that are unique for a given atomic system.

A simplified allowed-energy-level diagram for hydrogen is shown in Figure 7.22. A hydrogen atom excited to the $n = 3$ state may emit a $-1.9 - (-13.6) = 11.7$ eV photon and be left in the $n = 1$ or **ground state**, or it may emit a 1.5-eV photon and be left in the first excited state ($n = 2$). Note that if 13.6 eV or more is

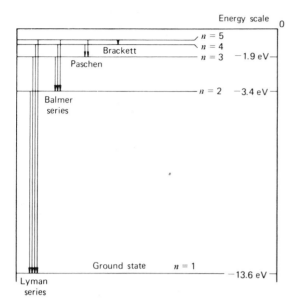

Figure 7.22 Simplified energy-level diagram for hydrogen. Lyman, Balmer, Paschen and Brackett are the names of series of spectral lines observed in the *uv*, visible, IR and IR, respectively.

absorbed by a hydrogen atom, it will be ionized; that is, the electron will be liberated. If the light coming from a hydrogen lamp is examined very carefully, it will be noted that the 11.7-eV photons are actually two groups of photons with energies differing by only about 10^{-4} eV. This "splitting" of the excited-state energy level is caused by the interaction of the spin angular momentum of the electron with its orbital angular momentum around the hydrogen nucleus. There are other small interactions that produce slight shifts and other splittings of the allowed energy levels of any atomic system.

If an energy source could be provided that could be adjusted to give exactly the right energy to ionize a ^{235}U system, but not a ^{238}U system, then isotope separation could be achieved, since the ionized atom is or molecules could be readily swept away from the neutral species by means of an electric or magnetic field.

Several techniques are being developed to use lasers for uranium isotope separation. But, because separated ^{235}U is one of the chief ingredients of nuclear weapons, much of this work is clouded in secrecy. The U.S. Government, however, in an attempt to interest private industry in isotope separation, disclosed some of the procedures a few years ago. There are also some European corporations interested in this area. We shall discuss one technique for two reasons: (1) it is reported in the literature, and, therefore, (2) it probably won't work (commercially).

This is the double-laser system shown in Figure 7.23. In this system, uranium atoms are vaporized in an oven and stream upward. About 45 percent of the atoms will be in the ground state, but about 27 percent will be in a **metastable** state only 0.077 eV higher in energy than the ground state. A metastable state is an energy state that is relatively long-lived; by that we mean long enough lived to be easily measurable. This particular metastable state is really just the ground state, split by the electron–nuclear interactions.

A xenon laser tuned to 378.1 nm excites only the ^{235}U atoms in the metastable state to a high-lying state having a short lifetime (235 ns). A strong krypton laser operating at 350.7 and 356.4 nm simultaneously is then used to further excite the ^{235}U atoms to ionization.

The ^{238}U atoms can and will absorb the krypton laser photons, but since these atoms are not in a high-lying excited state (they cannot absorb photons from the xenon laser), they will not be ionized. The ionized ^{235}U atoms are swept to a collector plate by a large negative electric potential. This system has actually worked. Using a 70-mW xenon laser and a 30-W krypton laser for 2 hr, the weapons laboratory at Livermore, California, obtained about 4 mg of 3 percent enriched ^{235}U.

It sounds easy, but there are problems. Mol-

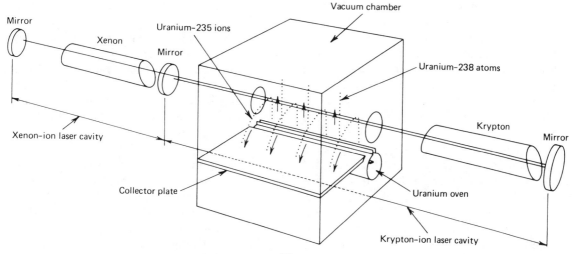

Figure 7.23 A simplified drawing of a double-laser ^{235}U enrichment scheme.

ten uranium is very corrosive. The oven in the above experiment probably didn't last more than 2 hr. At the high temperature required to vaporize uranium, many of the atoms of both isotopes will be thermally ionized and therefore attracted to the collector plate. The collector plate must be dissolved chemically to recover the uranium. Finally, operating expenses for the high-powered laser will not be trivial when scaled up to a commercial-sized plant.

There are other techniques involving compounds of uranium, such as UF_6. But these processes are even more complicated because of the much more complex nature of the excited-state system of compounds. Molecules have additional degrees of freedom not available to atoms: rotation and vibration. The payoff is large, however, so the research goes on.

There are nuclear proliferation arguments against laser enrichment technology. Laser systems are simple, relatively inexpensive, and are also relatively easy to build and conceal. The counterargument to that statement is that laser systems (at least, workable ones) are complicated, not easy to build, and it would be difficult to conceal the fact that a small system designed for nuclear reactor fuel enrichment had been upgraded to make higher enrichment materials. This counterargument loses some force if UF_6 laser enrichment processes are made to work commercially. It should be clear that this new technology does offer the hope of producing enriched uranium at a rather considerably reduced cost over either diffusion or the centrifuge.

There are other enrichment technologies, one of which is in use in Germany. This is the "Becker nozzle" technique. In this process the UF_6 gas is "sprayed" against a curved blade, with separation resulting from the centrifuge action of the gas flowing around the curve. This process has a separation factor α of about 1.01. This technology was recently sold to Brazil along with a reprocessing plant. It should not come as a surprise sometime in the decade of the 1980s when Brazil announces it has joined the nuclear club. (Brazil did not sign the Nuclear Nonproliferation Treaty.)

South Africa has announced it has developed a "new" enrichment system. But, from the few amplifications on that announcement, it seems likely that this new system is very much like the Becker Nozzle. Again, we should remember that South Africa did not sign the Nuclear Nonproliferation Treaty, faces an oil embargo from most of the OPEC nations, and has few native oil resources. It is in essence being forced to go nuclear.

Nuclear reactors in the United States and most of the rest of the world continue to burn up ^{235}U, which must be supplied at no small cost in an enriched form. The techniques for doing this enrichment are well established, but very inefficient. New processes are on the horizon that promise not only to make enriched uranium cheaper, thus extending our uranium resources, but also to make enriched uranium available to more countries, thus making life on earth a little more fragile.

Closing the Cycle

As indicated in Figure 7.19, the nuclear fuel cycle is supposed to be closed. We say *supposed to be* because, in fact, it has not yet

been closed. The missing link is the reprocessing step in which unused ^{235}U and the ^{239}Pu generated in the reactor are separated, concentrated, and refabricated into fuel to return to another reactor. There are a variety of reasons that this step has not yet been done successfully (some attempts have been made). To track down this story, let us recall that a 1000-MW reactor uses 3 kg of ^{235}U per day of full power operation. That means it also produces roughly 3 kg of radioactive by-products. There is also a nontrivial amount of transuranic nuclei produced. See Table 7.13 for the amounts and half-lives of the principal isotopes produced.

Thus the United States, as of August 1979, had produced about 565,000 kg of radioactive wastes per day; none of it was being processed. Wastes are, for the time being, stored at reactor sites after being removed from the core.

Reprocessing of reactor fuel is a necessity for extracting plutonium for weapons production. Since this has been done successfully for almost 40 years, it is difficult to understand why

TABLE 7.13   Principal Radioisotopes Produced in LWRs

		Yield (g/metric ton)	
	Half-Life (yr)	^{235}U Fuel	^{239}Pu Fuel
Actinides			
^{244}Cm	18.11	18	381
^{243}Am	7380	80	833
^{241}Am	432	33	251
^{237}Np	2.14×10^6	468	258
Fission products			
^{147}Pm	2.6234	107	96
Mixed Xi isotopes	—	5057	4547
^{144}Ce	284.4 days	304	264
^{137}Cs	30.17	1160	1198
^{134}Cs	2.062	165	175
Mixed Pd isotopes	—	1190	1770
^{106}Ru	368 days	146	239
^{103}Rh	56 min	344	437
^{99}Tc	2.13×10^5	792	795
^{90}Sr	29	508	307
^{85}Kr	10.4	27	17

ive and civil

the same procedures cannot be applied with the same success to civilian nuclear reactor wastes. The military fuel reprocessing technique is called the **Purex process**, which stands for plutonium–uranium extraction process.

The Purex process, as used in the one civilian reprocessing plant at West Valley, New York (no longer in operation), is shown in Figure 7.24. Fuel rods are chopped by a mechanical shear into small fragments. Some gaseous fission products are released; in early plants these were simply released to the atmosphere, but newer plants store these wastes. The fuel rod chunks are dropped into a nitric acid bath. The fuel—uranium, plutonium, and fission products—dissolves in the acid, but the fuel cladding does not.

The cladding, which is radioactive as well by virtue of neutron absorption and by contamination from the fuel, is removed and placed in containers for later burial—the radioactivity is at a relatively low level. The

uranium, $UO_2(NO_3)_2$, and plutonium, $Pu(NO_3)_4$, are leached out of the nitric acid solution by an organic solvent, tributyl phosphate diluted with kerosene.

This solvent extraction is done in a vertical countercurrent column; the acid solution flows downward and the solvent upward. At the bottom of the column the acid is neutralized, and the waste liquor is stored awaiting further processing. The solvent solution at the top of the column is sent through another column where a reducing agent changes the soluble tetravalent plutonium nitrate to a trivalent nitrate insoluble in the organic solvent but soluble in nitric acid. The uranium nitrate remains in solution in the organic solvent. Both the uranium and the plutonium are further purified and then converted into an appropriate form for shipment to a fuel fabricator.

This is the process that has been used for about 40 years to extract plutonium for reactor fuel rods for the purpose of making nuclear

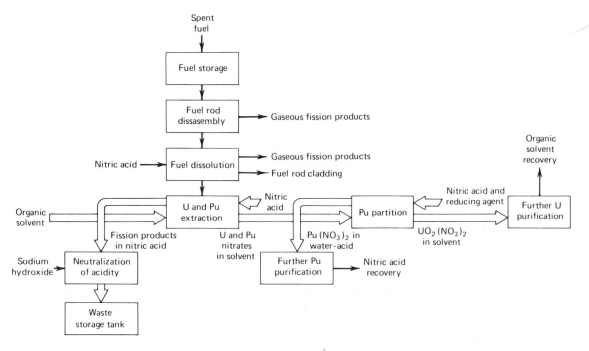

Figure 7.24 The Purex process for extracting uranium and plutonium from spent reactor fuel.

weapons. It is a technologically sound and proved process *for military wastes.* Civilian plants have not been able to use this process successfully. There are at least two major reasons why.

1. Civilian wastes have 5 to 10 times the residency in the reactor as military wastes.
2. Military plants are at least 50 percent "overdesigned" compared to civilian plants.

Higher residence times for civilian wastes means higher levels of radioactivity in the spent fuel rods. These higher levels require more shielding and make remote handling more difficult and expensive. High radioactivity levels also make the organic solvent breakdown much faster and result in a poor extraction efficiency for the U–Pu materials.

The West Valley plant was plagued with mechanical breakdowns caused by marginal construction. To be fair, we should say that since there had been no experience with civilian wastes at the time the West Valley plant was built some of these problems were probably not foreseeable. On the other hand, more than 60 percent of West Valley's operation was with military wastes under an AEC contract that amounted to an outright subsidy. The plant processed about 600 tons of fuel at a price of about $30/kg. Even with this, the plant could not operate in the black and eventually closed, leaving the state of New York a highly radioactive legacy of 600,000 gal of liquid waste residing in two stainless steel tanks.[16]

There were two other short-lived attempts at commercial nuclear fuel reprocessing. In the late 1960s General Electric built a reprocessing plant at Morris, Illinois. This plant was to use a new G.E. process, Aqualfor, which would have produced a solid uranium. The process was a technical failure, and the $64 million plant was written off in 1974. It is used today as a spent-fuel storage facility. A large reprocessing plant was virtually completed by Allied Chemical at Barwell, South Carolina. As it neared completion the operators began to read the handwriting written on the wall by G.E. and the operators of the West Valley plant, and consequently asked the federal government to underwrite their proposed operation. The government refused, and the plant today sits virtually complete and idle.

The government refused to underwrite nuclear fuel reprocessing primarily for one reason: reprocessing adds to the plutonium proliferation problem.[17] The policy has been that nuclear fuel should not be reprocessed; rather, it should be safely disposed of *as is.* The Carter administration in early 1980 announced that the federal government would take the responsibility for finding a safe method for disposal of this spent reactor fuel.

There are a number of points of interaction with the environment in connection with closing the cycle. These include

- Transportation.
- Handling.

Transportation is involved at all stages in the fuel cycle: ore, refined oxide, converted hexafluoride, and so forth. But it becomes a serious hazard only in the transporting of spent fuel from reactor sites to a reprocessing facility. If such plants were available, the number of shipments moving between reactors and reprocessing plants would today be of the order of 30 per day in the United States, averaged over the year. Even in the absence of reprocessing, if nuclear power expands, as many expect, there will be a very large number of spent fuel shipments daily from reactors to nuclear waste repositories of some kind.

[16]Litigation concerning the seat of responsibility for these wastes continues today.

[17]Requests by the Carter administration in mid-1980 for sales of enriched uranium to India cast doubts on the sincerity of its antiproliferation stance.

TABLE 7.14 Achievable Routine Emissions from Nuclear Facilities

Source	Isotope	Routine Release Rate (Ci/yr)	MPC[a] (Ci/m^3)	Relative Hazard[b] (m^3/yr)
Uranium mill[c]	Radon-222	90	3×10^{-9} A	3×10^{10}
	Uranium plus daughters	1,300	2×10^{-5} W	7×10^{7}
Enrichment complex[c]	Uranium	0.02	2×10^{-5} W	1×10^{3}
1000-MBWR	Krypton-85	315	3×10^{-7} A	1×10^{9}
	Xenon-133	28,000	3×10^{-7} A	1×10^{11}
	Hydrogen-3	6,000	3×10^{-3} S	2×10^{6}
	Iodine-131	1	1×10^{-10} A	1×10^{10}
Reprocessing plant[c]	Krypton-85	330,000	3×10^{-7} A	1×10^{12}
	Hydrogen-3	7,000	3×10^{-3} W	2×10^{6}
	Hydrogen-3	3,500	2×10^{-7} A	2×10^{10}

Adapted from J. P. Holdren, "Hazards of the Nuclear Fuel Cycle," *Bulletin of Atomic Scientist*, October 1974, p. 14, by permission of the Bulletin of Atomic Scientists, a magazine of science and public affairs, copyright © 1974 by the Educational Foundation for Nuclear Science, Chicago, IL 60637.
[a]"A" denotes MPC for air, "W" denotes MPC for water.
[b]For example, the first entry indicates 30 billion (3×10^{10}) cubic meters of air (A) per year is required to dilute the release of 90 Ci/yr of radon-222 to achieve its maximum permissible concentration (MPC) of three billionths (3×10^{9}) of a curie per cubic meter of air.
[c]Per 1000-MW reactor year, services.

The shipments are made in large, heavily shielded casks, each containing about 8×10^{16} Bq ($\sim 2 \times 10^{6}$ Ci) of radioactivity. Casks designed for truck shipment must be somewhat smaller, since most states have regulations placing an upper limit of about 25 tons on vehicles plus loads. Casks designed for rail shipment can be larger and more elaborate, in the sense that on-board cooling can be provided. These casks must be built to withstand some rigorous testing, but no amount of testing can ensure complete safety. Department of Transportation regulations require that the level of radioactivity measured at the *surface of the vehicle* carrying the cask must not be greater than 200 mrem/hr.[18] But the average

[18]The rem (1 mrem = 10^{-3} rem) is the amount of radiation of any kind that produces the same biological effect in man as 1 R (roentgen) of 250-keV X rays; that is, 100 ergs of energy deposited per gram of tissue. The rem has been replaced by the *sievert* (Sv) by international agreement; 1 rem = 0.01 Sv. See Chapter 14 for additional information.

annual background dose received in the United States is about 130 mrem, so this value could represent a significant dose for any prolonged exposure. Note that the level in the driver's compartment of a truck is not to exceed 2 mrem/hr; but even at this low level, a prolonged trip could result in a significant dose. We discuss the biological significance of these doses in Chapter 14.

With such a large number of shipments projected to be on the road in the last two decades of this century, it is inevitable that there will be accidents. And eventually an accident will result in some exposure to the general public, just as in the case of the Three Mile Island accident.

There exists the potential for significant releases of radioactivity from reprocessing plants. Certain restrictions are placed on these releases by the Nuclear Regulatory Commission. These restrictions take the form of Maximum Permissible Concentrations (MPCs) for various radionuclei. The published MPCs do not take into account the possibility of biological reconcentration of the radionuclei, but anyone releasing this material must demonstrate that reconcentration does not occur, or reduce releases accordingly. The achievable routine emissions and MPCs for various nuclear facilities are given in Table 7.14. At the one commercial reprocessing plant (West Valley, New York), annual releases

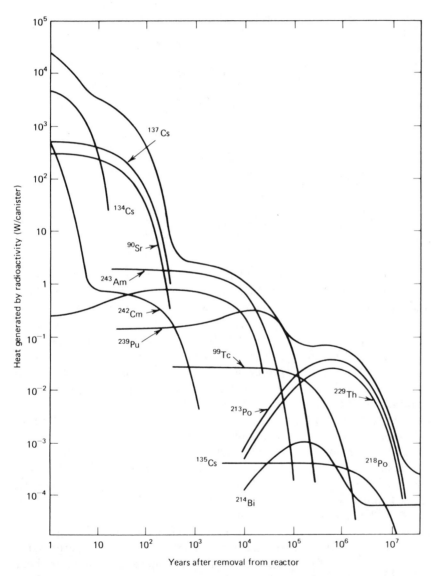

Figure 7.25 Heat generated by the more important fission products in spent reactor fuel as a function of time after removal. Total heat is indicated by the dark curve.

were as high as 22 percent of MPC values for liquid effluents and 7 percent for gaseous effluents.

Exposures to the public from the one reprocessing plant experience in the United States were no doubt within NRC guidelines. However, exposures to workers in the plant were rather high. In 1971, the last full year of operation, film-badge records indicate an exposure of 2366 person-rems for 67.9 metric tons of uranium reprocessed. Film-badge records can be misleading, since the exposures for workers employed on a contract with an outside firm would not appear in the plant's records. The general feeling in the industry is that the exposures to workers at the West Valley facility were much too high.

Finally, we note that attention has been called to the global accumulation of ^{85}Kr and ^{3}H. Projections based on continued expansion of nuclear power indicate that by the turn of the century the MPCs for these radionuclei may be exceeded if controls on their release are not initiated.[19]

The remaining unsolved problem in the nuclear fuel cycle is power-reactor waste disposal. Radioactive wastes are divided into two classifications: low level and high level. **High-level wastes** are defined[20] as the

aqueous raffinates that result from the operation of the first cycle solvent extraction system, or equivalent, and the concentrated wastes from the subsequent extraction cycles, in a facility for reprocessing irradiated reactor fuels.

Reactor spent-fuel rods fall into the high-level category. **Low-level wastes** are essentially

everything else, so long as the radioactivity does not exceed federal limits. These are primarily rags, contaminated water, irradiated metal, and other miscellaneous items. Low-level wastes are collected in barrels and buried. Some difficulty has been experienced with leaky barrels (Maxey Flats, Kentucky), but there should really never be a problem with this. Rigid control must be maintained over high-level wastes for two reasons: they generate substantial quantities of heat, and they are biologically toxic.

Since the Carter administration's policy with regard to reactor wastes was to store them

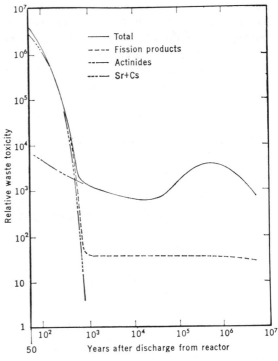

Figure 7.26 Relative toxicity of reactor-produced wastes as a function of time after removal. Note the large increase between 10^5 and 10^6 yr caused by actinide buildup. (From A. S. Kubo and D. J. Rose, "Disposal of Nuclear Wastes," *Science* **182**, 1205 (1973). Copyright © 1973 by the American Association for the Advancement of Science.)

[19]"The Safety of Nuclear Power Reactors (Light Water Cooled) and Related Facilities," *WASH-1250*, Washington, D.C.: Atomic Energy Commission, July 1973.
[20]"Alternatives for Managing Wastes from Reactors and Post-fission Operation in LWR Fuel Cycles," Volume 4, *ERDA 76-43*. Washington, D.C.: ERDA, 1976.

without reprocessing, we could now discuss the various options for storage. But it is worth digressing to consider some of the storage ramifications that accrue from reprocessing. We see in Figure 7.25 that after about 5 yr strontium and cesium isotopes account for most of the heat production, and from Figure 7.26 that the actinides (i.e., U, Pu, Np, Am, Cm, etc.) account for most of the toxicity after about 700 yr. In 700 yr the 30-yr half-life of ^{90}Sr will be reduced by a factor of 10^{-7}, making this storage problem within the realm of human consideration. Actinide half-lives are much longer, but the amount of material to be stored is vastly smaller. In fact, the actinides could be recycled in a reactor to be converted by neutron absorption to nuclides having much shorter half-lives with little loss of reactor efficiency. No doubt these arguments were considered when the decision was made to bypass reprocessing. Presumably the dangers of reprocessing with large amounts of enriched uranium and plutonium being present was a stronger argument against it.

Example 7.6 Trace the decay path(s) followed by the recycling of ^{241}Am.

$$^{241}\text{Am} + n \rightarrow$$

$$
\begin{array}{ccc}
^{242}\text{Am} + n & \rightarrow & ^{243}\text{Am} \\
\downarrow & & \downarrow \\
\tau_{1/2} = 100 \text{ yr} (\beta^-) & & \tau_{1/2} = 8000 \text{ yr} (\alpha) \\
\downarrow & & \downarrow \\
^{242}\text{Cm} & & ^{239}\text{Np} \\
\downarrow & & \downarrow \\
\tau_{1/2} = 163 \text{ days} (\alpha) & & \tau_{1/2} = 2.33 \text{ days} (\beta^-) \\
\downarrow & & \downarrow \\
^{238}\text{Pu} & & ^{234}\text{Pu} \\
\downarrow & & \downarrow \\
\tau_{1/2} = 90 \text{ yr} (\alpha) & & \tau_{1/2} = 24{,}300 \text{ yr} \\
\downarrow & & \\
^{234}\text{U} \ (\tau_{1/2} = 2.5 \times 10^7 \text{ yr}) & &
\end{array}
$$

All decays lead to uranium or plutonium isotopes, which would be extracted in the next reprocessing.

A variety of storage techniques have been suggested in this country and some preliminary work has been done, principally on salt mine storage. The various possible routes for reactor high-level wastes are indicated in Figure 7.27, with Route 0 being the one currently in favor.

Salt mines are preferred for waste storage for a very simple reason: salt is highly soluble in water. Therefore, the very existence of large salt deposits implies that these regions have not yet come in contact with groundwater since they were laid down a very long time ago. One must then extrapolate from that to conclude that these deposits are equally unlikely to contact groundwater in the future, and as a consequence would be appropriate waste storage sites. In a proposed salt mine storage facility, shown schematically in Figure 7.28, steel cylinders containing the **radwaste**, as it is called (each about 3 m long and 30 to 60 cm in diameter, containing either the concentrated waste from a reprocessing plant or the spent fuel from a reactor, depending on the route in Figure 7.27), would be buried in salt. The containers would be hot enough to cause "plastic" flow of the salt, which would then fuse around the containers. The steel would eventually corrode, and the radwaste would mix with the salt. But because of the heat and plastic flow character of the salt, the salt–radwaste system would be relatively immune to stress.

A series of tests was made at an abandoned salt mine near Lyons, Kansas, in the 1960s. Serious questions were raised by officials of the state when several hundred thousand gallons of water disappeared from a nearby working mine. Efforts then shifted to salt mines in New Mexico.

We shall not discuss in detail the other alternatives proposed. These are summarized in Table 7.15. Whatever method(s) is finally chosen, we should keep in mind not only the cost and short-term (i.e., operational) risks but also the long-term risks. In fact,

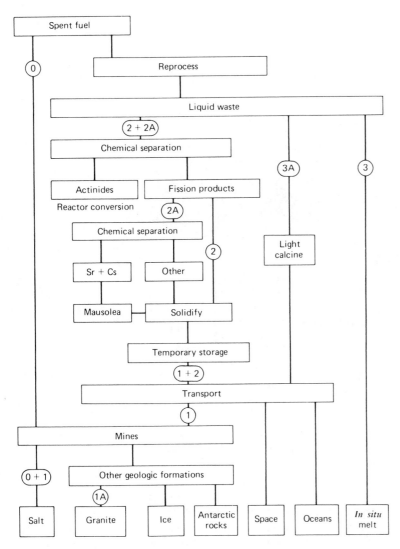

Figure 7.27 Routes for spent reactor fuel. (From A. S. Kubo and D. J. Rose, "Disposal of Nuclear Wastes," *Science* **182**, 1205 (1973). Copyright © 1973 by the American Association for the Advancement of Science.)

... there is a persuasive case for the subordination of both immediate risks and present costs to potential long-term hazards when selections are being made among alternative operations for the disposal of long-lived wastes.

... *Affordability is a flexible social and political decision.*[21]

[21]Gene I. Rochlin, "Nuclear Waste Disposal: Two Social Criteria," *Science* **195**, 23 (1977).

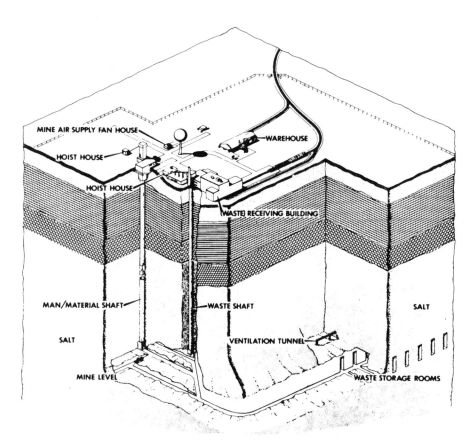

Figure 7.28 Artist's conception of a salt mine radwaste storage system. (From J. M. Dukert, *High-Level Radioactive Waste.* Washington D.C.: U.S. Energy Research and Development Administration, 1975.)

The costs of high-level waste disposal have not been determined, principally because a method has not been selected, studied, tested, and scaled up to operational size. Most projections predict a rather small incremental cost, something in the neighborhood of 0.01 to 1.5 mill/kWh. Others, while not predicting an exact number, believe it will be small. Salt mine storage has been used to a limited degree and with some success in West Germany. The Canadians have built above-ground concrete bunkers to store unprocessed spent fuel elements. But it is difficult to scale these operations up to the magnitude required for U.S. civilian nuclear power.

Finally, we should note that military radwaste in the form of concentrated liquids has been stored in large steel tanks at several locations around the country: Savannah River, South Carolina; Hanford, Washington; and Idaho Falls, Idaho. These tanks have corroded over the years, producing several major leaks; in general, they are not satisfactory for long-term storage.

We have now closed the nuclear fuel cycle, insofar as it is possible to do so. There remains one aspect to nuclear power yet to be considered in this section: the decommissioning of a power reactor. Every mechanical device has a finite lifetime; parts wear out. Reactors have

TABLE 7.15 Summary of Nuclear Waste Disposal Options[a]

Route	Option	Cost (mill/kWhe)	Advantages	Disadvantages
1	Salt mine	0.045–0.050	Most technical work to date; plastic media with good thermal properties occur in seismically stable regions	Corrosive media; highly susceptible to water; normally associated with other valuable minerals; difficult to monitor and retrieve wastes
1A	Granite	0.050–0.055	Crystalline rock; low porosity if sound; comparable to salt in thermal properties; retrievable wastes	Nonplastic media; presence of groundwater; difficult to monitor
2	Further chemical separation; recycle actinides	0.065–0.320	Reduced long-term toxicity; technology feasible; increases future options	Additional handling and processing; more toxic materials in fuel inventory; waste dilution due to processing; fission products remain
2A	Further chemical separation	0.140–1.100	Reduced long-term toxicity; reduced short-term thermal power; some reduction of fission product toxicity; increases future options	Additional handling and processing; more toxic materials in fuel inventory; waste dilution due to processing; storage and disposal of Sr and Cs extract; fission products remain
3	Melt *in situ*	0.011–0.016	*In situ* creation of insoluble rock–waste matrix; no transportation; reduce handling	Highly mobile wastes during 25-yr boiling phase; presence of groundwater; irretrievable wastes; proliferation of disposal sites; difficult to monitor
3A	Melt *in situ*, central repository	0.031–0.036	*In situ* creation of insoluble rock–waste matrix; short boiling period; no proliferation of sites	Presence of groundwater; irretrievable wastes; difficult to monitor
2	Antarctic rocks		Immobile water	Very narrow temperature limits; not a permanent geologic feature; difficult environment
2	Continental ice sheets		Immobile water	Cannot dispose of actinides; limited amount of ice; not a permanent geologic feature; difficult environment

Source: A. S. Kubo and D. J. Rose, "Disposal of Nuclear Wastes," *Science* **182**, 1205 (1973). *Copyright* © 1973 by the American Association for the Advancement of Science.
[a]Refer to Figure 7.27.

an additional wear mechanism: neutron bombardment. We have seen that maintaining a good neutron economy is essential in reactor operation; but, even so, a significant fraction of high-energy neutrons, the factor f in Figure 7.4, escape the core. Many of these will interact in the pressure vessel, dislodging atoms from the crystalline structure. This process over a period of time embrittles the steel, causing fatigue and eventually failure.

Reactor operators continually monitor this effect by bombarding samples of the original steel in the reactor core and periodically subjecting them to stress tests. After between 20 and 40 yr the reactor operator will be faced with a decision. The reactor is beginning to show signs of wear. Should the owners

1. Renew—repair, refit, rebuild.
2. Convert—burn fossil fuels.
3. Terminate—shut down the operation?

If the decision is made to terminate operation, three choices are available:

- Mothballing.
- Entombment.
- Dismantling.

In each of the three cases all mobile radioactivity (fuel rods, cooling water) must be removed and safely disposed of. Radiation must be monitored inside and outside the plant. **Mothballing** is usually for a specific period of time, to allow certain radionuclei in the pressure vessel activated by neutron bombardment to decay to lower levels of activity. After that, dismantling should follow. **Entombment** is the permanent sealing of the pressure vessel and structure. The entombment should survive long enough to permit the induced radioactivity to decay to acceptably low levels, although, as we shall see, this may require a very long time.

Only nine small reactors have been shut down in this country since the beginning of the nuclear age in 1957; these are given in Table

7.16. It should be noted that these are from 25 to 100 times smaller than the power reactors currently being constructed and that any conclusions we reach based on the experiences with these small reactors must be appropriately scaled, if possible, to the current generation.

The principal hazard to immediate dismantling of a reactor is the radioactivity induced in the iron, nickel, cobalt, and copper in the steel. Copper is usually an impurity in stainless steel, whereas cobalt and nickel are required. Stainless steel is required in the inner core structure of the reactor. The pressure vessel itself is usually a carbon steel, having a much lower content of cobalt and nickel. The radioisotopes and the amounts present in the Piqua Reactor 3 yr after shutdown and the amounts projected from this to be present in a 3485-MWt reactor immediately after shutdown are given in Table 7.17. These amounts are for the entire reactor, of course, but even taking this into account, it would be very easy to receive a maximum dose of radiation in a very short period of time.

Only one of the reactors listed in Table 7.15 has been dismantled: the Elk River Nuclear Reactor. To minimize radiation exposure to workers, the steel plates of the reactor vessel were cut up with a plasma torch under water. This added to the dismantling cost, which was $6.9 million compared to an original construction cost of $6 million. All of the other reactors have been mothballed, effectively entombed, for an indefinite period of time.

There are several environmental hazards in connection with decommissioning a reactor, aside from the radiation exposure to the workers. These include airborne radioactivity released during dismantling, particularly of the concrete containment structure, radiation exposure during transport of radioactive wastes, and contamination of local water supplies. There is no reason why, with proper care, these hazards cannot be minimized to the

TABLE 7.16 Decommissioned Reactors

Name and Owner	Location	Type	Capacity (MWe)	Start-up (year)	Shutdown (year)
Hallam Nuclear Power Facility, Sheldon Station (AEC and Consumers Public Power District)	Hallam, NE	Sodium graphite	75.0	1962	1964
Carolinas—Virginia Tube Reactor (Carolinas—Virginia Nuclear Power Associates, Inc.)	Parr, SC	Pressure tube, heavy water	17.0	1963	1967
Piqua Nuclear Power Facility (AEC and City of Piqua)	Piqua, OH	Organic-cooled and moderated	11.4	1963	1966
Boiling Nuclear Superheater Power Station (AEC and Puerto Rico Water Resources Authority)	Punta Higuera, PR	Boiling water integral nuclear superheat	16.5	1964	1968
Pathfinder Atomic Plant (Northern States Power Co.)	Sioux Falls, SD	Boiling water nuclear superheat	58.5	1964	1967
Elk River Reactor (AEC and Rural Cooperative Power Association)	Elk River, MN	Boiling water	22.0	1962	1968
Enrico Fermi Atomic Power Plant, Unit 1 (Power Reactor Development Co.)	Lagoona Beach, MI	Sodium-cooled, fast	60.9	1963	1973
Peach Bottom Atomic Power Station, Unit 1 (Philadelphia Electric Co.)	Peach Bottom, PA	High-temperature, gas-cooled	40.0	1966	1974
Saxton PWR	Saxton, PA	PWR	10.0	—	1973

TABLE 7.17   Radioactivity of Reactor Vessels

Radioactive Material	Half-Life (yr)	Activity in Piqua Reactors[a] (Ci)	Activity in Commercial-Size Reactor[b] (Ci)
Iron-55	2.60	51,000	34,600,000
Cobalt-60	5.26	1,100	730,000
Nickel-63	92	7.7	2,680
Nickel-59	80,000	0.049	171
Total short-lived		480	

[a]Activity 3 yr after shutdown of the Piqua reactor; total reactor exposure = 40 MW yr.
[b]Projected activity immediately following shutdown, based on Piqua.

extent that they will not pose a significant risk to the general public.

The costs of decommissioning have been a topic for debate for some time. While there has been very limited experience in decommissioning, (really one the Elk River plant), and while it is difficult to extrapolate upward to larger plants (current pressure vessels are 12.7 cm thick compared to 7.6 cm for Piqua), it does seem clear that decommissioning does not represent a very large investment on the part of the utility owners. The NRC currently requires utilities to set aside a fund of some sort to pay for decommissioning, even though the exact costs are not now and cannot be known. These costs are being borne by the consumers of the product of each nuclear plant and, in fact, it is a rather small part of their overall electrical energy price, about 1 percent.

Commercial nuclear power exists because a substantial number of electric utility owners perceive it to have an economic advantage over coal in many parts of the country. Whether such an advantage actually exists is the subject of much debate.[22] We cannot add productively to that debate here; but we do note that no new reactor orders have been placed in the United States since 1978. Whether there are economic or technical

reasons for this is not clear; however, the effect is the same in either case.

7.6   THERMONUCLEAR FUSION

The topic of fusion is perhaps out of place in a chapter on nuclear power, an existing technology. In some future edition of this book, fusion may well be entitled to its own chapter. For the time being, however, this seems an appropriate place to discuss it.

Thermonuclear fusion is in a unique position. It is the only energy technology that has been identified as a major energy source before it has produced a single joule of energy, even before it has been proved to be capable of producing energy in a useful fashion. Workers in the field have an almost religious con-

TABLE 7.18   The Advantages of Fusion Power

1. Effectively infinite fuel supply at low cost ($\ll 1$ mill/kWh).
2. Inherent safety, no runaway.
3. No chemical combustion products.
4. Relatively low radioactivity and attendant hazards.
5. No emergency core cooling problem.
6. No use of weapons-grade materials, so no

[22]See the references at the end of this chapter.

TABLE 7.19   Comparison of Energy Resources

Resource	Energy Content (TW yr[a])	Life Expectancy at Current Use Rate[b] (yr)
Recoverable oil	500	87
All recoverable fossil fuels	2500	438
^{235}U at \$220/kg (LWRs)	200	35
^{238}U at \$220/kg (breeders)	11,000	1,930
^{233}U at \$500/kg (breeders)	1,000,000	175,440
Lithium fusion	50,000	8,700
D–T fusion	500,000,000	8,770,000

[a]TW yr = terawatt-years. 1 TW yr = 10^{12} W yr.
[b]Current use rate estimated at 0.17 Q/yr or 5.7 TW yr/yr.

viction that fusion will prove to be the ultimate answer for our future energy needs. At first glance, fusion seems to be very attractive (see Table 7.18). The most important of these advantages is the first, the effectively infinite supply of raw material.

The raw materials for fusion can be obtained from seawater, and since these resources are projected to last a very long time—probably longer than the solar system will last—the supply is infinite. Fusion resources are compared with those for fission and fossil fuels in Table 7.19. We see that indeed fusion fuels from the ocean are anticipated to last more than 1 billion years, even at a rather high rate of energy consumption, and that solar radia-

tion is anticipated to last not a great deal longer. These resources dwarf both fossil and fission fuels in their capability to provide energy. Since the area of the ocean is 361×10^6 km^2 and its average depth is 3.8 km, its volume must be 1.37×10^9 km^3. This is approximately 40 million times the volume of 33 km^3 given in Table 7.20 as being equivalent in energy content to the fossil fuels. If we assume the extraction of only 10 percent of the deuterium from seawater, this still represents an amount of energy roughly *four million times that of the fossil fuels*. We now understand the tremendous attraction of fusion. But since we do not now have working fusion plants, what is the difficulty?

TABLE 7.20   Energy Obtainable from Seawater (D–D) Fusion

Volume of Water	Energy (thermal J)	Coal Equivalent (metric ton)	Crude-Oil Equivalent (bl)
1 ℓ	7.95×10^9	0.26	1.30
1 m^3	7.95×10^{12}	260	1300
1 km^3	7.95×10^{21}	260×10^9	1300×10^9
33 km^3	2.62×10^{23}	World's total supply of fossil fuels	

Physics of Fusion

The difficulty does not arise from the physics of fusion—this is really quite simple. Take two light nuclei and fuse them together. The product nucleus invariably has less mass than the original nuclei. If two light nuclei, the total mass number of which is less than about 60, are made to fuse together, the binding energy per nucleon of the resultant nucleus must increase. As a consequence, energy in the form of kinetic energy of the fusion products must be released.

Example 7.7 Calculate the energy release from the reaction

$$D + T \rightarrow n + {}^4He.$$

The total mass of the reactants is

$$m_D = 2.0149 \text{ u}$$
$$m_T = \frac{3.01605 \text{ u}}{5.03005 \text{ u}}.$$

The total mass of the products is

$$m_n = 1.008665 \text{ u}$$
$$m_{{}^4He} = \frac{4.00260 \text{ u}}{5.011265 \text{ u}}.$$

The mass difference is $(5.03005 - 5.011265) = 0.018785$ u. Since $1 \text{ u} = 931.395 \text{ MeV}$, $0.018785 \text{ u} = 17.496 \text{ MeV}$. Since $1 \text{ MeV} = 1.6 \times 10^{-13} \text{ J}$, $0.018785 \text{ u} = 2.799 \times 10^{-12} \text{ J}$ per fussion event.

A variety of fusion reactions have been proposed:

$$D + D \rightarrow p \rightarrow + T + 3.25 \text{ MeV} \quad (22{,}000 \text{ kWh/g}),$$

$$D + D \rightarrow n + {}^3He + 4.0 \text{ MeV} \quad (27{,}000 \text{ kWh/g}),$$
$$(7.31)$$

$$D + T \rightarrow n + {}^4He + 17.6 \text{ MeV} \quad (94{,}000 \text{ kWh/g}),$$

$$D + {}^3He \rightarrow p + {}^4He + 18.3 \text{ MeV} \quad (98{,}000 \text{ kWh/g}),$$

$$2H_2 + O_2 \rightarrow H_2O + H_2O + 0.000006 \text{ MeV}$$
$$(0.0044 \text{ kWh/g}).$$

The $D + T$ reaction has been extensively studied, not only because of its relatively high energy release per gram but also because it has a somewhat lower ignition temperature and a higher fusion cross section (refer to Figure 7.29).

Only the deuterium (D) occurs in nature; tritium (T) is radioactive, with an 18-year half-life. At first glance, it would not seem to be sensible to use a nonnatural resource, but the tritium can be regenerated *catalytically*. If the neutron (having 14 MeV energy) is absorbed in a blanket surrounding the reactor this way:

$$^6Li + n \rightarrow {}^3H + {}^4He + 4.8 \text{ MeV}, \quad (7.32)$$

not only are the 14 MeV of the neutron and the small positive energy release of Eq. 7.32 captured, but also the tritium is recovered, and the liquid lithium blanket can be used as a heat-transfer medium.

There are many other reactions. We know these reactions will work; they are the bases of thermonuclear weapons (H-bombs), and they are part of the energy-production mechanism in stars. This, of course, is the problem: to find a "box" capable of holding a miniature star!

High temperatures are required in order to get the nuclei close enough together so that the strong nuclear force of attraction can overcome the weaker electrical repulsion of the positively charged objects. A short calculation will show you that, if a fusion cross section of 1 barn is interpreted as a geometric shape and if we assume that the two nuclei must be within that shape in order to fuse, then this distance of closest approach must be of the order of 5.6×10^{-12} cm. This is actually not a large distance on the nuclear scale, being about equal to the actual size of the deuterium nucleus!

At the temperatures required for fusion—millions of degrees—the gas is not a gas in the usual sense at all, but a fully ionized **plasma**. Molecular bonds have been broken, all the electrons have been stripped away from each

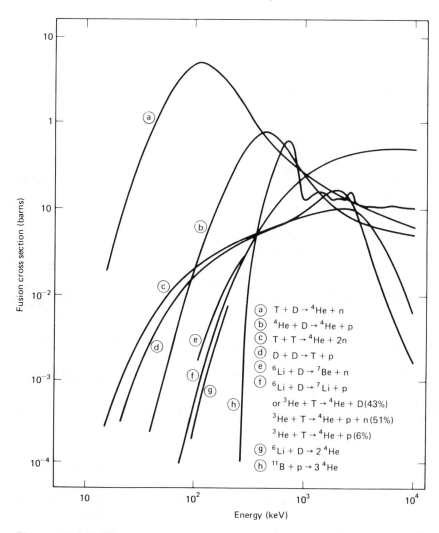

Figure 7.29 Fusion cross section versus average kinetic energy for several fusion reactions. (Used with permission from D. J. Rose and M. Feirtag, "The Prospect for Fusion," *Technology Review*, December 1976, p. 21. Copyright © 1976 by the Alumni Association of the M.I.T.)

atom, and the resulting plasma can be deflected by electric and magnetic fields. Accelerations of the plasma, changes in direction or speed, will cause radiation—Bremmstrahlung—in the form of high-energy ultraviolet photons or X rays. The motion of this plasma itself will generate magnetic fields, as does the motion of

any charged particle. These self-generated magnetic fields can interact with the plasma to cause effects, usually undesirable.

Because of these and other instabilities, current fusion reactors operate in fits and starts. So far none has been able to maintain the three required physical parameters high

TABLE 7.21   Tokamak Parameters[a]

Year	Confinement Time (s)	Ion Temperature T_i (K)	Density × Confinement Time $n\tau_E$ (s/cm^3)	Sustainment Time (s)
1955	10^{-5}	10^5	10^9	10^{-4}
1960	10^{-4}	10^6	10^{10}	3×10^{-3}
1965	2×10^{-3}	10^6	10^{11}	2×10^{-2}
1970	10^{-2}	5×10^6	5×10^{11}	10^{-1}
1976	5×10^{-2}	2×10^7	10^{13}	1
1978	5×10^{-2}	6×10^7	2×10^{13}	1
1980[b]	2.5×10^{-2}	8.2×10^7	3.7×10^{13}	1
Needed for a reactor	1	10^8	10^{14}	≥ 10

[a]Date prior to 1978 from B. Pease, Culham Laboratory, U.K.
[b]Data from the Princeton Large Torus. These data are for the highest temperature achieved in May 1980. Higher densities and confinement times have been obtained at lower temperatures.

enough, long enough. These are: temperature, density, and confinement time. For a given fusion reaction at a given temperature, there is a minimum density–time product required in order that **thermonuclear ignition** may occur. For $D + T$ at 10^8 K, this product must be at least 10^{14}—this is the so-called *Lawson criterion.*

Ignition refers to the onset of fusion. After ignition, the plasma will be heated by product α particles, and if feed material were supplied, the fusion process could continue. Researchers are coming tantalizingly close to achieving the Lawson criterion (see Table 7.21). The major obstacle to achieving ignition is containment. To date two very different techniques have been used: magnetic and inertial confinement.

Magnetic Confinement

The use of magnetic fields to confine the plasma has been the more extensively studied of the two techniques. Charged particles, of which the plasma is composed, experience forces when moving in a region of magnetic field. These forces, you will recall, are not directed along the field or the line of motion of the particles, but rather are perpendicular to both:

$$\mathbf{F} = q\mathbf{v} \times \mathbf{B}, \qquad (7.33)$$

where q is the charge of the particle, $\mathbf{v}$ is its velocity, and $\mathbf{B}$ is the magnetic field. The consequence of this relationship is that charged particles entering a magnetic field region will spiral along the lines of the field, unless by chance they happen to enter perfectly normal to the field, in which case they will move in circles, not spirals. This is shown in Figure 7.30.

If the magnetic field is not uniform, the spiral will decrease or increase in size depend-

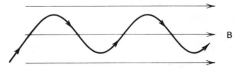

Figure 7.30 Path taken by a charged particle in a uniform magnetic field.

ing on whether the field is becoming stronger or weaker in the direction of the spiral. If the field becomes strong enough, the spiral motion of the charged particles will become tighter and tighter until the motion along the field line goes first to zero and then changes direction. The particle is "reflected."

Reflection would not be necessary if the charged plasma followed along the field lines that closed on themselves, as in a toroid. In either method, then, reflection or continuous lines, the plasma is contained and kept away from the interior wall of its enclosure by being forced to follow along lines of magnetic field.

Several different devices, some of the reflecting type, some of the continuous field type, have been studied over the years. A few are still under active consideration today. However, there is general agreement that one type has the greatest chance for success in the immediate future: the **tokamak**.

The tokamak effectively overcomes both the major disadvantages of the reflecting or pinch-type magnetic-confinement machines: the need to pulse the magnetic field to compress the plasma for heating purposes and the drift of ions from the plasma center to the walls. Pulsing means that (1) continuous operation is obviously not possible, (2) very large cyclic stresses are placed on the mechanical and heat-transfer parts of the machine, (3) currents in the magnets become very large, with concomitant heat losses, and (4) each fusion pulse must produce a very large amount of energy so that the time averaged amount can be economically large enough. In fact, some of the designs proposed, for example, the theta pinch, were shown to require more energy to pulse than could be produced! The ion-drift problem is caused by the fact that the field at the torus center is stronger than at the edges. Ions will not spiral along the field lines, then, but will drift toward lower field regions—that is, into the wall.

These problems were overcome, that is, it is

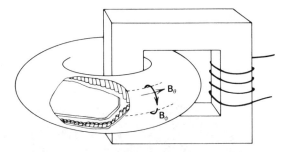

Figure 7.31 Simplified diagram of a tokamak magnetic fusion confinement device.

believed they will be overcome, in the tokamak, shown schematically in Figure 7.31. A transformer is added to the torus such that the plasma itself becomes the secondary winding. A varying primary current will induce an additional current in the plasma, giving rise to additional field lines, B_ϕ in Figure 7.31. These fields cause the drifting positive ions and negative electrons to remain in the plasma, rather than drift into the walls.

Unfortunately, the transformer functions only with varying primary currents, so we still have a pulsed machine. But these pulses can be quite long—seconds or even minutes, and, most importantly, the main confinement field is not pulsed. An added benefit with the transformer is that it causes *ohmic heating* of the plasma. That is, the induced secondary current generates heat in the plasma as would any current in any conductor. Unfortunately, the plasma resistance decreases considerably as its temperature increases, so it is not clear how much ohmic heating can be obtained.

The tokamak idea originated in the Soviet Union in the late 1950s and was liberally described by Soviet scientists in the open literature. But it was ignored in the United States until 1969. (There was tokamak work in Great Britain before 1969.) Several devices have been built, each larger than its predecessor, but none has achieved the conditions required for fusion (refer to Table 7.21). Two very large

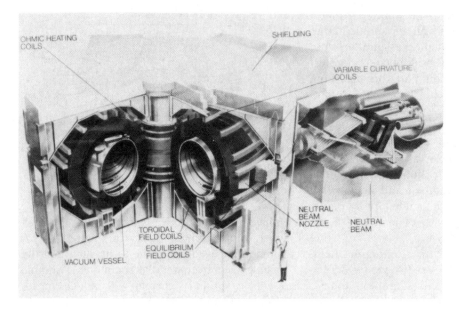

Figure 7.32 A drawing of the TFTR at Princeton Plasma Physics Laboratory, scheduled to go into operation in 1982. (Illustration courtesy of the Princeton Plasma Physics Laboratory.)

devices, the Tokamak Fusion Test Reactor (TFTR) at Princeton Plasma Physics Laboratory in New Jersey and the Joint European Torus (JET) at Culham, England, are currently under construction. They are scheduled to be completed in 1982 and 1983, respectively. These devices should be capable of producing, for the first time, thermonuclear ignition in the laboratory. A drawing of the TFTR is shown in Figure 7.32.

Inertial Confinement

In this scheme a small, hollow, glass pellet, called a *microballoon*, is filled with the fusion reactants (usually a D + T gas, although frozen mixtures have been used) under high pressure. An intense burst of photons from a laser or charged particles from an appropriate accelerator impinge uniformly on the surface of the pellet. Energy absorbed from the incident beam causes the outer layers of the pellet to ablate away at high velocity. The reaction force to this ablation crushes the pellet, raising the density of the remaining material to very high values. When the density has been increased by a thousandfold or so, the adiabatic increase in temperature of the reactants will be enough to cause thermonuclear ignition. The resulting thermonuclear blast wave will push forward faster than the materials can ablate away, thus "burning" most of the reactants.

Obviously, it cannot be that simple, or we would have working inertial-confinement fusion today. There is yet one major difficulty: coupling sufficient energy from the laser into the pellet via the plasma produced at the pellet surface. There are two ways to resolve this difficulty: produce more power in the laser or improve the pellet design to utilize the incident energy more efficiently. Both avenues are being pursued today.

Two types of lasers have been used in fusion work, the neodymium glass laser and the CO_2

gas laser. The glass laser produces relatively short-wavelength light (1.06 μm), is limited in power capability, must be pulsed, and has rather low efficiency, a few percent. The gas laser produces infrared (long-wavelength) radiation, about ten μm, can produce high power levels, can operate continuously, and is somewhat more efficient, perhaps 5 to 10 percent.

The multiarm Shiva neodymium glass laser at Lawrence Livermore Laboratory has produced 15 kJ of light energy and, by increasing the number of arms and the diameter of the final amplifying stages, the Nova successor to Shiva will, it is hoped, be able to produce a 100-kJ pulse by 1983. The Helios CO_2 gas laser at Los Alamos Scientific Larboatary has produced infrared beam energy levels in the neighborhood of 10 kJ, and the next-generation Antares, currently under construction at LASL, should be able to deliver 100-kJ beams.

Most workers in the fusion field now believe that at least 100 kJ will be required from the laser to induce thermonuclear ignition, even with developments in pellet design. The newer generation of lasers will be able to produce this amount of energy, but with rather low efficiency, so that even if ignition is achieved, "breakeven" is still far off. There are two types of breakeven: scientific and engineering. Scientific breakeven is when the thermonuclear output is equal to the energy input to the laser system. Engineering breakeven is when the energy extracted from the thermonuclear fusion is at least equal to the total energy input to the fusion reactor operation.

There is also a general belief that neither the neodymium nor the CO_2 laser will be appropriate for advanced fusion systems. The efficiency in both cases is much too low, substantial amounts of heat removal capacity are required to cool the high-power glass amplifiers, substantial time is required for cooling and adjustment between firings, and the wavelengths are too long.

The light wavelength must closely match the plasma frequency of the surface of the exploding pellet to optimize energy coupling into the plasma. This plasma frequency can be altered by changing the size of the pellet: the longer the wavelength of the laser light, the smaller the pellet should be. But with a small pellet, less than a few tens of microns in diameter, the shock wave travels too rapidly, and "hot" electrons, very energetic electrons, carry away energy that would otherwise be used for heating the plasma.

The shock wave can be slowed by shaping the laser pulse. A small prepulse can be used to create a small plasma layer that is then heated slowly by a time-changing main pulse. The pellet burn can also be controlled by layering the pellet with different materials. Figure 7.33 illustrates two types of multi-layered pellets. A pusher–tamper material is a high-Z element such as copper used to confine the fuel and shield the interior from X rays and hot electrons. This type of pellet has relatively low gain, that is, a low ratio of fusion energy output to laser energy input, and requires a rather simple pulse shape of about 100 ps in duration. Results with these pellets have yiel-

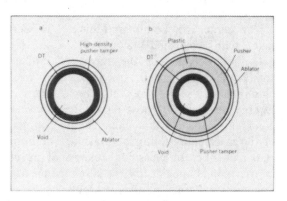

Figure 7.33 Cross sections of typical laser fusion fuel pellets: (*a*) "pusher–tamper" pellet with a thin ablator outer layer, (*b*) ablator pellet with a thin pusher layer. (From *Physics Today*, March 1975, p. 17.)

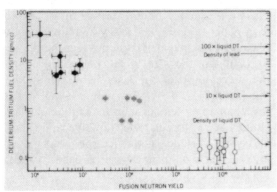

Figure 7.34 Results with fuel pellets similar to those in Figure 7.33. Ablative pellets are the six points on the left; pusher–tampr pellets are the open circles on the right. The center six points are with classified targets. (From *Physics Today*, November 1979, p. 21.)

ded high temperatures, but low compression densities; these are the open circles on the plot in Figure 7.34. An ablator material is a low-Z element that is used to generate an ablating plasma. This type of pellet has high gain but can generate hot electrons, so it requires laser pulse shaping. Results with ablator targets, the six points on the left in Figure 7.34, are characterized by low temperatures and high compression densities.

The center six points in Figure 7.34 were taken with "classified" targets. These are targets that, because of the supposed military significance of laser fusion, have been classified as secret by the Department of Energy. Presumably these targets have several layers of materials that enhance the coupling of the laser energy into the plasma. There are also a number of classified designs of asymmetric targets specifically for single beam use. Some information on these targets has been released[23]; they are fuel pellets mounted on plastic disks, each with a small plastic cap. The incident prepulse melts the cap, which flows

around the pellet, forming a uniform heat-transfer medium for the main pulse.

Single-beam asymmetric targets will obviously be required for an operating laser fusion reactor, since alignment with one beam will be hard enough, not to mention multiple beams! It would seem to be an appropriate step for the government to release information on these targets to stimulate work toward commercialization.

Even after scientific and engineering break-even are achieved by increasing the pellet gain or increasing the laser power, several substantial difficulties must be resolved before laser fusion reactors could be connected to your neighborhood power grid:

The laser must fire 10 or more times per second.
The pellet must be positioned to within 1 μm for each firing.
The first wall must last 10 to 20 years without requiring replacement.
The pellets must cost less than 1 cent each to manufacture.
The overall plant efficiency should exceed 40 percent.
The net energy produced should be comparable in cost to energy produced by other techniques.

These are formidable problems. And while it is not possible to say exactly when each of them will be solved, nevertheless ambitious projects continue. Obviously, an engineering test facility will not be constructed until thermonuclear ignition has been achieved. Most observers believe ignition will not be possible until the next-generation tokamak or laser facility has been completed, sometime after 1983. In an effort to speed up fusion development congress passed and President Carter signed into law the Magnetic Fusion Engineering Act of 1980. This act mandates the establishment of a fusion engineering device by 1990. This is somewhat before a Department of Energy panel of fusion experts believe such a device

[23]G. H. McCall and R. L. Morse, *Laser Focus* **12**, 40 (1974).

will be possible. Whether continued acceleration of the fusion energy program will be pursued in light of Reagan administration budgetary decisions has yet to be established at the time of this writing. Even if the program set out by the fusion act of 1980 were followed, commercialization of fusion could not result before 2010.

We have seen that fusion is at least a generation away from being a useful energy source and that in a generation fission may have ceased to be one. We have noted that an economy based on an increased reliance on nuclear power could be as precarious as one based on an increased reliance on fossil fuels. There are problems of radioactive wastes and heat generation with both fission and fusion, although the problems are somewhat different in the two cases. Fusion represents a potentially enormous resource; breeder reactors could do almost as well. If either one or both of these becomes a reality in the next generation, or if we are forced to increase our burning of fossil fuels, then heat production and management become even more important than they are now.

SUMMARY

The atomic *nucleus* consists of *protons* and *neutrons*. The chemical identity of an atom is determined by the number of protons in the nucleus. The total atomic mass depends on both the neutron and proton number.

Isotopes of an element are nuclei having the same proton number but different neutron numbers.

Radioactivity is the emission of a particle from the nucleus in order to change the proton-to-neutron ratio.

Fission releases energy because the *binding energy per nucleon* of the products is greater than that of the starting nucleus. Uranium

requires *thermal* neutrons in order that a **chain reaction** of fissions be sustained.

There are many different types of reactors, but only two are used commercially in this country: the **boiling water reactor (BWR)** and the *pressurized water reactor* (PWR).

The ease with which fission may be induced by a neutron depends upon the neutron energy and is measured by the *fission cross section*.

The second generation of neutrons released in a fission chain is related to the first generation by several fuel- and geometry-dependent factors. The relationship is expressed by the *multiplication factor* k_e.

The major task facing the reactor design is strict control of the neutrons. Neutron rate of change depends on three terms: production, absorption, and leakage. *Excess reactivity* is the amount of neutron flux beyond that required to maintain criticality.

Reactor control is an essential feature of reactor design. The reactor should *scram* in the event of an emergency. Reactivity is temperature dependent in a negative fashion; that is, if the temperature becomes too high the reactivity will drop below the value required for criticality.

A reactor that has been in operation for some time will still be physically hot long after it has been shutdown.

A *gas-cooled reactor* has been built, but it has not performed as well as water-cooled designs.

Breeder reactors provide a mechanism for using the substantial quantities of ^{238}U that are now being discarded. For a breeder to be successful the *conversion ratio* must be greater than 1.

For ^{235}U or ^{239}Pu reactors this means that thermal neutrons cannot be used. Fast neutrons are required.

The use of fast neutrons complicates reactor design. The heat density neutron flux must be higher. Higher fuel enrichments are also required. Fast reactors are more difficult to control, having a much smaller temperature coefficient of reactivity.

The design of breeder usually discussed in the United States is the liquid-sodium fast breeder reactor (**LMFBR**). This type of breeder has been successfully operated in France and the Soviet Union.

There are several design criteria and engineered safeguards for ensuring reactor safety. However, there have been a few reactor accidents in the plast 30 years. Only one of these (Three Mile Island) involved a commercial reactor of a type currently in use in the United States. The Three Mile Island accident provided that safety systems work; it also proved that safety systems are not immune from human error.

Uranium is difficult to find in concentrated ore bodies, because most salts are highly water-soluble. As a consequence, all inexpensive ore has been mined. Large deposits of more expensive ore still exist.

Light water reactors require ^{235}U *enrichment* of the order of 3 percent. Enrichment cannot be done chemically; a physical process sensitive to mass differences must be used.

The principle technique used to date is *gaseous diffusion*, which has a very small *separation factor*. Many stages and a great deal of power and area must be used. Future techniques may include *centrifuges* and *laser* enrichment. Both of these techniques have higher separation factors, but the capital costs may be high.

The original plan in the United States was to *reprocess* spent reactor fuel to recover the unused uranium and the plutonium produced by neutron absorption. However, no satisfactory commercial process has been discovered yet. Spent reactor fuel rods are currently being stored at reactor sites across the country.

Substantial amounts of radioactive by-products must be removed and stored to close the fuel cycle. Proposals for waste disposal include salt mine storage and above ground internment. No decision has been made by the federal government on a permanent solution.

Thermonuclear fusion may be the energy source of the future. It has a virtually unlimited fuel supply.

The physical processes behind fusion are rather simple: the binding energy per nucleon of the product nucleus is greater than that for the two original nuclei.

It is difficult to get the original nuclei to get close enough to each other to fuse, however. Two confinement techniques have been studied: *magnetic* and *inertial confinement.*

The *tokamak* magnetic device is believed to have the best chance of producing fusion energy in the immediate future, although it is generally conceded not to be a practical device for the long run.

Laser-beam bombardment of miniature glass fuel-containing spheres has also been attempted as an example of inertial confinement. This effort has also been unsuccessful. Current lasers are not large enough to couple enough power into the target in a short enough time period.

REFERENCES

General

Glasstone, Samuel, and Alexander Sesonke, *Nuclear Reactor Engineering.* New York: Van Nostrand, 1967.

Kaplan, Irving, *Nuclear Physics.* Reading, Mass.: Addison-Wesley, 1963.

Knief, R. A., *Nuclear Energy Technology: Theory and Practice of Commercial Nuclear Reactors.* New York: McGraw-Hill, 1981.

Tryor, J. C., and R. I. Vaughan, *An Introduction to the Neutron Kinetics of Nuclear Power Reactors*. Oxford: Pergamon Press, 1970.

Zweifel, P. F., *Reactor Physics*. New York: McGraw-Hill, 1973.

Breeder Reactors

Bupp, Irwin C., and Jean-Claude Derian, "The Breeder Reactor in the U.S.: A New Economic Analysis," *Tech. Rev.*, July/August (1974), p. 27.

Chow, Brian G., "The Economic Issues of the Fast Breeder Reactor," *Science* **195**, 551 (1977).

Cochran, Thomas B., *The Liquid Metal Fast Breeder Reactor: An Environmental and Economic Critique*. Baltimore: Johns Hopkins University Press, 1974.

Metz, Willian D., "European Breeders (I), (II), (III)," *Science* **190**, 1279 (1975); **191**, 368 (1976); **191**, 551 (1976).

Novick, Sheldon, "A Troublesome Brew," *Environment*, June (1975), p. 8.

Scott, R. L., Jr., "Fuel Melting Incident at the Fermi Reactor on October 5, 1966," *Nucl. Safety* **12**, 123 (1971).

Vendryes, Georges A., "Superphenix: A Full Scale Breeder Reactor," *Sci. Amer.*, March (1977), p. 26.

Reactor Safety

Castro, W. R., and Wm. B. Cottrell, "Preliminary Report on the Three Mile Island Incident," *Nucl. Safety* **20**, 413 (1979).

Cottrell, Wm. B., "Developments Pertaining to the Three Mile Island Accident," *Nucl. Safety* **20**, 619 (1979).

Gillette, Robert, "Nuclear Safety (I), (II), (III)," *Science* **177**, 771 (1972); **177**, 807 (1972); **177**, 970 (1972).

Hohenemser, Kurt H., "The Failsafe Risk," *Environment*, January/February (1975), p. 6.

Lanouette, William J., "No Longer Can the NRC Say . . . ," *Bull. Atom. Sci.*, June (1979), p. 6.

Leeper, Charles K., "How Safe are Reactor Emergency Cooling Systems?" *Phys. Today*, August (1973), p. 30.

Marshall, Eliot, "A Preliminary Report on Three Mile Island," *Science* **204**, 280 (1979).

Primack, Joel, and Franck von Hippel, "Nuclear Reactor Safety," *Bull. Atom. Sci.*, October (1974), p. 5.

Sheriden, Thomas B., "Human Error in Nuclear Power Plants," *Tech. Rev.*, February (1980), p. 20.

Study Group on Light Water Reactor Safety, "Nuclear Reactor Safety—The APS Submits Its Report," *Phys. Today*, July (1975), p. 38.

Wilson, Carrol L., "Nuclear Energy: What Went Wrong," *Bull. Atom. Sci.*, June (1979), p. 13.

Nuclear Fuel Cycle

Abajian, Vincent V., and Allan M. Fishman, "Supply of Enriched Uranium," *Phys. Today*, August (1973), p. 23.

Angino, Ernest E., "High-Level and Long-Lived Radioactive Waste Disposal," *Science* **198**, 885 (1979).

Bebbington, William P., "The Reprocessing of Nuclear Fuels," *Sci. Amer.*, December (1976), p. 30.

Carter, Luther J., "Radioactive Waste Policy in Disarray," *Science* **206**, 312 (1979).

Cohen, Bernard L., "The Disposal of Radioactive Wastes for Fission Reactors," *Sci. Amer.*, June (1977), p. 21.

Comey, David Dinsmore, "The Legacy of Uranium Tailings," *Bull. Atom. Sci.*, September (1975), p. 43.

Dahlberg, Richard C., "Weapons Proliferation and Criteria for Evaluating Nuclear Fuel Cycles," *Bull. Atom. Sci.*, January (1978), p. 38.

Gillette, Robert, "Plutonium (I) and (II)," *Science* **185**, 1027 (1974); **185**, 1140 (1974).

Harwood, Steven, Kenneth May, Marvin Resnikoff, Barbara Schlenger, and Pam Tames, "The Cost of Turning It Off," *Environment*, December (1976), p. 7.

Holdren, John P., "Hazards of the Nuclear Fuel Cycle," *Bull. Atom. Sci.*, October (1974), p. 14.

Jakimo, Alan, and Irwin C. Bupp, "Nuclear Waste Disposal: Not In My Backyard," *Tech. Rev.*, March/April (1978), p. 64.

Krass, Allan S., "Laser Enrichment of Uranium: The Proliferation Connection," *Science* **196**, 721 (1977).

Kubo, Arthur S., and David J. Rose, "Disposal of Nuclear Wastes," *Science* **182**, 1205 (1973).

La Guardia, T. S., "Nuclear Power Reactor Decommissioning," *Nucl. Safety* **20**, 15 (1979).

Lester, Richard K., and David J. Rose, "The Nuclear Wastes at West Valley, NY," *Tech. Rev.*, May (1977), p. 20.

Love, L. O., "Electromagnetic Separation of Isotopes at Oak Ridge," *Science* **182**, 343 (1973).

Nye, Joseph S., Jr., "Nuclear Nonproliferation and Energy Security," *Tech. Rev.*, December/January (1979), p. 48.

Olander, Donald R., "The Gas Centrifuge," *Sci. Amer.*, August (1978), p. 37.

Sefcik, Joseph A., "Decommissioning Commercial Nuclear Reactors," *Tech. Rev.*, June/July (1979), p. 16.

Speth, J. Gustave, Arthur R. Tamplin, and Thomas B. Cochran, "Plutonium Recycle: The Fateful Step," *Bull. Atom. Sci.*, November (1974), p. 15.

Nuclear Power Economics

Bupp, Irvin C., Jean-Claude Derian, Marie-Paule Donsimoni, and Robert Treitel, "The Economics of Nuclear Power," *Tech. Rev.*, February (1975), p. 15.

Fowler, John M., Robert L. Goble, and Christoph Hohenemsen, "Power Plant Performance," *Environment*, April (1978), p. 25.

McCaull, Julian, "The Cost of Nuclear Power," *Environment*, December (1976), p. 10.

Rossin, A. D., and T. A. Riech, "Economics of Nuclear Power," *Science* **201**, 582 (1978).

Shapley, Deborah, "Nuclear Power Plants: Why Do Some Work Better Than Others?" *Science* **195**, 1311 (1977).

Fusion

Clarke, John F., "The Next Step in Fusion: What It Is and How It Is Being Taken," *Science* **210**, 967 (1980).

Furth, Harold P., "Progress Toward a Tokamak Fusion Reactor," *Sci. Amer.*, August (1979), p. 51.

Kulcinski, G. L., G. Kessler, J. Holdren, and W. Häfele, "Energy for the Long Run: Fission or Fusion?" *Amer. Sci.* **67**, 78 (1979).

Metz, Willian D., "Fusion Research (I), (II), and (III)," *Science* **192**, 1320 (1976); **193**, 38 (1976); **193**, 307 (1976).

Stickley, C. Martin, "Laser Fusion," *Phys. Today*, May (1978), p. 50.

Miscellaneous

Bethe, H. A., "The Necessity of Fission Power," *Sci. Amer.*, June (1976), p. 21.

Hohenemser, Christoph, Roger Kasperson, and Robert Kates, "The Distrust of Nuclear Power," *Science* **196**, 25 (1977).

McIntyre, Hugh C., "Natural-Uranium Heavy-Water Reactor," *Sci. Amer.*, October (1975), p. 17.

Rose, David J., Partick W. Walsh, and Larry L. Leskovjan, "Nuclear Power—Compared to What?" *Amer. Sci.* **64**, 291 (1976).

Weinberg, Alvin M., "The Maturity and Future of Nuclear Energy," *Amer. Sci.* **64**, 16 (1976).

Periodicals

Annual Review of Nuclear Science, Annual Reviews, Inc., Palo Alto, Calif. *Proceedings of Internation Conferences of the Peaceful Uses of Atomic Energy*, United Nations, New York, N.Y.

Nuclear Science and Engineering, American Nuclear Society, La Grange Park, Ill.

Nuclear Safety, Department of Energy, Washington, D.C.

Nuclear News, American Nuclear Society, La Grange Park, Ill.

Nucleonics Technology, American Nuclear Society and the European Nuclear Society, La Grange Park, Ill.

Nuclear Fusion, International Atomic Energy Agency, Vienna.

Atomic Energy Review, International Atomic Energy Agency, Vienna.

GROUP PROJECTS

Project 1. Visit a nuclear power plant. Determine the operating characteristics of the plant; if possible, find out what the plant's capacity factor has been during the past year. (What percentage of the time did it operate?) How much spent fuel is stored on site?

Project 2. Determine what percentage of your local power company's generating capacity is nuclear? What provision does the utility have for obtaining power when the reactor is shut down? What plans does it have for obtaining additional generating capacity? How much is planned to be nuclear? How does this compare with the remainder of the country?

EXERCISES

Exercise 1. What is the minimum energy neutron for which the nuclear reaction below will occur?

$$^{32}S + n \rightarrow {}^{16}N + {}^{17}F.$$

Exercise 2.

(a) If the *mass excess* for an atom is defined as $\Delta_A = M_A - A$, show that

$$B = (\Delta_A - Z\Delta_H - (A - Z)\Delta_n) \times 931.502 \text{ MeV},$$

where Δ_H is the mass excess for hydrogen and Δ_n is the mass excess for the neutron.

(b) Calculate the binding energy per nucleon (B/A) for ^{208}Pb. Would symmetric binary fission of this nucleus yield an energy excess?

(c) Repeat (a) for ^{56}Fe.

Exercise 3. A living organism contained 75 μg of ^{14}C in the year AD 210 when it died. One part of the creature consisting of 25 percent of the original remains today. How much ^{14}C does it contain?

Answer: 15.13 μg (1982).

Exercise 4. The efficiency for counting a decay product of ^{40}K is 0.15 percent. How many atoms of ^{40}K must be present in a sample so that the quantity be known to 1 percent accuracy in 30 days of counting?

Exercise 5. In a Maxwellian distribution, the number of neutrons having a speed v is given by

$$n(v) = n_0 A v^2 \exp \left(-\frac{mv^2}{2kT} \right)$$

where n_0 is the number of neutrons per cubic centimeter, and

$$A = 4\pi \left(\frac{m}{2\pi kT} \right)^{3/2}.$$

(a) The average speed in a distribution is given by

$$\bar{v} = \frac{\int n(v)v \, dv}{\int n(v) \, dv}.$$

Using this and integrating between $0 \le v \le \infty$, show that for a Maxwellian distribution

$$\bar{v} = \left(\frac{8kT}{\pi m}\right)^{1/2}.$$

(b) The most probable speed, v_p, is the speed that is found by setting

$$\frac{dn(v)}{dv} = 0.$$

Show that for a Maxwellian distribution

$$v_p = \left(\frac{2kT}{m}\right)^{1/2}.$$

Exercise 6. The energy output of bombs is typically measured in equivalent tons of TNT. For the period between 1945 and 1963 (when a limited Test Ban Treaty was signed), 511 megatons of nuclear weapons were tested in the atmosphere or at the surface of the earth. Purely fission reactions accounted for 193 megatons of this total (fusion bombs the remainder).

(a) Using the relation, 1 megaton TNT = 132 MW yr (thermal), how many years must a nuclear reactor, with an electrical output of 800 MW and an efficiency of 32 percent, operate to produce the same number of nuclear fissions as have occurred in the nuclear testing?

(b) At 190 MeV thermal energy per fission, how many fissions occur per year?

Answer: 10.2 yr, 2.6×10^{28} fissions.

Exercise 7. Using the data in Table 7.3 find the *mean neutron lifetime* $\bar{\tau}$ for neutrons in a ^{235}U reactor. *Note*: For a multigroup reactor

$$\bar{\tau} = \frac{\sum\limits_{i=0}^{6} n_i \tau_i}{\sum\limits_{i=0}^{6} n_i}.$$

For ^{235}U, $n_0 = 99.35$, and $\tau_0 = 10^{-3}$ s.

Exercise 8. The moderator material in a reactor should be chosen so that the neutrons lose their initial kinetic energy rapidly by successive elastic collisions with the moderator nuclei.

(a) Show that if a mass m makes a head-on collision with another mass M that is initially at rest, then the relation between the KE_{before} and the KE_{after} is given by

$$KE_{after} = \left(\frac{m - M}{m + M}\right)^2 KE_{before}.$$

(*Hint*: Apply conservation of momentum and energy to the straight-line collision.)

(b) Consider deuterium and graphite as two possible moderators. Using the result from part (a), determine how many successive collisions a neutron must make with graphite nuclei (^{12}C) in order to lose as much energy as it would lose in one collision with a deuterium nucleus (^{2}H).

(c) Explain why light water makes a very effective neutron moderator. Why is heavy water better in some reactors?

Exercise 9. Assume that the heat in a reactor is produced only by the fission fragments, fast neutrons, and decay beta energies. From this estimate the radioactive waste that a 1000-MWe reactor produces per day of full power operation. Why does this not equal 3 kg?

Answer: 3.7 kg.

Exercise 10. Compare the fission cross section of ^{235}U at a neutron energy equivalent to 300°C to the geometrical cross section (radius $= 1.2A^{1/3}$ fm—1 fm $= 10^{-13}$ cm, or 10^{-15} m).

Exercise 11. What is the maximum permissible fraction of H_2O in the D_2O coolant of a CANDU reactor, if the total σ_a of the coolant is not to exceed 0.005 barns?

Exercise 12. In a neutron flux of 10^{14} neutrons/cm^2 s, what would be the fission rate in a typical UO_2 fuel pellet?

Exercise 13. If a sample of material is placed in a thermal reactor, the nuclei in the sample will absorb neutrons creating new, usually radioactive nuclei. If $n(v)v$ is the neutron flux (product of density and speed), $\sigma_a(v)$ is the neutron absorption cross section of a particular nucleus, and N the number of those nuclei present in the sample, the production rate of the new nuclei is given by

$$\frac{dN_a}{dt} = n(v)v\sigma_a(v)N.$$

(a) Show that this can be expressed in terms of the most probable velocity, v_p, using Eq. 7.9 as

$$\frac{dN_a}{dt} = nv_p\sigma_a(v_p)N.$$

(b) If the new nuclei are radioactive with a decay constant λ, show that the number of new nuclei present after an irradiation time t is

$$N_a = \frac{nv_p\sigma_a(v_p)N}{\lambda}(1 - e^{-\lambda t}).$$

Exercise 14. If a reactor contained only ^{235}U and no ^{238}U, what would be the k_∞? (Start from Eq. 7.10.)

Exercise 15. Show that η in Eq. 7.9 can be written

$$\eta = \nu \frac{\sigma_f}{\sigma_a},$$

where σ_f and σ_a are the fission and total ab-

sorption cross sections for uranium. Also show that for $\sigma_r(^{235}U)/\sigma_f(^{235}U) = 0.184$ and $\sigma_a(^{238}U)/\sigma_f(^{235}U) = 0.00474$,

$$\eta = \frac{2.5}{1.184 + 0.00474N_8/N_5},$$

where N_8 and N_5 are the number of nuclei per cubic centimeter of ^{238}U and ^{235}U, respectively, and $\sigma_r(^{235}U)$ is the nonfission capture cross section of ^{235}U. If in a uranium–graphite reactor η must be at least 1.63 for criticality, what value must N_8/N_5 have? What enrichment is required?

Exercise 16. How much UO_2 per cubic meter must be in a PWR having a neutron flux of 10^{14} n/cm^2 s in order that the thermal power density be 35 MW/m^3? Assume a total fuel volume of 800 m^3.

Exercise 17. The neutron absorption cross section for 0.06-eV neutrons in ^{235}U is 582 barns. What would the cross section be if the temperature of the fuel material were 2200° C?

Answer: 308.9 barns.

Exercise 18. A 700-MW reactor is operated at full power for 150 days. At that time, control rods are inserted to stop the fission reactivity. What is the core thermal power output 30 days later?

Exercise 19. What additional heat rate per kilogram is required in an PWR to cause the UO_2 fuel to melt? What percentage increase in fission rate could cause this? ($c_p = 0.25$ J/g° C.)

Exercise 20. Use the known values for half-lives and decays of nuclei heavier than ^{238}U to determine the chain of events for producing ^{249}Cf in a reactor.

Exercise 21. What is the doubling time for a 40 percent efficient 1000-MWe MSBR having $g_B = 0.05$ and $p_s = 50$ kW/kg? (Assume $t_c = t_r$.) What is the effect of tripling the total fuel cycle time? operating at 50 percent capacity? halving the reactor residence time? reducing g_B by 10 percent?

Exercise 22. Assume that in the year 2000 there are ten 5000-MWe LMFBRs that have been in operation since 1990. Estimate the total inventory of ^{239}Pu produced by the reactors by that time.

Exercise 23. One sequence of decays starting with ^{238}U leads to ^{206}Pb with ^{222}Rn as an intermediate product. Trace this sequence. Estimate the amount of ^{222}Rn in equilibrium with ^{238}U as a function of the initial amount of the latter. (*Hint*: Use the *Bateman Equations.*)

Exercise 24. Three parts per thousand uranium ore has the U_3O_8 extracted with 80 percent efficiency. What is the decay rate per cubic meter due to ^{238}U of the tails? What decay rate per cubic meter could be attributed to thorium?

Exercise 25. How many kilograms of feed are required to produce 10 kg of 95 percent enriched ^{235}U with a tails assay of 0.3 percent? What quantity of such tails are produced in order to satisfy the total requirements of 200 1000-MWe reactors over a 20-yr period? Assume 65 percent operation over that period.

Answer: 2309.8 kg feed per 10 kg ^{235}U; 6.58×10^8 kg.

Exercise 26. How many stages are required in a centrifuge cascade ($\alpha = 1.12$) to produce 3 percent enriched ^{235}U? What percentage enrichment do the tails contain?

Exercise 27. A D + T mixture is contained in a 600-μm-diameter fuel pellet at 100 atm pressure. If 5 percent of this material undergoes fusion, how much energy is released? If 50 percent of the released energy is absorbed by the first wall and if the first wall can withstand 1 kJ/m^2, what must the radius of the (circular) fusion chamber be?

CHAPTER EIGHT

WASTE HEAT MANAGEMENT

As we know, all energy conversion devices that use heat as an intermediate energy form must reject a large part of that heat. With the advent of nuclear power reactors, the public has become very conscious of this waste, although of course it has existed for many years with fossil-fueled plants. The Atomic Energy Commission (AEC), realizing when it began to promote commercial nuclear power in the early 1950s that the waste heat problem from reactors would be of concern, began to sponsor efforts to find beneficial applications for this discarded energy.[1]

There are certain obvious applications: space heating and urban snow removal, for example. Low-temperature steam is being used for heating in areas where transportation of the steam is not excessively expensive. Certain industries require heat; some chemical reactions occur more rapidly at elevated temperatures. Proposals to extend the shipping season in northern waterways using excess

heat from generating plants have been made. Finally, studies have been done on the feasibility of using excess heat from nuclear reactor power generators in seawater desalination plants.

These studies have had little success; the fact remains that most of the rejected heat from power plants must be absorbed by their surroundings: water, air, earth, and space. Of these four, only heat transfer to waterways and heat transfer to the atmosphere offer viable alternatives for large-scale operations. The absorption of heat by natural waters or the atmosphere produces local temperature changes that in turn affect both their biological and physical systems. We examine these local effects in this chapter; global effects are considered in Chapter 12. We also study several techniques for dispersing waste heat, as well as methods of minimizing heat production.

8.1 THERMAL EFFECTS

Waste heat is generated by many manufacturing processes, but by far the largest single source of waste heat in the United States is the electric power industry. As Table 8.1 indicates, more than 80 percent of all waste heat in 1964 was rejected by steam turbines. Of course, the cooling water requirements are larger today,

---

[1]It should be noted that, until the National Environmental Policy Act (NEPA) was tested in court in 1971 in the suit brought by opponents of the Calvert Cliffs, Maryland, nuclear plant, the AEC had argued that heat rejection from a power plant was a natural phenomenon, and so did not need to be accounted for in an Environmental Impact Statement. The court disagreed, and now heat discharge and its effects must be included in such statements for all projects under federal license or jurisdiction.

TABLE 8.1   Use of Cooling Water by U.S. Industry, 1964

Industry	Cooling-Water Intake (Billions of Gallons)	Percent of Total
Electric power	40,680	81.3
Primary metals	3,387	6.8
Chemical and allied products	3,120	6.2
Petroleum and coal products	1,212	2.4
Paper and allied products	607	1.2
Food and kindred products	392	0.8
Machinery	164	0.3
Rubber and plastics	128	0.3
Transportation equipment	102	0.2
All other	273	0.5
TOTAL	50,065	100.0

Source: D. A. Berkowitz and AM M. Squires, eds., *Power Generation and Environmental Change*, Cambridge: MIT Press, 1971.

since most sectors of the economy have grown since then. However, we should not expect the percentages to change radically.

Of course, not all electric power is produced by steam systems, but the percentage of electricity produced nonthermally, essentially by hydroelectric generators, is steadily declining, as we saw in Chapter 2.

In fact, as nuclear power generation proliferates and if geothermal power production should grow, the waste heat total will increase at an even greater rate because of the lower efficiencies of these conversion methods compared to fossil-fired steam systems. Thermal conversion efficiencies of fossil plants have improved over the years, but substantial improvement beyond the current value of about 40 percent is unlikely. Certainly, the current generation of light water reactors (LWRs) cannot be improved in terms of efficiency.

Figure 8.1 traces the history of steam electric system efficiencies. Current fossil plants are pushing the limits of technology; improved metallurgical techniques may make it possible to exceed current steam conditions of about 660°C and 270 atm, but surely the improvement will not be dramatic.

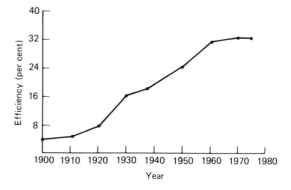

Figure 8.1 Average steam electric power plant efficiency. (From *Historical Statistics of the United States*. Washington D.C.: U.S. Department of Commerce, 1975; *Energy Perspectives 2*. Washington D.C.: U.S. Department of the Interior, 1976.)

In the years covered by Figure 8.1, the heat rate has improved at about 2.8 percent per year, but at the same time the total electric power generation increased at a rate of 7.2 percent per year. Of equal significance is the fact that in the past 20 years the maximum size of steam electric generators has increased from about 200 MWe to more than 1000 MWe.

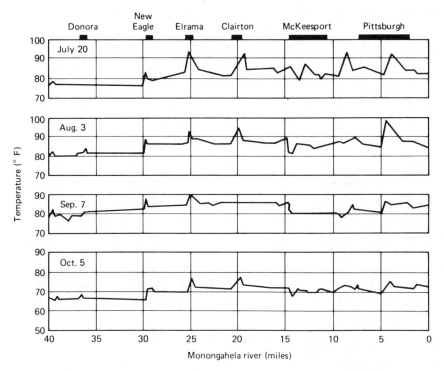

Figure 8.2 Temperature profile along the Monongahela River measured upstream from Pittsburgh. (Adapted from *Thermal Pollution—1968*, Hearings before the Subcommittee on Air and Water Pollution of the Committee on Public Works of the U.S. Senate, 90th Congress. Washington D.C.: U.S. Government Printing Office, 1968.)

The effect of this increase is to concentrate a much larger amount of waste heat in a small region. Consequently, environmental effects caused by waste heat will be much more evident now than they were a few years ago.

The primary vehicle for waste heat disposal today is water: rivers, lakes, ponds, or the ocean. The growth in electrical energy generation is threatening the stability of these natural waters. It is estimated that by the year 2000 electrical cooling needs could require two-thirds of our freshwater runoff. If electrical demands are allowed to increase for the next 100 years and are in fact met, the water requirement could not be satisfied without a substantial increase in the temperature of the water.

These projections do not take into account multiple use of the same water. Such concentration of heat sources along a river can alter the temperature profile above its normal value (see Figure 8.2). Also these estimates do not take into account the use of closed-cycle cooling towers as a substitute for direct cooling by water. The use of such cooling towers, while expensive, could nevertheless transform a problem of heated water into a problem of heated air. The heat problem still exists, but it may be easier to deal with. Surely, there is some upper limit to the amount of waste heat

TABLE 8.2 Water Properties as a Function of Temperature

Temperature (°C)	Vapor Pressure (mm HG)	Viscosity (cP)	Density (g/ml)	Surface Tension (N/m^2)	Oxygen Solubility (mg/l)	Nitrogen Solubility (mg/l)
0	4.579	1.787	0.99984	7.56	14.6	23.1
5	6.543	1.519	0.99997	7.49	12.8	20.4
10	9.209	1.307	0.99970	7.42	11.3	18.1
15	12.788	1.139	0.99910	7.35	10.2	16.3
20	17.535	1.002	0.99820	7.28	9.2	14.9
25	23.756	0.890	0.99704	7.20	8.6	13.7
30	31.824	0.798	0.99565	7.12	7.6	12.7
35	42.175	0.719	0.99406	7.04	7.1	11.6
40	55.324	0.653	0.99224	6.96	6.6	10.8

the earth can effectively handle; we shall look into this in Chapter 12.

Effects of Heat on Water Quality

Elevated temperatures affect virtually every physical and chemical property of water. Table 8.2 lists several relevant physical properties and their temperature dependences. The most dramatic change is in the vapor pressure; an increase in vapor pressure with increased temperature leads, of course, to increased evaporation. This implies an increased sedimentation rate for lakes and ponds. Oxygen solubility is also an important property, since it has a bearing on the level of animal activity a given body of water can support. All chemical reactions are accelerated by elevated temperatures. Reaction rate constants are temperature-dependent in the same manner as is the equilibrium constant referred to in Chapter 5. While generalizations are usually poor in this area, for many reactions of interest in the environmental field, reaction rates approximately double for each 10°C rise in temperature. In addition to affecting chemical reaction rates, temperature changes also impact upon ionic strengths, conductivities, and the dissociation and solubility of chemicals in water. This temperature dependence has im-

plications for the chemical requirements in water-treatment plants.

As well as having physical and chemical effects, temperature increases also affect

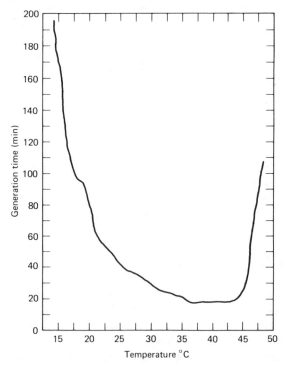

Figure 8.3 Generation time versus temperature for *Escherichia coli.*

microbiological life in water systems. These effects are most important in the consideration of sewage treatment. Growth rates of micro-organisms are sensitive to temperature. These organisms are complex collections of hydro-carbons and proteins whose various mutual interactions are affected differently by temperature changes. As a consequence, even though increased temperatures usually result in increased metabolic activity by micro-organisms, it is difficult to predict the exact temperature behavior an individual species will demonstrate. This is clearly borne out by the complex generation time curve for *Escherichia coli*, shown in Figure 8.3. Temperature also affects the bactericidal and viruscidal activity of chlorine. Temperature reductions of 10° to 20° require 1.5 to 2.5 times the chlorine content for 100 percent viral kill. Temperature increases affect not only the microbiological life in natural waters but also all aquatic life in general.

Effects of Heat on Aquatic Life

Temperature changes, especially increases, can have profound effects on aquatic life. In some cases the effects are beneficial; some increase in growth rate has been noted for some species. A temperature change may promote the growth of a desirable species over an undesirable one—although in practice the reverse often seems to happen.

On the other hand, the negative effects of temperature changes on aquatic life are many and varied:

Temperature requirements for certain fish vary for spawning, for the development of eggs and fry.
Tolerance to temperature changes varies with variety, age, size, and season.
Lethal highs and lows vary widely: $-11.8°C$ low for goldfish and $24°C$ high for pink salmon.
Sudden changes in water temperature can be lethal to certain varieties.
Certain fish can acclimate to slow temperature changes, within limits.
Fish can acclimate faster to temperature increases than to temperature decreases.
Fish acclimated to high temperatures will die quickly if suddenly placed in cold water.
Reduction in dissolved oxygen, increases in CO_2, or presence of toxic materials greatly reduces the tolerance of aquatic life to temperature changes.
Temperatures do not have to be lethal to kill a species—temperatures that favor competitors, predators, parasites, and diseases can destroy a species at temperatures far below lethality.
Increased temperatures speed bacterial metabolism, further reducing the dissolved oxygen available for fish.
Fish are not too bright and will swim into water having a lethal temperature.
Higher temperatures will produce lethargy in fish, reducing their ability to forage for food.
Increased temperatures favor blue-green algae growth, thus displacing the normal fish food, diatoms, in the water.

We could make a longer list; it is clear that the adverse effects of heat discharges into natural waters far outweigh any random beneficial effects. The utilization of waste heat for aquaculture must be done in a controlled fashion, with all aspects of the ecology of the water system studied in detail before the project is started.

8.2 DISPERSAL OF HEAT

Although large temperature increases resulting from waste heat discharges can be deleterious to aquatic life, it may be possible to design power plants to discharge the waste heat in a manner such that the temperature increase of the receiving water is within appropriate limits. If this is not possible, other methods of discharging the waste heat are available. To understand these options and the reasons for choosing a particular one in a particular situation, we need to study the physical mechanisms of heat transfer, and indeed what we mean by "heat" itself.

Heat on the Microscopic Scale

In Chapter 3 we described a transfer of energy by virtue of temperature difference. We called this transfer heat and we alluded to the average kinetic energy of atoms, molecules, and electrons of the substances. We were not very precise; we did not need to be at that time. But, since we want to talk about this transfer of energy from one substance to another, we need to take a closer look at the microscopic structure of materials and its relationship to what we call heat.

For gases a straightforward statistical mechanical analysis based on a Maxwell–Boltzmann distribution of velocities yields a relationship between the average kinetic energy of the gas molecules and the temperature:

$$KE_{av} = \frac{3}{2}kT, \qquad (8.1)$$

where k is Boltzmann's constant, 1.38×10^{-23} J/K. When the same analysis is applied to a solid, the total internal energy of a mole of substance is found to be

$$u_m = 3RT, \qquad (8.2)$$

where R is the product of k and Avogadro's number. Since the molar heat capacity c_v is the derivative of u_m with respect to temperature, this classical technique predicts that c_v should be a constant, in fact about 6 cal/mol K.

Measured heat capacities were found in the late nineteenth and early twentieth centuries to depart from this simple value (called the law of Dulong-Petit) at temperatures below about 250 K (see Figure 8.4). When the atomic nature of solids became firmly established, it was clear that the classical method of calculating atomic energies was incorrect. Not only are atomic energies quantized, but also in solids, because of the close spacing between atoms, there are cooperative interactions. Atoms do not act as isolated oscillators: rather, they are coupled together in a complicated three-dimensional structure.

Heat is conducted through a solid by means of the propagation of quantized vibrational modes along the crystal lattice of the solid. In

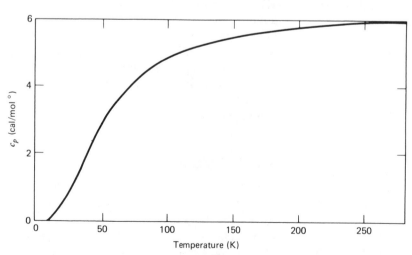

Figure 8.4 Heat capacity of metallic silver as a function of temperature.

this sense the temperature of a solid is still related to the average kinetic energy of its constituents. The vibration of the charged nuclei also gives rise to another heat propagation mode: radiation.

Heat-Transfer Mechanisms

Conduction and **radiation** represent the only physical mechanisms for the transmission of heat. A third category is often defined: **convection**. If a fluid moves across a hot surface, heat can be transferred to the fluid by conduction or radiation or both and can be moved by the fluid into a cooler region. The net result is a heat-transfer rate that is enhanced over that by simple conduction or radiation alone. Convection is a hydrodynamic problem that depends on the geometry of the problem as well as the fluid characteristics and the heat source. As a consequence, convection problems are a good deal more difficult to solve analytically than problems of conduction or radiation. In fact, one almost never solves them by any means other than empirical relationships developed by in-the-field measurements.

Conduction We find experimentally that the rate of heat transfer across a slab of material of thickness Δx, area A, having a temperature difference ΔT, is proportional to the temperature gradient:

$$\frac{\Delta Q}{\Delta t} \propto A \frac{\Delta T}{\Delta x}. \qquad (8.3)$$

We can generalize to infinitesimals and introduce a constant of proportionality, the **thermal conductivity**; the law of heat conduction then becomes

$$\frac{dQ}{dt} = kA \frac{dT}{dx}. \qquad (8.4)$$

Example 8.1 What is the heat loss rate by conduction through a wall 15 cm × 4.5 m ×

2.7 m, if the inside temperature is 20°C, the outside temperature is 0°C, and the average R-value of the wall is 10?

Use Eq. 8.4: $dQ = kA(dT/dx)$; apply to finite differences: $\Delta \dot{Q} = kA \, \Delta T/\Delta x$. "R-value" is $1/k$ in English units. Divide 62.3 by R to get k in joules per hour per centimeter per degree Celsius. Therefore, for $R = 10$,

$$k = 6.23 \text{ J/hr cm °C}.$$

Therefore,

$$\Delta \dot{Q} = \frac{(6.23 \text{ J/hr cm °C} \times 450 \text{ cm} \times 270 \text{ cm})(20 - 0)}{15 \text{ cm}},$$

$$\boxed{\Delta \dot{Q} = 1.0 \times 10^6 \text{ J/hr.}}$$

The thermal conductivity k for several materials is given in Figure 8.5. Thermal conductivity is usually quoted in the literature in English units. Note the conversion factor given in the example above.

Equation 8.4 assumes isotropic thermal properties for the material. It may be that for some substances the thermal properties are different for different directions. In general, substances that are optically anisotropic are also thermally anisotropic. For example, in calcite for a heat flux along the crystal axis $k = 293 \text{ mW/cm K}$ at 83 K; but for the flux perpendicular to the crystal axis $k = 54$ at the same temperature. Certain ceramics have been developed with strong thermal anisotropies, even at high temperatures, and have been employed as ablative heat shields for spacecraft reentry.

Once the geometry of a particular problem has been specified, Eq. 8.4 can be integrated to determine either the heat rate for a particular temperature difference or vice versa. For example, for a cylinder of length L, inner radius r_1, and outer radius r_2 (as in Figure 8.6),

$$\int_{T_1}^{T_2} dT = \frac{\dot{Q}}{k} \int_{r_1}^{r_2} \frac{1}{2\pi r L} \, dr, \qquad (8.5)$$

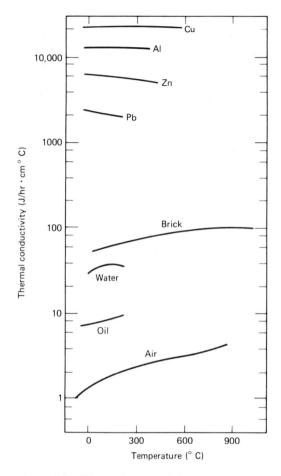

Figure 8.5 Thermal conductivity as a function of temperature for various substances.

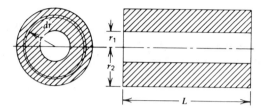

Figure 8.6 Radial heat flow in a cylinder.

there that the energy flux rate, that is, energy per unit area per unit time, from an arbitrary surface is given by

$$\dot{q}_r = \epsilon(T)\sigma T^4. \tag{8.7}$$

Of course, it is impossible to isolate any object completely; an object not only radiates energy at a rate characteristic of its temperature, but it also absorbs energy from its surroundings at a rate characteristic of the temperature of its surroundings. For an object of total area A, the net rate of heat transfer to the surroundings must be the difference between the rate of heat absorption and the rate of heat radiation:

$$\dot{Q} = A\epsilon\sigma(T_s^4 - T^4), \tag{8.8}$$

where T_s is the temperature of the surroundings, and T is the temperature of the object. Note that this assumes the emissivity and absorptivity of the object to be identical, which indeed they are.

Example 8.2 The absorptivity ϵ of a small body varies with temperature (°C) as

$$\epsilon = 1 - \frac{T^{1/2}}{100}.$$

Initially at 500°C, it is placed in an oven at 4000°C. Estimate the rate of energy absorption and emission.

$$\dot{Q}_a = A\epsilon\sigma T_s^4;$$

$$\frac{\dot{Q}_a}{A} = 0.36 \times 5.67 \times 10^{-8} \times (4273)^4,$$

$$\frac{\dot{Q}_a}{A} = 6.8 \times 10^6 \text{ W/m}^2$$

$$= \text{rate per area of energy absorption.}$$

and

$$T_2 - T_1 = \frac{\dot{Q}}{2\pi kL} \ln \frac{r_2}{r_1}. \tag{8.6}$$

This result is not directly applicable to situations of heated fluid flow down a pipe, since in those cases neither T_1 nor $\dot{Q}$ would be uniform throughout the length L. This geometry, for short L, could be used to determine k for various materials.

Radiation We have already studied some aspects of radiation in Chapter 6. It was noted

$$\frac{\dot{Q}_r}{A} = 0.78 \times 5.67 \times 10^{-8} \times (773)^4$$

$$= 1.58 \times 10^4 \text{ W/m}^2$$

$$= \text{rate per area of energy radiation.}$$

The object is obviously absorbing energy at a higher rate than it is radiating it away. It will become hotter until the two rates are equal:

$$\dot{Q}_a = \dot{Q}_r.$$

This will eventually occur when the object has the same temperature as the oven walls.

Radiation of heat energy is as important as conduction for many applications. In a real-world problem not only must heat loss rates by these two mechanisms be calculated, but also the effects of convection must be included, a much more difficult thing to do.

Convection The subject of convection could fill and has filled many volumes. We are interested here primarily in the transfer of heat through power plant condensers and steam or hot-water piping. As a consequence, rather than start from the fundamentals and work our way upward, we examine a few specific cases and develop whatever formalism we need.

It will be convenient to define a convection coefficient h in a manner similar to the definition of the conduction coefficient h in a manner similar to the definition of the conduction coefficient of Eq. 8.4:

$$\dot{Q} = hA \, \Delta T. \qquad (8.9)$$

For the moment, we defer the consideration of h. Consider next a fluid initially at temperature T_1 flowing through a pipe of cross section a, length L, and exiting at a temperature T_2. The inner wall of the pipe is maintained at temperature T_w; the pipe has length L. This is approximately the situation in a steam turbine condenser and, to a lesser extent, also is the situation in a long pipe transporting steam or hot water. Refer to Figure 8.7; the heat-transfer rate across a small surface area element dA

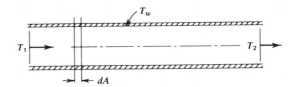

Figure 8.7 Heat convection for fluid flowing in a pipe.

is given by

$$d\dot{Q} = hdA(T_w - T), \qquad (8.10)$$

where T is the temperature of the fluid at the location of dA. The heat rate through this area must equal the heat-transfer rate of the fluid while passing over dA:

$$d\dot{Q} = mc_p \, dT, \qquad (8.11)$$

where dT is the temperature change in the fluid while passing over dA.

Setting these two heat rates equal and integrating from the inlet, $A = 0$ to the outlet, $A = aL$ (remember that A is the total surface area):

$$hdA(T_w - T) = mc_p \, dT, \qquad (8.12)$$

$$\frac{hdA}{mc_p} = \frac{dT}{T_w - T} = \frac{d(T_w - T)}{-(T_w - T)} = -d \ln (T_w - T), \qquad (8.13)$$

$$\int_0^{aL} \frac{hdA}{mc_p} = -\int_{T_w - T_1}^{T_w - T_2} d \ln (T_w - T), \quad (8.14)$$

$$\frac{haL}{mc_p} = \ln \frac{T_w - T_1}{T_w - T_2}, \qquad (8.15)$$

assuming h is constant along the pipe.

The total rate of heat loss through the pipe must also be given by Eqs. 8.10 and 8.11:

$$\dot{Q} = haL \, (\Delta T)_{av} = mc_p(T_1 - T_2), \quad (8.16)$$

where $(\Delta T)_{av}$ is the average temperature difference between the pipe wall and the fluid. From Eqs. 8.15 and 8.16,

$$(\Delta T)_{av} = \frac{T_2 - T_1}{\ln [(T_w - T_1)(T_w - T_2)]} = \Lambda. \qquad (8.17)$$

The quantity Λ defined above is called the logarithmic mean temperature difference, so

$$\dot{Q} = haL\Lambda. \qquad (8.18)$$

Now the trick is to determine h.

There are some interesting tricks for calculating h. First we need a few fundamentals. Fluid flow is usually characterized as being either *laminar* (streamlined) or *turbulent*. In the former case, the fluid velocity is always in the same direction; if confined to a pipe, the fluid has no radial component of velocity. In turbulent flow, while there is a net flow along the pipe, there is also a substantial fluctuating radial component to the velocity at any point. In either case there is a boundary layer of fluid next to the wall that has no turbulence and through which the heat is conducted. The convection coefficient h, then, must depend on the properties of the fluid flow that affect this laminar boundary layer. We can make some educated guesses about the factors that enter into this dependence, then use the technique of dimensional analysis to see how far we can get.

The thin film (for turbulent flow) or boundary-layer thickness and the heat transferred are thought to depend on the mass velocity of the fluid $G = \rho v$, the diameter of the pipe D, the fluid viscosity[2] μ, the thermal conductivity of the fluid k, and also the specific heat of the fluid c_p. We set

$$h = \alpha G^a D^b c_p^e \mu^f k^i, \qquad (8.19)$$

where α is an arbitrary constant and the exponents are to be determined.

---

[2]Viscosity is a measure of fluid "friction." It is defined for laminar flow as the constant of proportionality in

$$F = \mu A \frac{dv}{dy},$$

where F is the viscous force between laminae of area A, and dv/dy is the gradient of velocity perpendicular to v and normal to A. The SI unit of viscosity would be newton-second per meter^2. 1 N s/m^2 = 1 P (poise).

Dimensional analysis is a procedure by which we require the units of the left-hand side of Eq. 8.1 to equal the units of the right-hand side. The convection coefficient h by Eq. 8.9 has the units calories per second per square centimeter per degree, or $H/(tL^2T)$, where H stands for the units of heat, t stands for the unit of time, L stands for the unit of length, T stands for the unit of temperature and M stands for the unit of mass. Using this notation, Eq. 8.18 becomes

$$\left(\frac{H}{TL^2t}\right) = \left(\frac{M}{TL^2}\right)^a (L)^b \left(\frac{H}{Mt}\right)^e \left(\frac{M}{LT}\right)^f \left(\frac{H}{TLt}\right)^i. \qquad (8.20)$$

For equality to be established, the exponent of each unit on the left must equal the net exponent for the same unit on the right. An equation can be written for each unit:

$$\begin{aligned}
\Sigma H: & \quad 1 = e + i, \\
\Sigma M: & \quad 0 = a - e + f, \\
\Sigma L: & \quad -2 = 2a + b - f - i, \qquad (8.21) \\
\Sigma T: & \quad -1 = a - f - i, \\
\Sigma t: & \quad -1 = -e - i.
\end{aligned}$$

These equations must be solved simultaneously for the exponents. Unfortunately, however, there are more exponents than independent equations. There are five unknowns and three equations; we can solve for any three in terms of the other two. The choice of those two exponents is arbitrary; but it will be convenient if the choice leaves the expression in terms of one or more dimensionless factors, that is, a combination of parameters that is dimensionless. For this reason, in this case we choose to solve for b, f, and i in terms of a and e, and we obtain[3]

$$\frac{hD}{k} = \alpha \left(\frac{\rho v D}{\mu}\right)^a \left(\frac{c_p \mu}{k}\right)^e. \qquad (8.22)$$

---

[3]The first quantity in parentheses is called the Reynolds number, Re. For Re $\leqslant 2000$, the flow is laminar. Above Re $\geqslant 3000$ the flow is turbulent. Between these is an unstable transitional region.

Dimensional analysis has taken us as far as it can. In order to determine α and the exponents a and e we must appeal to experiment. For the special case of forced convection in circular pipes with turbulent flow, experiments yield

$$\frac{hD}{k} = 0.023 \left(\frac{\rho v D}{\mu}\right)^{0.8} \left(\frac{c_p \mu}{k}\right)^{0.2}. \quad (8.23)$$

Example 8.3 Use dimensional analysis to find an expression for the pressure drop per unit length in isothermal turbulent flow of an incompressible fluid at constant mass rate through a long, straight uniform pipe. The only physical variables that can enter into the expression are the length L and diameter D of the pipe, the velocity v, the density ρ, and the viscosity μ.

$$-\frac{dp}{dL} = f(D, v, \rho, \mu, g),$$

$$-\frac{dp}{dL} = \alpha D^a v^b \rho^c \mu^d g^e.$$

Substitute the dimensions for each variable:

$$\frac{F/L^2}{L} = \alpha (L)^a \left(\frac{L}{T}\right)^b \left(\frac{M}{L^3}\right)^c \left(\frac{M}{LT}\right)^d \left(\frac{ML}{FT^2}\right)^e.$$

All exponents must be dimensionless; therefore,

ΣF: $\quad 1 = -e$, $\qquad\qquad e = -1$,

ΣM: $\quad 0 = c + d + e$, $\qquad c = 1 - d$,

ΣT: $\quad 0 = -b - d - 2e$, $\quad b = 2 - d$,

Σl: $\quad -3 = a + b - 3c - d + e$, $\quad a = -1 - d$.

Since there are five unknowns and four equations, we must solve in terms of one exponent. We arbitrarily choose d:

$$-\frac{gD}{\rho v^2}\left(\frac{dp}{dL}\right) = \alpha \left(\frac{\mu}{Dv\rho}\right)^d.$$

In fact, experiment shows that $\alpha = 1$, and $d = -1$.

In this example we have treated force as a quantity having a separate unit. This was a matter of convenience, because we believed

there would be a dependence on g. In the use of dimensional analysis to determine h above, we did not anticipate a dependence on g, so it was convenient to make g a dimensionless quantity.

We are now in a position to return to Eq. 8.18 and calculate heat transfer through and along circular pipes. Let us now look at the techniques for transferring heat from the power plant exhaust to the biosphere.

Once-Through Cooling

The first method of large-scale heat rejection that we consider is once-through cooling. At present in the United States this method is most often used for the removal of heat from power plant condensers. Water is withdrawn from a river or estuary and pumped through the condenser tubes, where it picks up the latent heat of the steam condensing on the outer surface of the tubes. The water is then discharged back to the river or estuary. No effort is made to reduce the heat-loading effect on the receiving waters. The only concern is that there must be no recirculation of the warm water through the condenser tubes. From an economic viewpoint, once-through cooling is appealing simply because it is the least expensive method available provided, of course, a suitable site can be found. The costs of installation, operation, and maintenance are generally low.

It is possible to calculate the temperature increase of the cooling water that has passed through the condenser from the following equation:

$$\Delta T = \frac{\text{heat rejection rate to water}}{\text{water flow rate through condenser}}.$$
$$(8.24)$$

In this equation the specific heat of water, 1.0 in the appropriate unit system, has been omitted for simplicity.

This warmed water is then pumped back into the original body of water where, if there is thorough mixing, the temperature of the body of water will increase quickly to a mixed temperature given by

$$\Delta T_{mixed} = \Delta T p + T_{water},$$

where p is the percentage of total water flow pumped through the condenser and T_{water} is the original temperature of the water.

Under different conditions, there will be no mixing, and a thermal stratification will occur. That is, the warm water will simply form a layer on top of the dense, cool water. The criteria that deal with the formation of thermal stratification are complex and have not been completely determined, but the flow pattern of the stream or estuary that governs the advection, the placement of the discharge, and the channel characteristics that govern the turbulent mixing of layers seem to be the most important variables.

The federal government has not enacted comprehensive standards for temperature changes in natural waters induced by thermal discharges. The Water Quality Act of 1965 allows individual states to set standards along certain guidelines, subject to approval by the Department of Interior. Most states have obtained approval, but since the streams and lakes of the various states vary considerably in their temperatures, aquatic life, and seasonal environments, these standards also vary considerably. In only one case has the ΔT been allowed to exceed 10°F (that of Maryland). The National Technical Advisory Committee on Water Quality Criteria has recommended that ΔT be limited to 5°F, but, as we shall see, this constraint would impose a substantial burden on the heat-transfer systems.

Let us consider a 1000-MWe nuclear generating plant having an efficiency of 32 percent. The waste heat rate from this plant would be $68 \times 1000/32 = 2125$ MW. We can use Eq. 8.24 to determine the water flow rate required for a given temperature change:

Flow rate (kg/s)
$$= \frac{\text{heat rate (W)} \times (1/4.186)(\text{cal/J})}{\text{temperature change (°C)} \times 10^3 \text{ (g/kg)}}$$
(8.25)

For a ΔT of 5° and the waste heat rate above, the water flow rate required becomes: 1.02×10^5 kg/s, which would require a volume of about 10^2 m³/s, or about 3530 ft³/s.

Note that this flow rate is about one sixth the *total discharge* at the mouth of the Delaware, Hudson, or Sacramento rivers! These rivers are between 300 and 400 miles long, so assuming a linear increase in volume, this power plant located between 50 and 70 miles of the source would require *the total river flow*. These particular rivers are the 23rd, 22nd, and 21st largest in the country in terms of water discharge. Clearly, most streams and rivers are too small to support power generation plants in the 1000-MW class.

This analysis assumes that only one plant is located along the river. As we saw earlier in this chapter, that may not be the case. There can be persistent temperature changes along the course of a stream heavily used for cooling purposes.

There are other methods of heat rejection that we should study. A relatively simple way of disposing of the condenser heat load of a power plant is to use a man-made cooling pond or lake. An asset of the cooling pond approach is that it is inexpensive, especially in regions where land values are not too high. Cooling ponds can provide a community with considerable recreational facilities. Research is underway to determine the potential use of cooling ponds in aquaculture and fish farming.

When an engineer is designing a cooling pond, the most important factors to be determined are the surface area required to dissipate the heat for a given maximum surface temperature and the maximum downstream temperature (inlet temperature to the condenser for a closed cycle). The derivation of the equations to calculate the surface area requirement for a cooling pond is complex. The full treatment is included in the references for this chapter. Some engineers use a simple "rule of thumb" of sizing where 1 acre (0.405 ha)/MW plus 20 percent for surrounding land is used for fossil fuel plants and 2 acres/MW is used for nuclear plants. Whatever the method of size determination, it is clear that cooling ponds provide a workable solution to heat rejection problems.

When land values are high, spray ponds can be used to minimize acreage requirements. Spray ponds operate on the same principle as cooling ponds; however, the evaporation resulting from exposure to the air is enhanced by spraying the warm water into the air over the pond. A spray pond needs only about 5 percent of the area of a cooling pond because of the increased surface area of water in contact with the air. Both the residence time in the air and the relative velocity between water droplets and the air add significantly to the heat transfer. The spray nozzles, the design of which strongly regulates the performance of the pond, are usually located 2 to 3 m above the surface. Water losses in this process are usually very high; however, they can be minimized with the installation of louvered fences. Little or no research, either analytical or empirical, has been undertaken with regard to spray ponds.

Cooling Towers

One final area of heat-transfer techniques should be examined—that of cooling towers. Figure 8.8 shows several types. In order to understand the literature of this field a person needs to understand certain specialized terms. Cooling range, or *range*, refers to the number of degrees the water is cooled from inlet to outlet of the device. Wet-bulb temperature is the lowest water temperature that can be realized by evaporation into the sorrounding environment. It is measured by whirling a thermometer, the bulb of which is covered with a wet cloth; it represents the theoretical limit of cooling that a tower can achieve. The difference between the theoretical limit and the actual cooling accomplished by a tower is called the tower's *approach*, or approach to wet bulb. The approach is seldom closer to the theoretical limit than 3°C. More often it differs by 4° to 9°C. The term *heat load* refers to the amount of heat dissipated per unit time. The heat load is equal to the mass of the water circulated per unit time multiplied by the range.

Example 8.4 A particular cooling tower is designed to dissipate heat at a rate of 300 MW. If its cooling range is 8°C, what is the water use rate?

$$\text{Heat load} = \text{water mass rate} \times \text{specific heat} \times \text{range} = \dot{m} \, \Delta T,$$

$$\dot{m} = \frac{300 \times 10^6}{1 \text{ cal/g} \, °C \times 4.186 \text{ J/cal} \times 10^3 \text{ g/kg} \times 8°C}$$

$$= 8.96 \times 10^3 \text{ kg/s}.$$

There are various ways of increasing the heat load:

- Increasing the velocity of the air in contact with the water surfaces.
- Increasing the water surface exposed to the air.
- Lowering the atmospheric pressure.
- Raising the entering water temperature.
- Reducing the vapor content of the inlet air.

All of the above have an effect on the tower's performance or the ability of the tower to cool water. The cooling of a quantity of water from a specified hot water temperature to a specified cold water temperature at a specific wet-bulb temperature is the usual manner for expressing performance.

Cooling towers fall into two general classes—wet and dry. A wet cooling tower operates by bringing water into contact with the air. As mentioned earlier, the heat is trans-

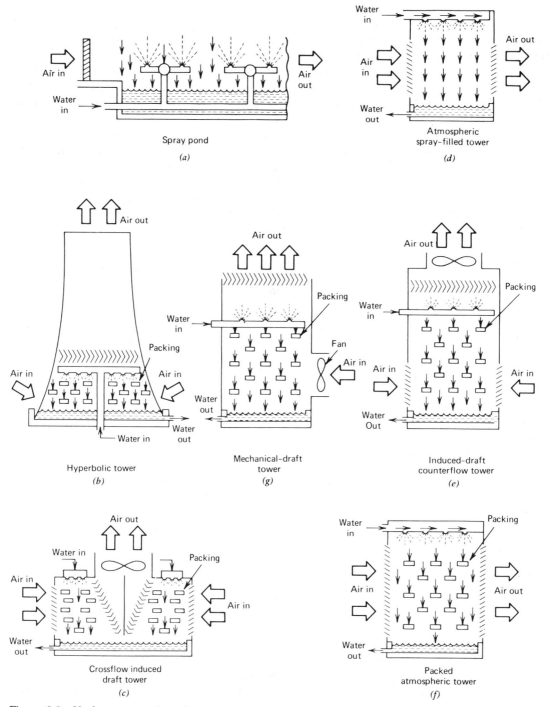

Figure 8.8 Various types of cooling systems. (*a*) Spray pond. (*b*) Hyperbolic tower. (*c*) Crossflow induced draft tower. (*d*) Atmospheric spray-filled tower. (*e*) Induced draft counterflow tower. (*f*) Packed atmospheric tower. (*g*) Mechanical-draft tower.

ferred mainly by evaporation (75 percent), while the remainder is cooled by sensible heat transfer. A factor affecting heat transference is the amount of contact area between water and air. As the water comes out of the distribution system, it falls onto the first of many layers of **fill**. The function of fill, which occupies much of the interior of a wet cooling tower, is to speed up the dissipation of heat by promoting the formation of water droplets and by increasing the wetted surface in contact with the air. The fill must be constructed so that it will provide low resistance to the flow of air and will maintain a uniform distribution of water and air. In most modern wet cooling towers, fill is of the splash type where the water splashes downward from one level of horizontal bars to another. In some smaller towers another type of fill is gaining acceptance. This type, called *film fill*, tends to spread the water into a thin film that spreads over a large surface area. At the end of its passage through the tower, the cooled water falls into a collecting basin at the bottom of the tower. From there it is pumped back to the place where it picked up the heat— for example, a steam condenser in a generating station—and begins another cycle.

Wet cooling towers can be further divided into three major types: atmospheric, natural draft, and mechanical draft. Atmospheric towers are designed so that they utilize the naturally occurring winds as the air flow. Water is sprayed down into the air stream, and the heat is carried away. These towers are usually built with a high broad side toward the prevailing wind. Louvers on the tower exterior are designed to keep drift losses to a minimum. Although these towers are simple and relatively trouble-free, they do have some important disadvantages. This type of tower is not built today for use in power plants because of the large amount of space required (large areas with no obstructions that would restrict air flow), a low cooling range, and relatively high material costs.

Natural-draft wet cooling towers are more sophisticated, and the movement of air in these towers is caused by a difference in density between the inlet and exit air streams. In other words, the air outside the tower has a slightly greater density than the air inside. A value of 0.008 kg/m^3 for this density difference is typical. The actual driving force is the difference in hydrostatic pressure between the two columns of air.

The air stream exit speed is important because the vapor plume must rise from the tower under all conditions. The efficiency of the tower is drastically lowered if air is recirculated. Typical exit speeds are 3 to 3.5 m/s. The small differences in density coupled with the pressure drop in the tower necessitates

Figure 8.9 A 152-m cooling tower under construction in West Virginia. This tower will handle the heat discharge from a 1300-MW coal-fired power plant. (Photograph courtesy of the American Electric Power Service Corporation.)

great heights for the towers. The latest designs used with large power stations are the huge hyperbolic units, with heights of 120 m being common. Figure 8.9 shows a natural-draft tower under construction. This tower, when completed, will handle the heat discharge from a 1300-MWe coal-fired plant using $2.27 \times 10^6 \, \ell/\text{min}$ of water.

Natural-draft cooling towers have several adantages over other types of cooling towers. First, they produce cooling effects similar to those produced by mechanical-draft towers without the use of mechanical parts and the power to run them. Also, when compared to atmospheric towers, their action is independent of wind velocity. They use much less ground space. In addition, these cooling towers operate under optimum conditions, in that the stream of air is in the opposite direction from that of the falling water with the coldest air meeting the colder water first, which ensures no loss of efficiency. But the great heights necessary and the large capital outlays are disadvantages of this method. Moreover, difficulties are encountered in the exact control of outlet water temperatures. Commonly quoted figures for optimum range and approach for such units are 14° and 10°C, respectively.

The third type of wet tower is the mechanical-draft tower. In mechanical-draft towers the air flow is produced by a fan. This

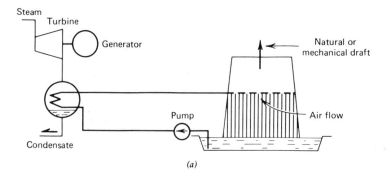

(a)

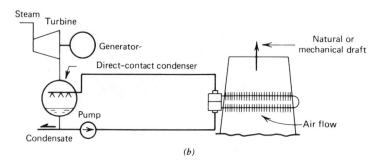

(b)

Figure 8.10 Two fundamental types of cooling towers. (*a*) Evaporative ("wet") cooling tower cycle. (*b*) Dry cooling to tower cycle.

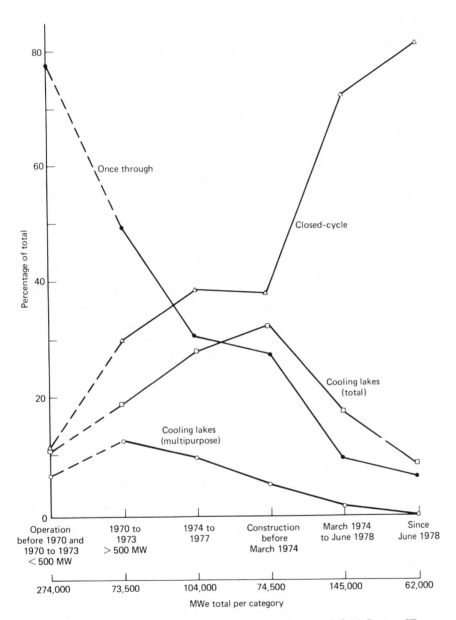

Figure 8.11 Types of condenser cooling systems in the United States. [From J. Z. Reynolds, "Power Plant Cooling Systems: Policy Alternatives," *Science* **207**, 367 (1980). Copyright © 1980 by the American Association for the Advancement of Science.]

fan may be placed either at the inlet (forced draft) or at the exit (induced draft). The use of a fan means that the air flow rate can be regulated in order to satisfy the particular cooling requirement. Mechanical-draft towers do not need the great heights required by natural-draft towers. These towers, because of this decreased height, have more problems regarding ground fogging and warm air recirculation. Although these towers have a lower initial cost than natural-draft types, this is offset by higher operating and maintenance costs over the life of the tower. The chief advantage of mechanical-draft towers is that a close approach and a high cooling range are possible.

All of the various types of wet cooling towers remove heat from the exhaust steam of a turbine by the process of evaporation. The second major grouping of cooling towers—dry towers—differs from this in that they are completely air-cooled, and sensible heat transfer only is used to dissipate heat. The dry cooling tower system is illustrated in Figure 8.10. In this system the turbine exhaust is condensed in a direct-contact spray condenser. Part of this condensate is returned to the boiler, while the other part is first sent through the cooling tower and then returned to the condenser. The actual cooling tower consists of an air-cooled heat exchanger inside a cooling tower chimney. This tower may be of either the natural-draft or mechanical-draft type.

One of the greatest advantages of the dry cooling tower is that there is no consumptive water usage. At present, water consumption from evaporation as a result of power generation is estimated to be 38 liters per day per person and is expected to increase more rapidly than the growth rate of power production.

It seems obvious that dealing with waste heat will cost the electric utility money, which of course it will extract from its customers in the form of increased electrical unit costs. It also is apparent that the cheapest method of managing the heat is to use once-through cooling, and that the other techniques—ponds, spray ponds, wet cooling towers, and dry cooling towers—become progressively more expensive.

Several studies on the incremental costs of various types of cooling methods predict costs ranging between 0.04 and 6 mills/kWh. Although these numbers seem large, especially the larger values for dry towers, we should remember that the cost of electricity delivered to customers these days averages more than 7 cents/kWh, or 70 mills/kWh.[4] At the most, even dry cooling towers would add only about 10 percent to the cost of the electricity. This does not seem to be too high a price to pay, and more utilities are beginning to pay it: see Figure 8.11. There are other alternatives: find some way to minimize heat production or find some useful application for the waste heat.

8.3 USEFUL APPLICATIONS OF WASTE HEAT

So far we have discussed several techniques for ridding ourselves of excess heat; all of these methods amount simply to throwing away something that in fact could be a valuable resource. After all, heat is required in many manufacturing processes; it is used to make our homes and offices comfortable in winter and summer. It seems paradoxical that at the same time we can both require heat for some purposes and yet be compelled to throw it away. Must it be this way? Is there no way to use the excess heat from power generation in industry or homes? Let us attempt to answer this question by examining first an idea we introduced in Chapter 3.

---

[4]Low rates in 1981 were about 4.5 cents/kWh; high rates (New York City) were in excess of 11 cents/kWh.

Heat Quality

The usability of heat is a function of the form of the transfer medium, its volume and temperature. These determine what is called the **heat quality**. High-quality heat is easy to use; low-quality heat, needless to say, is not. What is the difference? Figure 8.12 is a plot of the enthalpy per unit volume ($\Delta H/m^3$) and per unit mass ($\Delta H/kg$) versus temperature for water and steam. Note that 373.6°C is the critical point for water–steam. It is clear that water has a much higher heat content per unit volume than steam over most of the temperature range shown. Transporting heat as hot water would appear to be more economical than by using steam. If this is the case, why do we use steam at all? Why not run 340°C water through the turbine?

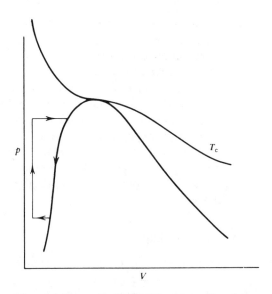

Figure 8.13 *p*–*V* diagram for water showing a complete cycle for liquid phase only.

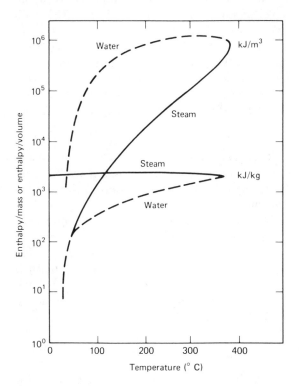

Figure 8.12 Enthalpy per unit mass and per unit volume for water and steam.

Let us set aside practical considerations for the moment and look again at the *p*–*V* diagram for water, Figure 8.13. In order for a Rankine cycle to operate with water *at all times* (as opposed to steam or steam–water mixtures part of the time), all parts of the process must occur to the left of the vapor dome and below the critical isotherm T_c, as shown. This results in a very small area for the integral, meaning little work output, and a rather small change in temperature between the inlet and outlet of the turbine. We could have predicted this from Figure 8.12; the change in enthalpy per cubic meter for water over a rather large change in temperature is very small. Consequently, the shaft work must also be small.

So much for turbines; how about transport of heat? Wouldn't water be preferable to steam? Not necessarily. Although water has a higher enthalpy per unit volume than steam, it has a lower value per unit mass. Remember also that as the steam condenses to water it gives up 540 cal/g as latent heat.

The density of water does not change much

as a function of temperature between 30° and 300°C; it decreases about a factor of 1.5. But saturated steam, on the other hand, *increases* a factor of 126. (In order to maintain saturation, the steam pressure must increase likewise from 1 atm at 100°C to about 200 atm at 360°C.) What this means is that at low steam temperatures a rather large volume is required to transport a given amount of heat. The higher the temperature, the smaller the volume required and, as a consequence, the smaller the capital investment.

Transportation of Heat

Let us design a steam pipeline to transport waste heat from a power generating station. How far could we transport heat, if we can afford a maximum loss rate of 20 MW using saturated steam through insulated pipe?

Equation 8.18 could be solved for this length L if we knew h and A. For these we need to assume a particular situation established by these parameters:

Pipe diameter,	$D = 91$ cm,
Exit temperature,	$T_2 = 100°C,$
Exit pressure	$= 87$ kPa
Inlet temperature,	$T_1 = 200°C,$
Exterior temperature	$0°C$ ($T_w = 50°C$),
Fluid viscosity of steam,	$\mu = 125\ \mu P,$
Thermal conductivity of steam	$k = \sim 24 \times 10^{-3}$ J/s m °C,
Mass velocity of steam,	$G = \rho v = ?,$
Specific heat of steam,	$c_p = 0.5$ cal/g °C.

To evaluate the mass velocity, $G = \rho v$, of the steam we need to know the heat delivery rate at the exit end of the pipe. Let us assume it to be 20 MW. (The reason for choosing many of these numbers will become evident later.) The

steam velocity to deliver heat at this rate for this particular pipe must be 15 m/s (remember, the steam carries latent heat as well as sensible heat).

Under these conditions the Reynolds number is

$$\text{Re} = \frac{\rho v D}{\mu} = 4.37 \times 10^5.$$

The value for the Reynolds number is suitable for fully developed turbulent flow; Eq. 8.18 is clearly applicable.

The second term in parentheses in Eq. 8.23 (called the Prandtl number) has a value of about 0.8 for all gases over a wide temperature range. Solving this equation for h yields

$$h = 1.5 \times 10^{-3} \text{ J/s cm °C.}$$

The quantity Λ can be evaluated with the given values of T_1, T_2, and T_w. The wall temperature T_w is not really very well known; it depends on the quality of the insulation between the pipe and the exterior. As it happens, Λ is not very sensitive to T_w, varying between 10.3 and 11.2 as T_w varies between 10° and 50°C. Then, solving for L, we have

$$L = 3.72 \times 10^6 \text{ cm (23.2 miles).}$$

This seems at first glance to be a reasonable distance—perhaps a little short. It would seem to be plausible to consider transporting heat in this manner; and it is, but what is the cost?

To answer this we must ask, from where does this heat come? Again, let's assume a particular situation: a nuclear power plant that, operating normally, would have a thermal efficiency of 32 percent. From this we have that the waste heat output rate Q_w would be given by

$$Q_w = \frac{68P}{32}, \tag{8.25}$$

where P is the output electrical power. But this is when the heat is exhausted at ambient temperature, say 25°C. Our heat-transport system requires steam at 200°C. To raise the exhaust to this temperature, we must lower the efficiency of the plant. If we assume the actual efficiency to be proportional to the Carnot efficiency, we find that raising the exhaust temperature to 200°C lowers the efficiency to 12.9 percent (assuming the high temperature to be 320°C). So that the waste heat production now would be given by

$$Q_w = \frac{87.1P}{12.9}. \qquad (8.26)$$

So 12.9 percent of our fuel energy goes into electricity; about 32 percent of the remainder, or about 29 percent of the total, goes into heat losses in transport. We have available about 60 percent of the total at the end of the pipe. At 32 percent efficiency, the electricity generation cost is about 1 cent/kWh. To recover this cost and produce useful heat with our system, the heat must cost 0.6 cent/kWh. This assumes that all the heat is used at the end of the pipe; in fact, probably less than 50 percent would be. So the actual cost of heat used would be in excess of $3.50/million Btu. This cost is far in

excess of the cost of heat produced directly on site for most applications.

Example 8.5 A certain industry requires 400 MW of electricity and 1500 MW of heat. Consider two methods of satisfying these requirements: (1) two plants, one producing electricity at $\eta = 0.3$ and one producing heat at $\eta = 0.7$, and (2) one plant producing electricity at $\eta = 0.129$. What is the difference in fuel requirements for these two methods?

(1) Fuel rate for
     electrical plant: 400/0.3 = 1333 MW
     Fuel rate for
     heat plant: 1000/0.7 = 2142 MW
                          Total: 3475 MW

(2) Fuel rate = 400/0.129 = 3075 MW. This plant would produce in excess of the 1500-MW heat requirement.

Again we see there are trade-offs. A system to use waste heat from a steam turbine would use less fuel, but would produce more costly heat. We should also mention that a special

TABLE 8.3 Applications for 2000 to 4000 MW of Heat

	Specific Application	Quantity
Central heating	Steam and hot water for residential, commercial, and industrial heating	For a city of 500,000 to 1,000,000 people
Central cooling	Evaporative cooling for residential and commercial needs	For a city of 500,000 to 1,000,000 people
Manufacturing	Electricity and heat for (typical mix):	
	Evap. salt	2775 tonnes/day
	Petrochemicals	60,000 bbl/day
	Kraft paper	500 tonnes/day
Desalination for municipal water	Wastewater recycling	To 230×10^7 ℓ/day
	Seawater distillation	To 230×10^7 ℓ/day
Agriculture	Arid land irrigation with distilled water	To 230×10^7 ℓ/day (128,000 ha)
	Greenhouse heating and cooling	To 400 ha
Transportation	Ice-free shipping lanes	6 to 12 km ice-free water
Aquaculture	Warm water and sewage for culture:	
	Shellfish	Unknown
	Fish	4,300,000 kg/yr (760 ha)

back-pressure turbine would have to be used, adding to the cost of the installation. Separate electricity- and heat-generating plants could be individually optimized. We shall return to this point.

It should be pointed out that the calculation we have made required many assumptions. In practice, empirical relationships developed in engineering studies would be used, and the resulting values of T_w, heat delivered, and so forth would probably be different from ours. Nevertheless, this simple calculation is useful for illustrative purposes.

The question then becomes one of whether or not a sufficient number of heat-requiring processes exist near enough to the reactor to use this large quantity of heat efficiently. There are many processes that require heat; see Table 8.3. We shall discuss one of these in detail.

Desalination

Several suggestions have been made for the application of waste heat from steam-electric generators, especially nuclear-powered generators. Uses such as aquaculture and agriculture, extending growing seasons, and freeing ice-locked rivers during winter have been suggested. But in fact none of these has been economical, practical, or technically feasible yet. Some use has been made in district heating, as we noted above. The only other truly promising area for waste heat application is desalination.

Cheap desalination would be a boon to arid, developing countries and is becoming almost a necessity for some of our coastal metropolitan areas. Water is required not only for household use (95 liters per 5-min shower, 10 liters per toilet flush), but also in enormous quantities for agriculture and industry. About 400 billion liters of freshwater is required daily for irrigation; 140 liters is required to produce one slice of bread, 14,175 liters to produce 1 lb of beef,

and 756,000 liters to grow 1 metric ton of alfalfa.

Industrial processing requires large quantities of water: 907,000 liters is used to produce 1 tonne of acetate, 2,495,000 liters to make 1 tonne of synthetic rubber, and more than 3,780,000 liters to process 100 barrels of synthetic oil from coal. In all, Americans currently use in excess of 7560 liters per day per person, and our total use rate is growing at a rate of 94,500 liters per minute.

It is little wonder, then, that commercially attractive desalination processes have been sought for some time. There are two basic processes to produce freshwater from brine, seawater, or brackish water[5]: remove the water from the salt, or remove the salt from the water. In the latter category, electrolysis and ion exchange have been used. These are more applicable to the purification of brackish water than seawater. For the large-scale commercial desalting of seawater, the principal technique is distillation. We shall examine it in some detail.

Water boils ordinarily at 100°C, of course, but if the pressure over the water is lowered, the boiling temperature is also lowered, as anyone who has tried to cook at higher elevations knows. Straightforward distillation at 100°C would be an uneconomic process because of the expense required to provide heat of that quality. Instead, in the multistage flash-distillation process, preheated seawater is exposed to stages of successively lower pressure. At each stage some of the water flashes into vapor. This vapor is condensed, using the seawater intake pipes to provide the low-temperature reservoir. The process is diagramed schematically in Figure 8.14.

In this type of operation the low pressure

---

[5]Freshwater by definition contains less than 1000 parts of salt per million parts of water; brackish water from 1000 to 35,000 ppm; seawater contains about 35,000 ppm salt, and brine contains more than that.

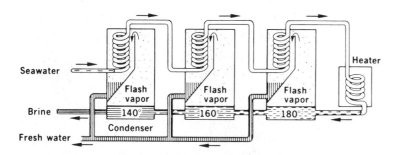

Figure 8.14 Flash–distillation desalination system. Temperatures are °F. (From Grace M. Arrows, *Nuclear Energy for Desalting.* Washington D.C.: U.S. Atomic Energy Commission, 1967.)

required is supplied by the condensation of the flash vapor on the cooling coils in the upper left of each chamber in Figure 8.14. We can calculate the performance of a single chamber by using energy conservation and making some reasonable assumptions.

Example 8.6 The brine is heated by waste heat from a power plant to a temperature of 180°F (82.2°C) (first chamber on the right). We must guess at the temperature of the cooling water entering from the left; let us assume the increase in temperature to be 10°C. And finally let us assume that the temperature of the freshwater as it is removed from the first chamber is 60°C. Then let $\dot{m}_1$, $\dot{m}_2$, and $\dot{m}_3$ represent the flow rates of brine into the chamber condensed flash vapor and brine out of the chamber, respectively. Conservation of energy yields

$$\dot{m}_1 h_1 = \dot{m}_2 h_2 + \dot{m}_3 h_3 + \dot{m}_1 c_p \, \Delta T$$

where the h's are the corresponding enthalpies. And it must be true that

$$\dot{m}_1 = \dot{m}_2 + \dot{m}_3.$$

We must be given $\dot{m}_3$; in other words, the system must be designed for a certain flow capacity. Then, eliminating $\dot{m}_1$,

$$\dot{m}_2(h_1 - h_2 - c_p \, \Delta T) = \dot{m}_3(h_3 - h_1),$$

$h_1 = 343 \text{ kJ/kg}; \quad h_2 = 251 \text{ kJ/kg}; \quad h_3 = 297 \text{ kJ/kg},$

$$\dot{m}_2 = 0.0011 \dot{m}_3.$$

A pilot plant of this type was built and operated successfully in San Diego, California, in the early 1960s.[6] This plant produced water at a cost of about 26 cents/kl, a price that is from two to three times too high to be economic. But cost is a relative thing; in some communities water must be trucked in at a cost of more than $1.85/kl. Desalination plants are clearly economic in such situations. There are clear benefits to be gained by scaling up the size of the plant.

Total Energy

We have seen that the fundamental problem in the use of waste heat from Rankine-cycle generators is that the heat quality is too low to warrant transportation. This problem can be solved by decreasing the Rankine-cycle efficiency and producing less electricity. Another possibility is to use some other heat engine that normally exhausts higher quality heat. If this heat is used on-site, both the

[6]In 1964 this plant was dismantled and reassembled at the U.S. Naval Base at Guantanamo, Cuba, when relations between the United States and Cuba were at a low ebb, and the Cuban government had threatened the continuity of the water supply of the base.

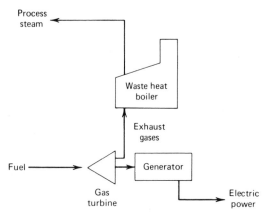

Figure 8.15 An example of electricity and heat cogeneration using a gas turbine.

quality and transportation problems are solved simultaneously.

We saw in Chapter 4 that the exhaust temperature from the gas turbine (Brayton cycle) is rather high, from 300° to 600°C. This exhaust can be used to produce high-pressure steam of reasonable temperature for process requirements. A system that would do this is diagramed in Figure 8.15. Diesel engines also exhaust at relatively high temperatures (300 to 450°C) and could be easily adapted for smaller-scale use.

A number of proposals have been made for industrial cogeneration of electricity and heat. In some places, large apartment complexes have been converted to this dual system. There are, however, economic and institutional barriers to the widespread application of the "total energy" concept.

Economies of scale are absent in small installations.
Heat and electricity are not necessarily needed simultaneously.
Power companies are reluctant to purchase excess electricity at appropriate rates.
Pollution controls have larger incremental costs on small installations.
Cogenerated electricity is usually not com-

petitive with central grid electricity, when average pricing is used.

These and other considerations have made the widespread application of industrial cogeneration or district heating very slow to evolve.

There are many other potential applications for the waste heat from steam-electric power plants, both fossil fueled and nuclear fueled. Not only is there a great need for pure water in many locations of the United States, not to mention the rest of the world, but also locating the power plant next to the ocean to provide cooling water would be desirable now because of the amount of waste heat to be dissipated from current designs. The only alternative to finding uses for this waste heat is to minimize its production.

8.4 MINIMIZING WASTE HEAT PRODUCTION

There are only two obvious ways to reduce our waste heat production: to reduce electricity production or to increase the efficiency with which we generate electricity. We discuss the possibilities for the former in Chapter 11. Here we look at the latter.

Topping Cycles

As we have learned, the only mechanism for increasing the efficiency of conversion of heat to mechanical energy is to increase the temperature spread between the inlet and outlet. The outlet temperatures are already as low as they can conveniently go, being limited by current technology. Although it must be said that industrial turbines lag five or so years behind turbines developed for military applications, and that research in the past few years has increased temperatures substantially, it should also be noted that even if inlet temperatures could be doubled, the efficiency increase would not double. Indeed, for a Carnot

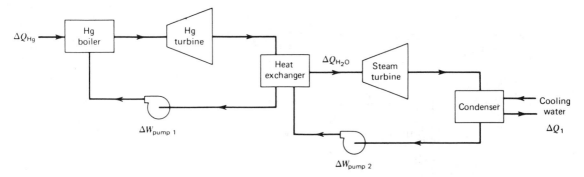

Figure 8.16 A mercury topping-cycle system.

engine, since

$$\eta = 1 - \frac{T_0}{T_i}, \qquad (8.27)$$

$$\frac{d\eta}{dT_i} = \frac{T_0}{T_i^2}, \qquad (8.28)$$

where T_0 and T_i are the outlet and inlet temperatures, respectively. Equation 8.28 indicates that the efficiency changes more and more slowly as the inlet temperature is increased.

That is not to say that efforts should not be made to increase operating temperatures. Any gain is valuable, but current steam-turbine designs have just about reached their limits. The use of saturated steam above about 260°C is very difficult because of the excessive pressure required. In fact, water is probably not the best substance, in terms of its thermodynamic properties, for Rankine-cycle operation. Water has a low critical temperature (647.4 K), requiring superheat to achieve the high operating temperatures necessary for high efficiency. Water has a high critical pressure (218.3 atm) requiring very expensive pipework to operate as saturated steam, or more extensive though lighter-weight pipework for superheat operation. Water has a low vapor pressure at condensing temperatures (0.0174 atm at 16°C), requiring expensive vacuum pumping for the condenser. Finally, water has a large heat capacity as a liquid, requiring substantial heat

addition at lower temperatures to raise the temperature to a reasonable working value.

Some of these disadvantages can be overcome by the use of a second turbine system at high temperatures employing a different working fluid. The exhaust heat from this system is used to produce steam to operate a steam turbine (see Figure 8.16). The **topping cycle** illustrated here uses mercury as the operating fluid. It has the advantages of having a high critical temperature, low critical pressure, somewhat higher pressure at the condenser temperature, and a low heat capacity as a liquid. It is also quite toxic in the vapor phase. Mercury does not "wet" steel surfaces, so that heat transfer is a problem; and it dissolves iron, requiring special steels for the pipework and the turbine. And finally, mercury is rather expensive in the quantities required.

A mercury–steam binary system was actually built by the Public Service Electric and Gas Company of New Jersey in 1949. This small station, 20 MW from the mercury turbine and 30 MW from the steam, had a combined thermal efficiency of 37 percent; steam-turbine generating plants were averaging 23 percent in the United States at that time. This plant is no longer in operation. There has been a substantial improvement in turbine generator systems since that time—the current average for new equipment is about 42 percent—a comparable increase in efficiency could be

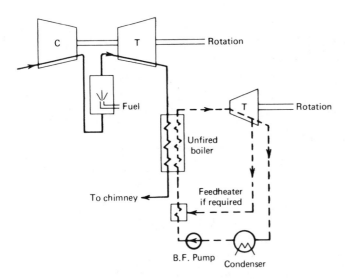

Figure 8.17 Combined gas/steam-cycle schematic.

achieved by the use of an appropriate topping cycle.

The New Jersey binary system required about 8 lb of mercury per kilowatt output; this represented an investment in the mercury alone of more than $400,000 (1963 dollars). No binary systems are under construction, and none is being planned. The (until recently) low cost of nuclear power plants and the lack of pressure to improve thermal efficiency have not helped create the appropriate climate for binary systems. As a consequence, the current economics of power generation may militate against Rankine binary cycles, even with the advantages of mercury as the high-temperature fluid. Instead, rather than a binary cycle, a **combined cycle** may be more practical.

This terminology is used to mean a combination of cycles: a rather high-temperature topping cycle, not Rankine, that exhausts heat into a steam generator for a Rankine-cycle turbine. The obvious choice for the topping cycle is the Brayton cycle. Our studies in Chapter 4 indicated that Brayton-cycle devices could accept heat at very high temperatures relative to Rankine cycles (up to about 1000°C)

and reject heat at temperatures that are still high enough to produce high-quality heat (about 400°C). The combined-cycle process is illustrated in Figure 8.17.

There are several advantages to the use of a combined cycle. First of all, combined-cycle generators can be located at the site of a coal-gasification plant that produces low-heating-value "power gas." Such a product does not have enough heating value to warrant transportation in pipelines, but can easily be used in an on-site gas turbine (see Chapter 6). In such a system, desulfurization of the coal becomes much easier at the gasification stage, so the products of combustion from the gas turbine would have very little impact on the environment. This use of coal to produce power gas would also have the beneficial effect of extending our oil and natural gas resources; these are the fuels most often used in Brayton-cycle power generation. Second, the overall efficiency of a combined cycle can be quite a bit higher than with a steam turbine alone. This is a consequence of the higher operating temperatures that can be achieved with a gas turbine. The consequence of the higher efficiency, approaching 50 percent, of course, is the reduction in heat rejection to the environment. Note that an improvement in efficiency from 40 to 45 percent will reduce the heat rejection by 10 percent. An additional increase in efficiency can be achieved with another, low-temperature turbine operating on the heat energy rejected by the steam turbine.

Tailing Cycles

In such a system the low-temperature turbine is often referred to as a **tailing cycle** or a bottoming cycle. The rejection temperature for a typical steam turbine is less than 100°C, but it can be made higher, as we know, by reducing the efficiency of operation. This is not particularly desirable; but if the rejection temperature is raised slightly to 150° or 200°C, then a variety of tailing cycle fluids can be used effectively to produce an overall increase in operating efficiency, with a subsequent reduction in heat rejection to the environment.

Fluids such as ammonia and isobutane have been proposed for tailing-cycle turbine operation. Prototype isobutane turbines have been built by Magma Energy, Inc., for use with low-temperature geothermal hot-water beds. Isobutane turbines are smaller and have fewer stages than steam turbines. This is because the physical characteristics of isobutane allow a greater enthalpy drop per stage. Since there is no water in the turbine, corrosion is not a problem, and special materials are not necessary. Isobutane is flammable, however, so special precautions are required.

The efficiency of a tailing cycle operating between about 150° and 30°C would not be very high—in the neighborhood of 12 percent. But this, added to the 40 percent of the normal steam cycle and the 10 percent or so of the topping cycle, would yield a combined efficiency of more than 60 percent. A system with this efficiency would waste about 33 percent less heat than a normal fossil-fueled steam turbine.

The reason tailing cycles are not in use, of course, is the expense. The current incremental cost of dealing with waste heat is far less than the incremental cost of a tailing turbine system. And this will continue to be the situation as long as heat rejection to rivers and lakes is essentially "free." Even the added costs of cooling towers are not really high enough to overcome the capital costs of tailing cycles. When the day comes that the "costs" of adding to the earth's heat burden are reckoned and charged to electric utilities, then tailing and topping cycles will seem a great deal less expensive than they do now.

SUMMARY

Even though there has been considerable improvement in the thermal efficiency of electric generating plants in the past 50 years, more and more waste heat will be generated in years to come.

Heat affects the plant and animal life in rivers and streams profoundly. It can cause death to some species.

Heat is transferred from one body to another by three mechanisms: *conduction*, *radiation*, and *convection*.

The rate of heat conduction depends on the geometry of the problem; it may be readily calculated when there are symmetries in the geometry.

Radiation follows the fourth-power temperature *law of Boltzmann–Stefan*.

Convection is quite complicated, as it is a combination of heat conduction and fluid transport. *Dimensional analysis* may sometimes be used to understand a particular problem.

Many power plants use *once-through cooling*, in which water from a river is used to condense steam. However, very large plants would heat the water too much. Cooling lakes and ponds are also in use.

Most new plants use a **cooling tower**. Several types of towers are in use. **Dry towers** reuse a fixed amount of water. **Wet towers** evaporate water to carry away heat.

Waste heat could be used advantageously. **High-quality** heat is difficult to obtain and difficult to transmit over long distances. Waste heat could be used to **desalinate** saltwater or brine.

The production of waste heat can be minimized by one or more of several systems: **combined cycles, topping cycles,** and **bottoming cycles.** These could produce an overall efficiency near 50 percent.

REFERENCES

Berkowitz, David A., and Arthur M. Squires (Eds.), *Power Generation and Environmental Change.* Cambridge, Mass.: MIT Press, 1971.

Clark, John R., "Thermal Pollution and Aquatic Life," *Sci. Amer.* **220**, 19 (March 1969).

Diamant, R. M. E., *Total Energy.* Oxford: Pergamon Press, 1970.

Karkheck, J., J. Powell, and E. Beardsworth, "Prospects for District Heating in the United States," *Science* **195**, 948 (1977).

Lee, Samuel S., and Subrata Sengupta (Eds.), *Waste Heat Management and Utilization.* Washington, D.C.: Hemisphere Publishing, 1979.

Merriman, Daniel, "The Calefaction of a River," *Sci. Amer.,* May (1970), p. 42.

Parker, Frank C., and Peter J. Krenkel, *Thermal Pollution: States of the Art.* Nashville, Tenn.: Vanderbilt University, School of Engineering, 1969.

Reynolds, John Z., "Power Plant Cooling Systems: Policy Alternatives," *Science* **207**, 367 (1980).

Sporn, Philip, *Fresh Water from Saline Waters: The Political, Social, Engineering and Economy Aspects of Desalination.* New York: Pergamon Press, 1965.

Williams, Robert H., "Industrial Cogeneration," *Ann. Rev. Energy* **3**, 313 (1978).

Woodson, Riley D., "Cooling Towers," *Sci. Amer.,* May (1971), p. 70.

_____, *Thermal Pollution.* Washington, D.C.: U.S. Government Printing Office, 1968.

GROUP PROJECTS

Project 1. Calculate the heat loss rate for conduction only in your house or apartment for an average winter day. Make estimates of the R-value for each outside wall, floor, and ceiling.

Project 2. Determine the method used by the nearest electricity-generating plant to dispose of its waste heat. Estimate the effects on the biosphere immediately adjacent to the plant. Have any permanent changes been noticed by residents of the area since the construction of the plant?

Project 3. For computer enthusiasts—Repeat Project 1 following David R. Jackson and John M. Callahan, "Energy Conservation with a Microcomputer," *Byte,* July (1981), p. 178.

EXERCISES

Exercise 1. An *E. coli* colony of 0.05 mg/ml is maintained at 40°C for 4 hr at which time the temperature is lowered to 15°C, where it is maintained for another 4 hr. How much *E. coli* exists at the end of the 8 hr?

Answer: 1373.04 mg/ml.

Exercise 2. A certain 900-MW nuclear power plant operates with 29 percent efficiency. The waste heat is to be carried away by running the entire Hudson River through the condenser (flow rate 50 m^3/s).

(a) What is the rate at which heat must be delivered to the river?
(b) What will be the temperature rise of the river?
(c) What would be the answers to (a) and (b) if the plant were fossil fueled with $\eta = 0.39$?

Exercise 3.

(a) Starting with the 1981 installed electrical generating capacity of 638,000 MW, calculate the figure for the year 2081, if the rate of increase is a constant 3 percent per year.
(b) Calculate the rate of waste heat production in 2081, if the average efficiency is 35 percent.
(c) If the total stream flow in the United States is 16×10^{14} ℓ/yr, calculate the average increase in temperature of cooling water in 2081, assuming all stream flow is used for all waste heat.

Exercise 4. An aluminum cylinder as in Figure 8.6 has $r_1 = 5$ cm, $r_2 = 25$ cm, and $L = 200$ cm. The outer temperature is maintained at 15°C, and the inner temperature is kept at 50°C by an electrical heater having a resistance of 10 Ω. What constant current should be provided?

Exercise 5. A 10-cm-thick wall having an R-6 average insulation is faced with 5-cm-thick silica bricks. What is the new R-value? What if a 1-cm dead air space were left between the wall and the bricks?

Answer: $R = 10.17$, $R = 675$.

Exercise 6. How much difference does ceiling insulation make in a typical midwest home? (Assume 15 cm of insulation, 450 m^2 of ceiling, ΔT of 30°C for 3 months.)

Exercise 7. If ceiling insulation costs $N/cm installed, home heating costs $M/MJ of loss, and the insulation should pay for itself in Y years, make a set of tables for $Y = 3$, 5, and 10; $N = 50$, 100, and 200; and $M = \$1$, \$2, and \$5. For your locality, what are reasonable values for N and M? If $Y = 5$, is 15 cm of additional insulation a bargain?

Exercise 8. A polished aluminum pan exposed to the sun on a beach reaches an equilibrium temperature of 90°C. What is the effective temperature at which it is absorbing energy?

Exercise 9. An orange on a tree in Florida has a temperature of 2°C. It exchanges energy with an open sky of effective temperature −45°C. What is its initial energy loss rate? Assuming a constant loss rate, what time would be required for the orange to drop to 0°C? (Assume $c_p = 20$ J/g °C.)

Exercise 10. If in Example 8.3 the flow were streamline instead of turbulent, the density would not be a factor. What would be the result?

Exercise 11. A sphere falling in a homogeneous fluid reaches a terminal velocity at which the weight of the sphere is balanced by the buoyant force and the frictional resistance of the fluid. Use dimensional analysis to find an expression for the terminal velocity. (Assume it depends on the first power of the density of the sphere.)

Exercise 12. A 1000-MWe nuclear plant is cooled by a 2500 acre cooling pond. Assume a vertical mixing height of 50 m and no excess horizontal mixing. If the air is originally at 20°C and 40 percent humidity, what fraction of the waste heat can be absorbed by evaporating water from the pond over an eight-hour period? If half the waste heat is conducted and radiated away, what is the final temperature of the pond? ($T_i = 20°C$; the average depth of the pond is 5 m.)

Exercise 13. If one percent of the waste heat from a 1000-MWe nuclear plant is used to evaporate water from a 2500 acre cooling pond and there is no replacement water, how much does the level of the pond drop in one day?

Exercise 14. In a natural-draft, wet cooling tower 150 m tall, what is the driving pressure on the interior air column, assuming a density difference between inside and outside air of 0.008 kg/m?

Answer: 11.76 Pa.

Exercise 15. If the cooling tower of Example 8.4 loses 2 percent of its water by evaporation and drift, what is the required replacement rate in liters per day?

Exercise 16. Recalculate the heat transmission distance that was developed in Section 8.3, Transportation of Heat, except use water with an inlet temperature of 100°C and an exit temperature of 50°C instead of steam. Assume all parameters remain the same except those that depend on the fluid.

Exercise 17. For a pipe diameter of 91 cm, exit temperature of 100°C, exit pressure of 87 kPa, inlet temperature of 200°C, exterior temperature of 0°C, $\mu = 125 \ \mu P$, $k = 24 \times 10^{-3}$ J/s m °C for steam, what maximum length is allowed if the heat loss through the pipe is to be no more than 10 percent of the 20 MW delivered?

Exercise 18. In Example 8.5, if the heat requirement were for 500 MW at 200°C, how would the fuel input comparison change?

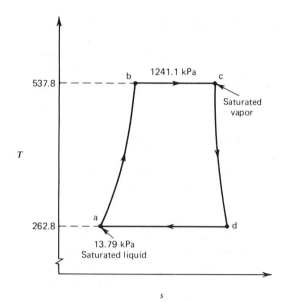

Figure 8.18 T–s diagram for the mercury turbine in a binary system (Exercise 20).

Compare also for 1500 and 500 MW of heat at 150°C. (Assume $T_H = 600°C$ for a Carnot system and scale the efficiency accordingly.)

Exercise 19. In Example 8.6, calculate the cooling-water flow rate required to provide 60°C condensate if the cooling-water temperature is 40°C and $\dot{m}_3 = 23,000 \ \ell/min$.

Exercise 20. The T–s diagram for the mercury turbine in a binary system is shown in Figure 8.18. Calculate the work per mole output and the efficiency of this cycle.

Exercise 21. If 10 percent of the coal-fired steam plant electrical-generating capacity were replaced with the equivalent electrical output of mercury–steam binary plants with overall efficiency of 52 percent, what annual coal savings could be realized?

PART FOUR

ENERGY UTILIZATION

CHAPTER NINE

ENERGY TRANSMISSION

In previous chapters we have examined the impacts of the various sources and forms of energy and conversion processes that are of value to our society. We know there are a number of techniques for transporting energy, either in a "raw" state as a fuel or as a "finished" product, as electricity or heat. In many cases geography or topography dictates the form of transport; in other cases there is a choice. In earlier years, the energy requirements of our country, the sizes of metropolitan districts, and the state of the technology were such that relatively small energy conversion plants could be built economically within cities; the only energy transport problem was with the fuel. Transmission of the "finished" product was usually not necessary or desirable.

These days the accretion of users into very large metropolitan districts, the economies of scale dictating ever larger units, the costs of transportation of fuel, and the uncertainties regarding the safety of certain types of generating plants have made energy transmission, both of the fuel and finished product, one of the most important considerations in plant siting. As a consequence, there are several questions we address in this chapter: What are the alternatives for energy transmission? What advantages and disadvantages

does each have? What are the physical bases for making a choice in a given situation?

9.1 ENERGY TRANSMISSION ALTERNATIVES

Energy can be transmitted either continuously, as in pipelines or electrical transmission lines, or in batches, such as in oil tankers or rail cars. Obviously, electricity is difficult to transmit in a batch, and coal is difficult (but not impossible) to transmit continuously. So not every transmission form is available in all situations.

In Table 9.1 several of the alternatives for both continuous and batch transmission are shown. In terms of relative costs, transporting Middle Eastern crude oil is cheap compared to transporting other forms of energy, even coal. However, if the coal or oil is being transported for the purpose of generating electricity, then its expense should be multiplied by about 3 to take into account the efficiency of the electrical generating process. This table does not take into account all the relevant criteria.

Criteria for Selection

There are several criteria that must be considered before a choice can be made for an

TABLE 9.1 Energy Form—Transmission Mode Matrix[a]

	Coal	Petroleum	Gas	Heat (Steam)	Electricity	Hydropower
Land—batch						
Rail		l				
Vehicle		X				
Land—continuous						
Aqueduct						X
Pipeline		l	l	X		
Transmission line					h	
Slurry pipeline	h					
Sea—batch						
Cargo Ship	l					
Tanker	l					
LNG Tanker			l			
Supertanker		l				

[a] l = relatively low cost, h = relatively high cost, X = not normally used for long distances.

energy transmission system:

- Cost per unit of energy delivered.
- Geographic limitations.
- Carrying capacity desired.
- Technical features.
- Environmental impact.

The cost per unit of energy delivered is a very important consideration in the choice of an energy transmission system for a given application. Some of these factors are rather more difficult to quantify. For example, conventional overhead electrical transmission lines require about 3 ha per linear kilometer of right-of-way. Ultrahigh-voltage lines (voltages in excess of 500,000 V) require more than twice as much. Such transmission lines are also rather unsightly, and there is some evidence that the large electric fields near ultrahigh-voltage lines can have deleterious biological effects. These effects, not yet thoroughly understood, are not being taken into account currently in power line siting or costing in most states.

Electricity can be transmitted in underground cables; but in most applications the cost of underground installations far outweighs any environmental considerations. The only exception to this general rule is the carrying of electrical power into large metropolitan areas, where the cost of above-ground right-of-way more than balances out the high installation costs for underground cable.

The choice of AC or DC for electrical transmission involves more than cost. Direct current transmission has several advantages over alternating current: higher average power for the same peak voltage, ease in tying together different generators, and reduced losses. On the other hand, AC can be easily stepped up or down in voltage and is easier to switch under load.

There are environmental considerations in the transmission of energy by batch as well. Probably everyone knows about the adverse effects of large oil spills on coastal flora and fauna. The general public may not be as familiar with the possible adverse effect on the global energy balance of an oil spill that covers a large ocean area.

Pipelines cause probably the minimum environmental disruption for a given amount of energy-carrying capacity, except in special situations (Alaska!). Although pipelines are generally used in the distribution of primary fuels—natural gas and oil—they can also be used to carry hydrogen. By this technique the electrical product of a large generating plant could be transmitted economically over long distances. (See Chapter 6.) Pipelines have also been used to transport solids as a slurry with some success.

Since electrical energy is about 15 percent of the energy in use today and is projected to account for more than 25 percent by the year 2000, we study it first.

9.2 ELECTRICAL ENERGY TRANSMISSION

Figure 9.1 shows the amount of transmission lines having voltages greater than 132 kV as a function of kilometers in place in the United States as of mid-1981. This represents a substantial commitment to electricity transmission, and several aspects of modern life may cause this commitment to increase. In 1980 electric utilities generated about 210×10^9 kWh that were not used by consumers, because they were lost in transmission and distribution. This power cost more than \$1 billion, equivalent in buying power to all the capital raised by utilities for electric transmission and distribution equipment. A substantial part of these losses could have been avoided by judicious choice of transmission systems or voltage. In order to be able to discuss the relative merits of one system over another, perhaps we should review some of the basics of electricity to make sure we understand all the terms.

Review of Fundamentals

Electric current, the motion of electric charge, can be generated either as a steady

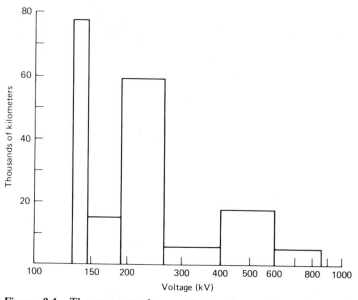

Figure 9.1 The amount of transmission lines with voltage over 132 kV in place in the United States in mid-1981. (Department of Energy Information Center.)

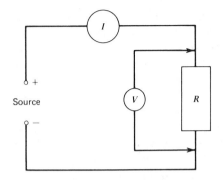

Figure 9.2 The voltage drop V across a resistance R carrying a current I.

one-directional flow of electrons or as a sinusoidally time-changing motion alternating in direction. We call these two options **direct current (DC)** and **alternating current (AC)**. (Actually, any sort of time-changing current could be generated, but it is convenient to consider sinusoidal currents, as these are the only ones used in practice.)

There is an empirical relationship between the current in a circuit and the potential difference (voltage) across the circuit, as shown in Figure 9.2:

$$V = IR, \tag{9.1}$$

where R is the equivalent resistance of the circuit.[1] This is, of course, the familiar **Ohm's law.**

The power required to sustain this current is just the work done per unit time. Now, the work done in moving a charge q C through a potential difference V is just qV, so the power is given by

$$P = \frac{qV}{t} \quad \text{or} \quad P = IV \tag{9.2}$$

[1]The circuit does not need to be a single component but, rather, can be a complex network of elements. An equivalent resistance can always be found for such a network using Kirchoff's rules.

In Figure 9.2 the energy is supplied by an external source and is dissipated by the circuit. This dissipation is in the form of heat. We can relate the power lost to the circuit resistance by eliminating V between Eqs. 9.1 and 9.2:

$$P_{\text{loss}} = I^2 R. \tag{9.3}$$

This relationship is known as **Joule's law.** Heat losses can be excessive if transmission lines of low voltage are used.

Example 9.1 A hydroelectric generating plant is required to generate electrical energy for a city 100 km distant, having a power load of $P_L = 0.48$ MW. The transmission lines carrying the current have a total resistance of $100\ \Omega$. The plant manager initially proposes to supply the electricity an an emf of 1200 V at the city, as shown in Figure 9.3a. What total generating capacity P_G must be installed in order to supply the city and account for transmission line losses? How would the answer change if the supply voltage were 200,000 V, as in Figure 9.3b?

You can easily find from Joule's law and Ohm's law that $I = P/V$. So the city load at 1200 V would demand $(0.48 \times 10^6)/(1.2 \times 10^3) = 400$ A. The current would generate heat in the

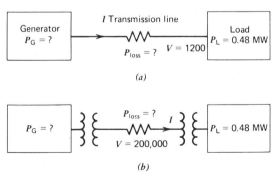

Figure 9.3 (a) A 1200-V transmission line between a generating station and a 0.48-MW load. (b) Step-up and step-down transformers used to transmit the power at 200,000 V.

transmission lines at the rate $(400)^2 \times 100 = 16\,\text{MW}$. Almost 34 times as much power will be lost in the lines as would be delivered to the city; clearly not a very economical operation. If, however, the voltage is stepped up with transformers to a high value, say 200,000 V, the power line losses will amount to only 576 W:

$$I = \frac{0.48 \times 10^6}{2 \times 10^5} = 2.4\,\text{A},$$

$$P_{\text{loss}} = (2.4)^2 \times 100 = 576\,\text{W}.$$

Transformers are always used to step voltages up and down to reduce the transmission line currents and thereby the losses; however, transformers require, AC, not DC. Of course, generators could be wound to generate high voltages directly, but insulation requirements and reduced current capacity would cause problems. And the difficulty in dealing with the high voltage at the load end would remain.

This is an important fact of life that we should understand before proceeding. A transformer consists of a core, usually laminated iron, around which are wound a primary and one or more secondaries, as shown schematically in Figure 9.4. The windings are insulated from the core and from each other so that there is no electrical connection between the primary and secondary. The iron core serves to trap the lines of the magnetic field generated by the primary and ensures that most of them go through the secondary windings. Since there is no contact between primary and secondary, how can a secondary current exist?

If the voltage source for the primary were a DC generator, there would be no secondary current. There will be no current unless there is an electromotive force, emf, generated in the secondary. The emf in any circuit is calculable from Faraday's law:

$$\mathscr{E} = -\frac{d\Phi}{dt}, \qquad (9.4)$$

where Φ is the total magnetic flux through the circuit. Clearly, the magnetic flux Φ must be changing in time in order for an emf and thereby a current to be generated. This is the reason alternating current generators must be used, if high-voltage transmission is desired.

Just for completeness, let us note that for a 100 percent efficient transformer the secondary and primary emfs are related by the ratio of the secondary to primary turns:

$$\mathscr{E}_s = \mathscr{E}_p \frac{N_s}{N_p}. \qquad (9.5)$$

Alternating current transmission also has

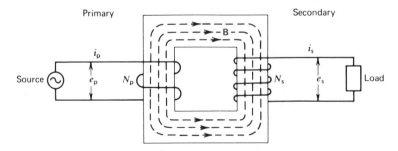

Figure 9.4 The secondary current i_s induced in the secondary windings of a transformer. Magnetic field lines B are generated by the alternating primary current i_p. The numbers of primary and secondary turns are N_p and N_s.

several characteristic disadvantages. In order to be able to make a sensible comparison of AC and DC, we must try to understand some AC circuit theory to see what causes these problems.

AC Power Transmission

In a circuit that has currents changing in time, there will obviously be changing magnetic fields. And, from Faraday's law, Eq. 9.4, there will be induced emfs. The minus sign in Eq. 9.4 indicates that the direction of the induced emf is such as to oppose the change in the magnetic flux that causes the emf. This opposition to changing currents in AC circuits acts similarly to DC resistance. We can make this idea quantitative by noting that for any circuit the ratio of the total number of magnetic field lines to the instantaneous current is a constant:

$$L = \frac{N\Phi}{i}. \qquad (9.5)$$

We call this ratio the self-inductance, or more usually **inductance**, of the circuit. Inductance is mesured in henrys in the metric system.

Inductors are coils of wire; they may be wound on iron cores to increase the $N\Phi$. Motors obviously are inductors; so are transformers, tuners in TV sets, and even long lengths of power lines. The AC resistance effect of an inductor mentioned earlier is called **inductive reactance**. It is given in terms of the inductance of the circuit and the frequency of the changing current:

$$X_L = 2\pi f L. \qquad (9.6)$$

However, we cannot simply replace R with X_L in Ohm's law. Why? With an alternating current does I represent the peak value of current or an average or what? Figure 9.5 is a drawing of a sinusoidal current, of frequency f Hz: $i = I_{max} \sin 2\pi f t$. The maximum value I_{max} occurs when $2\pi f t = \pi/2$. It is easy to see that

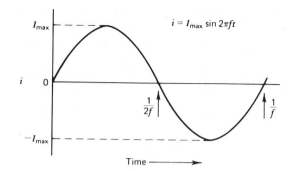

Figure 9.5 Graphical representation of an alternating current, $i = I_m \sin 2\pi f t$.

the average value is zero. Yet, we know that an alternating current will produce heat in a resistive circuit; your toaster is a good example of that effect!

Since this is the case, we can ask, "What is the value of a direct current that is equivalent in terms of the heat production in a resistive element to a given alternating current?" By averaging over one complete cycle, an enterprising student could show that a direct current given by

$$I_{eff} = \frac{I_{max}}{\sqrt{2}} \qquad (9.7)$$

has the equivalent heating power, averaged over one cycle, of the sinusoidal alternating of amplitude I_{max} and of any frequency. We call this value of current the **effective value**. Now we can generalize Eqs. 9.1, 9.2, and 9.3. Note that the voltages obtained in these equations will be effective values. If we use the I_{eff} given by Eq. 9.7 instead of I, these equations will apply to AC as well as to DC circuits with this important proviso: the resistances must be purely resistive.

All real resistances have some inductive reactance. So instead of R or X_L in Ohm's law, we must use another quantity that is called **impedance**. This is a kind of net result of resistance and reactance; it is defined by

$$Z^2 = R^2 + X_L^2. \qquad (9.8)$$

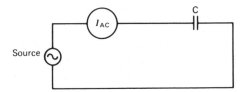

Figure 9.6 Pictorial representation of a capacitor.

But this neglects the effect of **capacitance**. In a circuit like that shown in Figure 9.6, even though there is no electrical contact across the device labeled C, an AC meter will give a reading.[2] Why? The current is changing direction periodicially; electric charge shuffles back and forth from one plate of the device to the other through the source. If the current were not changing, there would be some movement of charge immediately after the source were energized, but as soon as the potential difference across the device were equal to that of the source, the motion of charge would stop.

A device of the kind indicated by C in Figure 9.6 is called a **capacitor**. It is often in the form of parallel conducting plates separated by an insulating medium. For any particular arrangement of plates and at a given frequency, it is always found that the ratio of I_{eff} to V_{eff} measured across the capacitor is a constant, the capacitance:

$$C = \frac{I_{eff}}{2\pi f V_{eff}}. \quad (9.9)$$

The SI unit of capacitance is the farad (F). Thus, if we define the capacitive reactance as

$$X_C = (2\pi f C)^{-1}, \quad (9.10)$$

then for *any* circuit

$$V_{eff} = I_{eff} Z, \quad (9.11)$$

where the impedance is given by

$$Z^2 = R^2 + (X_L - X_C)^2. \quad (9.12)$$

[2]*Note*: AC meters read effective values.

This is the most general form of Ohm's law for AC circuits.

In an AC circuit that has reactances as well as resistances the current and voltage will not be in "sync." This is a consequence of the opposing effect of induced emfs in inductors and the opposing electric field set up in capacitors. The voltage will lead or lag the current by a phase angle ϕ. That is, if $i = I_{max} \sin \alpha$, then in general $v = V_{max} \sin(\alpha \pm \phi)$.

The instantaneous power in an AC circuit is just the product of i and v:

$$P_{inst} = V_{max} I_{max} \sin \alpha \sin(\alpha + \phi). \quad (9.13)$$

If this power is averaged over one cycle to eliminate time, the average power is given by

$$P = I_{eff} V_{eff} \cos \phi. \quad (9.14)$$

The $\cos \phi$ term is called the **power factor**. The watt-hour meter used by the electric utility measures the average power, including the power factor, but the utility must supply the peak power, which is $I_{max} V_{max}$. Obviously, the utility would like to maintain the power factor close to unity. The power factor can be written in terms of the resistance and reactance of the load:

$$\phi = \tan^{-1} \frac{|X|}{R}, \quad (9.15)$$

where $X = X_L - X_C$.

A factory that has a large number of electric motors will have a small power factor. Consequently, the power company will ask the factory owners to install correcting capacitors or suffer a power factor penalty. Long power lines will also have large inductive reactance; periodically, capacitors will be installed to offset this, even though high-voltage capacitors are expensive. The inductive reactance depends on the number and physical arrangement of conductors in the transmission line, but it can be as high as 0.15 Ω/km compared to about 0.09 Ω/km for the DC resistance of similar wires.

Example 9.2 A 160-km, 135-kV two-wire transmission line has an inductive reactance of 0.1 Ω/km and a DC resistance of 0.09 Ω/km. It is connected to a load that has a DC resistance of 120 Ω and a power factor of 0.707. What is the current (effective value) and power factor of the entire system?

The reactance of the load is found from $\tan \phi = X/R$. In this example $X = 120\,\Omega$. The total impedance is

$$Z = [(120 + 2 \times 0.09 \times 160)^2 + (2 \times 0.1 \times 160 + 120)^2]^{1/2}.$$

$$Z = 213\,\Omega;$$

$$X_L = 120 + 2 \times 0.1 \times 160 = 152\,\Omega,$$

$$I_{\text{eff}} = \frac{V_{\text{eff}}}{Z} = \frac{135,000}{213} = 633.8\text{ A}.$$

$$\text{Power factor} = \frac{152}{21} = 0.714.$$

Another peculiarity of alternating current is called the **skin effect**. A direct current will be distributed uniformly throughout the interior of a conductor. An alternating current will not be but, rather, will be concentrated in a surface layer, the "skin" of the conductor. This phenomenon is frequency-dependent, and is much more severe at high frequencies than at the 60 Hz normally found in the United States

in power transmission (50 Hz in many other parts of the world). But even at low frequencies the effect is significant. For example, if the DC resistance of a particular conductor is 0.022 Ω/km, at 60 Hz and 50°C its resistance is 0.030 Ω/km. Because of the inductive reactance of a transmission line, a current will be set up in the line—even with no load. This charging current, as it is called, produces Joule's law losses, because it is carried by the conductors.

Return for a moment to Eq. 9.13. Note that there are at least two times in every cycle when the instantaneous power is zero. This pulsating nature of the power can be particularly objectionable under certain circumstances. For example, a heavy current load will cause thermal stresses on the generator. It is desirable to maintain a more smoothly varying power load and yet still operate with alternating current. These apparently orthogonal requirements can be met with **polyphase systems**.

In most electrical transmission systems, three-phase generation is used, although proposals have been made to go to six phases to increase the load capacity of existing lines. Figure 9.7a shows schematically a generator with three identical windings, A′A, B′B, and C′C, physically spaced 120° apart on the armature, as indicated in Figure 9.7b. As the arma-

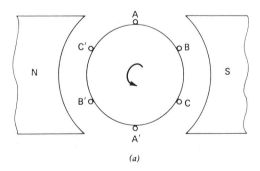

(a)

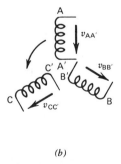

(b)

Figure 9.7 Schematic diagram of a three-phase generator, showing the relative phases of the three voltages.

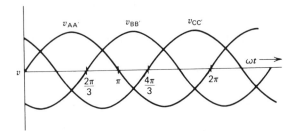

Figure 9.8 Graphical representation of the output of a three-phase generator.

ture rotates, emfs will be induced in the windings, as shown in Figure 9.8. Effectively, a three-phase circuit can be analyzed as three single-phase circuits.

The polyphase nature of most transmission line systems is the source of one of the major problems: tying together different generators. In general, the phase angle between the current and voltage in each phase will be different, depending on the loads on each phase. If the generator to be tied in does not have the correct phase angle in each current phase, it can appear as a load, rather than a source, to the existing generation network. This can result in excessive current in both generators and resulting automatic shutdown, for its own protection, of the entire system.

Switching a transmission line under load presents a major problem whether it is an AC or a DC line. With an AC line, there is at least one time during each cycle of each phase where the current is zero. But since mechanical switches require a finite time to operate, there is always an arc generated between the contacts of very sizable current. This arc must be quashed, before the contacts can be seriously damaged by melting. Modern designs use a high-pressure insulating gas, sulfur hexafluoride, to extinguish the arc.

High-Voltage Transmission

Much of the previous discussion of AC power lines is applicable not only to the large

high-voltage systems now so commonplace in this country but also to the lower-voltage distribution systems in use in our cities and towns. Here we will look only at those peculiarities of power transmission that result from the high voltage. Clearly, transmission at as high a voltage as is economically feasible is desirable to minimize Joule's law losses and deliver the largest possible percentage of the power generated to the consumer. Before discussing the question of how high that voltage should be, we must first answer the question of whether AC or DC should be used. We know that DC does not suffer from skin effects or reactive loading, but as a general rule, the cost of producing high-voltage DC is prohibitively high, except in situations where the high power load and very long distance of transmission make these costs comparable to the extra costs of losses peculiar to AC transmission. Figure 9.9 compares these costs. We see that DC transmission is comparable in cost to AC only for lengths exceeding about 1000 km and for power loads exceeding about 1000 MW. In

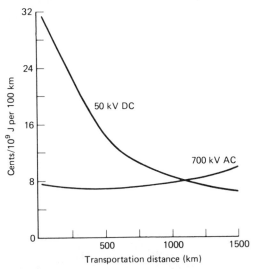

Figure 9.9 Cost of electric transmission versus distance for 50-kV DC and 700-kV AC systems.

fact, since there are very few places in this country where it is necessary to transmit power over such long distances, it is unlikely that DC will be used more extensively in the future than it is today. Of course, advances in technology that result in cheaper AC–DC and DC–AC converters could change this.

High voltages have risen from the early 69-kV lines to 135, 220, and 550 kV. Now there are several 750-kV lines in operation in this and several other countries. Figure 9.10 is a photograph of a 765,000-V transmission line in Indiana. There has been a great deal of research in the past decade to evaluate 1000- to 1500-kV power transmission. Over the years a number of physical and environmental problems have been found with high-voltage transmission lines; many of these will become much more severe at the higher operating voltages being comtemplated for the 1980s. These problems include:

- Corona discharges and consequent power loss.
- Electromagnetic interference (radio and television noise).
- Audible noise.
- Ozone and NO_x production.
- Electrific field gradients within the right-of-way.
- Load switching difficulties.
- Unsightliness.

Corona discharge occurs when the electric field intensity at the surface of a conductor exceeds the breakdown strength of the air. This is a very complex phenomenon that is not easy to describe in terms of physics principles because of its dependence on so many variables: air pressure, electrode, material, presence of water vapor or ionizing radiation, type of voltage, and conductor surface irregularities. Not only does it cause power losses,

Figure 9.10 A 765-kV transmission line in Indiana. (Photograph courtesy of American Electric Power Service Corporation.)

but also corona discharge is responsible for the next three items on the list above.

At electric field gradients above 15 kV/cm, corona discharge becomes important. The field gradient, of course, depends on the physical arrangements of the conductors relative to each other, to the supporting towers, and to the ground, so it is possible to design a transmission system to minimize corona. But it cannot be eliminated, even on fair weather days. In fair weather, corona discharge accounts for about the same amount of power loss as discharge across insulators; it is a rather small loss. But on bad weather days, rain or snow, or in the presence of particulates, ash or dust, corona loss can be very large. And the undesirable side effects can be quite objectionable, as well.

Corona discharge is characterized by rapid pulses of current having rise times in the microsecond range and repetition rates in the megahertz range. As a consequence, the frequency spectrum of electromagnetic waves generated by these discharges can cover a large portion of the communication bands, both radio and television. This interference is generally called EMI, electromagnetic interference, although sometimes it is broken down into two categories, RI and TVI, for radio and television interference. The EMI problem is compounded by the fact that although corona discharge is responsible for generating noise currents, these currents propagate down the transmission lines, which act as antennas. The actual EMI may occur many miles from the discharge.

There seem to be only two methods for minimizing EMI from high-voltage transmission lines: reduce corona discharge by conductor design, and choose the line siting as judiciously as possible to avoid regions where the signals interfered with, whether radio or television, are weak compared with the EMI. There are no quantitative levels specified by any regulating agency concerning electromagnetic emissions from power lines; however, most states do make the utility responsible for correcting EMI complaints.

Audible noise from power lines can be a serious problem, especially when the conductors are wet. Ten states as well as the Environmental Protection Agency have set quantitative standards for noise, which under wet conditions are difficult to meet. Audible noise reductions from existing power lines will be very difficult to achieve and appropriate siting of new lines is a necessity.

Electric field gradients in the right-of-way can be quite high. Measurements on existing 345-kV lines indicate gradients between about 3.8 and 5.5 kV/m and on 765-kV lines they have been measured to be near 11 kV/m. One state, New York, has set an upper limit of 1 kV/m for the field gradient at the edge of the right-of-way. Obviously, most high-voltage lines exceed this value. Three other states, Minnesota, North Dakota, and Oregon as well as New York, have set 5 mA (effective value) as the maximum current that would be set up in a short-circuited truck, vehicle, or piece of equipment in the ROW.

Example 9.3 If 1 mA can just be felt and an average body resistance—head to foot—is 10 MΩ, how tall would a person have to be in a 5-kV/m field to feel the current when grounded?

$$V = IR = 10^{-3} \times 10 \times 10^{6} = 10^{4}\,V,$$

$$h = \frac{10^{4}}{5 \times 10^{3}} = 2\,m.$$

These numbers are typical of the situation under a high-voltage power line.

Ozone and NO_x are produced in the corona discharge. These gases present significant health hazards, as we see in Chapter 13. However, under normal operating conditions the amounts generated are very small; certainly below the EPA standards, which are

0.08 ppm by volume not to be exceeded more than once a year for ozone and 0.05 ppm maximum annual arithmetic mean for NO_x.

The problem of unsightliness is inevitable. There is no physics involved in this, so there is very little for us to do, except indicate the magnitude of the problem. Towers for existing 135-kV lines are approximately 25 m tall; about 3 ha of right-of-way is required for each kilometer of line. 1500-kV lines would require towers are illustrated in Figure 9.11. If the demand for electrical energy continues to grow, even at the reduced rate of about 4 percent per year that was typical of the late 1970s, it is clear that many more miles of transmission lines will be required to fill that demand. It is easy to place blame on the power companies for building power plants and unsightly lines. But the electric utilities simply satisfy the demands of their customers. While it may have been true in the past that utilities helped to create that demand by their marketing strategies involving electric heat and air conditioning, today the demand seems to be coming from the consumers without external stimulus. As Pogo said, "We have met the enemy and he is us."

Most of the environmental regulations now required by the federal and state agencies were put into effect after many of the existing high-voltage power lines were built. The pressure of these regulations and the need for higher-voltage lines present a real challenge to designers of power transmission lines. It is possible to meet all the regulations most of the time; but it is not simple.

Superconducting Systems

An alternative to the transmission of electrical energy at high voltages using normal conductors is to make use of the fact that the

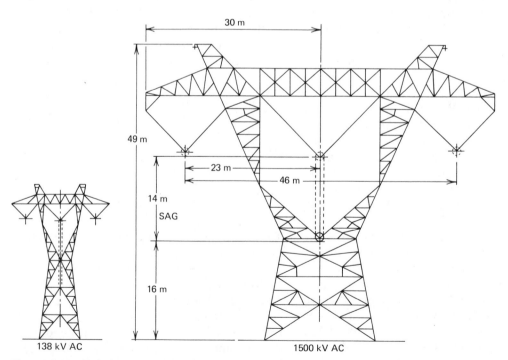

Figure 9.11 Existing 138-kV transmission line towers compared to the estimated size and shape of 1500-kV towers.

resistivity of a normal metal is a linear function of temperature over a wide range:

$$\rho = \rho_0[1 + \alpha(T - 20)] \ \Omega \text{ m}, \qquad (9.16)$$

where ρ_0 is the measured resistivity at 20°C, α is the temperature coefficient of resistivity, and T is the temperature in degrees Celsius. For copper $\rho_0 = 1.72 \times 10^{-8} \ \Omega$ m and $\alpha = 0.00393 \ °C^{-1}$, so using Eq. 9.16 at $-209°C$ the resistivity is 0.1 times its value at room temperature. (This assumes Eq. 9.16 to be valid to $-209°$; it is not.) This temperature is close to that at which nitrogen liquefies ($-195°C$); if the transmission lines in the preceding example were kept at liquid-nitrogen temperature, a factor of 10 in power savings could be realized. Of course, liquid nitrogen is not free! Energy would have to be expended to maintain the lines at that temperature.

Another alternative to high-voltage transmission of electric power is zero-resistance transmission. Certain materials, when cooled to very low temperatures, seem to lose all their electrical resistance. These are called **superconductors**.

Superconductivity was discovered in 1911 in Leiden by Kamerlingh Onnes, who found that the resistivity of mercury fell to as near zero as could be measured when the temperature was lowered to 4.15 K. Since that time many metals, most of them rather poor conductors at room temperature, alloys, and intermetallic compounds have been found to exhibit superconductivity below some temperature T_c, the *critical temperature*. A list of such elements and critical temperatures is given in Table 9.2. Note that all the temperatures are quite low, but that a few are in the liquid-hydrogen range. (Hydrogen liquefies at atmospheric pressure at 20.25 K.) In the superconducting state a current will pass through the material without Joule heating losses, so that even though the expense of maintaining the low temperature is quite high, there could conceivably be situations where superconducting power transmission

TABLE 9.2 Superconductivity of several materials

Material	Critical Temperature (K)
Titanium (Ti)	0.39
Cadmium(Cd)	0.56
Zirconium (Zr)	0.54
Zinc (Zn)	0.87
Indium (In)	3.40
Ti_2Co	3.44
Tin (Sn)	3.72
Mercury (Hg)	4.15
Vanadium (V)	5.38
Lead (Pb)	7.19
Niobium (Nb)	9.50
La_3In	10.4
NbN	16.0
V_3Ga	16.5
V_3Si	17.1
Nb_3Sn	18.05
Nb_3Ge	23.2

would be economically competitive. Unfortunately, life is not so simple.

It was found very early that if a superconductor were placed in a magnetic field, the superconductivity was destroyed for values of the field exceeding a critical value H_c and that the material became a normal conductor. The H_c values for pure metals were all very low and varied with temperature, as shown in Figure 9.12. This figure gives H_c in teslas, although the gauss is the cgs unit most familiar to workers in the field. The SI unit for magnetic induction (i.e., the total magnetic field), the tesla (T), is 10^4 gauss (G). This unit is not as commonly used; there are also several other units in the cgs system, none of which has any particular value. We shall use either gauss or tesla. In any case, the value of H_c for the metals in Figure 9.12 is quite small, and results in a very small current-carrying capacity.

Example 9.4 For a circular tin wire of 1 cm diameter, what is the critical current? (Use a temperature of 2 K.)

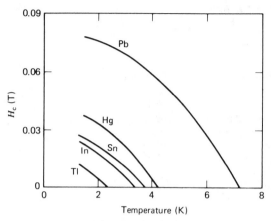

Figure 9.12 Critical magnetic field versus temperature for several Type I superconductors.

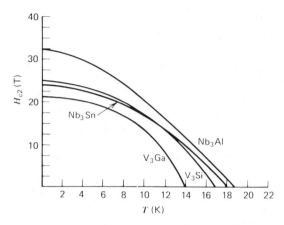

Figure 9.13 The critical field H_{c2} versus temperature for several Type II superconductors.

The magnetic field at the surface of a circular current-carrying conductor is given by

$$B = \frac{\mu_0 I}{2\pi R},$$

where R is the radius of the conductor.

$$I = \frac{2\pi RB}{\mu_0},$$

$$I = 2\pi (0.5 \times 10^{-2}) \times 210 \times \frac{10^{-4}}{(4\pi \times 10^{-7})} = 525 \text{ A}.$$

This current is far too small to be of economic value. If the line were being fed from a 135-kV source, this current would correspond to a power transfer of only 70.9 MW. The losses in a normal high-voltage transmission line delivering this power would not be worth the cryogenic investment required for the superconducting line.

All is not lost. The intermetallic compounds in Table 9.2 not only have high transition temperatures, but they also have very high critical fields; see, for example, Figure 9.13. Note the difference in vertical scale between Figures 9.12 and 9.13. But these materials behave somewhat differently than do the pure metals.

The difference is so striking that the first group has been called Type I superconductors and the other group Type II.

The major additional difference between the two types is apparent in what is called the Meissner effect. When a normal Type I superconductor is placed in a magnetic field at a high temperature, the lines of the field penetrate the metal, as shown in Figure 9.14a. If the temperature is lowered past the critical temperature for that material, the lines of the magnetic field are completely expelled from the interior of the sample, as in Figure 9.14b. It is as if the metal becomes completely diamagnetic, aligning all the electrons in the metal so

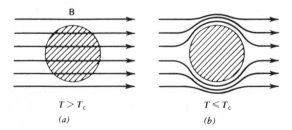

Figure 9.14 The Meissner effect. (a) A superconducting sphere at a temperature above T_c is penetrated by magnetic field lines. (b) At a temperature $\leq T_c$, the field lines are expelled.

that the resulting interior field is antiparallel to the applied field. This will be true so long as the surface field is below H_c, and the temperature is below T_c.

In a Type II superconductor, penetration of an applied field is complete even below the critical temperature so long as the applied field is larger than a critical value H_{c2}. If the applied field is between H_{c2} and a lower value H_{c1}, the penetration is partial and varies smoothly. For field strengths below H_{c1} field lines are completely expelled, just as in Type I superconductors. For most Type II materials, H_{c1} is rather low, not very different from H_c for Type I materials. However, H_{c2}, which is the vertical variable in Figure 9.13, is very much larger by several orders of magnitude. Very high fields can be achieved—Figure 9.15 is a photograph of a superconducting magnetic built by Inter-magnetics General Corporation that will produce fields up to 16.5 T (165 kG).

Unfortunately, this partial penetration of the flux into the material causes a problem that is particularly severe with alternating currents, even at 60 Hz. This problem is losses. These are not I^2R losses, since the resistance is zero

Figure 9.15 A commercially available superconducting magnet that can achieve fields as high as 16.5 T (165 kG). (Photograph courtesy of Intermagnetics General Corporation.)

below H_{c2}, but the effect is the same. Power is lost in the transmission line, and heat is generated.

The mechanism for these Type II superconductor losses requires the complete quantum-mechanical treatment of superconductivity to be understood. We do not require such completeness here; the references at the end of this chapter can supply it. We note that these losses can be reduced by choice of conductor size and configuration and by careful surface treatment of the conductors. Losses vary considerably with temperature and magnetic field intensity. At a constant temperature of 6 K they range from a small fraction of a microwatt per square centimeter of conductor outer surface at a surface field of 0.1 T to several hundred microwatts per square centimeter at 0.2 T for a Nb$_3$Sn ribbon. As can be easily calculated, these losses can be very large and will more than offset the absence of I^2R losses. A DC transmission system would experience only about half the Type II superconductor losses as the AC system; even this amount may be too large in most situations.

An additional serious problem must be satisfactorily solved, that of superconductor stability. Transmission lines are subject to sudden changes in loads caused by equipment failure, lightning, or conscious load shifting. Sudden surge currents could drive the superconductor to normal resistivity, and the resulting heat would at the very least evaporate all the helium coolant and at the worst melt the conductors. One way to overcome this is to back the superconductor with an appropriate thickness of normal conductor, such as aluminum or copper. If the superconductor suddenly goes normal, the regular conductor backing will handle the load, some heat being generated. However, one difficulty is that it would take several seconds to restore superconductivity. Normal high-voltage lines can restore service within 25 cycles of a 10-times-normal current surge.

Another potential problem has to do with the availability of helium. Unless superconductors with much higher critical temperatures are found, helium is the logical coolant. But helium is found in the atmosphere in quantities too small to make extraction from the atmosphere economic. Up to 1972 helium had been extracted from natural gas streams and stored by the Bureau of Mines. In 1972 this helium conservation program was terminated. It is not unreasonable to ask whether there is enough "cheap" helium to support an expanded superconducting power transmission network.

Table 9.3 is an estimate of the U.S. helium reserves of two kinds: rich reserves (greater than 0.3 percent He) and lean reserves (less than 0.3 percent He). Secure reserves are those being held by the Bureau of Mines. Insecure reserves are those in privately owned natural gas fields. None of the latter are being used for helium recovery today, and the likelihood that significant helium can be recovered after the year 2000 is very small. This is because the likelihood that natural gas will be recoverable from those same fields is also very small. There are about $80 \times 10^8 \, m^3$ of helium in storage now, and estimates are that more than 40,000 km of superconducting transmission lines could be built with only 75 percent of this amount being required (very little should be lost in operation).

Many superconducting applications have been studied and are still being studied. These include also generators and magnets. High-field magnets are the one area in which superconducting technology holds a significant edge over normal construction techniques. But it does not appear likely that superconductivity can make such of an impact on electricity transmission problems. Very large loads, in excess of 1000 MW, and very long distances, in excess of 1000 km, will be required before the added costs of the cryogenic refrigerators can be offset. The one area of power transmission

TABLE 9.3 U.S. Helium Reserves

	Reserve Quality	Reserve Amount (GCF)[a]
Rich Reserves (>0.3% He)		
Secure reserves		
1. Bureau of Mines storage program	60% He	28
2. Tip-top field	>0.3%	>15[b]
Total secure, rich, proven reserves		43
Insecure reserves		
3. U.S. natural gas fields, predominantly Texas panhandle, Oklahoma, and Kansas (Alaska excluded)	>0.3%	105 (1971)
Total rich, proven reserves		148
Lean Reserves (<0.3%)		
Insecure reserves		
4. U.S. natural gas fields, predominantly Texas and Gulf Coast (Alaska excluded)	0–0.3% (mostly 0.006%)	130 (1971)
Total proven reserves		278

Source: E. B. Forsyth, ed., *Underground Power Transmission by Superconducting Cable*. Brookhaven National Laboratory Report BNL 50325, March 1972.
[a] 1 GCF = 10^9 SCF (Standard Cubic Foot).
[b] May be much greater.

that could be well served by superconductivity in the future is underground cable transmission.

Underground Electric Power Transmission

Underground cables are used only as a last resort, when the cost of overhead right-of-way is excessive. Underground cables are expensive and difficult to install, hard to maintain, and almost impossible to upgrade to increase capacity. Nevertheless, in high-population-density urban areas there are more than 3600 km of high-voltage underground power transmission cables today, and many more are projected for the future.

Current practice is to use high-pressure oil-filled (HPOF) cables; several alternatives are just becoming available:

■ Compressed gas insulation.

■ Cryogenic transmission.
■ Superconducting transmission.

The last of these is not yet available, and, as discussed earlier, may not be practical for some time. Let us look at some of the unique problems of underground cables to see if we can determine to what extent this means of energy transmission may contribute in the future.

The first obvious problem is insulation. Because of the high cost of right-of-way, air cannot be used as the insulating medium. The conductors must be quite closely spaced; until recently, underground cables were insulated with a paper wrapping that was impregnated with a mineral oil. The thin, paper-wrapped cables are tightly fitted into a sheath; then three cables, one for each phase, are pulled into a 0.9-km-long pipe, which is then filled with high-pressure oil. The cables are spliced every half-mile, so that a manhole and splicing station are required at half-mile intervals. The characteristics of underground cables at several voltages are shown in Table 9.4.

The rather small thickness of the paper insulation should be noted. Even at 69 kV the thickness is only about three times that of ordinary electrical tape. Notice also that the losses are very high compared to those from superconducting cables or from overhead lines. And there is one other entry to be noted: the critical length.

This length is significant only for AC systems. It comes about because of the charging current, which does not exist in DC transmission lines. The charging current, as discussed earlier, is the current in a transmission line without a load resulting from the reactive components of the line. While the inductive reactance of an underground cable will be only somewhat larger than that for an overhead line, the capacitive reactance will be much larger, by several orders of magnitude. This is a consequence of the close spacing of the conductors and the presence of a dielectric medium between them. The capacitive reac-

tance increases with the length of the conductors; therefore, the reactive load and the power lost to the reactive load increase as well. The reactive loading does not directly absorb power, but the currents induced by the reactive load do cause Joule's law losses in the conductor. The ultimate limit to the amount of power a cable can carry is determined by the ability of the environment around the cable to absorb heat. When the length of the cable is increased to the point that the charging current equals the thermal loss maximum current, then no net power can be extracted from the cable—the critical length has been reached. Of course, this critical length can be extended by using inductive compensation at the ends or at intermediate points in the cable. But this adds considerably to the expense of the system.

Cable heating comes about not only from Joule's law losses in the conductors, but also from heating of the dielectric. Nominally insulators, dielectrics nevertheless have leakage resistance. The small resulting leakage current can generate a significant amount of heat. These dielectric losses at voltages higher than 345 kV are unacceptably high with paper insulation. Therefore, to operate at higher vol-

TABLE 9.4 Some Characteristics of Oil–Paper Dielectric Pipe Cables

| | Voltage Rating | | | |
	69 kV	138 kV	230 kV	345 kV
Permissible average loss (W/ft)	6.66	6.96	6.99	7.30
Dielectric loss (W/ft)	0.54	1.36	1.40	2.68
Insulation thickness (in)	0.285	0.505	0.835	1.025
Thermal rated capacity[a] (MV A)[b]	105	200	330	440
Critical length (miles)	55	41	38	26

Source: P. H. Rose, "Underground Power Transmission," *Science* **170**, 267 (1970), copyright 1970 by the American Association for the Advancement of Science.
[a]The thermal capacity depends on the thermal conductivity of the surrounding soil; an average value has been used here.
[b]The volt-ampere is a measure of power that takes into account only average values for voltage and current; the power factor is disregarded.

tage, different dielectrics are required that can be thinner to give better heat transfer, yet are able to withstand the resulting higher dielectric stress. Newer synthetic materials such as Mylar, polyethylene, and nylon have such capabilities and are being used. Several gases are also being studied.

Sulfur hexafluoride, SF_6, has been used for several years as the insulating gas in the high-voltage terminal of van de Graaff generators and similar very-high-voltage electrostatic particle accelerators. The breakdown strength as a function of pressure for SF_6 and several other gases is shown in Figure 9.16. With gases the dielectric losses are considerably reduced, so that the critical length is extended. At 500 kV it is about 550 miles compared to 17 miles for a paper-insulated cable. Gases also conduct heat better because of convection currents, and since they require larger-

diameter pipes for insulation, there is a larger heat conduction surface to the surrounding medium. But since the pipes are larger, more expensive trenches must be dug.

Another proposal for underground power transmission is the use of refrigeration to capitalize on the fact that the resistivity of conductors decreases with decreasing temperature. This is unrelated to the phenomenon of superconductivity discussed earlier; it is simply the fact that conduction electrons make fewer collisions as the temperature of the metal lattice is lowered, reducing the vibrational energy of the atoms. The change in resistivity can be quite dramatic, as we see in Figure 9.17. Here the resistivity of pure alu-

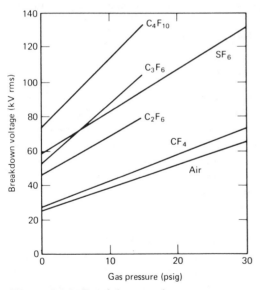

Figure 9.16 Breakdown voltage versus gas pressure for several insulating gases used in underground high-voltage cables. [From P. H. Rose, "Underground Power Transmission," *Science* **170**, 267 (1970). Copyright © 1970 by the American Association for the Advancement of Science.]

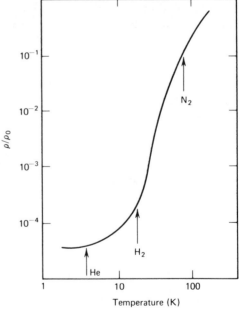

Figure 9.17 Resistivity of pure aluminum as a function of temperature. ρ_0 is $2.44 \times 10^{-6}\,\Omega$ cm. The arrows mark the boiling points of several gases. [Adapted from P. H. Rose, "Underground Power Transmission," *Science* **170**, 267 (1970). Copyright © 1970 by the American Association for the Advancement of Science.]

minum is plotted as a function of temperature. The boiling points of helium, hydrogen, and nitrogen are indicated by arrows. In the region below about 40 K, the resistivity is strongly dependent on impurities, and may be an order of magnitude higher than indicated.

However, the maintenance of temperatures near the boiling point of hydrogen is a very expensive process. The results of one study are plotted in Figure 9.18. This indicates that the best cost per unit load-kilometer may occur at liquid-nitrogen temperatures, which are much easier to maintain.

There are some Type II materials that are superconducting at or near the liquid-hydrogen boiling point. An interesting suggestion has been made to use transmission lines of these materials to transmit not only electrical power, but also liquid hydrogen fuel! If an efficient means of producing hydrogen along with elec-

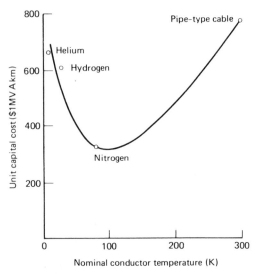

Figure 9.18 Unit capital costs versus conductor temperature for several types of underground high-voltage cable systems. [From P. H. Rose, "Underground Power Transmission," *Science* **170**, 267 (1970). Copyright © 1970 by the American Association for the Advancement of Science.]

tricity can be developed, then this dual transmission system could be highly effective. There are no serious plans underway at present.

As with all cryogenic or superconducting systems, even though some studies may show benefits, few entrepreneurs are willing to risk the investment in an essentially untried technology. Only an increasing power load that present technology cannot satisfy will bring these newer ideas on. It is difficult to see that such power loads will develop before the turn of the century, if then.

9.3. TRANSMISSION OF FLUIDS

Fluids are transmitted either in batch form in tankers or trucks or continuously in pipelines. There is not much physics to the former, but significant environmental impact. There is a great deal of physics and some environmental impact in the continuous transmission of fluids (or solids) in pipelines.

Tanker Transport

The United States receives about half of its crude oil from overseas suppliers. Europe obtains about 95 percent of its petroleum requirements from the Middle East, and Japan is almost totally dependent on foreign supplies. The worldwide picture of crude oil transport is shown in Figure 9.19. As U.S. supplies continue to dwindle (and we saw in Chapter 2 that they must), the dependence on the tanker lifeline will grow even stronger. More tankers will be built; even larger tankers will be used, since it is more economical to do so. And more tanker accidents can be expected. The economies of scale dictate the size of oil transport ships. Tankers are rated in terms of *deadweight tonnes* (dwt); this is the cargo carrying capacity in metric tons. The tankers of World War II were smaller than 50,000 dwt. By 1965, 471 tankers had been built in the 50,000-

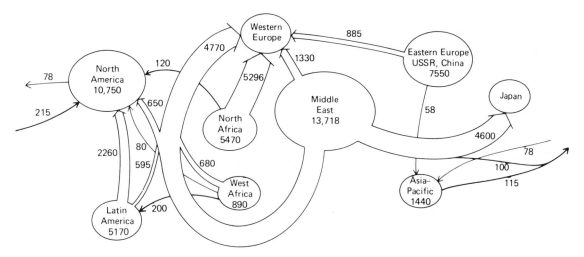

Figure 9.19 World-wide crude oil distribution in 1980. The major supiliers and their average daily production in thousands of barrels are shown with the average daily delivery to the major consumers of petroleum. Note that most of the connections are by means of oil tankers.

to 199,999-dwt class. Tankers between 200,000 and 300,000 are arbitrarily called **very large crude carriers (VLCCs)**. In the 1960s 131 VLCCs were placed in operation. Tankers larger than 300,000 dwt are called **ultralarge crude carriers (ULCCs)**. The largest of these are the *Globtik Tokyo* at 476,000 dwt, the *Esso Atlantic* at 509,000 dwt, and the *Bellamya* at 541,000 dwt. These ships are so large that no port in the United States can accommodate them. They require more than 23 m of clear depth when fully loaded, but they extend only 5 or 6 m above the surface. As a consequence they have been called "oilbergs." Other larger ULCCs are planned, and if the current world-wide economic recession ameliorates, these will no doubt be built.

While these behemoths are indeed very expensive to build, the transportation costs per barrel of oil with them are considerably smaller than with smaller ships. For example, the fuel requirements for a 32,000-dwt ship are about 75 tonnes per day, but for a 500,000-dwt ship only 330 tonnes per day are necessary. Clearly, as long as the demands of our society

force us to look overseas for our crude oil supplies, these large, cumbersome ships will be a necessity.

Many of the accidents reported with VLCCs are caused by operator error. It is difficult to realize, but a VLCC traveling at 16 knots[3] requires 4.8 km or 22 min to come to a crash stop. The engines are placed in a full reverse; during this time the ship cannot be steered. An even larger number of tanker losses were reported to have been the result of "structural failure"; this means the ship simply came apart. Structural failure seems to be a common occurrence for tankers as they approach 15 years of age.

Of the 6800 or so tankers in existence at the end of 1978, more than 580 were VLCCs or larger. Many of these were built in the early 1960s. We can expect a rash of "structural failures" to begin occurring in the 1980s and to continue for several years.

Nevertheless, far more oil is lost in routine

---

[3]One knot = 1 nautical mile/hr = 1.15 statute miles/hr = 1.85 km/hr.

TABLE 9.5 Sources of Crude Oil and Petroleum Products Introduced in the Ocean Per Year

Source	Amount (tonnes)	
Tanker operations		
LOT cleaning and ballasting	84,500	
Non-LOT cleaning and ballasting	575,250	
Miscellaneous operations	229,500	
Total tanker operations		889,250
Other ship operations		
Bunkers	9,000	
Bilges, cleaning, ballasting	292,500	
Total other ship operations		301,500
Offshore production		
Total		118,100
Natural seepage		
Total		600,000
Vessel accidents		
Tankers	124,100	
Other vessels	49,000	
Total vessel accidents		173,100
Total all sources		2,081,950

Adapted from W. B. Travers and Percy R. Luney, "Drilling, Tankers and Oil Spills in the Atlantic Outer Continental Shelf," *Science* **194**, 791 (1976), copyright 1976 by the American Association for the Advancement of Science.

operations than in accidents. These are the cleaning and ballasting entries in Table 9.5. The unloading of tankers is a rather inefficient process; a substantial amount of crude oil is left at the bottom and on the sides of the tanks. If the tanker is to take on a load of grain for the return trip, as is often the case, this oil must be cleaned out. Usually, seawater is sprayed into the tank, then discharged—the oil goes with it. If the tanker is to return to the Arabian Gulf or other oil port empty, then the oil tanks must be filled with water to provide ballast and a safe passage. Again, this ballast is frequently discharged with leftover oil.

The oily water need not be simply discharged. Rather, it can be collected in a central tank and the oil be allowed to segregate to the top of the water—the **load on top (LOT)** procedure. Then the oil can be drained off and only a small amount be discharged with the water. Unfortunately, there is no real control once a ship leaves port, and it is sad but true that most ships do not take the extra effort to minimize oil discharge.

Another possibility is to provide segregated ballast tanks (SBTs); that is, separate tanks that are used only to hold ballast water for return trips. These extra tanks would be placed around the oil tanks to provide another line of assurance in the event of a grounding accident piercing the hull. Extra tanks mean higher costs, however, something that ship owners and operators are not eager to accept. The United Nations sponsors the Intergovernmental Marine Consultative Organization (IMCO). The U.S. members in this organization have tried to implement stricter regulations with respect to marine oil pollution, but have not been successful. Whether or not the current administration will continue with such efforts may depend in part on how many additional tanker spills actually affect the American coastline in the next few years. If the past can be used as a guideline, we must expect it to happen.

In the past decade there have been several well-publicized oil tanker accidents in which large amounts of crude oil have spilled into the ocean near land. The worst of these was when the *Amoco Cadiz* went aground off the English Channel in March of 1978. The entire cargo of 216,000 tonnes[4] of crude oil was lost.

However, we should place this in perspective. Table 9.5 indicates sources of crude oil and petroleum product losses and the average amounts for the early part of the past decade. We see that the total release is much greater than the contribution from accidents. Part of this release from tanker operations can and should be avoided. Part cannot be; this is natural seepage, which is a continuous and widely distributed process.

The oil from a spill is not widely distributed but, rather, is concentrated in a relatively small region of ocean and sometimes ocean shore. The biological effects of oil in the ocean are not clearly understood but are significant. Many thousands of animals are killed by a single large spill, and an unknown amount of damage is done to the biological support system of the ocean. In areas where there is a vigorous cleansing action—waves, tides, and so on—the effects of a spill appear to dissipate quickly.[5] At other locations where the wave action is small and the water temperatures are cool the year around, a spill may take 10 years or more to dissipate.[6]

There are also physical problems associated with oil spills. As we see in Chapter 12, the air–ocean interface is responsible for the injection of a considerable amount of matter into the atmosphere—more than 3×10^{11} metric tons annually. The injection mechanisms involve both evaporation and spray production. Oil slicks affect both, by reducing water vapor evaporation and by coating the minute sea spray bubbles with oil. This oil can be transported many miles from the oil spill.

The transportation of crude oil accounts for about half of the petroleum released to the marine environment annually; of this amount, only about 10 to 20 percent is attributable to accidents. If oil were not transported in such large quantities by tankers, considerable reduction of the amount of environmental damage would result—although it is difficult to see that permanent environmental effects have been observed to date as a result of oil loss.

Pipelines

In a pipeline the fluid (which may be a liquid, such as oil, or a gas, such as methane) flow is laminar. Pumping is required to overcome the frictional effect of the viscosity. In Example 8.3 we looked at the isothermal flow in a pipe; this is usually the situation in a pipeline. However, in that example the flow was assumed to be turbulent. In the case of laminar flow, the pressure gradient does not depend on the mass of the field. A relationship between the pressure gradient and flow volume can be easily obtained using the method of dimensional analysis developed in Chapter 8.

Example 9.5 Use dimensional analysis to find an expression for the pressure drop per unit length in isothermal laminar flow of an incompressible fluid through a long, straight, level uniform pipe. The only physical variables that can enter into the expression are the diameter of the pipe D, the velocity of flow v,

---

[4]Because crude oils vary so much in composition it is not possible to relate mass and volume in a simple fashion. For comparison purposes, however, it can be assumed that in average 7 bbl = 1 metric ton.

[5]*Jacob Maesrk* spill, January 1975 at Leinoes, Portugal.

[6]*Metula* spill, August 1974 at Straight of Magellan, Chile.

and the viscosity μ.

$$-\frac{dp}{dL} = f(D, v, \mu),$$

$$-\frac{dp}{dL} = \alpha D^a v^b \mu^c.$$

Substitute the dimensions for each variable:

$$\frac{(ML/T^2)(1/L)}{L} = \alpha L^a \left(\frac{L}{T}\right)^b \left(\frac{M}{LT}\right)^c.$$

All exponents must be dimensionless; therefore,

$$\Sigma\ M: \quad 1 = c,$$

$$\Sigma\ T: \quad -2 = -b - c, \qquad b = 1,$$

$$\Sigma\ L: \quad -2 = a + b - c, \quad a = -2,$$

$$-\frac{dp}{dL} = \alpha\,\frac{v\mu}{D^2}.$$

This can be written in terms of the volume rate of flow V, where $V = \pi D^2 v / 4$:

$$V = \frac{1}{\alpha}\left(\frac{\pi D^4}{4\mu}\right)\frac{dp}{dL},$$

The multiplicative factor α has been determined from experiment to be 2, so that for a finite length L the volume flow given in terms of the pressure difference between the ends is

$$\boxed{V = \frac{\pi D^4 (p_1 - p_2)}{8\mu L}.}$$

The power required to transmit a given volume flow rate in a level pipe can be calculated using the result of this example. The power is the work per unit time, and the work is the pressure difference times the flow rate:

$$P = \frac{\Delta W}{\Delta t} = (p_1 - p_2)V, \qquad (9.17)$$

and from the example above,

$$P = \frac{8\mu L V^2}{\pi R^4}. \qquad (9.18)$$

Since the power required varies as the flow rate squared, increasing the net throughput of a pipeline is best accomplished by increasing its diameter (presumably by adding pipes in parallel) rather than by increasing the volume rate.

The viscosity of fluids varies considerably with fluid and temperature. For methane over most of the temperature range likely to be encountered in a pipeline, the viscosity is about $100\ \mu P$. For crude oil, the viscosity varies so much that a generalization cannot be made. Some crudes, the so-called heavy crudes, are barely fluid at room temperature; these do not even begin to flow until temperatures above $30°C$ are reached. Other, light crudes have viscosities in the $1000\text{-}\mu P$ range at room temperature. Of course, at lower temperatures, the viscosity increases substantially. It is for this reason that the crude oil in the Alaskan pipeline must be heated.[7] And it is the heated pipeline that causes concern among environmentalists.

Trans-Alaska Pipeline

Oil was discovered in Prudhoe Bay, Alaska, north of the Arctic Circle, in 1968. Since this area is ice-free only 6 months of the year, the owners of the discovery decided to construct an oil pipeline overland to Valdez, Alaska, an ice-free year-round port on the Pacific Coast. The pipeline route is shown in Figure 9.20. There was an immediate response from interested conservation groups, and the project was held up by federal court mandate. Construction did start after 3 years of hearings and, finally, congressional action. On July 28, 1977, oil began to flow down the pipeline. The total

---

[7]In this pipeline the oil is self-heated by friction part of the year.

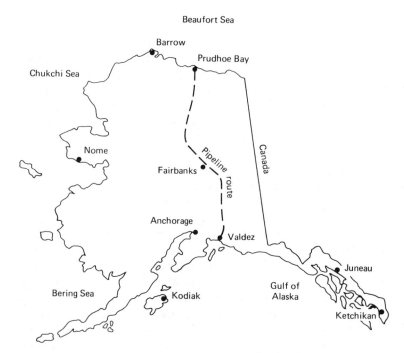

Figure 9.20 The route of the trans-Alaska pipeline from Prudhoe Bay to Valdez, a total of 1288.8 km.

project cost was in excess of $12 billion, a figure that makes it the single most expensive construction project ever attempted by mankind, dwarfing even the Panama Canal! The political ramifications of the trans-Alaska pipeline may take years to sort out. These we set aside and look instead at some of the less controversial technical features.

The pipeline is 1288.8 km long and is designed to handle 2.0 million barrels of oil per day. About half the distance covered is in *permafrost* country. Permafrost means simply that, aside from a surface layer a few inches thick that may thaw during the summer, the remainder of the ground is permanently frozen. The oil is heated to an average temperature of 60°C and is pumped by 12 pumping stations located along the line. The maximum oil pressure inside the line is not supposed to exceed 1180 lb/in.^{2}, although the pressure

exceeded that during early tests. One section of the line ballooned out and ruptured; the welds did not fail. The high pressure was caused by line blockage.

The pipe is made from 1.27-cm steel in 12.2-m lengths. These lengths were welded, some prior to placement, some afterward. Many of the welds were found to be defective and were redone. Since operation has begun, there has been no major leakage from defective welds, although we should note that the operators of the pipeline, Alyeska Pipeline Service Company, agree that only major leaks could be easily detected.

Opposition to the pipeline was based on many concerns:

- Damage to the tundra.
- Disruption of animal migrations.
- Erosion of permafrost.

■ Effects of large spillage.
■ Proliferation of large tankers.

The first three of these have been minimized by the choice of construction techniques. The pipeline is above ground in the permafrost region. It is insulated with gravel and insulating blankets. Stands support the pipeline at a height of from 2 to 3 m in locations where caribou migrations are known to occur. Access roads were not build by scraping away the tundra, but rather were made by bringing in gravel or an artificial gravel made of polyethylene chunks. In the few years of operation so far, it is safe to say that those first three ill effects have not materialized.

The long-range effects of a large oil spill either on the land or in the Pacific near the shore are not known. Many studies are under way, not only because of the pipeline, but also because of offshore drilling in lower Alaska between Juneau and Anchorage.

The fact that the pipeline terminates at Valdez on the seacoast and does not go overland (i.e., across Canada) to the United States may present a much greater environmental hazard in the long run than the pipeline itself. The port of Valdez is difficult to get into and out of with the gigantic tankers of the current generation.

The question of what to do with Alaskan oil is one of those political difficulties mentioned earlier. Alaskan oil has displaced Middle Eastern oil in those refineries that can handle it, but many cannot refine this oil. Most West Coast facilities are designed for Indonesian crude, a "sweeter," lower-sulfur oil. Proposals to construct a pipeline from a California port of entry to Texas have been stymied in red tape. The pipeline owners must obtain Presidential approval with congressional veto power in order to ship Alaskan oil outside the United States. As a consequence, excess oil is being stockpiled on the West Coast, and the trans-Alaska pipeline is operating at only about half capacity.

There are proposals in the wind to build a natural gas pipeline from Prudhoe Bay to the Midwest across Canada. There is another proposal to pipe natural gas from Venezuela to Texas. These would also be monumental undertakings, but technically possible. The next few years of operation of the trans-Alaska pipeline will determine the extent to which additional large projects are built in the future.

Pipelines have been used to transmit coal as a water slurry. A 50 to 60 percent by volume powdered coal slurry has about the same pumping characteristics as the water. However, the coal must be dried before it can be burned; this can require as much as 500 MJ/tonne. The slurry could be made with oil produced at the mine from the coal. An oil slurry would be more viscous, requiring higher pumping power, but the oil could be burned along with the coal. A mixture of, say, 50 percent coal, 30 percent oil, and 20 percent water would have a viscosity closer to water than to oil and would be directly combustible. Alternatively, methanol could be used instead of oil; it would have a low viscosity and be combustible. To date, only water slurries have been used. Expanded use of western coal may make slurry transport to Midwestern markets more attractive.

In this chapter we have discussed the transportation of energy. We have seen that, for the immediate future, current techniques will probably continue to be used. Whether or not any of the technical innovations that we described will be applied in the future depends critically on the future demand for energy. If the demand curve departs from the historical growth pattern and begins to level off, we may never see superconducting systems and the like. Then we may never have to worry about helium depletion, effects of ultrahigh voltages, and the ramifications of additional pipelines. If the demand curve continues to climb, then energy transmission technology will become very interesting.

SUMMARY

Energy is transmitted in many forms, but there are two general types: *batch* and *continuous*. Several criteria can be used to select a particular energy transmission form including cost, geographic limitations, carrying capacity, and others.

Electrical energy transmission is one of the most important forms. Most electrical transmission lines are AC, so that high voltages may be used to minimize *Joule's law* losses. However, AC transmission involves *reactance* losses, which DC transmission does not have. *Inductance* and *capacitance* cause the current and voltage to become out of phase.

Polyphase systems are used to minimize thermal stresses on generators.

Very-high-voltage systems have special problems: corona discharge, electromagnetic interference, audible noise, and others.

Superconductors having very low or no electrical resistance can be used to minimize heat losses. Type II superconductors can carry very large currents, but there are some losses. One difficulty with superconductivity is maintaining the very low temperatures, required. There may not be enough helium to support widespread superconducting transmission systems. In some places underground cables are required. These are more expensive because insulation problems are more difficult to solve. Underground cables also have a maximum effective length because of capacitive reactance.

Tankers are used to transport oil because of the great economy of scale that can be achieved with them. Tankers having a carrying capacity greater than 300,000 *dead-weight-tonnes* are called *ultralarge crude carriers* (ULCCs).

Tankers cause many oil spills annually. However, more oil is deposited in the ocean by natural seepage.

The volumetric flow rate in a pipeline can be related to the pressure difference, length, and viscosity of the fluid by the use of dimensional analysis.

The trans-Alaska pipeline has a maximum flow rate of 2×10^6 bbl/day. It represents a monumental construction task. None of the environmental fears associated with the project have come to pass.

REFERENCES

Barthold, L. O., and H. G. Pfeiffer, "High Voltage Power Transmission," *Sci. Amer.*, May (1964), p. 39.

Cook, E., "The Helium Question," *Science* **206**, 1141 (1979).

Fishlock, David, *A Guide to Superconductivity.* New York: American Elsevier, 1969.

Foner, Simon, and Brian B. Schwartz, *Superconducting Machines and Devices.* New York: Plenum Press, 1974.

Graneau, Peter, *Underground Power Transmission.* New York: John Wiley, 1979.

Kittel, Charles, *Introduction to Solid State Physics*, 5th ed. New York: John Wiley, 1976.

Rieder, Werner, "Circuit Breakers," *Sci. Amer.*, January (1971), p. 76.

Rose, P. H., "Underground Power Transmission," *Science* **170**, 267 (1970).

Roston, James P., *810 Miles to Valdez.* Englewood Cliffs, N.J.: Prentice Hall, 1977.

Shonle, John I., *Environmental Applications of General Physics.* Reading, Mass.: Addison-Wesley, 1975, pp. 235–246.

Schwartz, Brian B., and Simon Foner, "Large-Scale Applications of Superconductivity," *Phys. Today*, July (1977), p. 34.

_____, *EHV Transmission Line Reference Book.* New York: Edison Electric Institute, 1968.

———, *Transmission Line Reference Book—345 kV and Above*. Palo Alto: Electric Power Research Institute, 1975.

GROUP PROJECTS

Project 1. How are oil, natural gas, coal, and electricity transported to your community? What is the cost to the carrier in cents per kilojoule? Which produces the greatest environmental impact? What are the trends in your area for the next 10 to 20 years?

Project 2. Determine the maximum and average electrical load of your community. What are the characteristics of the transmission line(s) that deliver(s) this energy? How far is the electricity transported? What are the losses?

EXERCISES

Exercise 1. A particular crude oil pipeline is 750 km long and has a capacity of 2000 metric tons/hr. If this oil (4.3×10^7 J/kg) is used in a steam plant to produce electricity ($\eta = 0.4$), what electrical power output could be produced? Assume the same power is to be carried the same distance by a 500-kV three-phase electrical transmission line, each phase having 10 aluminum conductors each of 350 mm^2 area. What are the losses in this line? (Assume each phase carries one-third the total current.) Compare the available output power in these two cases.

Answer: 3824 MW, $P_{loss} = 39$ MW/phase, oil/electricity = 0.97.

Exercise 2. A hydroelectric plant with a head of 30 m is designed to generate 200 MWe. An earthquake diverts the river that had been used to fill the reservoir, so that an aqueduct must be built to transport the water. What carrying capacity in liters per minute should it have? If the water velocity in the aqueduct is not to exceed 1 m/s, what should be the cross-sectional area of the aqueduct? Compare this flow rate to several rivers.

Exercise 3. A tanker ran aground, spilling 57,000,000 ℓ of crude oil. If the energy cost of recovering this oil is 20 kWh per tonne, what net energy can be obtained from the recovered oil? If the original cargo was 220,000,000 ℓ, by what percentage did the eventually delivered oil increase in cost? (Assume 75 percent recovery of the spilled oil.)

Exercise 4. If $f = 60$ Hz, what inductance is required to product a reactance of 100 Ω? Would this be a physically large or small device?

Exercise 5. If $f = 60$ Hz, what reactance would be produced by a 0.01-μF capacitor? Would you be more likely to find this capacitor in a power-factor-correcting bank or in an audio amplifier?

Exercise 6. In most European countries the power line frequency is 50 Hz. What would be the effect of using appliances such as electric razors designed for 60-Hz operation on 50-Hz current? How about electric clocks? (Assume voltages are compatible—usually they are not!)

Exercise 7. Find the frequency in terms of L, C, and R for which the circuit in Figure 9.21 is purely resistive; that is, $Z^2 = R^2$. At this frequency, what is the power factor?

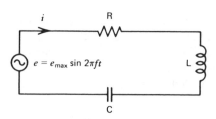

Figure 9.21 Drawing for Exercise 7.

Exercise 8. A particular inductive load has a power factor of 0.4. In terms of the initial X_L, what capacitance must be added to make the power factor 0.9? In terms of its own operating costs, why does the utility want each customer to have unity power factor?

Exercise 9. Consider a city of 10,000 households that is served by a power plant 8 km away. If each household uses 1200 W of power at 120 V, how much current must be delivered to the entire city? If we require that no more than 10 percent of the power load of the city be wasted in heating the transmission lines, what diameter of copper wire must be used? How is this changed if the power is transmitted at 120 kV? Given that copper costs $1.10/kg, what saving is involved in using high-voltage transmission lines?

Answer: 10^5 A, 1.19 m, 10^2 A, 1.19 mm savings 8.48×10^7.

Exercise 10. The New York City peak demand is about 7600 MW: this is supplied by several different generators. But assume it is delivered by a single transmission line 100 km long. What would the diameter of the copper wire have to be, if the power loss is not to exceed 2 percent of the load at 375 kV? How much total heat would be absorbed by the atmosphere in the New York area per day (include generator heat rejection, as well)?

Exercise 11. A transformer is required to step up 110 V (AC) to 600 V for a printing press that is a 6-kW load. Assume power losses of 5 percent of the load in the transformer. What should be the minimum size of the primary wire so that the primary losses are less than 10 percent of the load? (Assume 1000 primary turns of copper wire, each turn of radius 4 cm.)

Answer: minimum wire area = 0.219 mm^2.

Exercise 12. Assume that the iron core of the transformer ($m = 110$ kg) of the previous exercise absorbs 75 percent of the heat generated and is thermally isolated. How hot will it become in 10 hr of operation? ($T_i = 23°C$.)

Exercise 13. A 400-km transmission line has a total DC resistance of $8 \Omega/\text{km}$ and an AC resistance of $9 \Omega/\text{km}$. The load has a resistance of 9000Ω and an inductive reactance of 1000Ω. At what transmission voltage does the line loss equal 1 percent of the power delivered to the load?

Exercise 14. A transmission line serving a particular city has losses of 3 percent of the load when delivering the normal load. On a summer evening the load increases to 150 percent of normal. What are the transmission losses as a percentage of normal load under this condition? (Assume voltage remains constant.)

Answer: losses = 6.75% of load.

Exercise 15. The total cost of a 135-kV copper transmission system can be written in terms of the cross-sectional area of the conductors a as

$$\text{cost} = A + Ba + \frac{C}{a},$$

where A represents the fixed costs, B is the cost per unit area, and C approximates the $I^2 R$ losses. Find the a in terms of A, B, and C that minimizes cost. If $C = \$600/\text{mm}^2/\text{km}$ and $B = \$500/\text{mm}^2/\text{km}$, what is the maximum current that can be carried over a 150-km, two-wire minimum cost line? The power loss is not to exceed 1 percent of the load. What is the power delivered?

Exercise 16. A typical corona discharge dissipates 50 W/m. About 80 percent of this goes to heat, and most of the remainder goes to light. Assume about 1 percent of the heat is used to form ozone. How much ozone is produced in 1 day in a 150-km line consisting of three phases? Each phase has a bundle of six cables, and each cable loses 50 W/m to corona.

Answer: 3.29×10^7 g/day.

Exercise 17. Assume the energy for corona discharge is distributed as in the previous exercise. If the ozone concentration at the right-of-way surface of a 138-kV, three-phase transmission line is not to exceed 0.08 ppm by volume, what maximum corona discharge rate in watts per meter is permitted? Assume all ozone created over an 8-hr period accumulates. How could this calculation be improved?

Exercise 18. An oil tanker with a total mass of 300,000 tonnes steaming at 16 knots reverses its engines for a crash stop. The ship requires 22 min to come to rest. What is the constant reverse thrust provided by the engines? What power output is required from the engines?

Exercise 19. At a speed of less than 4 knots a VLCC cannot be steered. Estimate the maximum sideways thrust developed by the rudder. How does such a ship maneuver at low speeds?

Exercise 20. Assume the pumping stations in the trans-Alaska pipeline are evenly spaced and that the pressure drops by a factor of 1.4 between stations. If the flow rate is to be 2×10^6 bbl/day and $\mu = 10.1$ P, what is the diameter of the pipeline?

Answer: 1.5 m.

Exercise 21. At 1.5×10^6 bbl/day with a radius of 61 cm, compare the energy required for pumping with the energy transported.

Exercise 22. There are 62 remotely operable block valves in the trans-Alaska pipeline. Assuming they are evenly spaced, at a flow of 2×10^6 bbl/day, if a major break occurred, what is the maximum amount of oil that would be spilled?

Answer: 1.46×10^5 bbl.

Exercise 23. What could be the volume flow rate of a coal slurry in a 30.5-cm pipeline with pumps spaced 150 km apart, if $p_1 = 27$ atm and $p_2 = 13.5$ atm? How much net energy per day is available? Neglect pump energy; $\mu = 10$ P.

Exercise 24. If the slurry in Exercise 23 were made with an oil having $\mu = 1500 \, \mu$P, how would the answers change? Assume 10 pumps. How much net energy is available per day? Do not neglect pump energy.

Exercise 25. On average each tonne of bituminous coal is mixed with 760 ℓ of water in a coal slurry.
(a) How many liters of water are required to meet the needs of a 1000-MW generating plant each day?
(b) If this water comes from a lake with a surface area of 260 hr, how many centimeters will the level drop each day in supplying the slurry pipeline's water needs?

Exercise 26. Natural gas in one 50-cm pipeline is transmitted at 2.76 MPa. At what pressure would hydrogen have to be transmitted in order that the rate of energy delivery be the same?

Answer: 8×10^6 Pa.

CHAPTER TEN

ENERGY STORAGE

There are at least three areas in which improved energy storage could have an impact:

- Electricity generation.
- Transportation.
- Domestic and commercial applications.

Effective energy storage would enable electric utilities to shift a large part of the demand away from intermediate and peaking plants to baseload plants (see Figure 10.1). The latter are usually newer nuclear or coal-fired plants that operate continuously at high efficiency. Intermediate plants are often older, less efficient plants or natural-gas-burning turbines. Peaking plants are usually gas combustors or diesels. By placing more baseload plants on line and storing the energy generated during the demand "valleys," the "peaks" could be handled without requiring the services of older, less efficient plants. Not only would a substantially increased efficiency result from this shift, but also a reduced reliance on the more valuable fossil fuels would come about. Improved storage would also enable intermittent energy sources such as solar and wind power to be tied into a grid more effectively.

Shifting to another energy storage system for automobiles from the current chemical means (i.e., gasoline) would pay double dividends. Not only would the overall petro-leum usage be reduced, but also the dispersed air pollution source represented by the Otto-cycle engine would be eliminated. Of course, if battery cars replaced current models, there would still be pollution at the power plants required to generate the electricity to charge the batteries. But these plants would be large, central pollution sources, from which the emissions would be much easier to control (see Chapter 13).

Providing thermal energy storage for home and industry would enable consumers to take advantage of lower, off-peak electricity rates, if such were available in this country as they are in Europe. This would have the additional benefit of leveling off the peaks and valleys of the daily load cycle of Figure 10.1. Utility companies could reduce their investment in older, inefficient intermediate plants even more.

There are a large number of techniques for storing energy; most of them were known at the turn of the century. Only a few are in use today. These can be improved, and some that are marginally economic or speculative today could become realistic in the next few decades. Energy storage is obviously a legitimate concern of an environmental scientist, who should be capable at least of being able to make quantitative comparisons between the various types.

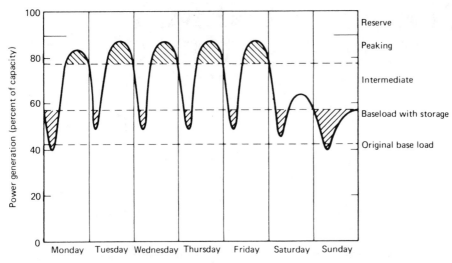

Figure 10.1 The effect of storage on the daily electrical load curve of an electric utility is to reduce the intermediate and peaking requirements by increasing the baseload. The demand curve "peaks" are to be supplied by stored energy that is generated during the demand "valleys."

10.1 ENERGY STORAGE ALTERNATIVES

Figure 10.2 is a comparison of the relative "storage" capability of many different physical systems. We do not think of many of these as energy storage devices, because for many of them we do not know how to put energy in effectively or how to recover it in a useful way. Nevertheless, this figure does help to put the various options in proper perspective.

Every energy storage system can be characterized by several physical properties. Among the most important of these is the **energy density**. Two units are in common use for energy density: joules per kilogram and watt-hours per pound. This characteristic is particularly important for electricity-generating plants. A high energy density means that a large amount of energy can be stored in a relatively small volume. Energy density is important for transportation, but also important is the **power density** of a storage system. An automobile

may have to accelerate rapidly; this requires the expenditure of energy at a high rate, and if very massive storage systems are required to supply this power, the automobile will be too heavy to drive economically.

Example 10.1 Compare the energy density of gasoline to lead–acid batteries.

From Chapter 2:

$$E/m) \text{ gasoline} = 4.6 \times 10^7 \text{ J/kg} \quad \text{(Table 2.1)}$$

Chapter 5:

$$E/m) \text{ Pb–acid} = 15 \text{ Wh/lb} \quad \text{(Table 5.4)}$$

$$15 \frac{\text{Wh}}{\text{lb}} = \frac{15 \times 60 \times 60 \text{ W s}}{2.2 \text{ lb/kg}} = 2.4 \times 10^4 \text{ J/kg}.$$

E/m for a lead–acid battery is rather considerably less than that for gasoline. But remember that the mass of the engine is also large, so that a comparison of E/m for the entire power plant would not be as unfavorable for batteries.

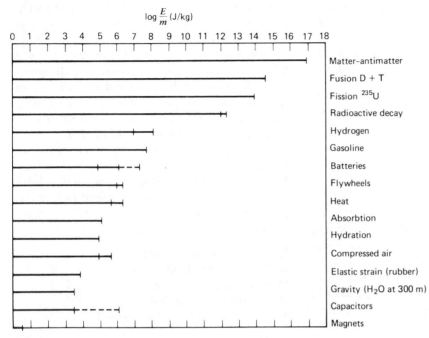

Figure 10.2 Comparison of the energy storage per unit mass for several systems. (Data courtesy of Dr. A. Klein, Melbourne University.)

The physical form of the storage medium often determines how and when it can be used. Gases are difficult to store; they have a low energy-to-volume ratio, except under pressure. Liquids also require a large volume for a given energy. But then, large volumes may not be a drawback for some applications, such as electricity-generating plants.

The chemical stability of the storage medium is quite important. Water does not easily change chemical composition under pressure, but it does evaporate, if left exposed. Flywheels can be disastrously unstable in some circumstances. Media that are flammable present special problems. As we mentioned in Chapter 5, one drawback with batteries is the deterioration of the electrodes after many charge–discharge cycles.

Perhaps the most important consideration is the cost. Cost calculations are complicated by the fact that there are several factors. An energy storage system can be thought of as two separate units: an energy conversion subsystem and an energy storage subsystem. The unit capital costs associated with the former decrease with system storage capacity, and the unit costs associated with the latter increase with the time required to discharge the system. We could write the total cost of a system as

$$C = C_c + C_s t, \tag{10.1}$$

where C_c is the cost of the conversion subsystem in dollars per kilowatt, C_s is the cost per unit of storage capacity in dollars per kilowatt per home, and t is the maximum discharge time of the system. If C_c and C_s are known or can be estimated for various systems, then cost comparisons can be made. Unfortunately, "hard" numbers are not easy to obtain for many of the less-developed techniques; as a consequence, utility companies are understandably unwilling to invest in untried technologies.

We examine several storage systems in detail to see what the current state of the technology is and how it might change.

10.2   MECHANICAL ENERGY STORAGE

Pumped Hydrostorage

We mentioned briefly this method of energy storage in Chapter 2. It is one of two mechanical systems in use today (see Figure 10.3). During the off-peak hours, overnight, electricity from a baseload plant is used to power a pump that lifts water from the lower reservoir to the upper. When a peak load in excess of the baseload capability is required, the water is returned to the lower reservoir through the turbine, thereby generating the required peak-load power. The largest pumped hydrostorage facility in the world is located near Ludington, Michigan. The average water head of 85 m can generate more than 2000 MW at full power. At full charge, the reservoir can be used to generate about 15 MWh. This plant, reservoir plus turbines, cost 300×10^6 in the

late 1960s and early 1970s, comparing extremely favorably with the costs of new nuclear or fossil-fired plants.

Pumped hydro accounts for about 2 percent of all U.S. electrical generation, but the prospects for expansion of current facilities seem limited. Since most large electric-generating plants that could benefit from pumped hydrostorage are located close to large metropolitan areas, suitable bodies of water and appropriate topography are very difficult to find. There have also been serious complaints from environmentalists, who are concerned about possible side effects of pumped hydrostorage.

There are several points of concern:

- Effects on aquatic life.
- Unwarranted land acquisition.
- Increased weight load.
- Instability of earthen dams.
- Generally undesirable aesthetics.

Certainly, these are legitimate worries, but the general consensus in the industry is that these problems are far less severe than difficulties in finding geographically suitable sites.

As a consequence, an alternative to normal above-ground pumped hydrostorage has been suggested: underground caverns. Water is stored in a small upper reservoir. It is allowed

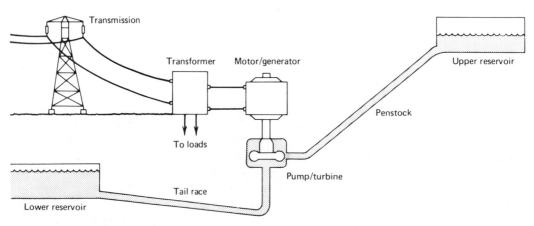

Figure 10.3   Schematic diagram of a pumped hydrostorage system.

to pass through the pump/turbine to a lower underground cavern, which may be more than a thousand meters below the upper reservoir. This higher water head means a smaller volume of water is required for a given energy storage. Suitable caverns can be drilled in the rock underlying much of the northeastern, north-central, and western United States. Drilling technology exists, and suitable high-lift pumps can be built. Engineering studies are underway in several locations, and the first underground pumped hydrostorage facility may soon be built.

Compressed Gases

This energy storage form has been studied for almost 40 years, but it has only recently become a reality (in Huntorf, West Germany). Refer to Figure 10.4. In this system, off-peak electricity is used to turn a compressor system to compress air in an underground cavity. This cavity can be a natural cavern, a disused mine, or a cavity drilled for this purpose. When the stored energy is required, the high-pressure air is used to turn turbines that drive generators to replace the electrical energy. Of course, this is a simplification, but the net result is that the

overall efficiency of storage is about 70 percent, acceptably high for most applications.

Example 10.2   At the Huntorf storage plant each kilowatt-hour of electrical output requires 0.8 kWh of electrical input plus 5600 kJ of heat input at the turbines. What is the actual efficiency of storage?

$$1 \text{ kWh} = 3.6 \times 10^6 \text{ J},$$

$$\text{Input} = 0.8 \times 3.6 \times 10^6 + 5.6 \times 10^6$$
$$= 8.48 \times 10^6 \text{ J},$$

$$\text{Output} = 3.6 \times 10^6 \text{ J},$$

$$\text{Efficiency} = \frac{\text{output}}{\text{input}} = 42.5\%.$$

This is not particularly high, but it does represent a significant improvement in the utilization of the fossil fuels used to supply the heat. Normally 10 to 12 MJ/kWh is required.

The reason that additional heat energy must be supplied in this system is that heat must be removed at the time of compression, or the air temperature would rise to very high levels. This may have deleterious effects on the cavity walls. In the case of the Huntorf plant, the

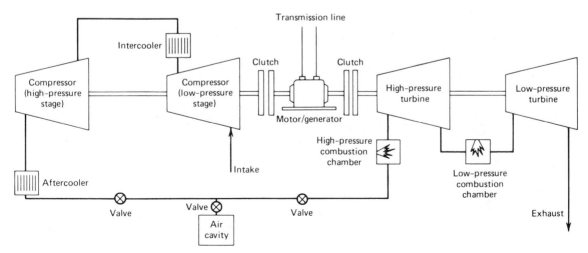

Figure 10.4   A simplified compressed-air storage system.

cavity has been leached from a salt deposit, so excessive temperatures would be risky. Engineering studies have been made of the possibility of "saving" the heat extracted during compression by means of a recuperator and adding it back at the time of expansion. This may considerably reduce the need for external heat. It will certainly reduce the required heat input per kilowatt-hour output.

Example 10.3 Assume air is compressed adiabatically to 50 atm, starting from STP. What would be the final temperature?

From Chapter 4: $T_p^{(1-\gamma)/\gamma} = \text{const.}$

Use $\gamma = 1.4$:

$$293 \times 1^{(1-1.4)/1.4} = T \times 50^{(1-1.4)/1.4},$$

$$T = 893.6 \text{ K},$$

$$\boxed{T \approx 620°\text{C.}}$$

A great deal of work remains before compressed air storage becomes widespread in the utility industry. The geological conditions for storage need to be established and methods of constructing caverns examined. Contamination of the air and resulting corrosion of turbine blades need to be reduced. The effect of thermal cycling on the cavity walls needs to be thoroughly reviewed. In short, the commercial applicability of compressed-air storage has not been established.

Compressed gases are also used on a much smaller scale to store energy, principally in transportation. We saw in Chapter 6 that one of the drawbacks to the use of hydrogen in automobiles is the problem of storage, either as a gas or as a liquid. Clearly, storage under high pressure is required to increase the energy density. The limitation is the weight of the storage vessel. Of course, lightweight and strong materials have been developed in the space program, but these are often too expen-

TABLE 10.1 Comparison of Several Materials

Material	Yield Stress (MPa)	Density (g/cm³)
Aluminum 5083	124	2.70
Beryllium	310	1.85
Carbon fiber	2,760	1.5
Copper	69	8.9
Glass filament (Type S)	4,830	2.48
Inconel K, annealed	690	8.2
Magnesium alloy	124	1.74
PRD-49 (Kevlar)	3,590	1.44
Silica, fused	13,790	2.16
Steel, stainless 301 heat-treated	200	7.8
Titanium 6A14V	830	4.5

sive for general industrial and commercial use (see Table 10.1).

Example 10.4 Show that the circumferential stress on a thin-walled cylinder at interior pressure p, radius r, and height h is given by

$$\sigma = \frac{pr}{h}.$$

Consider an imaginary bisection of the cylinder as shown in Figure 10.5. In equilibrium

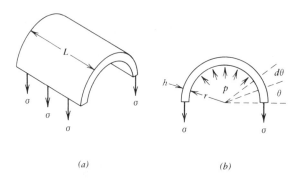

Figure 10.5 (*a*) Bisection of a cylinder of length L, indicating the circumferential stress σ. (*b*) End view, θ is the integration angle.

$\Sigma F = 0$, so that

$$2\sigma hL = \int_0^\pi pr \sin \theta L \, d\theta$$

or

$$\sigma = \frac{pr}{h}.$$

Similarly, we could show that the tensile stress in a spherical vessel is given by

$$\sigma = \frac{pr}{2h}.$$

We can use the result of Example 10.4 to show that the ratio of the masses of tank and fuel for thin-walled spherical tanks is given by

$$\frac{m_t}{m_f} = \frac{3\rho_t p}{2\rho_f \sigma}, \qquad (10.2)$$

where ρ_t and ρ_f are the tank and fuel densities, p is the interior pressure, and σ is the maximum allowable stress. Remember that stress is the normal force per unit area on a cross section of the material.

In the space program, cost per unit of energy storage was less important than weight per unit of energy, since the total liftoff weight determined the thrust required, and therefore the overall cost of the mission. For practical use, compressed-gas storage is not currently economic and does not seem to have much hope of being so in the future.

Flywheels

Flywheels have been used for energy storage in small applications for some time: in engines, in hand-powered potter's wheels, for ship stabilization, in watches, and a variety of other applications. Research is currently underway in the application of flywheels, both to transportation and to electricity storage at the generator and near the point of use. There have been a number of advances in flywheel construction, which may make this type of energy storage economically viable in the near future.

There are a number of advantages offered by flywheel energy storage:

- High efficiency, 80 to 90 percent.
- Quiet.
- Free of pollution.
- Reasonable power density.
- Rapid charge.
- Location near point of consumption.

At the same time, we should point out that high energy densities are difficult to achieve and that costs so far have been excessive.

A steel-flywheel-operated Oerlikon bus was in use in Altdorf, Switzerland, from 1953 to 1969. The 1500-kg flywheel could store only enough energy to drive the bus between stops, about 1 km, where it would require 2 minutes to run-up the flywheel. Since 1970, flywheels have improved in terms of weight and energy storage. To discuss these improvements, let us first look at the basic physics behind flywheels.

A flywheel stores kinetic energy of rotation; the amount of energy stored is simply

$$E = \tfrac{1}{2}I\omega^2, \qquad (10.3)$$

where I is the moment of inertia of the object, and ω is the angular velocity of rotation. Since the moment of inertia of an object through a given axis is just

$$I = \int r^2 \, dm, \qquad (10.4)$$

it follows that a large mass located a great distance from the axis of rotation and spinning at a large rotational velocity will store a large amount of energy, and the larger each one of these quantities is, the better. Obviously, these variables cannot be increased without limit. Eventually the "centrifugal" forces of rotation cause tensile forces in the material, which will exceed the maximum allowed stress. Then the material fails, often catastrophically.

Obviously, materials with high tensile strengths are required. We see in Table 10.1 that lighter, somewhat exotic materials have considerably higher tensile strengths than the commonplace metals that have been previously used in flywheel construction. Metals have higher densities, but curiously enough, this is not the best arrangement. By examining the forces on a rotating hoop, we can show that the maximum energy density is given in terms of the maximum stress allowed and the density:

$$\frac{E_{max}}{m} = \frac{\sigma}{\rho}. \qquad (10.5)$$

With this in mind, consider two materials with equal maximum allowed stress values, but considerably different densities. The stress value for the denser one will be reached at a proportionately lower value of stored energy density. "But," you may ask, "how about total energy stored?" The total energy stored is the product of the energy density and mass. We can see from Eq. 10.5 that for the same maximum stress, the ratio of masses required to store the same amount of total energy is equal to the inverse of the ratio of densities. Thus, for a given *mass*, a high-strength, low-density material is far superior to a standard high-density material.

Several of the materials listed in Table 10.1 have been used in modern, so-called superflywheels. The fibers—carbon, glass, or fused silica—are particularly interesting as they have anisotropic physical properties. Their tensile strength along the direction of orientation of the fibers is several orders of magnitude greater than in a transverse direction.

Many different flywheel configurations have been used (see Figure 10.6). Composite flywheels using fibers, wires, or filaments have had energy densities as high as about

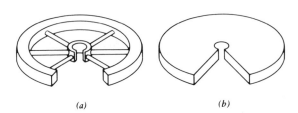

Figure 10.6 Two flywheel designs: (*a*) rim, $I = mR^2$; (*b*) disk, $I = mR^2/4$.

20 Wh/kg, although commercial versions of these cannot be produced because of the costs of the exotic materials. A better number for flywheels currently available would be closer to 10 Wh/kg. This is still competitive with lead–acid batteries in terms of energy density, and, depending on which report you read, it may also be competitive in terms of costs. Figure 10.7 shows a low-cost flywheel made of steel tire wire. This flywheel could store about 1 kWh when rotating at 15,000 rpm. The fiber flywheels also have the advantage that when they fail, they tend to unravel into a tangled mass of threads, rather than into high-speed and dangerous projectiles as do metal flywheels.

The failure mode is particularly important in the transportation application of flywheels. These units would be small, needing about 30 kWh, and requiring only a few hundred kilograms of flywheel. The total mass of the drive system would probably be in excess of 200 kg. Such a unit would have a range of about 300 km at 90 km/hr. This should be compared to a comparable lead–acid battery system that would have a mass in excess of 1000 kg for the same performance. This flywheel could also be "charged" in about 5 min; the batteries would require several hours. However, to charge a 30-kWh flywheel in $\frac{1}{12}$ hr (5 min) would require a 360-kW motor!

Flywheels, whether for vehicles or stationary storage systems, will have to be mounted in a vacuum to reduce friction losses.

Figure 10.7   A flywheel made with steel tire wire. This flywheel has an energy density of about 10^5 J/kg at 15,000 rpm. (Photograph courtesy of Dr. D. W. Rabenhorst of the Applied Physics Laboratory of Johns Hopkins University.)

They will need special support bearings that can function without lubrication in a vacuum. Magnetic bearings have been proposed and appear to be a satisfactory solution. It is estimated that an automobile flywheel could remain charged for as long as 10 to 12 months.

Stationary flywheels located at the point of consumption can be weightier and not built to such exacting standards as those in automobiles. A flywheel storage system coupled with a solar-power converter for domestic use could be an important method of saving energy in the next decade, if costs can be reduced.

In the late 1970s flywheel system costs were in the neighborhood of $50/kWh; this is too high to be competitive with pumped hydro-storage for central power generation, but it is in the ball park for competing with automobile battery systems. Clearly, research in this area is justified and may pay off large dividends.

10.3 CHEMICAL ENERGY STORAGE

Chemical energy is the storage of energy as chemical potential of reacting species that can be made to react with a net release of energy. This energy storage form has several obvious advantages:

- High energy density.
- Easy conversion to other energy forms.
- Potential for reversibility.
- Established handling network.

Chemical energy in the form of petroleum products supports virtually our entire vehicular transportation system. And, although the chemical reactions are not reversible in a practical way, the ease of handling and relatively low cost of these fuels have made consideration of other chemical storage forms unattractive. This kind of chemical energy does not represent storage in the same sense as pumped hydro. The formation of these fuels required an energy input that would be difficult to duplicate today. In that sense, few chemical reactions represent reversible storage.

The conversion of electrical energy to chemical energy by reason of reactions at battery electrodes is a true energy-storage mechanism. We have discussed this in some detail in Chapter 5 and will not repeat it here. We simply recall that, with the exception of the high-performance batteries not now readily available, this energy storage form cannot compete with other methods either for central electricity-generating plants or for vehicular applications.

There are other chemical forms of energy storage that do have potential as future storage systems. We examine them in some detail in this section.

Reversible Reactions

Chemical reactions that require large amounts of energy to proceed in one direction and release that energy when the reaction is reversed can be used to store energy. Reversible chemical reactions usually store thermal energy and therefore could be discussed in Section 10.5. We make a distinction, although a somewhat artificial one, because these reactions result in a complete change of the chemical identities of the reactants. Several reactions have been suggested as a means of storing or transporting energy from a solar or nuclear power system. The reverse of catalytic methanation has a reasonable reaction rate at high temperatures:

$$CH_4 + H_2O \rightleftharpoons CO + 3H_2. \qquad (10.6)$$

Using the technique discussed in Chapter 6, the energy storage capability of this reaction can be calculated.

Example 10.5 The chemical reaction of Eq. 10.6 proceeds at 200°C. What is the energy storage per mole of reactants?

Following the techniques of Section 6.1, we look up the enthalpies of the reactants (Eq. 6.3), taking into account the fact that the reaction occurs at an elevated temperature (Eq. 6.4):

$$\Delta h_R + \Delta h(CH_4) + c(CH_4)\,\Delta T$$
$$+ \Delta h(H_2O) + c(H_2O)\,\Delta T$$
$$= \Delta h(CO) + c(CO)\,\Delta T$$
$$+ 4\Delta h(H_2) + 4c(H_2)\,\Delta T,$$

$$\Delta h_R + (-57.02) + \frac{200-25}{1000} \times 8.54 + (-57.80)$$
$$+ \frac{200-25}{1000} \times 8.02$$
$$= -26.416 + \frac{200-25}{1000} \times 6.96$$
$$+ 4 \times 0.0 + 4 \times \frac{200-25}{1000} \times 6.89,$$

$$\boxed{\Delta h_R = 117.28 \text{ kcal/mol.}}$$

The Δh values are in kilocalories per gram-mole.

It is not possible to use any reaction fully for energy storage. The reaction will not go completely to the right or left; there will be products other than those desired at each end of the process. Heat will be lost in transporting or storing the products, and some loss of material can be expected. Nevertheless, reversible chemical reactions of the type of Eq. 10.6 are currently being considered for use with solar power plants.

Hydrogen

We have examined hydrogen as a means of transporting and distributing energy in Chapter 6, where we also discussed methods of hydrogen production. Since it is clear that the production of hydrogen, by whatever means, requires an energy input, much of which can be recovered by means of fuel cells (Chapter 5) or combustion (Chapter 6), we concern ourselves here not with the production of hydrogen but, rather, with its storage. In particular, we examine the storage of hydrogen in a form suitable for use in automobiles.

Hydrogen can be stored as a gas, as a liquid, or as a chemical compound. We know already that compressed-gas storage has limitations, especially when applied to vehicles, because of the poor container-to-gas weight ratio. Storage as a liquid also presents difficulties. Hydrogen liquefies at $-252.87°C$; a substantial amount of energy is required to produce this liquefaction. Long-term cryogenic storage without large losses is difficult, and safety problems are very serious, indeed. A compact and safe alternative for the storage of hydrogen is presented by a particular class of compounds, metal hydrides.

When hydrogen is forced under pressure to come into contact with a pure metallic surface, a substantial amount of the hydrogen "dissolves" into the metal in the form of atomic hydrogen (H). These hydrogen atoms occupy the space between metal atoms, called interstitial space. At sufficiently high pressures the ratio of hydrogen atoms to metal atoms will exceed 1; in fact, it can be in excess of 2. This occurs when the chemical compounds known as **hydrides** are formed.

Hydrides can be made with any pure element or with most binary alloys. However, for the purpose of energy storage, certain characteristics are required of the host material. It should

1. Form hydrides at relatively low pressures.
2. Decompose the hydride at relatively low temperatures.
3. Be abundant and cheap.
4. Hold up under many charge–discharge cycles.
5. Not present any safety hazard.

None of the metal hydrides studied so far meets all these criteria. Most of them are highly combustible, burning with very hot fires. However, most of them have rather high energy-to-volume ratios, compared to both liquid and gaseous hydrogen (see Table 10.2).

Unfortunately, the weight required for energy equality with pure hydrogen is excessive for most automotive applications. However, we should point out that this energy-to-weight ratio is of the same order of magnitude as that for batteries that would be suitable for automotive use. This, coupled with the fact

TABLE 10.2 Properties of Some Metal Hydrides

Hydride	Hydrogen Storage by Weight (%)	Energy Density (J/g)
MgH_2	7	9,916
$MgNiH_4$	3.2	4,477
$FeTiH_{1.95}$	1.75	2,469
Liquid hydrogen	100	141,838
Gaseous hydrogen	100	141,838

that hydrogen-fueled engines would burn virtually without polluting emissions, makes the metal hydride storage system particularly attractive as a future energy device. No doubt research will continue in this area.

10.4 ELECTRICAL ENERGY STORAGE

Energy is required to produce static electric fields, static magnetic fields, or electromagnetic fields. If such fields could be established and then maintained without the addition of energy, an effective energy storage system could be developed. There are methods of doing just that, which we discuss in this section.

Electrostatic Energy Storage

If parallel conducting plates of surface area A separated a distance d in a vacuum are connected to a battery of emf V, as in Figure 10.8, the electric charge in this circuit will redistribute itself so that after a period of time the following will be true:

There is an amount of charge $\pm q$ on the two plates.

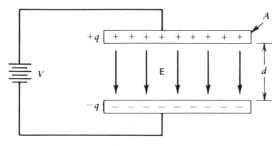

Figure 10.8 A parallel-plate capacitor of surface area A, separation distance d, each plate charged with q, having a potential difference between plates of V and a consequent electric field **E**.

There is a potential difference, V, equal to the emf of the battery across the plates.

There is a uniform electric field, **E**, between the plates (neglecting edge effects).

Work has been required to rearrange this charge, so we can think of the electric field in the space between the plates as a convenient energy storage reservoir.

The amount of energy stored can be related to the strength of the electric field, which itself depends on the physical distribution of charge. Since we would like to be able to compare various electrostatic storage devices, we develop some of the formalism required to determine the energy stored. We have presented some of these ideas already in Chapter 9; but we risk repetition here for clarity.

The work done to bring a charge dq to an object at a potential V is

$$dW = V \, dq. \tag{10.7}$$

the total work to charge an object is therefore

$$W = \int V \, dq. \tag{10.8}$$

For any particular arrangement of electrodes it is always found experimentally that the ratio of charge to potential is constant. It is convenient to use this ratio to characterize that particular arrangement of electrodes. We call it the **capacitance**, and such a device is called a **capacitor**. The unit of capacitance is the farad; one coulomb per volt is one farad:

$$C = \frac{q}{V}. \tag{10.9}$$

Using this definition in Eq. 10.8 yields

$$W = \frac{CV^2}{2} \tag{10.10}$$

for the energy stored in a capacitor. Since capacitors may vary in size and shape, the energy density is more instructive than the energy itself. For a parallel-plate capacitor the energy density is given by

$$u = \frac{W}{Ad} = \frac{CV^2/2}{Ad}. \qquad (10.11)$$

The capacitance for a given arrangement of electrodes can usually be calculated. For a parallel-plate capacitor it is relatively easy to do, but we shall not dwell on the technical details. For this case

$$C = \frac{\epsilon_0 A}{d} \qquad (10.12)$$

And since for parallel plates

$$E = \frac{V}{d} = \frac{q}{Cd}, \qquad (10.13)$$

then

$$E = \frac{q}{\epsilon_0 A}. \qquad (10.14)$$

If the space between the plates is filled with a nonconducting dielectric medium, then ϵ_0 (8.85×10^{-12} F/m), the permittivity of free space, is replaced with ϵ, the permittivity of that particular material. Using Eq. 10.13 in Eq. 10.11, we have

$$u = \frac{\epsilon E^2}{2}. \qquad (10.15)$$

The energy density will be in joules per cubic meter when the electric field is expressed in volts per meter. Although derived for a parallel-plate capacitor, this result is true in general.

A capacitor has the property that, if it is properly insulated and charge is not allowed to leak away, it will remain charged after the voltage source is removed for a significant period of time given by the time constant of the capacitor–leakage path combination,

$$\tau = RC, \qquad (10.16)$$

TABLE 10.3 Dielectric Properties of Some Insulators

Insulator	Dielectric Constant	Dielectric Strength (kV/mm)
Vacuum	1.00000	∞
Air	1.00054	0.8
Nylon	3.5	59
Paper	3.5	14
Pyrex glass	4.5	13
Polyethylene	2.3	50
Teflon	2.1	60
Titanium dioxide	100	6

where R is the resistance in ohms of the leakage path. The energy stored in the capacitor can be recovered by connecting the capacitor to a circuit in which it can "dump" its stored charge. Energy storage in this fashion is quite efficient, so long as excessively long storage times are avoided.

The amount of energy that can be stored depends on the size of the capacitor and the type of insulating material. Insulators are usually classified not by the permittivity ϵ but, rather, by the ratio of ϵ to ϵ_0, called the dielectric constant κ. We should note also that every insulator will break down at some critical electric field strength gradient, referred to as the dielectric strength. Both the dielectric constant and dielectric strength are given for several substances in Table 10.3.

Example 10.6 What would be the area of a parallel-plate capacitor having a separation distance of 5 mm and a paper dielectric that could store 3.6×10^3 kJ?

With a paper dielectric the maximum electric field gradient according to Table 10.3 is 14 kV/mm; so the maximum applied voltage would be $14\,kV/mm \times 5\,mm = 70\,kV$. From Eq. 10.11

$$C = \frac{2uAd}{V^2} = \frac{2 \times (3.6 \times 10^6)}{(70 \times 10^3)^2},$$

$$C = 0.0015\,F.$$

From Eq. 10.8,

$$A = \frac{dC}{\epsilon_0} = \frac{(5 \times 10^{-3}) \times (1.5 \times 10^{-3})}{8.85 \times 10^{-12}},$$

$$\boxed{A = 8.47 \times 10^5\,m^2.}$$

This is an unreasonably large area; clearly, no set of plates this large could be practical.

The amount of energy storage requested in Example 10.6 is only 1 kWh. It seems that capacitive energy storage cannot be used on the scale required for central electricity-generating plants. However, this technique can be used to deliver reasonable amounts of electrical energy to a load in very short time periods, of the order of microseconds; the power capability is therefore quite large. When the weight of a capacitor required to store a large amount of energy is determined, it is found to be excessive. The energy per unit weight for capacitor storage is particularly small, as is apparent from Figure 10.2.

Inductive Energy Storage

The energy density in magnetostatic systems is not normally very different from electrostatic systems, but there are circumstances under which it can be usefully large.

If an air core coil, called a solenoid, is connected to a battery, as shown in Figure 10.9, after a time the following will be true:

1. There will be a steady current I through the circuit.
2. There will be a steady magnetic field in and around the solenoid.

Work is required to establish this field because initially an electromotive force is generated that opposes the current. This emf depends on the rate of change of the current, which in turn depends on the physical characteristics of the winding. It is found experimentally that the ratio of the opposing emf to the rate of change of the current is a constant for a given winding configuration. This ratio is called the in-

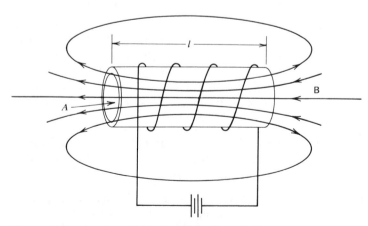

Figure 10.9 A solenoidal inductor having winding length l m and a cross-sectional area A m².

ductance:

$$L = \frac{|\mathscr{E}|}{di/dt}. \qquad (10.17)$$

An inductance of 1 H generates 1 Vs/A. The power required to maintain that current, in the absence of resistance, is

$$P = \frac{dW}{dt} = \mathscr{E}i = Li\frac{di}{dt}, \qquad (10.18)$$

where dW/dt is the rate at which energy is stored by the inductive circuit.

The energy is found by integrating Eq. 10.16:

$$W = \int dW = \int Li\,di = \frac{1}{2}Li^2. \qquad (10.19)$$

For the particular case of a solenoid of cross-sectional area A and length l, the energy density is

$$u = \frac{Li^2/2}{Al}. \qquad (10.20)$$

As with capacitors, the physical arrangement of the winding determines the inductance. This is often somewhat complicated to calculate; many introductory texts give the details. Instead, we present the result we need. For a solenoid the inductance is

$$L = \mu_0 n^2 lA, \qquad (10.21)$$

where n is the number of windings, and μ_0 is the permeability of free space (1.26 × 10^{-6} H/m). There is a relationship between the current and the magnetic field:

$$B = \mu_0 in. \qquad (10.22)$$

The energy density can thus be written in terms of the magnetic field,

$$u = \frac{1}{2}\frac{B^2}{\mu_0}. \qquad (10.23)$$

This relationship is analogous to Eq. 10.15 for capacitors. We can think of the energy as being stored in the static magnetic field. However, there is a very significant difference

between the two cases. For the capacitor, when the charging voltage is removed, as long as the capacitor plates are well insulated the charge will remain. For the inductor, when the current is shut off, the magnetic field collapses, the stored energy being fed back into the circuit.

Example 10.7 What is the energy density in a solenoidal inductor that carries a current of 10 A and has 50 turns?

From Eq. 10.19

$$u = \frac{Li^2/2}{l} = \frac{\mu_0 n^2 lAi^2/2}{Al},$$

$$u = (1.26 \times 10^{-6}) \times 50^2 \times 10^2 \times 0.5,$$

$$\boxed{u = 0.1575 \text{ J/m}^3.}$$

This is not a very large amount of energy density, a very small fraction of 1 kWh/m³. Since energy density scales linearly with inductance, we see that a rather large solenoid would be required to store a significant amount of energy in the magnetic field.

We could conceive of a circuit like that shown in Figure 10.10 where, after a current were started in the solenoid, a switch could be thrown to produce a closed circuit containing only the solenoid. When energy output was desired, the switch could be returned to the original position. This is possible because of the "electrical inertia" contained in an inductor. Just as in a capacitor there is an asso-

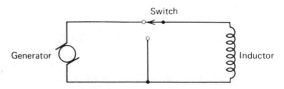

Figure 10.10 A circuit that would enable a solenoid to be "charged" with a current from a generator, then switched to a short circuit.

ciated time constant given by

$$\tau = \frac{L}{R}, \qquad (10.24)$$

where R is the resistance of the inductor. For very large inductors, this time constant could be large enough that the current would not change appreciably during the switching time.

This technique would "trap" the current, but with normal conductors it would not store the energy very effectively. There will always be Joule's law losses because of the resistance of the wire. In time all the stored energy would be dissipated as heat in the windings. Only if the resistance were very small, preferably zero, would this system of energy storage be practical.

Proposals have been made to use super-conducting solenoids and toroids for energy storage. These would be very large units, solidly fixed in bedrock to withstand the mechanical forces caused by the large magnetic fields generated by the stored currents. Type II superconductors, as discussed in Chapter 7, would have to be used because of their large critical fields. As a consequence, some losses would be experienced. But long-term storage beyond 10 or 12 hr would not be required.

Costs for superconducting storage are estimated to be rather high in comparison with other, proven techniques. A large part of the cost goes to the cryogenic cooling system. If materials with significantly higher critical temperatures can be found, large savings can be realized. Additional research in these areas, which is currently under way, could change this dramatically in this decade.

10.5 THERMAL ENERGY STORAGE

Finally, we come to the question of storing energy directly as heat and recovering it from storage in the same form. This is an important question, not only for solar power plants, but also for domestic and commercial users. It has been the custom in Europe for many years for utilities to offer both domestic and commercial consumers lower rates if they used electricity off the peak hours. This has not been the case in the United States until recently, but it is likely to become much more widespread in the future.

Sensible Heat

In this storage system, heat is used to raise the temperature of a large quantity of some material, which is thermally isolated to reduce heat losses. This technique can be used with low- or high-quality heat. It has been used for many years in Germany to store high-temperature steam generated during off-peak hours. The major problem with high-temperature storage is the high cost of the storage device, usually a pressurized steel vessel, but an underground cavern could be used. For use with power plants, rather large volumes are required.

Example 10.8 What volume of storage vessel is needed to store 65 MWh of thermal energy as sensible heat in saturated steam at 200°C?

From Appendix B,

$$h = 279.3 \text{ kJ/kg},$$

$$\text{Quantity of steam} = \frac{(65 \times 10^6) \times 60 \times 60}{279.3 \times 10^3} \text{ kg}.$$

But at 200°C the specific volume is 0.127 m³/kg, so the volume required is

$$V = \frac{0.127 \times 3.61 \times (65 \times 10^9)}{279.3 \times 10^3},$$

$$\boxed{V = 1.064 \times 10^5 \text{ m}^3.}$$

This is equivalent to a cylindrical tank 60 m in diameter and 40 m tall! Of course, this example assumes no losses; in reality, at least 30 percent more heat would have to be stored for short-term (6 to 8 hr) retrieval.

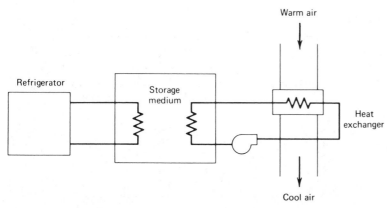

Figure 10.11 A coolness storage system. The storage medium is cooled with refrigeration at night. Warm building air flows through a heat exchanger, depositing heat in the storage medium during the day.

High-quality heat storage has the advantage that the energy density is rather high, approaching that of batteries. A major disadvantage is the difficulty in thermally insulating the storage vessel. Low-quality heat storage does not usually require an expensive pressure vessel, but, on the other hand, it does require a much larger volume for a given quantity of energy. Several U.S. power companies are cooperating with the Department of Energy in testing several sensible-heat-storage systems for industrial and domestic customers. These involve ceramics, pressurized water, and ice-cool storage.

Sensible heat storage can also be used to store "coolness." In Figure 10.11 off-peak electricity at night is used to cool a reservoir by means of a refrigeration system. During the day, building air is cooled by a heat exchanger, transferring heat to the reservoir. This system has been used in some large office buildings; architects are reluctant to employ it routinely, however, because of uncertain costs.

Latent Heat Energy Storage

Another way to store heat is to cause a phase change in a substance; that is, to cause a material to change from solid to liquid or from liquid to vapor. This isothermal change, depending on the direction of the change, requires or liberates an amount of heat L J/g. This heat is called the latent heat of phase change. Certain specific changes are also referred to heat of fusion, heat of condensation, heat of vaporization, and so forth. For most pure substances these heats are not particularly large. For example, in water the heat of fusion is 334.9 J/g, and the heat of vaporization is 2260 J/g. Nevertheless, latent heats are always larger than heat capacities, so phase-change heat storage, when possible, is always more effective than sensible heat storage.

Many other substances have higher heats of phase change. Lithium has a heat of vaporization of 19.74 kJ/g; unfortunately, lithium boils at 1315°C, much too high to be of value in an energy storage system. The dual requirements of high latent heat plus reasonable phase-change temperature rule out most substances.

There is a class of materials, however, called **hydrates**, that satisfies these requirements. Hydrates are chemical compounds in which water molecules occupy definite lattice sites in the crystalline structure. Rather large amounts of energy are required to melt these com-

TABLE 10.4   Properties of Some Hydrates

Hydrate	Melting Point (°C)	Heat of Fusion (kJ/kg)	Specific Heat (kJ/kg °C) Solid	Liquid	Density (g/cm³)
$Al_2(SO_4)_3 12H_2O$	88	260	0.46	1.03	1.65
$NaC_2H_3O_2 3H_2O$	58	264	0.60	1.00	1.30
$LiNO_3 3H_2O$	30	306	0.58	0.94	1.44
$Na_2SO_4 10H_2O$	18	186	0.54	1.00	1.51
$1(Na_2SO_4 10H_2O) + 1.5(NH_4Cl)$	11	162	0.41	0.77	1.48

pounds. Upon melting, the water coalesces, and the compound dissolves in this *water of hydration*, as it is called. Several of these compounds are listed in Table 10.4. Many have melting points in the appropriate range and high enough heats of fusion to be interesting. Note also that these are relatively common materials and do not present any particular hazards.

Example 10.9   What is the energy per unit mass storage capability for a volume of $Na_2SO_4 10H_2O$ effectively equal to 3 m³, if the temperature range is from 15° to 50°C?

$$Q = c_{solid} m (T_{fusion} - T_{low}) + L_{fusion} m + c_{liquid} m (T_{high} - T_{fusion}),$$

$$m = \rho V = 1.51 \times 3 \times (10^2)^3 = 4.53 \times 10^3 \text{ kg},$$

$$Q = [0.54 \times (18 - 15) + 186 + 1.0 \times (50 - 18)] \times 10^3 \times (4.53 \times 10^3),$$

$$Q = 994.9 \times 10^6 \text{ J},$$

$$\frac{Q}{m} = 219.6 \times 10^3 \text{ J/kg}.$$

a substantial amount of energy!

These hydrates can also be used effectively in coolness storage systems. A number of questions need to be answered, however, before widespread use in either heating or cooling is possible.

Do these materials withstand many thermal cycles?
Can appropriate heat-transfer surfaces be developed?
Can costs be kept down?

Many home solar heating systems are being built today using hydrates for thermal storage. As experience is gained, we can expect to see many more.

In this chapter we have examined several of the energy storage systems that are in use or have been proposed for use, not only for central electricity-generating plants, but also for automobiles and for homes. We have presented some of the basic physics required to analyze and compare these systems and thereby to make sensible judgments on the applicability of one or another in a given situation. Energy storage can help as we try to solve our energy–environment dilemma, but it may be many years before advanced storage systems can contribute in a large way, particularly with vehicles.

SUMMARY

Effective electrical energy storage would smooth out the peaks and valleys in the demand curve.

Pumped hydrostorage has been used and could be used more to store off-peak energy. There are environmental concerns with such storage systems.

Compressed-gas storage in underground caverns has also been used in Europe. This system is somewhat lower in efficiency than pumped hydrostorage. High-pressure storage in cylinders does not have a good gas-to-cylinder weight ratio.

Flywheels have been used in automobiles with little success. Developments in new materials may make this storage system more feasible in the future.

Certain *reversible chemical reactions* have been suggested as a means of energy storage, particularly for solar power systems.

Hydrogen can be stored as a liquid or as a *metal hydride*. Liquid storage has many problems. Hydride storage is promising; much work is being done on it.

Electrical energy can be stored in a *capacitor* or in an *inductor*. Storage densities are rather low.

Heat can be stored directly as steam or as *latent heat* of phase change. Some compounds have very high latent heats and are promising as storage media.

REFERENCES

Hohenemser, Kent, and Julian McCall, "Wind-up Car," *Environment*, June (1970), p. 14.

Kalhammer, Fritz R., "Energy-Storage Systems," *Sci. Amer.*, December (1979), p. 56.

Kalhammer, Fritz R., and Thomas R. Schneider, "Energy Storage," *Ann. Rev. Energy* **1**, 311 (1976).

Millner, Alan R., "Flywheels for Energy Storage," *Tech. Rev.*, November (1979), p. 32.

Post, Richard F., and Stephen F. Post, "Flywheels," *Sci. Amer.*, December (1973), p. 17.

Reilly, J. J., and Gary D. Sandrock, "Hydrogen Storage in Metal Hydrates," *Sci. Amer.*, February (1980), p. 118.

Robinson, Arthur L., "Energy Storage (II): Developing Advanced Technologies," *Science* **184**, 884 (1974).

Tam, S. W., C. A. Blomquist, and G. T. Kartsoines, "Underground Pumped Hydro Storage—An Overview," *Energy Sources* **4**, 329 (1979).

GROUP PROJECTS

Project 1. Make a chart of kilowatt-hours versus time for your home, apartment or dormitory over two 24-hr periods: a weekday and a weekend day. From that estimate the power demand versus time curve. Estimate the savings that could be obtained by variable rate structures and several levels of energy storage.

Project 2. There are at least three energy storage systems in and around most homes. Identify as many as you can. Find other places in the home where energy storage could be used.

EXERCISES

Exercise 1. Calculate the distance an automobile could travel on the energy stored in 1 m^3 each of gasoline, Pb–acid battery, and hydrogen gas at 20 atm. Assume 4.8×10^6 J/km required.

Answer: 5.7×10^3, 60, and 48 km, respectively.

Exercise 2. Compare the energy per molecule in electron volts per molecule for energy storage in fission, gasoline, and a coiled spring (0.04 Wh/kg).

Exercise 3. The coefficients for Eq. 10.1 for

TABLE 10.5 Coefficients for Eq. 10.1 for Various Energy Storage Systems

System	C_c ($/kW)	C_s ($/kWh)
Advanced batteries	80	49
Compressed air	290	5
Flywheels	80	100
Hydrogen	600	20
Pumped hydro	250	15
Thermal storage	450	18

several storage systems are given in Table 10.5. For what discharge time would thermal storage costs equal those for advanced batteries?

Exercise 4. By what factor would the capital costs of a hydrogen system of 5000 MW capability need to be reduced in order that the total cost of 10-hr storage equal that of a similarly sized compressed-air system?

Exercise 5. At what water flow rate would the Ludington, Michigan, pumped hydro reservoir generate 2000 MW? Estimate the weight load per unit area added by this reservoir. (Assume the average depth is 10 m.)

Answer: 2.67×10^{13} m^3/s and 9.8×10^4 N/m^2.

Exercise 6. What would be the volume of an underground reservoir having a 600-m head that would generate the same power and energy as the Ludington, Michigan, pumped hydro system?

Exercise 7. The air at the Huntorf plant fills a (3×10^5)-m^3 cavity to a pressure of 6900 kPa. Assume adiabatic compression. How much heat must be removed to maintain a constant temperature?

Answer: 5.15×10^{12} J.

Exercise 8. Derive Eq. 10.2. Show that the tank-to-fuel mass ratio for a gas of molecular weight A_g and temperature $T(K)$ in a spherical tank is given by

$$\frac{m_t}{m_f} = \frac{3\rho_t RT}{2A_g\sigma},$$

where R is the universal gas constant.

Exercise 9. Compare the tank-to-fuel ratio (spherical tank) for aluminum 5083, beryllium, and 301 stainless steel, each at a pressure corresponding to the yield stress. Compare the energy stored in each case for butane gas. ($T = 20°C$.)

Exercise 10. If the 1500-kg, 1-m-diameter flywheel of the Oerlikon bus were replaced with the same mass of gasoline, assuming the gasoline engine weighs the same as the electric motor, how would the range of the bus between "charge-ups" change? (Assume 3600 rpm for the flywheel.)

Answer: Bus now travels 10,360 km instead of 1 km.

Exercise 11. Derive Eq. 10.5.

Exercise 12. Compare the maximum stored energy for 1000-kg flywheels of aluminum, of fused silica, and of steel.

Exercise 13. What mass of Type S glass filament flywheel would be required to deliver 40 kWh for domestic use? Assume 95 percent efficiency and a factor of 3 safety margin in terms of stress. If $\omega = 15,000$ rpm and the shape is as shown in Figure 10.6(a), what is the diameter of the flywheel?

Exercise 14. At what rotation speed would failure occur for a solid stainless-steel disk? a fused silica rod? (Assume $R = 1$ m in each case.)

Exercise 15. Calculate the moment of inertia of the flywheel shown in cross section in Figure 10.12.

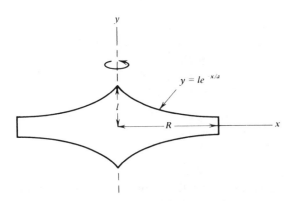

Figure 10.12 Drawing of flywheel in cross section for Exercise 15.

Exercise 16. The reversible chemical reaction of Eq. 10.6 is used to store energy at 200°C with the reactants pressurized to 20 atm. What is the energy density in joules per cubic meter?

Answer: 2.76×10^7 J/m^3.

Exercise 17. Consider this reversible chemical reaction:

$$CH_4 + H_2O \rightleftarrows CO + 3H_2.$$

What is the energy storage per mole of reactants?

Exercise 18. How would the result of Exercise 17 change if 10 percent of the time the reaction were

$$2CH_4 + H_2O \rightarrow C_2H_5OH + 2H_2?$$

Answer: $\Delta H_R = -27.77$, reduces total ΔH_R by 14.4 percent.

Exercise 19. If the capacitor of Example 10.6 were fully charged and the resistivity of the dielectric were 10^{15} Ω cm, what would be the total charge and the time constant?

Exercise 20. A 2-F capacitor is charged with a 1000-V supply. It is discharged through a 0.01-Ω load. What is the average power delivered in the first three time constants?

Exercise 21. A 10-μF parallel-plate capacitor with a titanium dioxide dielectric 1 mm thick ($\rho = 10^{13}$ Ω cm) is fully charged by a 100-V potential difference. What is the energy storage after 60 days? Assume no external leakage paths.

Answer: E after 60 days $= 4.95 \times 10^{-2}$ J, a 1 percent loss.

Exercise 22. An inductor is to be built in a confined space so that the length can be no more than 1 m and the diameter no more than 1 m. It is designed to carry 100 A in the winding. What is the maximum number of closely wound turns of copper wire, if the total DC resistance is not to exceed 0.1 Ω? What is the energy density in this inductor?

Exercise 23. A 100-H inductor is in a circuit as in Figure 10.10. If the total resistance of the circuit is 0.02 Ω and the charging current was 400 A, how much energy is contained in the field after 10 hr?

Exercise 24. A superconducting solenoid is made of Nb$_3$Sn at 3 K; it is 100 m in diameter and 50 m long. It has 100 turns. What maximum current can be stored? How much energy does this correspond to?

Exercise 25. Remember the expression for the force on a current, $F = BiI$. Use this to calculate the radial thrust on the solenoid in Exercise 22, if $I = 10,000$ A.

Answer: 3.96×10^6 N per turn.

Exercise 26. If the tank of Example 10.8 is made of stainless steel and if a safety factor of 2.5 is required, how thick should the walls be?

Exercise 27. What volume of sulfur is required to store one heating season's worth of heat? (Find the heating degree-days for your area; assume 50 percent losses.)

Exercise 28. What thermal energy storage density is available with liquid sodium at 300°C? What volume flow rate at this temperature is required to carry heat at a rate of 3000 MW?

Answer: 362.7 kJ/kg, 9.40 m^3/s.

Exercise 29. Compare the volume required to store a given quantity of heat as saturated steam at 200°C to that for saturated water at 30°C.

Exercise 30. The bottom of an elevator shaft is filled with water as a coolness storage medium. It is 4.5 × 4.5 × 6 m. If the temperature of the water is always such that 4°C < T < 15°C, how much energy can be absorbed? Estimate the number of average-sized offices that can be accommodated.

Exercise 31. Repeat Exercise 30, except instead of water use the $Na_2SO_410H_2O + NH_4Cl$ mixture.

Exercise 32. What mass of sodium hydroxide would be required to store 2000 MWh over the temperature range 10° to 400°C?

CHAPTER ELEVEN

EFFICIENCY IN ENERGY USE

In preceding chapters we have discussed sources of energy, transformations of energy, handling energy—in general, energy use from the points of view of the technical processes and the environmental interactions involved. In this chapter we would like to discuss almost the opposite: various ways of *not* using energy. Since the early 1970s it has become painfully evident that it is energy, not money, that makes the world go around, and that it is in our own best interests to find ways of doing what we do with less energy. What do we do?

Figure 11.1 shows a very general flow diagram that represents most of our activities. We take raw materials that have a low level of organization (ore, stone, timber, etc.) and add a considerable energy investment to produce primary materials (metals, lumber, paper, etc.). These materials are much more organized in a thermodynamic sense. Substantial additional organization and somewhat smaller energy in- vestment are required to produce finished sys- tems from these materials (houses, cars, ap- pliances, etc.). These finished systems are operated for their various lifespans, requiring the addition of sometimes large amounts of energy. This large amount is indicated by the breaks in the horizontal lines. Eventually, the systems reach the end of their useful lives, at which time they are usually discarded to a scrap heap. The scrap heap is at a lower level of organization, indicating a substantial loss of available work in the discard process. From the scrap heap some recycling can be done, requiring some increase in organization, but not as much as that required to come from raw materials to primary materials. Most of the discards are simply allowed to decay, however, a process that lowers the organization even below that of raw materials toward a level corresponding to total dispersal. Of course, useful material can be recovered from this total dispersal level, as indicated by the dashed line, but the energy investment required is higher.

It is clear from this figure how to go about saving energy: improve the processes of benefication and assemblage, minimize energy requirements in the operation of systems, design systems that have long lifetimes and do not require frequent discard, and recycle as much as possible. Some of these requirements are mutually exclusive with the goals of certain types of businesses; some of these are being done today. But rarely do we find decisions being made on the basis of the best thermo- dynamic solution to a given problem.

We should keep in mind the definition of the Second Law efficiency, which was introduced

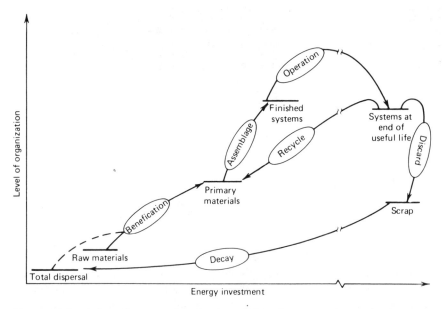

Figure 11.1 A flow diagram of society's activities in terms of organization and energy investment.

in Chapter 3.

$$\epsilon = \frac{B_{min}}{B_{act}}, \qquad (11.1)$$

where B_{min} and B_{act} are the minimum and actual amounts of available work required and used, respectively, in a particular task. Note that this efficiency may be increased in three ways:

■ Change the device in use.
■ Improve the device.
■ Change the task.

As an example, consider home heating. Most houses use forced-air furnaces, which we know have low Second Law efficiencies. This system can be improved by preheating the exterior air with room return air. Replacing furnaces with heat pumps would provide a significant increase in efficiency. But changing the task would be even better. After all, the object of home heating in general is to heat the occupants, not the house. So why not use a

microwave system that would heat only the *people*, requiring far less energy than any system, furnace or heat pump, that heats the *air*? There may well be arguments against this proposal on other grounds. But this example illustrates that there are always several options that should be considered in every situation.

It will be convenient to approach the question of energy conservation from the point of view of energy end use, since statistics are more easily obtained on this basis. We note from Figure 1.4 that there are three major end uses for energy: residential and commercial, industrial, and transportation. As a rough rule of thumb each of these uses about 25 percent of the total energy produced in the United States. The remaining quarter goes to energy conversion losses, which we have discussed in some detail in preceding chapters. In this chapter we study these three end uses for energy to see whether energy can be saved and to what extent physical science can contribute in this effort.

11.1 DOMESTIC AND COMMERCIAL ENERGY CONSERVATION

The practice of energy conservation has a great deal in common with the weather: everybody talks about it, but nobody does anything about it. This at least is a step forward; until recently no one was even talking about it. But even with its newfound vitality, energy conservation is more than simply turning down thermostats. Domestic and commercial energy use must be examined from the ground up. What are the present energy requirements? What policies could be easily implemented to save energy? What construction practices are wasteful of energy? How could we operate our homes more efficiently? Finally, have we fallen into the trap of substituting apparent convenience for efficiency?

Policy Influences

Table 11.1 shows the distribution of dwelling types in the United States in 1970 and the prediction for 1990. Clearly, single-family dwellings constitute the largest category. It may be thought that this distribution of dwelling types is largely the result of consumer

TABLE 11.1 Distribution of Dwelling Types in the United States[a]

	1970 Units		1990 Units	
	($\times 10^6$)	(%)	($\times 10^6$)	(%)
Mobile home	2.1	3.1	7.4	7.8
Single-family detached	44.6	65.9	59.4	62.6
Low density (duplex)	11.0	14.8	14.8	15.6
Low-rise, townhouse	6.5	9.6	8.8	9.3
High-rise	3.3	4.9	4.5	4.7

[a]Data from *Final Task Report—Residential and Commercial Energy Use Patterns 1970–1990*. Washington D.C.: Federal Energy Administration, 1974.

demand, rather than policy influences. However, policy does have an effect. A homeowner may deduct mortgage interest on his federal income-tax return; a renter has no equivalent deduction. This policy, which has considerable effect when interest rates are high, coupled with a traditional desire for home ownership by a large fraction of Americans, produces dwellings that have a high surface-to-volume ratio. A large volume would imply a large number of occupants; a small surface would imply reduced heat losses. Using this reasoning, it can be concluded that high-rise apartments are rather more energy-efficient than single-family detached dwellings, which is indeed the case. If the trends indicated in Table 11.1 come to pass, about 70 percent of all dwellings will be single-family detached or mobile homes. These have traditionally been the least energy-efficient of all buildings, particularly the mobile homes. The reason this has been true in the past has always been believed to be cost. The technology certainly exists to construct homes that are very energy efficient. A superinsulated home in central Saskatchewan requires very little heat other than passive solar for the entire winter season. Again we see that here, as in so many other aspects of the energy–environment dilemma, economic rather than technical considerations are the driving forces.

Consequently, when we are examining various improvements to dwellings from the standpoint of energy benefits, we should also consider economic benefits. In fact, we can generalize cost–benefit analysis into three considerations:

■ Energy benefit.
■ Economic benefit.
■ Social benefit.

We could propose that fitting every single-family detached dwelling with aluminum siding would save 10 percent of our annual heating bill. Fine, but how much energy would it cost to refine the aluminum required for this task? Would this be a net energy savings to the country? Similarly, if every window in the United States were triple-glazed, energy savings would result. How much? Would each homeowner be able to recover the additional cost of the added window in a reasonable period of time? Unless the double-glazing were done very poorly, the answer must be no. Finally, would the energy expenditure required if every man, woman, and child in the country were to use an electric toothbrush be justified? The result could be strikingly better dental hygiene; perhaps the energy cost would be worth it!

Energy conservation is a tangled web of interlocking, socioeconomic factors. The problem almost defies straightforward analysis. A large number of studies and reports have been made. As a consequence, some policy decisions will be made in the next few years. For example, the Buildings Energy Performance Standards (BEPS) program will soon become a reality. This program was established in 1976 as part of the Energy Conservation and Production Act, but it has run into bureaucratic red tape, as well as considerable opposition from the construction and utilities industries.

The objections of the construction industry to BEPS are that it would make the already high costs of housing even higher and that the standards do not provide a "recipe" for the builder. Rather, BEPS require that the finished house perform according to a set of standards. Performance standards are more flexible than construction recipes and are certainly preferred by architects, but they are definitely harder from the standpoint of the builder, who usually cannot worry about computer verification of design acceptability. The electric utilities oppose BEPS because they incorporate a weighting system that will penalize electric resistance heating.

Some sort of building standards will emerge, even though they will not be as stringent as many conservation groups would like. They will be a first step, and perhaps when builders have had an opportunity to assess the impact of energy standards, a second round of regulation can be made more effective. We must realize that *fully half of the buildings that will be standing in the year 2000 have not yet been built.* If these can be constructed with energy efficiency, as well as cost effectiveness, in mind than a large step toward the efficient use of energy will have been taken.

Construction

The objective of a home builder is to construct a building that meets the applicable construction codes, has some attractiveness to potential buyers, and minimizes his costs. This objective is not in the best long-range interests of the country, since minimizing first costs usually does not minimize life-cycle costs.

Life-cycle costs include not only the cost of construction of the house but also the costs associated with heating, cooling, and maintaining the house over its lifetime. These costs clearly depend intimately on the construction practices used in the first place. Heating, for example, can be achieved either by large quantities of expensive insulation along with a smaller, cheaper heating plant or with a minimal amount of insulation and a larger, more expensive heating plant. The latter usually minimizes first costs while the former does not.

Table 11.2 gives the approximate first costs for home construction in 1980. The shell accounts for about half of the energy-related costs. How can the optimum insulation be determined? We can parameterize the costs of the shell in terms of a surface-area-dependent cost a_1 and a cost that depends on the quality of the insulation used, p_1, so that

$$\text{Shell first cost} = (a_1 + p_1 R_0)A, \quad (11.2)$$

TABLE 11.2 First Costs of a Typical Midwestern House in 1980

	Dollars	Percent
Energy related		
Shell	8,000	20.1
Heat and cooling system	4,500	11.3
Lighting	800	2.0
Hot water	150	0.4
Appliances	1,350	3.4
Total	14,800	37.2
Other		
Interior finish, plumbing, etc.	25,000	62.8
Total first cost	39,800	100

Adapted from *Final Task Report—Residential and Commercial Energy Use Patterns 1970–1990*. Washington D.C.: Federal Energy Administration, 1974.

where R_0 is the net R-value for heat transfer through the shell. Recall the R-value was defined in Chapter 6 as the reciprocal of the heat-transfer coefficient; it has units of 1/(British thermal units per hour per foot per degree Fahrenheit). Multiply the R-value by 0.534 to obtain joules per second per meter per degree Celsius.

The first cost of the heating plant also has an R-dependent term, but in this case it is an inverse dependence, since the larger the thermal resistivity of the shell (large R_0), the smaller the heating plant must be.

$$\text{Heating plant first cost} = a_2 + \frac{p_2 \Delta T A}{R_0},$$
$$(11.3)$$

where p_2 is a unit cost, ΔT is the design inside–outside temperature difference, and A is the shell surface area as before. The total first cost is then the sum of these two costs:

$$\text{First cost} = (a_1 + p_1 R_0)A + a_2 + \frac{p_2 \Delta T A}{R_0}.$$
$$(11.4)$$

First costs can now be minimized by differentiating Eq. 11.4 with respect to R_0 and solving for the optimum R_0. This yields

$$R_0)_{\text{opt}} = \left(\frac{p_2}{p_1} \Delta T\right)^{1/2}. \qquad (11.5)$$

This result does not include annual operating costs. To take these into account we must include a term in the life-cycle cost that includes the heating requirements H, the efficiency of the heating plant η, the unit per area energy cost p_3, and a factor f that takes into account the increase in costs of energy over the lifetime of the house:

$$\text{Heating operating costs} = \frac{p_3 H A f}{R_0 \eta}. \quad (11.6)$$

When the life-cycle costs, Eqs. 11.4 plus 11.6, are minimized, a different value of $R_0)_{\text{opt}}$ is found:

$$R_0)_{\text{opt}} = \left(\frac{p_2 \Delta T + p_3 H f/\eta}{p_1}\right)^{1/2}. \quad (11.7)$$

If this value of R_0 were used in the original construction, the first cost would be increased by significant factor. In particular, the rate of increase of energy costs through the factor f can have large effects.

Example 11.1 Calculate heating operating costs, given 3000 degree days required, 2 percent/yr increase in energy costs over 30 yr, 260 m^2 of shell surface area, effective R-value of 1.5, furnace efficiency 0.6, and energy per unit cost of \$2.70/10^9 J.

The heat required is given by

$$\frac{3000 \text{ degree days}}{\text{yr}} \times \frac{24 \text{ hr}}{\text{day}} \times \frac{10^6 \text{ J/hr}}{\text{degree day}},$$

$$H = 7.2 \times 10^9 \text{ J} \times 30 \text{ yr} = 2.16 \times 10^{12} \text{ J},$$

Energy cost factor $f = (1.02)^{30} = 1.8$,

Heating operating cost

$$= (2.70 \times 10^{-6})(2.16 \times 10^{12}) \frac{1.8}{1.5 \times 6}$$

$$= \$1.17 \times 10^4.$$

Note:

1. The number of degree days of heating required for any day is the difference between 65°F and the high and low temperature of the day.
2. p_3 is a per unit area cost. Here $2.70 = p_3 A$.
3. 10^6 J/hr per degree day is an average, which may not be correct for well-built houses, higher occupancy, and so forth.

There is little incentive for a builder to use a higher effective R-value; this would simply make his product more expensive and more difficult to sell. If the buyer could realize an immediate benefit by buying a life-cycle-built home rather than accumulate the benefits over a 30- or 40-yr period, there could be such an incentive. This might have to take the form of property or income-tax credits.

There may be ways to increase the effective R-value without increasing the cost. If 2×4 in. wall studs spaced 16 in. apart were replaced by 2×6 in. studs spaced 24 in. apart the amount of material required would be the same. There would be no loss of structural support, but the amount of space for insulation would be increased by 50 percent and the standard R-11 insulation would be increased to R-16. Other ways to increase the effective insulation of the house include reduction of window area, double-glazing, and reduction of air infiltration.

This latter factor should be investigated in more detail. Outside air must not only be warmed to room temperature, but it must also be humidified if comfort is to be maintained. Outside air at 4.5°C having a humidity of 60 percent (it could be quite a bit lower) will have a humidity of only 20 percent when the air is warmed to 20°C. Typically, 40 percent humidity is considered optimum. However, to warm and humidify the air requires about as much energy as is lost by conduction through an average-size house.

Example 11.2 A house having a volume of 425 m^3 has one complete air exchange per hour. The outside temperature is 4.5°C with 60 percent humidity, and the interior is to be maintained at 21°C with 40 percent humidity. How much water is required to accomplish this?

Amount of water in air at 40°:

Vapor pressure of water at 4.44°C

$$= 6.27 \text{ mmHg for 60 percent humidity,}$$

$$p = 0.6 \times 6.27 = 3.76 \text{ mmHg,}$$

$$n = \frac{pV}{RT} = \frac{3.76 \text{ mmHg} \times (4.25 \times 10^8 \text{ cm}^3)}{62,400 \times (4.5 + 273.2)}$$

$$= 92.2 \text{ mol.}$$

Note that the gas constant R can be written in a variety of units.

Amount of water required at 21°:

At 21° vapor pressure is 18.65 mmHg.
For 40% humidity $p = 0.4 \times 18.65 = 7.46$.
Number of moles $= 182.9$.

$$(182.9 - 92.2) = 90.7 \text{ moles additional water required}$$

$$= 1.633 \text{ kg.}$$

Air infiltration can be minimized by careful construction practices. Modern homes have from 0.5 to more than 2 complete air exchanges per hour. A figure nearer 0.2 should be adequate for most homes. In office buildings or in places where smoking is allowed, higher exchange rates would be necessary. Homes could be built with near zero exchange rates, although this would be undesirable. Fresh air is needed in a home, even if some energy expenditure must be made. However, recall that each individual in a room is equivalent to a 100-W heat source!

Fireplaces present an energy problem not only in construction, because infiltration cracks are usually more abundant around them than in other places, but also in operation. Fireplaces left open to a room will draw air from the room, heat it, and deposit it outside. The warm air in the house is replaced by a higher rate of infiltration from outside. Virtually all fireplaces that are open to the house cause a net loss of energy. Glass firescreens that seal against the fireplace prevent the drawing of air from the room, but allow radiant heat from the fire to penetrate the room almost as efficiently as if there were no screen.

We have not mentioned the energy costs of the actual construction itself. The greatest expense in construction is labor; with few exceptions the energy cost of providing the materials required is relatively small. Most of the materials used in single-family detached dwellings are from renewable sources—wood. However, many commercial buildings and some homes use a great deal of concrete, which is a very energy-intensive material, as we see in the next section. Concrete became a popular construction material in the 1930s because it was easy to form and had low labor costs compared to the most used material at that time, stone. Neither stone nor concrete is a particularly good heat insulator used by itself, at least in the thicknesses usually encountered in construction.

We should perhaps modify our life-cycle cost procedure to take into account the energy expense of the construction materials. This would add a term of the form

$$\text{Materials energy cost} = p_4 A, \quad (11.8)$$

where A is the surface area, and p_4 is a per unit area cost of the materials. Note that there must be an interaction between a_1 and p_4 since the choice of construction material dictates not only the materials energy cost but also the fixed labor costs represented by a_1. Currently, this term is included in the calculation of p_2 and is not a very large fraction of it. But the cost of producing concrete may increase radically in the next decade as fossil fuel prices continue to climb.

Operation

We have discussed those aspects of building construction which are significant in terms of energy, particularly insulation. In this section we inquire about the activities that go on in buildings. Domestic and commercial energy uses are summarized in Table 11.3. From this we see that the major part of this energy goes to space heating, as we might have guessed.

Space Heating A number of aspects of the heating question need to be studied.

What are the characteristics of the various heating systems?
What are the patterns of space heating use?
What potential energy savings can be realized?

There are several heating systems in use today: electric resistive, electric heat pump, forced air (fuel burning), and hot water (fuel burning). These are drawn schematically in Figure 11.2. We can calculate the First Law efficiency for each one readily. But to calculate the Second Law efficiency, we have to know the minimum available work B_{min} required for the heating task.

The heating task has been solved by several means, as given above. We can generalize

TABLE 11.3 Residential and Commercial Energy Use

Energy Use	Percent of National Total
Space heating	18
Water heating	4
Cooking	1
Refrigeration and air conditioning	5
Other	4

Source: *Patterns of Energy Consumption in the United States*, prepared by Stanford Research Institute. Washington D.C.: U.S. Office of Science and Technology, 1972.

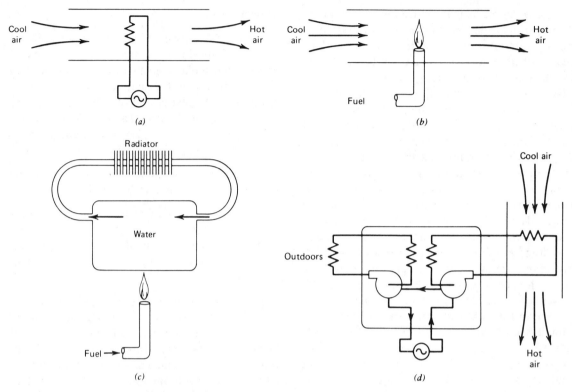

Figure 11.2 Several techniques for supplying home heat. (*a*) resistive hot air, (*b*) gas (or oil) fired hot air, (*c*) gas (or oil) fired hot water, and (*d*) electrically operated heat pump.

these solutions in the form of a pair of devices, as shown in Figure 11.3. A *task device* converts the driving energy to the heat delivered to the room. A *supply device* converts the primary energy, which may be in the form of a fuel or electricity, into driving energy; at this stage there may be wasted heat. In Table 11.4 we show several combinations of task and supply devices. Not all combinations are viable; these are indicated by NA, not applicable.

The minimum available work required in every case must be the available work from the combustion of the fuel. As we saw in Chapter 6, that value is very nearly the heat of combustion of the fuel:

$$B_{min} = |\Delta H|. \qquad (11.9)$$

Is there a device that would enable us to use this amount of work to produce heat? The fuel cell, discussed in Chapter 5, could come very close. As a consequence, many of those who are advocating the use of the Second Law efficiency as a measure of quality use the available work from the fuel cell as the B_{min} for Second Law calculations. The values of ϵ in Table 11.4 have been determined by comparison with the fuel cell. An interesting conclusion from this table is that resistive space heating, with the electricity supplied by a fuel cell, is the most energetically efficient method of using fossil fuels. Unfortunately, practical fuel cells using hydrocarbon fuels are not yet available.

Example 11.3 What is the Second Law

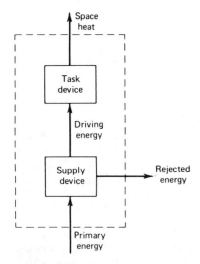

Figure 11.3 Generalization of a space-heating device, consisting of a supply device and a task device (see Table 11.4).

From Eq. 11.10,

$$\epsilon = \frac{Q_1}{W}\left(1 - \frac{T_2}{T_1}\right),$$

$$\epsilon = \frac{2.0 \times 10^7}{5 \times 10^3 \times 60 \times 60}\left(1 - \frac{273 - 1}{273 + 45}\right),$$

$$\boxed{\epsilon = 0.16.}$$

The efficiency seems low, but, compared to some alternatives yet to be examined, it is not so bad. The COP for this device is

$$\omega = \frac{2 \times 10^7}{5 \times 10^3 \times 3.6 \times 10^3} = 1.11.$$

This COP is so low because the lower temperature reservoir is also too low for efficient operation. At about 4°C the COP in a properly operating heat pump would be closer to 3.0.

---

efficiency of a heat pump operating between −1 an 45°C with a heat deposition rate of 2.0×10^7 J/hr and an electric power consumption of 5 kW?

One possibility that is available today is to use a diesel engine to drive the compressor of a heat pump. Diesel engine efficiencies are reasonably high, and the efficiency of the overall process can be increased by using the

TABLE 11.4 Task–Supply Device Matrix for Space Heat[a]

Supply Device	Task Device		
	Resistive Heat	Fuel Burning	Heat Pump
None[b,c]	$\epsilon \approx \eta$ (0.35)	$\epsilon = \eta\left(\frac{T_2}{T_1}\right)$ (0.08)	$\epsilon = \omega\left(1 - \frac{T_2}{T_1}\right)$ (0.15)
Fuel cell	$\epsilon \approx \eta$ (0.55)	NA	$\epsilon = \omega\left(1 - \frac{T_2}{T_1}\right)$
Diesel engine[d]	NA	NA	$\epsilon \approx \epsilon_1\epsilon_2 + (1 - \epsilon_1)\left(1 - \frac{T_2}{T_1}\right)$ (0.2)

[a]The Second Law efficiency in terms of measurable quantities is given, along with a typical value in parentheses.
[b]η is the First Law efficiency, ω is the coefficient of performance.
[c]T_2 is the low temperature, T_1 the high temperature.
[d]ϵ_1 is for the engine, ϵ_2 is for the heat pump; assume engine waste heat used for heating.

heat rejected by the engine. A drawback to this arrangement is that diesel engines are large and bulky; they would not be suitable for individual homes. However, apartment complexes would be ideal for the application of such a hybrid arrangement. This is a logical extension of the cogeneration schemes discussed in Chapter 8.

We should also note that by far the most common method of providing space heat—forced–air heating by fossil-fuel burning—is also by far the least energetically desirable technique. It is also particularly inefficient, even from the First Law viewpoint, when all the losses are taken into account. First, the temperature at which air is delivered to the room is seldom higher than about 22°C, a loss of quality. A large part of the heat generated in the combustion is lost through the duct walls or through leaks. A large part of the heat is delivered to unused areas and spaces between wall studs. Overall, less than 75 percent of the fuel energy is actually used for space heating.

If a house is well constructed, it will have a large **thermal relaxation time** τ. This is the time required for the temperature difference between the exterior and interior of a house to drop to $1/e$ of its initial value. Naturally, this time constant varies considerably from house to house. Since air heats rapidly, thermostatic control of a forced-air system during the night and during unoccupied periods is simple to accomplish. A typical furnace can supply heat at a rate of about 17 kW. Heat at this rate will raise the temperature of a 430-m³ house (a typical value) about 1.9°C/min. Therefore, if the furnace were turned on for an hour or so prior to the time a family wishes to begin the day, the house should become comfortable rather quickly after a night of falling temperature (refer to Figure 11.4). The total amount of heat lost by a house during a time period t while the furnace is off is given by

$$Q = K\,\Delta T_0 \tau (1 - e^{-t/\tau}), \qquad (11.10)$$

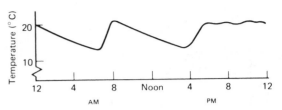

Figure 11.4 Interior temperature in a well-insulated home with an outside temperature below 10°C. At 6:30 A.M. the furnace is turned on for $1\frac{1}{2}$ hr, then turned off. It is turned on again at about 3:30 P.M. and kept on until about midnight.

where K is the effective thermal conductance of the house. You can quickly convince yourself that considerable savings can be realized if the furnace is turned off completely at night.

Additional savings can be obtained by heating only those portions of a house that have occupant zone heating. Direct resistive radiant heating and hot water heat (Figure 11.2c) lend themselves to zone control because of the ease of dividing these heating systems into several circuits that can be individually thermostated. Finally, we should also note that all of the comments given in this section for space heating also apply to water heating. Considerable savings can be made by applying additional insulation, by turning the temperature control down when hot water is not required and by selecting efficient devices initially.

Air Conditioning Comfort cooling of homes and commercial buildings has been the fastest rising end use of electricity in this country over the past two decades, having a growth rate averaging more than 16 percent per year. It is estimated that in 1980 air conditioning accounted for more than 4 percent of all the energy use in the United States. We saw in Figure 1.10 that not all air-conditioning units have equally good performances, so a considerable savings in energy could be obtained by simply choosing the correct unit.

Beginning the late 1970s, manufacturers were required to label air-conditioning units with an energy efficiency ratio, EER, as a guide to prospective purchasers. Unfortunately, since the EER is expressed in a hybrid set of units, a direct comparison with the First Law coefficient of performance (COP) can be made only by dividing the EER by 3.414. The EER is the heat-removal rate in British thermal units per hour divided by the power consumption in watts. Thus, a unit with an EER of 6, a typical value, has a COP of 1.75, a not particularly high number.

High-performance air conditioners are expensive to construct; they require larger compressors, larger heat-exchange surface areas, and longer-lasting materials. There is usually a linear relationship between first cost and EER. However, if life-cycle costs are used instead of first costs, then these higher EER units are almost always more economical. Unfortunately, many purchasers do not or cannot look at it this way.

Air conditioning is clearly not the luxury it once was. But this one electricity end use has shifted the peak load for electric utilities from the winter, where it was a few years ago, to the summer. This one load taxes the utilities most severely, particularly on stormy summer nights, when air conditioning is in demand, and lightning is playing havoc with the high-voltage transmission systems. Until the end of the 1970s electric utilities were advertising air conditioning rather heavily. It soon became very clear that the utilities could not handle the demand they had helped to create, so they stopped such advertising. Many concerned citizens are advocating a return to natural, un-air-conditioned comfort. Indeed, if houses were better constructed with attention paid to siting and many passive solar features were built in, the air-conditioning requirements of the country, at least for dwellings, could be reduced substantially. In commercial buildings, if the common practice of overcooling and then reheating with resistive heat to bring the temperature up to something comfortable were eliminated, the load could also be reduced. In some office buildings in the winter, heat is provided for the building perimeter while the central core containing elevators, duplicating machines, computers, and the like is air conditioned. The unfortunate fact is that it is cheaper to do it this way than to provide the heat-exchange systems necessary to enable the rejected heat from the machines to serve as space heat for the people. Finally, we note with sadness that many modern office buildings are built with windows that *cannot* be opened to allow for natural ventilation.

Whether air conditioning is moral or immoral is not the issue here. Air conditioning is a necessity for a very large portion of our populace. High technology occupations could not exist without it. Virtually everyone who lives and works in a humid summer climate is more productive, efficient, and innovative in an air conditioned environment. It is essential for developing nations to bring their products on the world market in a competitive fashion. As a consequence we are concerned here with the efficiency of air conditioning and the alternative methods for producing high efficiency devices.

We have already noted the coefficient of performance for an air conditioner (refrigerator):

$$COP = \omega = \frac{Q_2}{W}, \qquad (11.11)$$

where Q_2 is the heat extracted from the cold reservoir, and W is the work input required. The minimum work required to extract Q_2 units of heat from a reservoir at T_2 and deliver it to a reservoir at T_1 is given by

$$B_{min} = Q_2\left(\frac{T_1}{T_2} - 1\right). \qquad (11.12)$$

The Second Law efficiency is the minimum

work divided by the actual work, or

$$\epsilon = \frac{B_{\min}}{W} = \frac{Q_2(T_1/T_2 - 1)}{W} = \omega \left(\frac{T_1}{T_2} - 1 \right). \quad (11.13)$$

For typical units ϵ ranges between 0.01 and 0.07. These values are comparable to those achieved with forced air heating. We can conclude that considerable improvements can be made in the construction and application of air conditioners. Note that all of the previous comments on building construction relative to heating also apply for air conditioning, since it is, after all, basically the heat infiltration rate—through the walls, through leaks, or because of occupancy—that causes an air-conditioning load.

Recall that in Chapter 4 we discussed some alternatives to vapor-cycle refrigeration. These were various absorption or adsorption devices that, in concept at least, can operate with the low-quality heat available from the sun. These devices must be improved to the point of widespread applicability if substantial energy savings in air conditioning are to be made. These coupled with enlightened construction practices could reduce the air-conditioning load to a small fraction of its present value, at least for dwellings.

Lighting Lighting accounts for about 5 percent of the total energy use in the United States. Of the total amount, 20 percent is used in homes, 40 percent in public and commercial buildings, and about 10 percent in out-of-doors applications (street lights, advertising, security lights, etc.). Lighting in the homes tends to be restrained, if inefficient, while lighting in schools and offices tends to be excessive, but somewhat more efficient.

In order to discuss lighting intelligently we need to know the units of measure. Unfortunately, while light intensity can be defined in terms of watts per square meter or similar unit, this does not take into account the fact that the human eye responds differently to

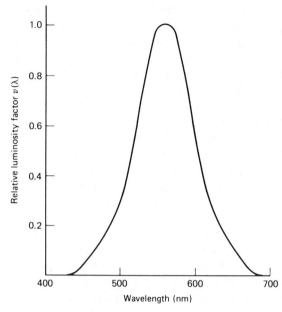

Figure 11.5 A standard measure of the human eye response to light as a function of wavelength.

different wavelengths of light. To do this a new unit has been devised, the lumen [abbreviated lm—1 lm = 1 cd sr (candela-steradian)]. The luminous flux[1] Φ in lumens of a light source that has a power distribution $P(\lambda)$ W/(unit wavelength interval) is given by

$$\Phi = 680 \int_0^\infty P(\lambda)v(\lambda)\,d\lambda, \quad (11.14)$$

where $v(\lambda)$ is the "relative luminosity factor," a standard measure of eye response. This function is plotted in Figure 11.5. Note that this is a standard curve; each individual eye may have a different response function.

The term **efficacy** is used by lighting engineers as a measure of the efficiency of a source of light. It is the luminous flux in lumens divided by the power input in watts. For a typical source, this power input can be obtained by integrating $P(\lambda)$. For a 100-W

[1]Flux is an unfortunate word here, since Φ does *not* measure intensity per unit area.

incandescent bulb the efficacy is about 18 lm/W; for a 40-W fluorescent fixture it averages about 60 lm/W.

The luminous flux is a measure of the light energy per unit time. A measure of the actual "flux" in the physics sense is called the **illuminance**; it is measured in lumens per square foot or lumens per square meter. A lumen per square foot is also called a foot-candle. Standards for lighting in schools and offices are usually given in terms of lumens per square foot. How much lighting is required? In terms of construction standards this is simple to answer. From 1950 to about the late 1970s the required illuminance had risen at the rate of about 43 lm/m^2 yr. In the early 1950s, therefore, typical illuminance levels were 200 lm/m^2, but by the late 1970s they were nearly 1000 lm/m^2. At this point, because of a newly found interest in energy conservation, architects and engineers became more task-oriented and again asked, "How much light is actually required?"

The required illuminance depends on the task, since there is a connection between illuminance, size of the object to be seen, and distance of observer from the object. This connection is indicated in Figure 11.6. Here the

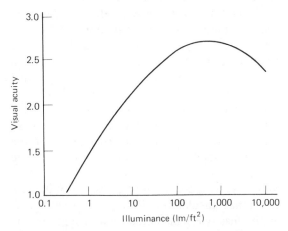

Figure 11.6 The minimum visual acuity that can be achieved as a function of illuminance.

visual acuity is plotted as a function of the illuminance. The visual acuity is $1/\theta$, where θ is the angular gap in minutes of arc that can be discerned in a printed circle. This plot assumes that the light reflected from the object is the same as that incident upon it and that there is no difficulty with visual backgrounds. We can easily use this figure to determine required light levels.

Example 11.4 What illuminance is required for comfortable reading of a typical textbook?

The smallest letters on a printed page are about 1.3 mm tall. To read comfortably at a distance of 50 cm it would be desirable to be able to resolve distances about equal to the width, say 0.13 mm. This is an angle of $\tan^{-1}(0.013/50) = 0.89'$. Visual acuity = $1/0.89 = 1.12$. From Figure 11.6, approximately 4.3 lm/m^2 is required. This is from 100 to 300 times less than is found in most office buildings, but is nearer the illuminance you would have at home in your easy chair.

It is clear from Figure 11.6 that high illuminance levels are not required for most tasks to be found in schools and offices and that levels above about 400 lm/m^2 do not significantly improve the visual acuity for most tasks. Yet, we find very high levels in use throughout the country. This level of illuminance does not contribute to worker productivity, but it does provide a much higher heat load on the building air conditioning, since, as we have seen, most buildings do not use this rejected heat.

Heat loads from lighting can be minimized by using the more efficient fluorescent lights in both homes and offices. These lights use the UV radiation from an electric discharge set up in a mercury vapor to cause a coating on the inner walls of the discharge tube to fluoresce in the visible range. Usually, one or more additives are placed with the mercury to give the

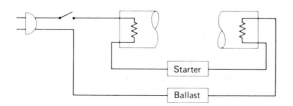

Figure 11.7 Wiring diagram for a fluorescent lighting fixture.

light a more natural color. These lights require a starter and a ballast, as shown in Figure 11.7. The starter is used when the circuit is initially energized to heat filaments in each end of the tube in order to vaporize some of the mercury. The starter automatically opens after a time to present a high potential across the tube so that the discharge can begin. The ballast is simply a series resistor to limit the current drawn by the discharge. Fluorescent tubes will age, causing a reduction in light output, and dust accumulations on the outside will also have an effect. Perhaps the largest single loss of light in fluorescent systems is in the light fixture. These tubes are usually covered by some sort of plastic diffuser so that the bare bulb will not be visible. These diffusers, and the lack of a downward light reflector, often absorb more than half of the light.

Lighting in an office building can be calculated in terms of electrical watts per square meter of usable space. On this basis many modern buildings have between 20 and 50 W/m² of equivalent lighting installed. This is considerably more than can possibly be needed.

Example 11.5 A building has 25 W/m² of equivalent lighting using 40-W fluorescent tubes. What is the average illuminance?

The efficacy for 40-W fluorescents is 60 lm/W. Therefore the illuminance is $(60 \text{ lm/W}) \times (25 \text{ W/m}^2) = 1500 \text{ lm/m}^2$. As we have seen, this is much larger than required for most tasks.

Other improvements that can reduce the lighting load include more extensive use of natural lighting, improved fluorescent fixtures operating at frequencies higher than 60 Hz, automated lighting controls to turn off lights in unoccupied spaces (it always pays to turn off the light when leaving the room, virtually without regard to the length of absence), and reduced levels of outdoor advertising lighting.

Appliances Household appliances account for about 7 percent of the national energy consumption, more than any individual category except space heating. Some energy savings is possible in the use of appliances, but only a shift in product usage will bring about large savings. For example, gas ranges are rather inefficient in terms of the heat actually absorbed by the food compared to the heat generated in the unit. Electric ranges are somewhat more efficient in this respect, but the much higher price for the electricity more than offsets this advantage. Microwave ovens are certainly more efficient from this point of view, but are not usable for all types of cooking. There are also inhibitions in some householders to the use of microwave devices.

Table 11.5 lists several appliances and their energy requirements. This list, along with what we have previously seen, tells us that any house equipped with two or more of the following has a high energy demand that will be difficult to reduce with simple conservation measures: electric heat, electric range and oven, electric air conditioning, electric clothes dryer, and electric hot-water heater. These appliances are so energy-intensive (and so energy-expensive!) that the remainder of the household load is "down in the noise." Some appliances can be made more efficient; hot-water heaters, for example, can be built with additional insulation. Refrigerators are currently less well insulated than could be the case. Frost-free refrigerators and freezers achieve this freedom by periodically turning on

TABLE 11.5 Energy Requirements for Several Household Appliances

Appliance	Energy Requirement (kWh)	Appliance	Energy Requirement (kWh)
Bed covering (blanket)	0.4/day	Oil furnace pump	0.25/hr
Broiler	1.3/hr	Radio	0.08/hr
Carving knife	0.9/hr	Hi-fi stereo	0.1/hr
Clock	1.4/month	Range	98/month
Clothes dryer	5.0/hr	Small unit	1.6/hr
Coffee maker	0.84/hr	Large unit	2.7/hr
Cooker (egg)	0.49/hr	Oven unit	3.2/hr
Corn popper	0.49/hr	Refrigerator	
Deep fat fryer	0.8/hr	12 ft^3	70/month
Dehumidifier	0.22/hr	12 ft^3 frost-free	100/month
Dishwasher with heater	1.2/hr	Refrigerator-freezer	
Without heater	0.3/hr	14 ft^3	100/month
Fan, furnace	0.3/hr	14 ft^3 frost-free	140/month
Floor polisher	0.3/hr	Roaster	65/hr
Food blender	0.3/hr	Sewing machine	0.7/hr
Food freezer		Slow cooker	
15 ft^3	90/month	Low	0.075/hr
15 ft^3 frost-free	140/month	High	0.15/hr
Food mixer	0.13/hr	Skillet	0.6/hr
Food disposal	0.4/hr	Sun lamp	0.28/hr
Hair dryer	0.4/hr	Television	
Heater, portable	1.3/hr	Black and white	0.15/hr
Heating pad	0.06/hr	Color	0.18/hr
Hot plate	1.25/hr	Toaster	1.1/hr
Humidifier with fan	0.12/hr	Toothbrush, electric	0.007/hr
Ice-cream freezer	0.19/hr	Vacuum cleaner	0.5/hr
Iron, hand	0.75/hr	Water heater (80 gal)	500/month
Microwave oven	1.6/hr	Washing machine	0.35/hr

a resistive heating strip to evaporate the ice that has been deposited on the interior. By choosing the best device and by doing a little manual labor some reductions in energy use can be achieved.

Home Computerization A new generation of computers became widely available in the late 1970s that will have a profound impact on American society. These **microcomputers**, as they are called, will soon be found in most households. They will be able to do a large variety of tasks, some of which will be important in saving energy. We have already mentioned two in this chapter: thermostat and lighting control. The list of tasks is limited only by one's imagination, because the hardware (i.e., equipment) and software (i.e., computer programs) necessary to do most of them already exist or can be easily obtained. For example:

1. Sense open windows, either directly or indirectly by detecting increased heating or cooling requirements beyond those normally found for the existing temperature difference.

2. Sense dripping or running water faucets by prolonged water flow.
3. Control hot-water heater to take advantage of off-peak rates and reduce the temperature during periods of little use.
4. Sense electrical appliances left on inadvertently.
5. Control zone-heat distribution to reduce heat to unoccupied areas.
6. Sense by temperature differences the reduction in performance of furnaces or air conditioners caused by clogged filters.

There are of course many other non-energy-related funtions that home computers can and do perform that we have not listed here. The investment in hardware to do the tasks above is less than $1500 in 1981 and likely to become even smaller as more units become available. These devices can pay for themselves in a few years of operation, and now that energy-related expenditures can be deducted from income reported to the IRS, we can expect to see them being used in larger numbers in the decade of the 1980s.

11.2 INDUSTRIAL ENERGY CONSERVATION

Industry accounts for more than 40 percent of the total energy used in the United States. The six largest industries in terms of energy use and in order of use are primary metals; chemicals and allied products; petrochemicals and allied products; food; paper; and stone, clay, glass, and concrete products. These six industries account for about 65 percent of the total industrial energy use. The distribution of energy end use in industry is given in Table 11.6; we see that heat in one form or another is by far the largest energy form in industrial use. To discover ways to save energy in industry we must examine several of the most important processes.

TABLE 11.6 Industrial End Use of Energy

End Use	Percent of National Total
Process steam	17
Electric drive	8
Electrolytic processes	1
Direct heat	11
Feedstock	4
Other	<1

Source: *Patterns of Energy Consumption in the United States*, prepared by Stanford Research Institute. Washington D.C.: U.S. Office of Science and Technology, 1972.

Industrial Processes

In many manufacturing processes heat is required to drive chemical reactions or for drying. Some of these are particularly inefficient. For example, cement production in the United States typically requires about 12×10^8 J/bbl, while in Europe, where preheaters are used, the figure is closer to 6×10^8 J/bbl.

If air to be used in combustion is preheated, substantial fuel savings can result. Figure 11.8 shows a heat recuperator using stack gas heat

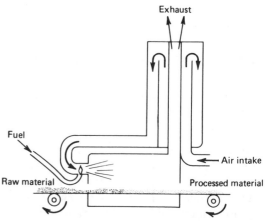

Figure 11.8 A schematic diagram of a heat-recuperating system for a drying oven. Intake air is heated by exhaust air in the stack.

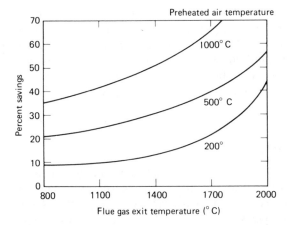

Preheated air temperature

Figure 11.9 A plot of the heat savings that can be achieved in a processing system similar to that in Figure 11.8 as a funtion of exhaust gas temperature.

from an oven to preheat the combustion air. The effectiveness of this system is indicated in Figure 11.9. If a particular process requires 1370°C and if the combustion air can be heated to 540°C, Figure 11.9 indicates that a fuel savings of about 30 percent can be achieved.

Example 11.6 Verify the above statement.

We can compare the heat required to raise air from 20° to 1370°C, Q_2, with that required to raise it from 540° to 1370°C, Q_1, using $Q = mc\,\Delta T$.

$$\frac{Q_1}{Q_2} = \frac{mc\,\Delta T_1}{mc\,\Delta T_2} = \frac{830}{1350} = 0.62.$$

Preheating clearly does result in a fuel savings!

Ovens are only one of several different methods of providing direct process heat, but in many of the others a similar type of recuperation process could be developed to preheat the combustion air.

Some industrial processes use electrical resistive heating elements. These could be replaced with direct fuel combustion, giving about a factor of 3 increase in the overall

efficiency of utilization of this part of the fuel. However, some electrical processes cannot be replaced; electric furnaces are still refining steel and aluminum. In these cases the very high temperatures required can only be obtained from an electric arc. There does not seem to be an economic replacement for electric arc furnaces on the horizon.

A large part of the industrial heat requirement is in the form of steam. Steam production is relatively efficient in modern plants. However, research has shown that improvements in combustion can be expected in the future.

There are several important manufacturing processes that require physical separation of a product stream into components. Some of these involve a macroscopic separation, such as metals from nonmetals in a municipal waste facility. We cannot easily apply the ideas of thermodynamics to such processes. Clearly, though, physics can be applied. Some of the separation techniques are very sophisticated. For example, one process separates dark glass from uncolored glass by means of the difference in magnetic susceptibility.

Microscopic separation can be analyzed, since we can calculate the minimum amount of work necessary to do the job. This is simply the reverse of a diffusion process described in Chapter 6 when we discussed combustion. Using diffusion as an analogue, it must be that the amount of energy required to separate two gases originally at concentrations x_i and temperature T at a constant pressure is given by

$$B_{\min} = RT\left[n_1 \ln\left(\frac{1}{x_1}\right) + n_2 \ln\left(\frac{1}{x_2}\right)\right], \quad (11.15)$$

where n_i is the number of moles of each species.

One common separation is that of oxygen from air. Typically this is done by reducing the temperature of the air to the point that the oxygen liquefies. The average industrial process for doing this requires about $1.5\times$

10^6 J/kg of oxygen produced. Equation 11.15 can be used to determine the minimum energy that should be required; of course, finding a technique that uses only this amount of energy may be very difficult.

Example 11.7 Find the minimum separation energy for oxygen in air at 27°C. Assume air to be composed of only oxygen and nitrogen in the ratio of 2 to 7.5. Then $x_1 = 2/9.5 = 0.215$, and $x_2 = 7.5/9.5 = 0.78$. Then

$$B_{min} = RT \left[2 \ln \left(\frac{2}{0.215} \right) + 7.5 \ln \left(\frac{7.5}{0.78} \right) \right]$$

$$= 2 \times 300 \times 21.36 \text{ cal},$$

$$B_{min} = 53,600 \text{ J} = 26,800 \text{ J/mol O}_2,$$

$$\boxed{B_{min} = 1670 \text{ J/kg O}_2.}$$

Cryogenic separation thus appears to have a Second Law efficiency of only $(1.67 \times 10^3)/(1.5 \times 10^6) \approx 0.0011$, a rather small value.

A molecular sieve process using 3 atm air has had some success, although the energy requirement is even larger than that of the cryogenic technique. The fact that the Second Law efficiency of current methods is so low leads us to believe that other techniques can be found, so research continues.

Other industrial processes could be studied in a similar manner; however, the largest energy end use in industry is, as we have seen, heat. The only way to improve that utilization is to operate at higher temperatures, not always desirable or possible; preheat the combustion material, perhaps requiring high capital outlay; or use a different fuel, if such is available. The fact is that until recently it was almost always cheaper to do things inefficiently than to worry about energy losses. In fact, manpower has been displaced by machines primarily because manual labor became much more expensive than mechanical labor, even though the mechanical equipment is much more energy-intensive. Even today it is probably still more desirable from an individual manufacturer's point of view to be energy-intensive, but for the entire nation it does not make good sense to carry on with business as usual.

Recycling

In recent years there has been a great deal of interest in recycling of discards as an energy conservation measure. Yet the United States recycles only about 20 percent of its potentially reusable materials. The easy reason for this is that it is not profitable to do so; the hard answer is that it is a very complex issue. Two major difficulties must be overcome:

1. Institutional and social problems.
2. Technical problems.

Among the former is the claim that freight rates discriminate against scrap materials in favor of virgin ores. Rates are supposed to be set by the Interstate Commerce Commission so that the equivalent values of service pay equivalent rates. But iron and steel scrap rail rates average two and one-half times that of iron ore. This puts scrap at a definite disadvantage, and adversely affects the environment as well. Processing of virgin ore "costs" considerably more in terms of fuels and effluents than does processing scrap. This is true for almost all kinds of scrap metals, not just iron and steel. Scrap is very abundant (see Figure 11.10). But even more would be used if there were no economic penalty.

Another institutional complaint has to do with the depletion allowence. Mine operators are allowed to take a 15 percent deduction in their gross income because of the depletion of their resource; scrap operators can take no such deduction. Scrap dealers are also concerned about legislation being pushed hard by some environmental groups. This legislation

Figure 11.10 Shredded iron and steel scrap ready to transport to an electric-arc steel reducing furnace. (Photograph courtesy of the Institute of Scrap Iron and Steel, Inc.)

would ban the sale of throwaway containers and require a deposit on all bottles and cans. From the point of view of the scrap operator, this would put him out of business—the amount of scrap available would decline markedly, as is evidenced by the experience in Oregon and other states that have initiated bans on disposable containers.

Another problem has to do with labeling. Many industries are required to label products as to virgin material content. Motor oil made from reprocessed oil must be so labeled and does not command the consumer respect that virgin oil does, even though there is no difference in performance.

There are also technical barriers to recycling. Automobiles contain large quantities of copper and aluminum, which, if allowed to be mixed with the iron in processing, would produce a lower-quality steel, usable only for reinforcing rods used in building construction.

The "tin" in tin cans is only a thin layer plated on a basically steel can. The seam also contains solder made from tin and lead. These cans cannot be easily recycled. Steel production is done primarily in basix oxygen furnaces (BOF), which can take only a small percentage of scrap, less than 30 percent of the full load. Electric arc furnaces can operate with 100 percent scrap, but these account for only about 15 percent of the total U.S. steel production. The BOF does not use externally supplied fuel but rather relies upon exothermic oxidation of carbon, silicon, and manganese by oxygen blowing through molten pig iron. Scrap is deficient in these elements, and unless preheated, delays the conversion process so that a lower overall productivity is achieved for an individual furnace.

The technical situation with other metals is more favorable. The energy requirements to produce ferrous metals, aluminum, and copper from ore and from urban scrap are indicated in Table 11.7. Substantial energy savings can be made, particularly from aluminum. In this case, while impurities are important in determining the usability of scrap, many aluminum products are not so sensitive to impurity content; for example, recycled beverage containers can be used to produce either more containers or household foil. Recycling of aluminum will become much more important in

TABLE 11.7 Approximate Energy Requirements for Producing Finished Metal

	Energy to Produce 1 tonne from	
	Ore (GJ)	Scrap (GJ)
Ferrous metals	60	12.0
Aluminum	330	11.0
Copper	100	6.0

the next decade as some of the products produced in the early 1950s come to the ends of their useful lives. This is a result of the almost exponential increase in aluminum production in this country in the past 30 years.

There are some materials for which there is little advantage (indeed there may be some disadvantages) to recycling scrap as opposed to using virgin ore. These are glass and rubber. About 1.2×10^6 J of thermal energy is required to produce a single 0.5-ℓ beverage bottle; the same energy is required to recycle scrap glass into a bottle. However, scrap glass varies substantially in composition and color and is in general not suitable for high-quality glass production. Even small amounts of impurities, either metallic or organic, can cause difficulties. Less than 5 percent of the glass produced in this country is eventually recycled. (And the "half-life" of glass in the biosphere is very long!)

The difficulty with recycling rubber tires is in the separation of the rubber from cords or wires molded into the tires. Recapping of tires is a desirable alternative to complete recycling, but still the overall amount of rubber recycled total is less than 25 percent. Textiles as well are not recycled much, less than 4 percent of the total production. This is primarily an economic problem, not a technical one. Many textiles degrade quickly in sanitary landfills, so do not present the long-term problem that glass does.

In summary, recycling could contribute a great deal more to the campaign to conserve energy in this country than it does at present. To do this many of the institutional and technical barriers will have to be withdrawn.

11.3 TRANSPORTATION

Some ancient societies never invented the wheel; perhaps they were lucky. Modern transportation has enabled us to bridge con-

tinents in hours, to live many miles from our jobs, and to enjoy the produce and goods of many different lands. But it has also enabled us to spend many hours per week in our cars commuting to and from work, to breathe air proved to be toxic to laboratory animals, and to pay out an ever-increasing amount of our income, much of it overseas, on the fuel to accomplish all this.

Transportation uses about 25 percent of all the energy produced in the United States. A savings of even a few percent in this category can have a great effect on our overseas balance of payments, on the production of polluting effluents and on our resource depletion rate. The goal of this section is to examine our transportation systems and to find ways of becoming more efficient at them.

Aspects of the Transport Problem

We must transport people and freight. The methods that have evolved over the years for doing this have been based on speed, convenience and economy, *in that order.* Until recently, very little consideration was given to energy use; but now that energy has a heavy impact on economy, we find that as a nation we have suddenly become energy-aware in dealing with transportation.

This awareness has taken several forms. There have been increased expenditures on federal, state, and local levels on mass-transit studies and systems. The federal government has mandated higher performance standards for automobiles, and consumers have shown a willingness to buy such cars, even though they were available initially only from foreign manufacturers.

Much of the discussion of transportation has been concerned solely with the transport of people. This reflects the fact that most of the energy spent on transportation goes to passenger traffic (refer to Table 11.8). Most of the discussion in this section as well concerns the

TABLE 11.8 Transportation Energy Consumption Patterns Estimated for 1980[a]

	Intercity		Urban
	Freight Percentage of Total	Passenger Percentage of Total	Passenger Percentage of Total
Air	0.2	10	
Truck	20		
Rail	34	1	
Auto		87	97
Bus		2	3
Total energy (10^{15} J)	3.4	6.0	9.5

[a]Based on data in E. Hirst, *Energy Consumption for Transportation in the U.S.* Oak Ridge: Oak Ridge National Laboratory Report ORNL-NSF-EP-15, 1972.

movement of people, but we also consider some new alternatives for freight transport.

The difficulties in providing workable mass-transit systems for Americans involve more than simple technical barriers, although those certainly exist. Many cities have no prior history of mass transit, so that when tunnels must be dug, as in Washington, D.C., it becomes a very expensive proposition. Most cities, such as Boston, were not originally planned with mass transit in mind, so a simple, workable system is very difficult to institute. One of the few systems built in the past several decades is the Bay Area Rapid Transit system (BART) in the San Francisco Bay area. This multibillion dollar project required 8 years to construct (1964–1972) and now operates on more than 70 miles of above-ground and underground tracks. A BART train is shown in Figure 11.11.

Most of the transit studies have been concerned with either the "hardware" (vehicle type) or the "software" (networks) along with the economics of a particular design. As a consequence we now have designs for super-

trains with magnetic levitation and linear induction motors, people movers having small monorail-driven cars, and other technological innovations. The network studies have yielded information on optimum timing, pick-up on demand, route network theory, and other valuable ideas for implementing mass transit. However, none of these studies cuts to the heart of the mass transit problem: the commuter. Even with the high price of automobiles and gasoline, the fact is that most Americans actually *want* to own and use cars. It is a matter of pride. The car gives us freedom to go and to do. People see their cars as reflections of themselves. A transit system does none of this. Thomas Lisco says it very well:

It is preposterous to expect anyone with a reasonable income, an attractive home and air conditioned office to submit to rush hour jostling, crowding and standing in transit cars that combine inadequate heating and nonexistent air-conditioning with poor riding characteristics. People willing to pay a high premium for their comfort at home and in the office will pay just as much for comfort traveling between them. And when they are faced with the choice between being herded like cattle into a rush hour transit vehicle and driving a well-engineered, quiet and comfortable automobile the decision is an easy one.[2]

This is the usual failure of transit systems. But often, as well, particularly in smaller communities, the transit system fails to offer an economic advantage to a potential user. For an individual given the choice of driving 10 miles round trip or paying $2.00/day (or more) on a transit system, the choice is clear: drive. Most drivers will not consider the amortization or maintenance costs of their cars in coming to a decision. They will argue that those costs must be paid even if the car sits in the garage; they believe that only the fuel costs should be

[2]Thomas E. Lisco, "Mass Transportation: Cinderella in our Cities," *Public Interest*, Winter (1970), pp. 52–74.

Figure 11.11   The MacArthur Station of the Bay Area Rapid Transit (BART) system in Oakland, California, showing a BART train.

compared to the transit system cost. In addition, most communities are forced to require the transit system to pay for itself. It cannot do this because it doesn't have a large enough ridership, in part because its fares are too high. A vicious circle is created that is difficult to break without intervention from some level of government.

Again, as has been the case so often in our energy–environment studies, we see that sociological, psychological, economic, and political considerations far outweigh the technical aspects of the problem. There are, however, some technical features that we should consider. The mathematical analysis of traffic flows in recent years has been successful in providing transit system designers with data helpful in optimizing designs. Several radical departures in transit vehicle designs have been suggested; we should examine the technical aspects of these proposed vehicles. And finally, since there does not seem to be any way to abandon the automobile, we should look at it in detail to see what factors affect its performance and how it may be improved in the future.

Traffic Flow

The question confronting applied mathematicians was, "Can traffic flow, q vehicles/s, be written as a function of some parameters λ_i that can be controlled by designers of streets and highways so that q can be optimized?" In other words, if $q = q(\lambda_i)$, find λ_i such that

$$\sum_i \frac{dq(\lambda_i)}{d\lambda_i} = 0. \qquad (11.16)$$

The average traffic flow q can be defined as

$$q = kv, \qquad (11.17)$$

where k is the concentration in vehicles per unit length of highway, and v is the speed. Some success in answering this question has been achieved[3] by assuming the concentration k can be written as the sum of the concentrations of moving and stopped cars.

$$k = k_r + k_s. \qquad (11.18)$$

This has been called a "two-fluid model" because it is in a sense similar to Bose–Einstein condensation, in which at low temperatures a gas separates into two components, one being ground-state molecules and the other molecules in excited states.

In Bose–Einstein condensation the thermal energy of the excited molecules is proportional to a power of the fraction of those molecules present. Similarly, we can propose that the speed of the moving vehicles should be proportional to the fraction of vehicles that are moving.

$$v_r \propto f_r^n \qquad (11.19)$$

or

$$v_r = \beta(1 - f_s)^n, \qquad (11.20)$$

where β is a constant of proportionality having units of speed, f_s is the fraction of stopped vehicles, and n is a number to be determined empirically.

The average speed of all vehicles on a highway depends on the average speed of the running cars and the fraction of running cars:

$$v = v_r f_r, \qquad (11.21)$$

so that Eq. 11.21 becomes

$$v = \beta(1 - f_s)^n f_r. \qquad (11.22)$$

In order to obtain a final expression that depends only on parameters related to the highway, we need a last piece of information: at vehicle concentrations above some critical value k_{max} traffic jams will occur. We expect the fraction of stopped cars to depend on this jamming concentration in a fashion that reflects the design of the road system. Let

$$f_s = \left(\frac{k}{k_{max}}\right)^p, \qquad (11.23)$$

where p is a measure of the "quality" of the design.

Equation 11.22 can be solved for f_s and that result can be used with Eq. 11.23 to derive the concentration k:

$$k = k_{max}\left[1 - \left(\frac{v}{\beta f_r}\right)^{1/(n+1)}\right]^{1/p}. \qquad (11.24)$$

This can then be used in our starting equation for traffic flow, Eq. 11.17, to produce an equation that represents a family of curves relating the traffic flow with average speed:

$$q = k_{max}v\left[1 - \left(\frac{v}{\beta}\right)^{1/(n+1)}\right]^{1/p}. \qquad (11.25)$$

For a given set of β, k_{max}, n, and p the speed required for a maximum flow can be determined.

Example 11.8 If $p = 1$ and $n = 2$, in terms of β and k_{max} what value of v maximizes q? Set $dq/dv = 0$. Then

$$\frac{dq}{dv} = k_{max} - \frac{4}{3}\left(\frac{k_{max}}{\beta^{1/3}}\right)v^{4/3} = 0,$$

$$\boxed{v = \left(\frac{3}{4}\right)^3 \beta.}$$

There is no dependence on k_{max}.

---

[3] I. Prigogine and R. Herman, *Kinetic Theory of Vehicular Traffic*. New York: American Elsevier, 1971.

If the parameters of Eq. 11.25 can be easily related to the physical design of highway system, then it should be possible to determine traffic patterns in advance of construction, minimize construction costs, and save fuel. Unfortunately, the clear-cut relationships one would hope to find have not yet emerged; work continues by many researchers. Assuming our society continues to believe this effort is worth supporting, we can expect to see results in the near future.

Mass-Transit Systems

The objectives of mass-transit studies have changed somewhat since the late 1960s, when many were initiated. In those days, getting automobiles off the streets to alleviate air pollution was more important than saving energy. Of course, then as now, producing a system that would be attractive and save the consumer money was also very important. The high-technology solutions that we referred to earlier are products of these studies. One example is the "people mover."

This was to be a series of small, four-passenger cars moving along a computer-controlled, usually elevated guideway. These cars would move at low speeds along a one-way path and would enable users to travel virtually from their homes to offices in reasonable times. However, in terms of passenger distance per unit of fuel, these could not possibly compete with the higher-passenger-density devices, like trains and buses. It is not clear they could ever have become economically competitive, either. The few full-scale demonstration projects that were built were never satisfactory—from a technical or an economic standpoint. No new ones have been constructed in 10 years, and none is on the drawing board, although studies continue to be made.

There are really three kinds of transit systems we must consider; they are quite different in terms of performance and vehicles required:

- Urban passenger transport.
- Intercity high-speed transport.
- Intercity low- to medium-speed transport.

The alternative vehicles to provide for these transport requirements are listed in Table 11.9. We see that not every vehicle could be used in each of the three types of transit systems.

From the standpoint of minimizing overall energy use, passenger-kilometers per liter is an appropriate measure of merit. Even though the automobile is rather efficient compared to the other vehicles in terms of liters per kilometer, it carries very few passengers. We have probably been excessively generous in assigning an average value of four passengers. The typical commuter trip averages slightly more than one passenger. Of course, by the same token, neither is an intercity bus that has only a few passengers operating very efficiently.

From Table 11.9 we see that only a few vehicles can provide the speed most people desire for intercity business travel, namely, airplanes. Airplanes, however, are by far the most energy-inefficient transportation vehicles. A new generation of wide-body planes is now on the drawing board; these promise to be somewhat better in terms of passenger-kilometers per liter, but they will never approach the values obtained by trains and buses. And, in the case of a TACV (tracked air-cushioned vehicle) the speed is not all that bad.

Example 11.9 Compare the travel times between mid-city Chicago and Los Angeles via regular jet and TACV (tracked air-cushioned vehicle).

The road-atlas distance between Chicago and Los Angeles is 3305 km. An airplane would be able to fly a more or less straight-line distance, which is about 2200 km. A TACV would have to travel somewhat farther, say, 2500 km. The travel times in the two cases would be

TABLE 11.9 Transportation Vehicle Alternatives

	Speed (km/hr)	Number of Passengers	Fuel Usage (ℓ/km)	Passenger-km/ℓ
Supersonic transport	3200	120	35.3	3.4
Regular jet (707)	480	130	9.4	14
Jumbo jet (747)	640	220	14.1	16
TACV	320	100	5.6	18
Overnight train	145	100	4.7	21
Air bus	640	3501	11.8	30
Automobile	80	4	0.1	40
High-speed train	160	250	4.0	62
Commuter train	60	500	5.9	85
Intercity bus	50	70	0.4	175
Suburban train (two-deck)	60	1400	3.4	412
Bicycle[a]	10	1	2.59×10^{-4}	3.85×10^4

[a]Fuel usage is the required energy equivalent.

Plane: $\dfrac{2200}{480} = 4.6$ hr,

TACV: $\dfrac{2500}{320} = 7.8$ hr.

But to these times must be added the travel time to mid-city at each end. In Chicago this would be about an hour, and in Los Angeles about 0.7 hour. However, this would be the case for the plane only, since the train would leave and arrive at mid-city. Therefore, the ratio of plane to TACV time would be (4.6 + 1.7)/7.8 = 0.81. With the rather considerably greater comfort possible in a TACV as compared to a plane (not to mention the difference in food!), it doesn't seem unreasonable to assume that a large segment of business travelers would opt to a fast, reliable TACV or equivalent, if such were available.

Conventional train speeds are limited by the ability of the cars to remain on the tracks at high speeds, around curves, and over minor bumps and misalignments. This limitation is fundamental and stops the tracked train at about 250 km/hr. The Japanese National Railway Tokaido line travels routinely at 209 km/hr, but maintenance must be done on the entire track each night to forestall track alignment problems. British Rail's High Speed Trains (HST) will travel at 240 km/hr, but they have not yet been placed in routine service. These are lighter and smaller than conventional trains.

The only way to increase the speed of a surface vehicle beyond about 250 km/hr to a useful value in terms of mass transit is to eliminate the contact between the vehicle and the track by some sort of suspension system. Three have been studied: air cushion, magnetic levitation, and dynamic levitation.

The TACV is an example of an air-cushion-supported system; it is shown schematically in cross section in Figure 11.12. There are other designs that differ in the configuration of the guideway. The air is forced through channels in the car body and ducted through an "air bag," the cushion, into the guideway. Air pressure supports the weight of the vehicle and propulsion can be provided by a variety of

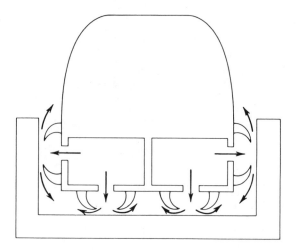

Figure 11.12 A schematic drawing of a tracked air-cushioned vehicle (TACV).

techniques: rocket boosters, fans, or linear induction motors. The primary drawbacks to TACVs are the need for secondary suspension to smooth out irregularities in the guideway, wear and tear on contact points, which inevitably occurs, significant drag caused by the fact that the expelled air does not move relative to the vehicle, some lack of stability at high speeds, and excessively high-quality guideway requirements. The British, French, and

Americans were at one time engaged in air-cushion vehicle research, and several test vehicles were built. But the limitations of this support technique have been realized, and research has turned in other directions.

There are two varieties of magnetic levitation: attractive and repulsive. These are illustrated in Figure 11.13. In attractive levitation, electromagnets energized in the vehicle cause an attractive force between the vehicle and steel rails. While in motion a sensor measures the air gap and provides a feedback signal that is used to correct the electromagnet current. About 2 kW of power per tonne of vehicle is required to maintain levitation for up to about 2 cm of air gap between magnets and rail. However, this average power must be much higher as the amplitude and frequency of required adjustments increase. As a consequence, very accurately aligned and smooth guiderails are required. It has become clear in recent years that the trade-off between cost and speed is not a very good one, and that attractive magnetic levitation cannot be used in practice.

Repulsive levitation requires very large magnetic fields to maintain a reasonble separation distance between the vehicle and the guideway. The only "practical" method of

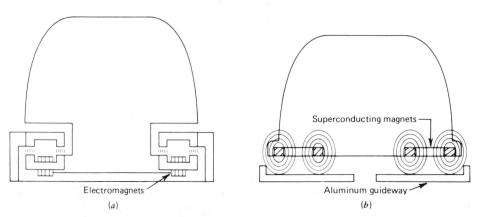

Figure 11.13 Magnetic levitation. (*a*) Attractive support, (*b*) repulsive support.

producing these fields is to use superconducting magnets in the vehicle. Estimates are that about 37 kW/ton would be required. Both magnetic suspension schemes suffer from eddy current losses. Eddy currents are induced in conducting materials by changing magnetic fields that penetrate them. These currents produce I^2R losses, and this lost energy must be supplied by the magnetic field.

An interesting alternative to air-cushion and magnetic levitation is dynamic levitation. In this scheme an air cushion is formed by the vehicle in a kind of low-altitude flight. The cushion, called a ram cushion, can be formed by the motion of the vehicle or by a turbine-powered fan. In this system the guideway becomes very simple and need not be particularly massive, as shown in cross section in Figure 11.14. Hinged "wings" are used to adjust the flow of air from beneath the vehicle to maintain the correct operating pressure. The ram air cushion can be developed either by a forward-mounted fan powered by a gas turbine, or by forward motion caused by a linear induction motor. In the former case the estimate is that only about 3 kW/tonne would be required! This fact, coupled with the relatively simple and inexpensive guideway, makes ram air cushion highly competitive with current transportation, even short-haul aircraft. Except for the very large initial capital investment

required, such a system would be practical today; as fuel costs continue to rise, we may see ram air-cushion systems coming into use for certain heavily traveled intercity corridors, such as Washington–New York and Los Angeles–San Francisco. The various suspension schemes are compared in Figure 11.15.

The seriousness of the transportation energy picture has triggered some exotic ideas. For example, there is the gravity vacuum transport (GVT). This is a train that is pulled by gravitational forces downward through an evacuated tunnel. Inertia carries the train upward to the next station, as shown in Figure 11.16. It has been estimated that a total length of travel of 12.1 km could be accomplished in 3 min, with a

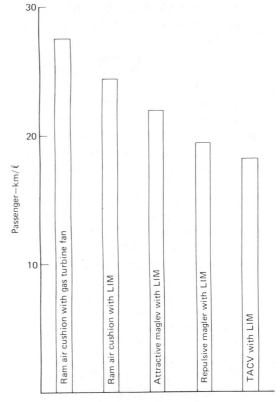

Figure 11.15 A comparison of the performance of several TACV suspension schemes in terms of passenger-kilometers per liter of fuel.

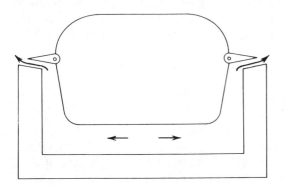

Figure 11.14 A schematic diagram of a tracked air-cushion vehicle supported by a ram cushion.

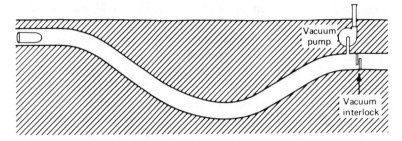

Figure 11.16 The gravity vacuum transport (GVT) scheme. A passenger vehicle falls in an evacuated tunnel to gain speed, then coasts up to the next station.

vertical drop of about 900 m. Since speeds in excess of 240 km/hr would be reached, the question of rails versus guideway plus suspension would have to be solved. Of course, the biggest problem is drilling the required tunnel; the technology does not exist to do this economically.

Another idea is the development of extra-large lighter-than-air craft, or dirigibles (the original French word means "steerable"). These craft would be similar to the famous Zeppelins of the 1930s, including the infamous Hindenberg, except that they would be three to four times larger and would use helium not hydrogen. Such airships, while slow, being capable of speeds in the range of 160 to 320 km/hr, would nevertheless have the ability to carry very large loads, which cannot now be easily accommodated on normal aircraft, trains, or trucks. They would also permit site-to-site loading and delivery. They could even launch and retrieve aircraft. And transoceanic travel could be very luxurious. Costs, however, are completely unknown; few entrepreneurs are willing to risk capital under these circumstances, and the government is not particularly interested in this technology, having been burned once on the SST. The great airships may yet return, but for now it doesn't appear likely.

The development of high-speed surface transportation will help relieve some of the

energy pressure on our society, but no real progress will be made until either the use of private automobiles decreases significantly, which seems unlikely, or the automobile itself is changed into a rather more efficient vehicle than it has traditionally been, a more achievable goal.

Automobiles

Automobile travel accounts for more energy use by far than any other single activity in this country. Figure 11.17 charts the growth of automobile mileage as a function of time after World War II. Except for a short period following the oil embargo of 1973–1974, automobile mileage has increased steadily. When this is coupled with the average performance of automobile engines (discussed in Chapter 4), it becomes clear that the first statement of this paragraph may indeed be correct. It also becomes clear that there are only two alternatives for changing this situation:

1. Reduce the number of miles driven.
2. Improve automobile performance.

Since automobile ownership and use is as much, perhaps more, of a psychological and sociological problem as it is a technical one, there may be very little that can be done to reduce auto usage. We have dicussed some of the strategies for the first of these: mass tran-

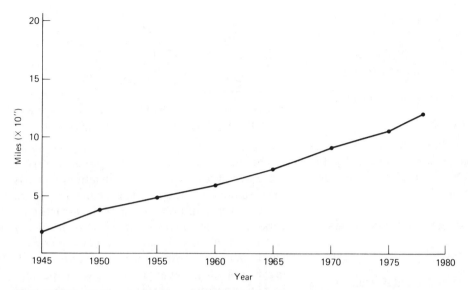

Figure 11.17 U.S. passenger automobile mileage since 1945.

sit, people movers, and others. The success of this approach has varied from slight to none. To be fair, very few major mass-transit systems have been built in the past 70 years. Time may show these to be highly successful. And since gasoline demand has been much more elastic than imagined, improved performance can do a great deal to reduce our overall energy demand.

Automobile performance can be improved in several ways:

- Improve aerodynamics.
- Improve accessories.
- Reduce weight.
- Improve current power plants.
- Replace current power plants.

Some of these improvements represent rather large changes; some are small. It is a fact of life that only small changes are easily implemented in our economy. As a consequence, in recent years we have seen improvement in weight (i.e., size), some improvement in accessories, and some improvement in the power plant itself. We have

not seen improved aerodynamics, especially in trucks (this has an impact on styling), nor have we seen a movement to radically different power plants.

We can easily discover the penalty we are paying for not implementing some of these improvements. The Otto-cycle efficiency that we calculated in Chapter 4 was about 50 percent. We knew then, however, that real engines do not achieve this idealization; we said perhaps half of this figure. In Chapter 4 we treated the internal combustion engine as a thermodynamic device. Here we have to connect that device to the transmission, drive train, and tires of a real automobile to discover the real efficiency of conversion of fuel to propulsion.

To do this let us examine the forces on a car in level motion, Figure 11.18. The forward force F_t is provided by the friction between tires and road surface. It must be true that

$$\sum F = F_t - F_d = m\,a. \qquad (11.26)$$

In level motion at constant speed the ac-

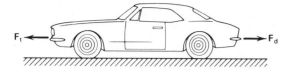

Figure 11.18 Forces acting on an automobile in motion.

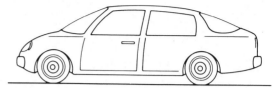

Figure 11.19 An automobile shape that would be very nearly perfect in terms of aerodynamic drag.

celeration is zero, so the forward force provided by the engine is just balanced by the total drag force. This total drag is a combination of friction in the tires and aerodynamic drag.

The aerodynamic drag force is approximated by

$$F_{ad} = 0.65 C_d A v^2 \text{ N}, \qquad (11.27)$$

where A is the projected frontal area of the vehicle, and C_d is a drag coefficient that depends upon shape. This coefficient is listed in Table 11.10 for several shapes and types of automobiles.

While it is not possible for an automobile to have the best possible shape, that of a teardrop, it is possible to produce shapes that are substantially better than those now available. Figure 11.19 is a sketch of such an automobile; this shape auto would have a drag coefficient of only 0.2, better than a factor of 2 improvement over most automobiles now on the market. This would not mean a factor of 2 improvement in losses, because other effects are present, of course.

TABLE 11.10 Aerodynamic Drag Coefficients

Shape	C_d
Teardrop shape (best possible shape)	0.03
Potential automobile	0.20
Porsche 904	0.35
Jaguar XK-E	0.40
Typical U.S. auto 1960–1980	0.50
Typical U.S. station wagon 1960–1980	0.60
Ordinary truck	0.7
Square flat plate (worst possible shape)	1.3

The major difficulty in attempting to market more aerodynamically efficient automobiles is the problem of styling. The shape in Figure 11.19 is very different from typical American cars, although not so different from many European models. Whether or not Americans would accept such a radical departure in styling is not clear, even if the new shape delivered improved performance. Streamlined European cars, like the Peugeot, have not sold that well in this country; but initial cost and servicing difficulties may be more important than life-cycle operating expenses. Perhaps this is not a good measure of consumer acceptance.

Example 11.10 How much energy is expended against aerodynamic drag in 1 hr at 100 km/hr by a truck?

From Eq. 11.27

$$F_{ad} = 0.65 C_d A v^2,$$

and

$$\text{Power} = Fv,$$

$$\text{Energy} = \text{power} \times \text{time},$$

$$E = F_{ad} v t = 0.65 C_d A v^3 t$$

$$= 0.65 \times 0.7 \times (3 \times 2.7) \times \left(10^2 \times \frac{10^3}{60 \times 60}\right)^3$$

$$\times 60 \times 60 \qquad (A = 3 \times 2.7 \text{ m}^2)$$

$$\boxed{E = 2.8 \times 10^8 \text{ J} \qquad \text{for 1 hr.}}$$

Since the energy content of diesel fuel is about 1.1×10^8 J/gal, this would require about 2.5 gal. The energy required to overcome tire resistance would be somewhat less, say equivalent to 1.9 gal. Overall, 4.0 gal of diesel fuel would be required for 100 km (62 miles), or 14.1 miles/gal.

Rolling resistance comes about from tread slippage on the road, air resistance, and internal flexing losses. It can be calculated using experimentally derived parameters, but a parameterization is correct only for a specific type of tire. This parameterization is reasonably good for new tires and speeds above 20 m/s:

$$F_r = \left(\frac{M}{1000}\right)\left[118 + 1.1v + 0.2(v-20)^2\right] \text{N},$$

$$(11.28)$$

where M is the mass of the vehicle in kilograms, and v is the speed in meters per second. Figure 11.20 is a plot of the rolling resistance, the aerodynamic drag, and the constant speed power required for a large luxury car ($M =$

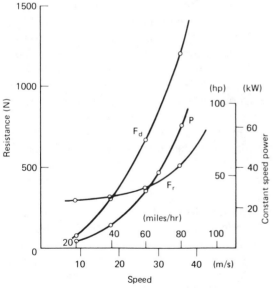

Figure 11.20 Rolling resistance F_r, aerodynamic drag F_d, and constant-speed power P required for a 2270-kg car as a function of speed. F_r and F_d use the left-hand vertical axis; P uses the right-hand axis.

2270 kg) as a function of speed. The constant speed power is simply the product of the speed and the total force, $F_r + F_d$; this is the power that must be delivered to the tires simply to maintain a constant speed in level motion.

We see from this figure that to maintain a high constant speed in level motion a significant amount of power must be delivered to the wheels. This is *not* the power developed by the engine, as some power is lost. The "brake horsepower" (bhp) is the amount of power delivered to the flywheel by the engine. In level, constant-speed motion the bhp must be about 3 percent higher than that required at the wheels to account for losses in the transmission and axles. In addition, required and optional accessories can absorb a great deal of power. The air pump, cooling fan, and water pump required by the engine will absorb from 5 to 15 hp, depending on engine speed. Air conditioning and power steering can each account for between 4 and 8 hp, depending on speed and temperature difference. For 30 m/s (67 miles/hr) level motion about 37 kW (51 bhp) is required at the wheels. This could translate to as much as 48 kW (65 bhp) required at the flywheel, when these losses are taken into account.

This discussion so far has not taken into account accelerations, coasting and idling, and energy lost by braking. When these effects are averaged into a typical driving trip, the power required at the flywheel, on average, is about 1.5 the power required to drive the wheels.

The variation in available power as a function of engine speed gives an interesting insight into automobile operation. We have mentioned three engine devices that require power that varies with engine speed: air pump, cooling fan, and water pump. There is another "device," piston friction. Even in a well-lubricated engine there is piston friction that increases with piston speed and with temperature. If we call the power lost to this effect **friction power**, then Figure 11.21 shows that

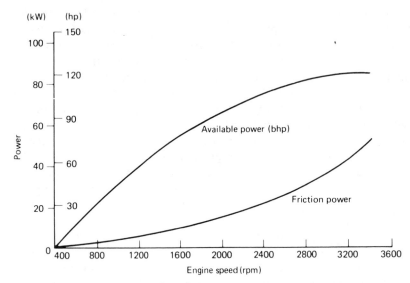

Figure 11.21 Available power (bhp) and friction power versus engine speed in revolutions per minute (rpm).

the available power does not vary linearly with engine speed, because of the power lost to friction.

We can think of available power (bhp) as being the difference between the power developed by the engine—**indicated power** (ihp)—and the friction power (fhp). Then a **mechanical efficiency** can be defined in terms of these:

$$\eta_{mech} = \frac{bhp}{ihp} = 1 - \frac{fhp}{ihp}. \qquad (11.29)$$

A plot of this mechanical efficiency versus engine revolutions per minute (rpm) would have a maximum at some intermediate value of rpm. This is a result of the fact that at low rpm the indicated power developed by the engine is small, while the friction power tends to a low constant value. At high rpm the indicated power is high, but then so is the friction power. This mechanical efficiency is directly proportional to the torque developed by the engine.

The ideas presented so far have been for "full throttle"; that is, they assume no restriction in air flow to the carburetor. The reduction

in engine rpm while maintaining a high mechanical efficiency is accomplished by gearing down with the transmission. The solid line in Figure 11.22 for full throttle indicates that the transmission must be continuously variable in gear ratio to produce the continuously variable curve. Recall that engine torque and mechanical efficiency are directly related. If engine speed is reduced by constricting air flow in the carburetor—letting up on the throttle—then the situation is quite different.

An automobile traveling at full throttle in high gear on a level road would be operating at point A on Figure 11.22. Slowing down by constricting the air flow, but staying in high gear, would cause the operation to shift to point B. At each point along this dashed line the power required at the wheels is reduced and therefore so is the power required at the flywheel, the bhp. In fact, since the available power is the product of engine torque and engine speed, we would expect the available power to consist of a family of hyperbolas on this plot, as indicated by the light lines.

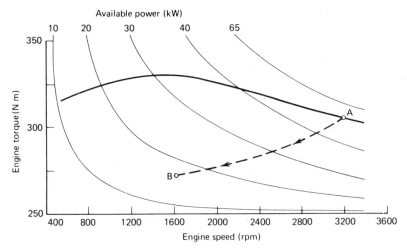

Figure 11.22   Engine torque versus engine speed in revolutions per minute. The solid curve is for "full throttle" and continuously variable transmission.

Point B is in a region of lower mechanical efficiency; the fuel consumption of the automobile must increase. The overall efficiency of operation must then decrease, since less energy is being produced, but more fuel is being used.

Operation at partial load, which is the situation most of the time in urban driving, accounts for a great deal of the power losses in the automobile. One way of overcoming some of these losses is to make sure that the engine is always matched as closely as possible to the load. Manual transmissions help, but automatic transmissions have some very nice features that enable operation at high speeds in an efficient manner. A better design would be one that has a continuously variable transmission under microcomputer control, which could select the best position on the diagram (Figure 11.22) and adjust the automobile accordingly. Such systems may be available in the near future as options.

Partial-load losses coupled with high-speed losses produce vehicles for which the fuel economy is a very sensitive function of speed. Overall automobile performance taking into account partial-load losses and drag losses has a maximum between 35 and 45 miles/hr (56 and 72 km/hr), as indicated in Figure 11.23. The upper curve is an estimate for a 1980 mid-sized car; obviously, smaller cars would do some-

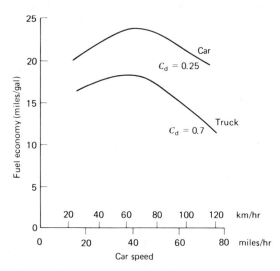

Figure 11.23   Fuel economy (in miles per gallon) versus speed for a typical 1980 car and truck.

what better and trucks rather considerably worse, depending on weight.

It is clear that the 55-miles/hr federal speed limit is a compromise between performance and convenience. It must also be clear that there are savings to be made in energy consumption by adhering to the speed limit. The statement by many, particularly truck drivers, that their vehicles operate most efficiently at high speeds is really quite absurd.

The question of weight is very significant. It has triggered a trend in the late 1970s and early 1980s that may be indicative of the future of the automobile industry in this country. In 1975 the federal government passed the Energy Policy and Conservation Act, which enabled the Department of Transportation to set performance standards for automobiles sold in this country beginning in 1978. The standards for the first three years were set, and in 1977 standards were set for all years up to 1985. These performance standards are listed in Table 11.11. Through the 1981 model year, all U.S. automobile manufacturers have been able to meet these standards, which apply to the "fleet" produced by each automaker, that is, a weighted average over all sizes of cars produced. Auto manufacturers made a large number of improvements in engines and accessories in an effort to meet the standards, but in actual fact only one improvement was

TABLE 11.11 Mandated Automobile Performance Standards

Year	Required Miles per Gallon
1978	18.0
1979	19.0
1980	20.0
1981	22.0
1982	24.0
1983	26.0
1984	27.0
1985	27.5

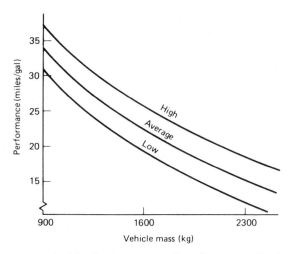

Figure 11.24 Fuel economy (in miles per gallon) versus vehicle mass for high-, average-, and low-weight cars, corresponding roughly to full-sized, mid-sized, and compacts.

able to turn the trick: reducing automobile weight.

There is a very clear correlation between performance and weight, as shown in Figure 11.24. Subcompacts are in the weight range 900 to 1300 kg, compacts about 1300 to 1600 kg, mid-size from about 1600 to 1900 kg, and full-sized cars are heavier. The curves in Figure 11.24 are only averages; the weight of a particular model may be less or more than that indicated for its class. Weight can be and has been reduced by substituting plastic and aluminum for steel, where possible. But in order to achieve the weight reductions required to meet the mandated fleet averages, it has been necessary to produce *and sell* smaller cars. This is fortuitous, of course, since in the past few years many Americans have recognized the virtue of driving smaller, more fuel-efficient cars. But even with very clear message from consumers in the 1970s and with the determination of the federal government in the form of mandated fuel efficiencies, it seems to have required a virtual depression in auto sales to convince the automakers that the produc-

TABLE 11.12 Some Conservation Strategies

Area of Strategy	Potential Savings (%)	Notes
		Homes, Buildings
Space heating	5–8	Insulation; heat pumps cut electric heating needs by 50%; gas heat pump is a possibility
Air conditioners	>1	Save peak power, fewer brownouts; insulation, design, window improvements reduce heat load
Home appliances	2	Fluorescent lights, better motors, and insulation in refrigerators; insulation in water heaters, electric igniters replace pilot lights
Design of buildings	5	Includes redefining lighting levels and tasks; total energy systems, conserving window systems; orientations that reduce energy needs
Solar heating/cooling	10	If 40% of today's heating and cooling were solar; economics depends on cost of glass, storage, and alternative fuels
		Industry
Process heat	5–12	Once through the facility is sufficient, with insulation, leak plugging; more sophisticated treatment requires redesign, pipes, cascading high-temperature processes with lower-temperature demands
Total energy: cogeneration of electricity and heat at factory	3–5	Energy independence to factories; siting communities near industry is a possibility
Returnable bottles, use of recycled materials	1–3	Many institutional problems as "no deposit, no return" becomes ingrained
		Transportation
100% shift to 40% lighter cars	5	
More careful driving cycle	1	Appreciable savings in energy cost of building car, refining oil; less congestion, less wear and tear, less pollution, with less traffic
Improved technical efficiency of autos	5	
Switch one half of urban passenger miles to bus	2	
Improved load factors in rail, bus, plane, mass transit	2	Savings in freight and other passenger modes come mainly from fuller utilization of existing routes and higher load factors
Freight mode mix, improved technical efficiency	2	
		Other
More durable, repairable, and recyclable goods	?	Substitutes quality work for endless throwing away
Urban design	?	Live near work, district heating, etc.
Changes in consumer preferences	?	"Output juggling"—vacation near home, ride a bike, work in the garden

Adapted, with permission, from L. Schipper, "Raising the Productivity of Energy Utilization," *Annual Review of Energy* 1, 455 (1976), copyright © 1976 by Annual Reviews Inc.

tion of smaller cars is in their own best interest. Some things change slowly.

The change to a new power plant will also come very slowly, if at all. There are few alternatives, and these do not have the characteristics that are appealing to consumers. Electric cars, while available today, are expensive, have limited driving ranges without recharge, are very small, and accelerate slowly. There can be no doubt that electric vehicles, if a large enough number could be put into service, would help alleviate the urban air pollution problem. Of course, pollutants would be generated at the power plants where the electricity is produced; but these central plants are far easier to clean up than the individual mobile pollutant sources represented by automobiles.

Progress is being made in convincing the American public to use smaller, more fuel-efficient automobiles and in convincing the automakers to build them. Far more progress will have to be made in providing a workable mass transit alternative to the automobile in urban areas. Interstate transport of passengers should be handled by high-speed trains, but a commitment of the size of the space program is required to provide the physical facilities necessary to do it. There is no sign that the public is ready to make such a commitment.

Changes in lifestyle can easily save a great deal of energy, and there is evidence that Americans are becoming more aware of this. On the whole there is reason to be optimistic about the future of energy-conservation efforts in this country. Now that the energy fuels are selling for prices that more clearly reflect their real worth and that come a great deal closer to paying for their environmental effects, their hidden elasticity has become apparent. People now have a vested interest in saving energy: it saves them money! We can hope that many of the ideas we have discussed in this chapter will be acted upon in the very near future. These are summarized in Table 11.12.

SUMMARY

Conserving energy is a process of making the actual work expended in a process approach as closely as possible the minumum work required.

Policy often dictates the extent of energy efficiency. The construction industry, for example, could produce more energy-efficient homes. However, there is no government policy to encourage it. Often first costs are more important to builders and buyers than life-cycle costs.

Space heating can be thought of in terms of a task device and a supply device. The Second Law efficiency is very different for different combinations of devices. There are a number of strategies for improving the efficiency of space heating.

A considerable amount of energy is expended on lighting. Most offices have much more light than is necessary.

Industrial processes can also be made more efficient. The use of waste heat to preheat an air stream can result in considerable savings.

Recycling has many institutional and technical problems. Some materials, such as glass, offer little savings. Others, such as aluminum, can be recycled at substantially lower energy cost.

Transportation accounts for the lion's share of energy use in the United States. One way to reduce the energy requirements is to study traffic flow mathematically to help design better highways. The "two-fluid" theory has had some success.

Since most transportation energy is expended by commuters, improved mass-transit systems would seem to be a logical solution. However, there are many difficulties in instituting them. Most automobile drivers will not switch to a less-comfortable, more-restrictive system.

There are few economically viable mass-transit alternatives. **Tracked air-cushion vehicles** of several designs have been considered, but none are in use.

The automobile itself can be improved in order to save energy. Several effects reduce the efficiency of the automobile. Principle among these are tire friction and aerodynamic drag. Substantial improvement can be made by reducing the weight of autos.

REFERENCES

General

Berg, Charles A., "Energy Conservation through Effective Utilization," *Science* **181**, 128 (1973).

Berg, Charles A., "A Technical Basis for Energy Conservation," *Tech. Rev.*, February (1974), p. 15.

Hirst, Eric, and John C. Moyers, "Efficiency in Energy Use in the United States," *Science* **179**, 1299 (1973).

Lincoln, G. A., "Energy Conservation," *Science* **180**, 155 (1973).

Roberts, Fred S. (Ed.), *Energy Modeling and Net Energy Analysis*. Chicago: Institute of Gas Technology, 1978.

Ross, Marc H., and Robert H. Williams, "The Potential for Fuel Conservation," *Tech. Rev.*, February (1977), p. 49.

Schipper, Lee, and Joel Darmstadter, "The Logic of Energy Conservation," *Tech. Rev.*, January (1978), p. 41.

Williams, Robert H. (Ed.), *The Energy Conservation Papers*. Cambridge, Mass.: Ballinger, 1975.

————, *Efficient Use of Energy*. New York: American Institute of Physics, 1975.

Domestic and Commercial Buildings

Hirst, Eric, and Bruce Hannon, "Effects of Energy Conservation in Residential and Commercial Buildings," *Science* **205**, 656 (1979).

Snell, Jack E., Paul R. Achenback, and Stephen R. Petersen, "Energy Conservation in New Housing Design," *Science* **192**, 1305 (1976).

Stein, Richard G., "A Matter of Design," *Environment*, October (1972), p. 17.

Thompson, Grant P., *Building to Save Energy: Legal and Regulatory Approaches*. Cambridge, Mass.: Ballinger, 1980.

Industrial Energy Conservation

Berg, Charles A., "Conservation in Industry," *Science* **184**, 264 (1974).

Bever, Michael B., "Recycling in the Materials System," *Tech. Rev.*, February (1977), p. 23.

Cannon, James, "Steel, the Recyclable Material," *Environment*, November (1973), p. 11.

Harwood, Julius J., "Recycling the Junk Car," *Tech. Rev.*, February (1977), p. 32.

Kakela Peter, "Railroading Scrap," *Environment*, March (1975), p. 27.

Rose, David J., John H. Gibbons, and William Fulkerson, "Physics Looks at Waste Management," *Physics Today*, February (1972), p. 32.

————, *Resource Recovery and Utilization*. Philadelphia: American Society for Testing and Materials, 1975.

Transportation

Anderson, J. Edward, *Transit System Theory*. Lexington: D. C. Heath, 1978.

Anderson, J. Edward, Richard D. Doyle, and Raymond A. MacDonald, "Personal Rapid Transit," *Environment*, October (1980), p. 25.

Barrows, Timothy M., "Suspension Concepts for High-Speed Ground Transportation," *Tech. Rev.*, July/August (1975), p. 31.

Grey, Jerry, George W. Sutton, and Martin Zlotnick, "Fuel Conservation and Applied Research," *Science* **200**, 135 (1978).

Height, Frank, and Roy Cresswell (Eds.), *Design for Passenger Transport*. Oxford: Pergamon Press, 1979.

Herman, Robert, *Theory of Traffic Flow*. New York: Elsevier, 1961.

Hirst, Eric, "Transportation Energy Policies," *Science* **192**, 15 (1976).

Kolm, Henry H., and Richard D. Thornton, "Electromagnetic Flight," *Sci. Amer.* **229**, 17 (1973).

Laithwaite, E. R. (Ed.), *Transport Without Wheels*. Boulder: Westview Press, 1977.

McCormick, J. Byron, and James R. Huff, "The Case for Fuel-Cell-Powered Vehicles," *Tech. Rev.*, August/September (1980), p. 54.

Pierce, John A., "The Fuel Consumption of Automobiles," *Sci. Amer.*, January (1975), p. 34.

Rice, Richard, "Toward More Transportation with Less Energy," *Tech. Rev.*, February (1974), p. 45.

Rosenthal, Barbara, "The Auto Option," *Environment*, June/July (1977), p. 18.

Trzyna, Thomas N., and Joseph R. Beck, *Urban Mass Transit: A Guide to Organizations and Information Resources*. Claremont: California Institute of Public Affairs, 1979.

Vittek, Joseph F., Jr., "Is There an Airship in Your Future?" *Tech. Rev.*, July/August (1975), p. 23.

Journals

High Speed Ground Transportation Journal, Institute for Transportation, Durham, N.C.

Institute of Transportation Engineers Journal, Institute of Transportation Engineers, Arlington, Va.

Logistics and Transportation Review, University of British Columbia, Vancouver, British Columbia.

Transportation, Elsevier, Amsterdam, The Netherlands.

Transportation Journal, American Society of Traffic and Transportation Engineers, Chicago.

Transportation Research, Pergamon Press, New York.

GROUP PROJECTS

Project 1. Examine the elasticity of demand for gasoline. Interview at least two groups of consumers. At what price level will people begin to cut back on automobile usage? What will they substitute? What are their attitudes toward small cars? transit systems?

Project 2. Devise a technique for measuring the effective R-value for a house or apartment. What is τ for this dwelling? Plot measured temperature and calculated temperature versus time for different heating strategies. Compare heat losses for these strategies.

Project 3. Estimate the lighting in watts per square meter for several classrooms and offices. Compare with requirements. Estimate annual savings by reducing illuminance to reasonable levels in these rooms. Do the same for a typical home or apartment.

Project 4. Determine the performance versus weight relationship between cars of the current model year. Develop criteria for excluding certain models from the group. Estimate the savings of this group over those in 1977 before mandated performance standards went into effect.

Project 5. For computer enthusiasts: Develop a program for minimizing the distance traveled in performing a set of typical Saturday morning chores. See Richard R. Parry and Howard Pfeffer, "The Infamous Travelling Salesman Problem," *Byte*, July (1981), p. 252.

EXERCISES

Exercise 1. Estimate the total energy cost to supply every single family dwelling in the United States with aluminum siding. If this saves 10 percent of the home heating energy, what is the net energy balance over 20 years? (Aluminum siding is typically 0.061 cm thick.)

Exercise 2. Using Figure 6.33 and Table 6.21 estimate the heating savings expected from triple-glazing compared to double-glazing for a typical single-family detached home.

Exercise 3. If you lived in a $15 \times 9 \times 3$ m cave where the temperature was a year-round constant $13°C$ and you wanted to maintain a temperature of $21°C$, at what rate would heat have to be supplied? Assume an R-value of 10 for 1 m effective thickness of the cave walls.

Answer: 40.2 MJ/day.

Exercise 4. If $p_1 = \$2.22/m^2$ unit thermal resistivity and $p_2 = \$2.37/m^2$ unit thermal resistivity $°C$, what is the optimum R-value for $\Delta T = 30°C$ neglecting life-cycle costs? If $A = 450$ m^2, $a_1 = \$10$ and $a_2 = \$2000$, what is the first cost? Use the data in Example 11.1 and recalculate $R_{0)opt}$ and first cost including life-cycle costs. Compare total life-cycle costs for the two cases.

Exercise 5. Estimate p_3 for a typical house for resistive heating, oil, and natural gas in your area, and find the life-cycle heating cost in each case. Is p_2 a constant?

Exercise 6. In Example 11.2, how much total energy is required for each air change?

Answer: 4.6 MJ/air change.

Exercise 7. If the outside air is at $-5°C$ with 80 percent humidity, how much energy per unit volume is required to bring this air to $20°C$ at 40 percent humidity?

Exercise 8. A typical fireplace 7 m tall with a circular cross section heats the air inside it, say 1 m^3, to an average temperature of about $200°C$. Estimate the air outflow from the fireplace. (Assume the temperature outside is $0°C$.) Hint: refer to Example 9.4.

Answer: 1.33 m^3/s.

Exercise 9. Some people (the author among them) like to prewarm their coffee cups so that the coffee will not cool so quickly. Compare the energy requirements for this prewarming, if two alternatives are available:

1. Warm water from the tap. The water must run about 2 minutes before it is warm enough, at the rate of 1 cup per 4 s.
2. Microwave oven. A cupful of water heats sufficiently in about 45 s at 1600 W.

Also compare the actual costs; use rates appropriate for your area.

Exercise 10. The B_{min} for a Diesel-engine-driven heat pump is given by

$$B_{min} = (Q_2 + Q')\left[1 - \left(\frac{T_2}{T_1}\right)\right],$$

where Q_2 is the heat delivered by the heat pump, and Q' is the heat rejected by the engine. Derive the expression for ϵ for this arrangement in Table 11.3.

Exercise 11. Assume the interior temperature of a house falls from $20°$ to $3°C$ during an 8-hr period for which the furnace is off. If $K = 2.9 \times 10^5$ J/hr $°C$, how much heat is lost? Plot T versus time. If the furnace had been required to maintain a temperature of at least $15°C$, how much energy would have been required? ($\tau = 5$ hr)

Exercise 12. A heat pump with a COP of 3 is required to deliver heat at the rate of 15 kW. It is driven by a diesel engine having an η of 0.3. What is the rate of fuel usage of the engine, if its work output is exactly the amount required by the heat pump? What is the rate of heat rejection from the engine?

Exercise 13. If a certain heat pump has $\epsilon = 0.1$, at what temperature ratio does its COP become unity? In other words, at what temperature difference does it become useless? (Assume $T_1 = 20°$ C.)

Answer: $-9.3°$ C.

Exercise 14. Assume the first cost of an air conditioner is given by

First cost in dollars $= 0.02B + 10$ EER,

where B is the heat rate in British thermal units per hour. Calculate the life-cycle costs over a 10-yr period averaging 1000 hr/yr. Assume electricity costs 5 cents/kWh. Choose several room air conditioners ($B = 6000, 8000$, and 12,000) for EER $= 5, 6, 8$, and 10 and several central units ($B = 20,000, 22,000$, and 24,000) for EER $= 5.8, 6.0$, and 6.4.

Exercise 15. A 6000 Btu/hr 117-V air conditioner with an EER of 6.2 has $\epsilon = 0.04$. If this value were to be increased to 0.06, how much would the electrical current demand decrease for the same Q_2?

Answer: A reduction of 2.8 A.

Exercise 16. If air conditioning at $\epsilon = 0.05$ accounts for 4 percent of the total energy use in the United States, what annual savings would be obtained if the average ϵ could be raised to 0.07?

Exercise 17. What is the luminous flux of a light source for which

$$P(\lambda) = 10 \exp\left[-\left(\frac{600 - \lambda}{100}\right)^2\right]?$$

Exercise 18. Assume sunlight to be described by a blackbody with $T = 5800°$ K. Find Φ for the sun.

Exercise 19. The night response of the eye is a curve similar to that in Figure 11.5 except the peak is shifted to about 510 nm. Calculate the luminous flux of the full moon (assume the moon reflects 0.1 percent of sunlight without changing the frequency distribution).

Exercise 20. What is the approximate percentage increase in visual acuity if the illuminance is doubled from 970 to 1940 lm/m^2?

Exercise 21. What illuminance is required to resolve 0.0025-cm objects at a distance of 50 cm?

Exercise 22. A task requiring a visual acuity of 2.4 is to be illuminated by a single 100-W incandescent bulb. How far should this bulb be from the work surface?

Exercise 23. A 40-W fluorescent fixture usually has a 13.5-W ballast. What energy savings over a 12-hr day could be realized if a 5-W ballast were used instead?

Exercise 24. An office is uniformly illuminated with 40-W fluorescent tubes providing 1000 lm/m^2 illuminance. What is the total electrical power load for these lights? What power reduction could be achieved if the uniform illuminance were reduced to 100 lm/m^2 and each of 10 desks in the office given a single 40-W tube?

Exercise 25. If the starting current in a 40-W fluorescent is 1.6 A, and start-up requires on the average 5 s, for what length of time is it more economical to leave the light on rather than to turn it off?

Answer: 23.5 s.

Exercise 26. The efficacy of a 75-W fluorescent tube is about 75 lm/W. What energy savings over a 10-hr day could be obtained by providing a uniform $600 \, lm/m^2$ illuminance for an office with 75-W tubes compared to that for 40-W tubes?

Exercise 27. If every man, woman, and child in the United States were to use electric toothbrushes, how would the total power demand change?

Exercise 28. If a stew takes 10 hr to cook in a slow-cooker (low) and 3 hr on an electric range (large unit at 0.2 of full range), which method is more economical?

Exercise 29. "Permanent press" shirts require no ironing if dried in a dryer. But, if dried out of doors, each must be ironed (about 4 min each). Compare the two techniques for 10 shirts requiring 20 min of dryer time.

Exercise 30. Calculate the minimum work required to separate helium from air. Assume only He, O_2, and N_2 in the air.

Exercise 31. What is the percent savings in fuel used to heat air as a function of the percent increase in input air temperature? (That is, develop an expression for the fuel requirements for a given output temperature as a function of input temperature.)

Exercise 32. The production of aluminum in the United States since 1950 can be approximated by

$$P = a + b(t - t_0) + c(t - t_0)^2,$$

where P is the annual production in 10^6 tons, $t_0 = 1945$, $a = 1.0$, $b = 0.1$, $c = 0.0006$, and t is the year. Show that if half the aluminum produced in 1950 were recycled in 1990, less than 5 percent of the 1990 production would be obtained.

Exercise 33.
(a) Derive Eq. 11.24.
(b) Show that a reasonable interpretation of β in Eq. 11.20 is that it is the maximum average speed of moving cars.

Exercise 34. Define the total trip time per unit distance, the running time per unit distance, and the stopped time per unit distance for $T = T_r + T_s$, where $T = 1/v$, $T_s = f_s T$ and $T_r = f_r T$. Also define $T_{max} = 1/\beta$. Show that

$$T_s = T - T_{max}^{1/(n+1)} T^{n/(n+1)}.$$

Exercise 35. If all of the intercity rail passengers and two thirds of the intercity air passengers in the year 2000 travel on ram air-cushion devices powered by turbines, what energy savings could be expected? Extrapolate from 1980 using a 2 percent/yr growth rate.

Exercise 36. At what mid-city distance does the total TACV travel time equal regular jet total travel time?

Exercise 37. What cargo could be carried by a helium-filled airship having a volume equivalent to a cylinder 30 m in diameter and 800 m long, if the structural mass is 5 metric tons/m of length?

Exercise 38. An 2270-kg automobile can climb a 4 percent grade at 95 km/hr. What force can it deliver to the tires at that same speed in level motion?

PART FIVE

ENVIRONMENTAL DEGRADATION

CHAPTER TWELVE

GLOBAL ENERGY BALANCE

In the preceding parts of this textbook we have discussed some of the interactions between society and its environment and the reasons for those interactions. Many of the results of our dependence on energy are obvious: air pollution, water pollution, thermal pollution, and noise pollution, to name a few. The short-range, local consequences of these effects have received much attention; similarly, we must worry about more long-range, global processes that can have a profound effect on mankind. Among the more serious questions in this regard is that of the earth's heat balance. If we can apply physics to this as well as to other problems, perhaps we can discover a means of ameliorating them. All of these basic environmental problems can be grouped together logically under the heading of "environmental degradation," since this title more clearly describes the situation. The fact that we dump polluting materials on the earth is not in itself dangerous; the earth has remarkable recuperative powers. What must concern us is how far the degradation of the earth's recuperative processes can be carried before those processes begin to fail. The approach we take is to attempt to determine the equilibrium situation and from that estimate the effect of human activities in terms of the factors upon which equilibrium depends. Per-

haps the most important concern for mankind, just now is the question of the earth's energy balance, since it is a very fragile balance indeed.

12.1 NATURAL ENERGY SOURCES AND SINKS

Figure 12.1 is a diagram of the natural energy sources and sinks for the earth. Note that there are only two sources: the sun and the earth's natural radioactivity. Of course, if human activities were to result in the deposition of an amount equivalent to a significant fraction of either of these, they too would have to be counted. The energy input drives the two heat engines, the atmospheric and ocean currents, to transfer heat from the equator to the polar caps. At equilibrium the solar energy absorbed by the earth should eventually be radiated to space; we may write this:

$$(1-k)\frac{S}{4}(1-\alpha) = \sigma\epsilon T^4, \qquad (12.1)$$

where S = solar flux—8.37 J/cm^2 min, k is the fraction absorbed by the atmosphere, α is the fraction reflected by the earth (albedo), σ is the Stefan–Boltzmann constant—5.67×10^{-8} J/m^2 K^4, ϵ is the effective emissivity of the earth, and T is the average temperature in

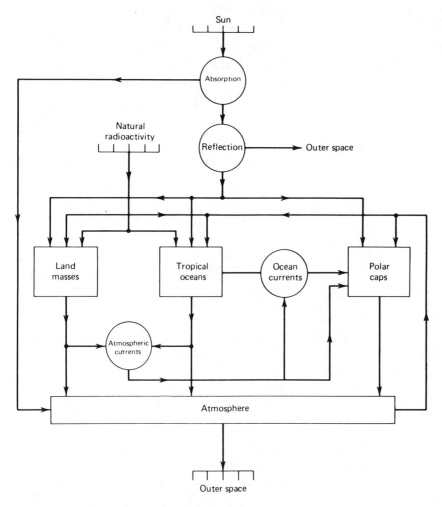

Figure 12.1 Natural energy sources and sinks.

kelvins. (Note: the $\frac{1}{4}$ term appears because the solar flux must be averaged over night and day and all the surface area.)

Example 12.1 What is the effective radiating temperature of the earth?

For the earth the product $\sigma\epsilon$ is about 5.8×10^{-12} J/cm^2 s K^4, $k = 0.25$, and $\alpha = 0.15$.

$$0.75 \times 0.8 \times 10^{-3} \times 0.85 = (1.4 \times 10^{-12})T^4,$$

$$\boxed{T = 245.7 \text{ K} = -27.5° \text{ C.}}$$

Of course, this is not the temperature of the earth. But it is the approximate temperature of that part of the earth that radiates to the sky, namely, the stratosphere.

We believe that at present there exists a balance between the amount of heat arriving at the earth given by the left-hand side of Eq. 12.1 and that being radiated away to space, the right-hand side. That is to say, we think the earth is becoming neither cooler nor hotter—it is in equilibrium.

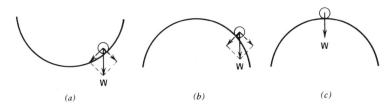

Figure 12.2 Three types of equilibrium: (*a*) stable, (*b*) unstable, and (*c*) neutral.

There are three kinds of equilibrium, and they are illustrated in Figure 12.2. In **stable equilibrium**, restoring forces exist that would tend to restore equilibrium if a perturbation occurred. In **unstable equilibrium**, forces exist that tend to take the system farther away from equilibrium. In **neutral equilibrium**, there are no restoring forces, and no motion of the system so long as there is no perturbation; but any change will cause instability.

Because of the rather large mass of the earth and the slowness with which heat transfer occurs, it is not possible to judge the state of equilibrium by simply measuring the temperature at various locations over a short time period—accurate meteorological records have been kept for only about the past 100 years. Geologic records indicate large climatic variations, oscillations around a stable situation, perhaps. Is the earth still in a climatic cycle, or have the oscillations damped out? If the latter is true, how much change in heat exchange with the rest of the universe would be required to trigger instability? Would a small displace-

ment away from thermal equilibrium result in restoring forces or greater displacement? Would an increase in the earth's average temperature cause more cloudiness, thereby increasing the albedo and thereby decreasing solar energy absorption? Or would it increase the CO_2 content of the atmosphere, causing more IR absorption, thereby increasing the earth's temperature? Perhaps small perturbations are allowed and are recoverable, but large ones are not, as shown in Figure 12.3. But how large is large? The answers to these questions are not known. It has been estimated that a decrease of 1.6 percent in solar radiation reaching the earth's surface would result in a freezing of all the oceans; there are three means by which such a change could result: (1) decrease in solar radiation leaving the sun, (2) increase in absorptivity of the atmosphere, or (3) increase in reflectivity of the earth. It is not believed that any one of these or any combination would produce a change as high as 1.6 percent, but we really do not understand the overall picture well enough to conclude that with absolute certainty.

Energy Sources

The largest single source of energy on the earth is, of course, the solar radiation flux of 1.4 kW/m^2 (measured at the top of the atmosphere). This value is, of course, an annual average; because of the earth's elliptical orbit the figure will vary seasonally. There are also long-period variations to the earth's orbit.

Figure 12.3 A possible description for the equilibrium of the earth. Stable equilibrium will result so long as the displacements from equilibrium are small.

There is a cyclic variation in the ellipticity with a period of about 105,000 yr, a variation of the direction of the major axis with a period of about 21,000 yr, and a variation of the tilt of the axis relative to the orbital plane with a period of about 40,000 yr. These variations are small and even combined without a nonlinear feedback mechanism cannot produce a major change in the received solar radiation; it is interesting, however, that the frequencies of these variations coincide roughly to the glacial cycles. There are short-period fluctuations in the sun as well; the 11-yr solar sunspot cycle is a familiar example. The effect of this cycle on the solar constant is not known at present. There are some very interesting speculations about solar cycles and terrestrial climate (see the references for this chapter). It may even be that the current stage of solar activity may be rather new, beginning at the end of the Maunder Minimum[1] in about 1750. Satellite data or a solar monitoring station on the moon may shed light, so to speak, on this question.

The solar flux is distributed in wavelength approximately like that given by a blackbody of temperature 6000 K, since this is the temperature of the photosphere of the sun. There are missing wavelengths in the solar spectrum as observed at the surface of the earth because of the absorption bands of the constituents of the atmosphere, principally O_3 (ozone), CO_2, and H_2O.[2] The spectrum is reproduced in

[1]From 1645 to 1715 very few sunspots were reported. This has led to many speculations about solar activity and terrestrial weather during that period. See, for example, John A. Eddy, "The Maunder Minimum," *Science* **192**, 1189 (1976).

[2]Do not confuse these absorption *bands* with Fraunhofer *lines* caused by atomic absorption in the sun's photosphere. Molecules and atoms absorb energy by similar mechanisms, predicted by quantum mechanics, but the essential difference is that allowed molecular absorption wavelengths cover entire bands. In fact, O_3 effectively absorbs all wavelengths shorter than about 0.3 μm, the ultraviolet region, keeping earthlings from being fried each time they step out of doors.

Figure 5.6. There is only one other natural source of heat for the earth, and that is heat from within the earth resulting from radioactive decay of naturally occurring radioisotopes.

Measurements indicate a temperature gradient of about 30° C/km exists near the surface of the earth; this implies a net outward heat flux. While the figures vary widely from one location to another, the generally accepted average value for this heat flux is 6.28 μJ/cm^2 s. The origin of this flux is believed to be radioactive decay in the earth's crust. Heat-gradient measurements indicate a significant difference between continent and subocean heat fluxes. These differences have led to the development of a differentiated mantle model of the earth, illustrated in Figure 12.4. In this model the radioactive isotopes are assumed to have separated out during the early stages of the earth's formation, leaving the lower mantle and core relatively free of heat-producing substances. As a result, the bulk of the measured heat flux is thought to be the result of radioactive decay in the crust. The total amount of energy flow upward at the earth's surface resulting from internal heat is, however, much smaller than that which the solar flux provides.

Example 12.2 Assume the 6.28 J/cm^2 s heat flux is a consequence of a small molten core. Estimate what the temperature of that core would have to be.

The heat conduction rate through a spherical shell of outer radius r_0 and inner radius r_i with temperatures T_o and T_i, respectively, is given by

$$\dot{q} = k(4\pi r_i r_o)\frac{T_o - T_i}{r_o - r_i},$$

$$T_i = T_o - \frac{\dot{q}(r_o - r_i)}{k(4\pi r_i r_o)}.$$

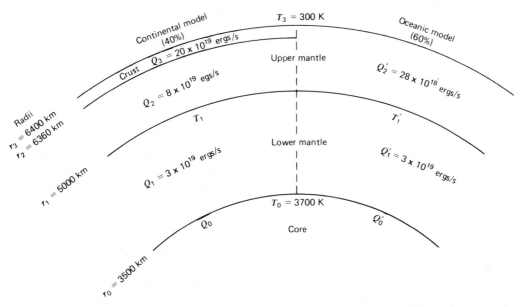

Figure 12.4 Thermal model of the Earth, based on the hypothesis of a differentiated mantle. (From F. D. Stacey, *Physics of the Earth.* New York: John Wiley & Sons, 1969.)

Assume

$$r_i \approx 0.1 \quad r_o, \quad r_o = 6.4 \times 10^6 \text{ m (radius of earth)},$$

$$k = 2.0 \text{ cal/m hr}^\circ \text{C (average value for stone)},$$

$$T_o = 15^\circ \text{ C},$$

$$T_i = 15$$

$$-\frac{(-15 \times 10^{-6} \times 100 \times 100 \times 60 \times 60)(0.9 \times 6.4 \times 10^6)}{2 \times 4 \times \pi \times 0.1 \times (6.4 \times 10^6)^2},$$

$$\boxed{T_i = 3{,}000{,}000^\circ \text{C}.}$$

The center of the earth is not a plasma! It would have to be at that temperature. We conclude that the heat gradient at the surface cannot be caused by a molten core.

Satellite measurements indicate about 37 percent of the incident solar radiation is reflected directly without change in wavelength. This figure for the **albedo**, as it is called, is not a constant but depends on the surface features of the earth. Clouds and snow or ice account for 80 to 90 percent of the albedo; oceans for less than 5 percent. Land features account for the remainder, with desert areas being more effective than forested regions. The fact that snow and ice are so effective as radiation reflectors makes the polar regions very important in terms of global weather and climate patterns.

The Arctic ice covers about 5 percent of the northern hemisphere and has an annual variation of about 10 percent of this amount. In the Antarctic, about 8 percent of the hemisphere is covered, but the variation is about 75 percent. This difference can be accounted for by several factors. Antarctica is a land mass surrounded by deep oceans in a hemisphere that is largely ocean. The Arctic area is a shallow ocean surrounded by land masses in a hemisphere that is largely land. Because of this difference the Antarctic is important in determining global weather, although surprisingly enough there appears to be very little interchange of ocean or atmospheric currents across the equator. At any rate, because of the high reflectance of the polar ice, these regions have a positive feedback effect. That is, if a perturbation of the incoming energy through either the S or k term in Eq. 12.1 occurred such that the energy were reduced, the polar ice would advance. This advance would increase the α term, producing an even greater decrease in the net available energy. Estimates are that a decrease in solar radiation at the surface of the earth of 1 percent would produce an advance of sea ice by 10° latitude and a total planetary temperature decrease of 2.8° C. Similarly, a 1.5 percent solar radiation decrease would advance the sea ice by 18° latitude and produce a temperature drop of 5° C. These figures should be compared to the estimated difference in temperature between the glacial and interglacial periods of 5° C. It would appear that in the past the earth has come dangerously close to the point from which stable equilibrium cannot be restored. There must be other processes that have a negative feedback effect to counteract the positive effect of the sea ice.

Energy Sinks

The earth's atmosphere can be divided into roughly two zones (refer to Figure 12.5). The region from the surface of the earth to an altitude of about 12 km is called the **troposphere**. In this region most of the earth's weather features appear, and the majority of the atmospheric mixing occurs. The remainder of the atmosphere above about 12 km is known as the **stratosphere**; in this region there are few clouds and pressures are of the order of 10 to 100 millibars (1 bar, the air pressure at sea levels, equals 101.325 kPa). There are no precise boundary lines between the stratosphere and the troposphere, and the rather diffuse separation layer, called the **tropopause**, varies in altitude with latitude, being highest at the equator. The composition of the atmosphere near sea level, except for water content, is given in Table 12.1. The water content of the atmosphere varies seasonally, daily, and hourly at any location and varies with altitude, as well. Over the tropics the concentration of water vapor relative to air is about 0.04 by mass; in the stratosphere it is about 3×10^{-6}.

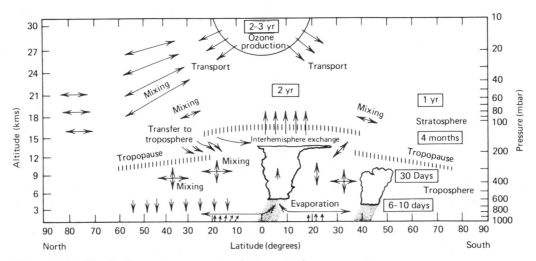

Figure 12.5 Distribution and movements of atmospheric components.

TABLE 12.1 Average Composition of Air (Near the Surface of the Earth)

Substance	Percentage by Volume
Nitrogen	78.09
Oxygen	20.93
Argon	0.93
Carbon dioxide	0.03
Neon	0.0018
Helium	0.0005
Krypton	0.0001
Hydrogen	0.00005
Xenon	0.000008
Ozone	0.00005[a]

[a]Varies considerably.

The low value for ozone in the troposphere should be multiplied by about a factor of 20 to obtain the upper stratosphere concentration.

About 3 percent of the solar radiation incident upon the upper atmosphere is absorbed in the production of ozone, and by the ozone thus produced. Wavelengths shorter than about 220 nm are virtually completely absorbed by the atmosphere. However, the absorption capability of the atmosphere is a rapidly varying function of wavelength, thus at 400 nm about 70 percent of the radiation is absorbed.

We can understand this absorption in a more quantitative way. We know that electromagnetic radiation is absorbed in matter exponentially (see Chapter 14):

$$I = I_0 e^{-\mu z}, \qquad (12.2)$$

where I is the intensity of radiation, μ is the coefficient of linear absorption, and z is the penetration distance. It is convenient to define a *mass attenuation coefficient*, which avoids the necessity of detailed knowledge of the nature of the absorbing substance:

$$\mu(\text{cm}^{-1}) = \frac{\sigma\,(\text{cm}^2/\text{atom})N\,(\text{atoms}/\text{mol})\rho(\text{g}/\text{cm}^3)}{A(\text{g}/\text{mol})},$$

$$(12.3)$$

$$\frac{\mu}{\rho}\,(\text{cm}^2/\text{g}) = \frac{\sigma N}{A}. \qquad (12.4)$$

In these expressions σ is the cross section for absorption, the factor μ/ρ is most easily found for photons of various energies by referring to compilations of computations.[3]

Example 12.3 μ/ρ for 400-nm photons in air is about $75 \times 10^3\,\text{cm}^2/\text{g}$. If 90 percent of the absorption occurs in the stratosphere, what is the average air density in this region?

Assume the stratosphere to be about 10 km thick.

$$0.1 = e^{-\mu \times 10},$$

$$\mu = 2.3 \times 10^{-1}(\text{km}^{-1}) = 2.3 \times 10^{-6}\,\text{cm}^{-1},$$

$$\rho = \frac{\mu}{75 \times 10^3} = 3.1 \times 10^{-11}\,\text{g}/\text{cm}^3.$$

The air density at sea level and STP is $1.293 \times 10^{-3}\,\text{g}/\text{cm}^3$. The stratosphere is clearly quite rarefied!

There are other photoinduced chemical reactions of importance in the troposphere, especially those related to the production of photochemical smog; but these reactions do not contribute significantly to the absorption of solar energy. In Figure 5.6 a substantial part of the solar spectrum near 1400 and 1900 nm is missing at the surface of the earth. This absorption is by CO_2 and water vapor, which gases are especially sensitive to the infrared portion of the spectrum and absorb all but a few "window" wavelengths. This absorption takes place regardless of the direction from which radiation impinges on the atmosphere.

The energy radiated by the earth to space has a wavelength distribution approximated by that of a blackbody having a temperature of 300 K, except, of course, for the missing ab-

[3]See, for example, G. R. White, "X-Ray Attentuation Co-efficients from 10 keV to 100 MeV," *National Bureau of Standards Report* 1003, 1952.

sorbed wavelengths. The energy radiated by the earth must penetrate the atmosphere; in doing so a great deal of it is absorbed. This absorbed energy must eventually be reradiated; but the reradiation takes place isotropically. In other words, roughly half of it is directed back down to earth! This effect by the atmosphere, passing short wavelengths, absorbing and reradiating long wavelengths, is referred to as the **greenhouse effect**, because this is the action of a botanical greenhouse. Clouds, consisting largely of water droplets, absorb and reradiate in the infrared quite readily, almost as well as a blackbody. At the same time clouds reflect incoming solar radiation. It is not clear which effect predominates, reflection or absorption; there may even be a variation with altitude. It may be that cloud formation provides a negative feedback mechanism; that is, decreased insolation promotes cloud formation, which traps more outgoing radiation, preventing a temperature decrease.

The remaining natural constituents of the atmosphere are the aerosols: particles and liquid drops in suspension. The natural sources for these aerosols are sea spray, windblown dust, volcanoes, and the conversion of naturally occurring gases to particles. Particulates having diameters less than about 100 nm are found primarily in the troposphere, where they appear to have a residence time of less than 2 weeks; these aerosols have primarily terrestrial origins. The larger-diameter particles from 100 to 1000 nm, are found chiefly in the stratosphere, peaking near 18 km, where they have a residence time of 2 or more years. The stratospheric particles appear to be the result of nucleation of trace gases, particularly those involving sulfur, although volcanic eruptions have been known to produce stratospheric particulates.

In particular, the eruption in March 1963 of Mt. Agung in Bali had a substantial effect on the stratosphere that has since come to be called the "Agung effect." Antarctic observers noticed a steady decrease in the *normal-incidence* solar radiation in December of 1963, until by mid-February the normal-incidence solar radiation was only 15 percent of its average value over the previous 10 years. This was attributed to the deposition of volcanic material in the stratosphere by the Agung eruption. The normal-incidence depletion continued to be observed for several years afterward; by 1968 it was 98 percent of normal. During this period of time, the temperature of the atmosphere was found to increase 6° to 7° C initially and remain 2° to 3° C higher for several years. At the same time, no change in tropospheric temperature was noted. Several years of brilliant sunsets can be attributed to Agung: Other weather phenomena can be caused by volcanic eruptions. Boston had a killing frost *each month* of 1816, a direct result of the eruption of Mt. Tambora in Indonesia in 1815.[4] Surprisingly enough, after the Agung eruption the total solar flux at the Antarctic remained only slightly below normal, the total being the sum of normal incidence and scattered (diffuse) radiation. From this we are led to the conclusion that particulates in the stratosphere do not significantly affect the total heat balance of the earth. They may indeed cause local fluctuation in atmospheric parameters, however. It is not clear whether or not this same conclusion can be drawn about the smaller aerosols that are normally found in the troposphere.

12.2 MECHANISM FOR RADIATION SINKS

Radiation incident upon the atmosphere, either the short-wave solar flux or the long-wave terrestrial flux, is attenuated by two

---

[4]It is estimated that about 60 km³ of mountaintop was tossed into the air by this eruption. See H. Stommel and E. Stommel, *Sci. Amer.*, July (1979), p. 176, for additional information.

mechanisms: absorption and scattering. Since both these effects are atomic phenomena, it may not be possible to describe them satisfactorily using the methods of "classical physics." Instead, the techniques of quantum mechanics must be applied.

We shall not apply these techniques in all their mathematical glory! Instead, we shall use the results of such calculations for the specific physical systems and interactions of interest to us. The intent of our coverage in this section is to indicate how physics contributes to the understanding of our environmental problems in a very fundamental way. For those interested in the details of these calculations we have included in the references for this chapter a number of useful texts.

Absorption

A given atom will ordinarily be found in its lowest energy state, called the *ground state.* Additional energy cannot be given to this atom in an arbitrary amount. The energy states available for hydrogen have been measured and calculated; the smallest increment of energy a hydrogen atom in the ground state can accept is 10.20 eV. The next smallest is 12.09 eV, and so on. A hydrogen atom originally in the ground state that has absorbed 10.20 eV energy is said to be in the first excited state. Other possible energy states are likewise numbered. The excited-state structure for hydrogen is shown in Figure 7.22. The energy indicated in the left-hand column is the energy measured relative to the ground state, so that 13.6 eV is required to ionize the hydrogen atom. All atoms have quantized energy-level sequences; in general, they are very complex in terms of the number and spacing of allowed levels. It is usually too difficult to calculate these levels exactly for any individual element, but we can be certain that no two elements have identical energy-level sequences. This fact is very useful.

An atom in an excited state rarely remains in that state very long. Instead, it releases the excess energy by emitting a photon of electromagnetic energy and returns to the ground state. The energy and the frequency of the photon are related:

$$E = hf, \qquad (12.5)$$

where h is Planck's constant, $h = 6.62 \times 10^{-34}$ J s. Therefore, the emission spectrum, that is, the frequencies of radiation emitted by excited atoms, can be used to identify those atoms. The wavelengths of photons emitted from hydrogen are indicated in Figure 7.22. The lines H_α, H_β, ..., result from transitions between various excited states and the first excited state; these lines are in the visible range and are named after the Swiss school teacher Balmer, who devised a recursion formula for predicting their wavelengths in the late nineteenth century, before atomic structure was understood. The other series, Lyman Paschen, and so on, are in the ultraviolet or infrared. Likewise, the absorption spectra can be used for identification purposes. If white light (light containing all frequencies) is incident upon a cell containing hydrogen gas, the light passing through the gas will be found to be deficient in those frequencies characteristic of hydrogen.[5]

The emission and absorption of radiation by molecules (two or more atoms bound together by electrical forces) are more complicated. Molecules have more degrees of freedom than atoms; they can rotate and vibrate. As a consequence they can absorb and emit energy by these additional mechanisms. Of course, as we would expect in atomic systems that have spa-

[5]Helium was first discovered on the sun by identification of the Fraunhofer lines, the absorbed wavelengths in the sun's spectrum. Lines corresponding to most known elements at the time (1868) had been identified, but the remaining could not be; so they were attributed to a new element, which was then named after the sun, *helios* in Greek.

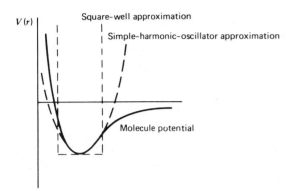

Figure 12.6 Molecular potential diagram. The simple harmonic-oscillator approximation used to calculate the energy levels is indicated by the dashed curve.

tial constraints, the rotational and vibrational energies for a given molecule are quantized.

A typical molecular potential is shown in Figure 12.6; a simple-harmonic-oscillator approximation, valid near the equilibrium position, is also shown. Quantum mechanics can be used to solve for the equally spaced energy levels that result, giving

$$E_n = (n + \tfrac{1}{2})\omega, \quad n = 0, 1, 2, \ldots, \quad (12.6)$$

where ω is the "classical" frequency of vibration, $\omega = (k/m)^{1/2}$. If the effective spring constant k for a typical molecule is evaluated the energy-level differences obtained are rather small, typically $0.1 \, eV$. As a consequence, molecules, in addition to having electronic excited states irregularly spaced, but averaging in the tens of electron volts, can have bands of vibration states built on each electronic excitation.

The existence of rotation complicates the allowed energy spectrum further. The classical energy of rotation is given by

$$E = \frac{L^2}{2I}, \quad (12.7)$$

where L is the angular momentum of rotation, and I is the moment of inertia. To evaluate the

energy of a rotating atomic or molecular system Schrödinger's equation, appropriately modified for the problem, must be solved. When this is done, the resulting allowed energy levels are given by

$$E_l = \frac{l(l + 1)\hbar^2}{2I}, \quad (12.8)$$

where $l = 0, 1, 2, 3, \ldots$ is the angular momentum of the rotating system, I is the moment of inertia, and $\hbar$ is $h/(2\pi)$. For a typical molecule the energy levels are spaced about $10^{-2} \, eV$, about $1/10$ the vibration spacing. A molecule in a particular electronic excited state and in a particular vibration mode can also be rotating with a rotational energy, E_l.

We see that for molecules the excited-state structure appears as an almost solid band of allowed energies, many states separated by very small energy differences (see Figure 12.7). Because the energy spacings of these levels are so small, the wavelengths of radiation emitted when the molecule changes from one of these allowed states to another are very long; in fact, these wavelengths are in the infrared region. This is the reason carbon dioxide and water vapor are such good absorbers in the infrared.

Scattering

The other mechanism by which the atmosphere and its constituents remove energy from the impinging radiant flux is scattering. Scattering is clearly an atomic phenomenon, and so we would expect to have to use the language of quantum mechanics to examine the process in detail. This is certainly corrent; but if we wish only to make predictions of the gross behavior of the atomic scatterers and not attempt to predict the detailed results, the methods of classical physics may suffice. At any rate, we know how to apply classical physics, so let us do it. Then, at the end, we can see how far from reality our calculations are,

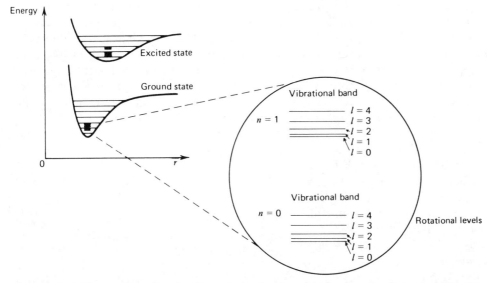

Figure 12.7 Electronic-, vibrational-, and rotational-energy levels of a diatomic molecule. The rotational levels are shown in an enlargement of the $n = 0$ and $n = 1$ vibrational levels of the ground electronic state.

and make appropriate adjustments, if required, to account for the quantum nature of matter.

In addition, we assume we know the actual scattering mechanism to be as follows: radiant energy in the form of an electromagnetic wave impinges on a scatterer consisting of a collection of positive and negative charges bound together by Coulomb forces. The radiant energy causes this charge collection to oscillate in simple-harmonic motion, the Coulomb force being the restoring force. The charges thus accelerating and decelerating radiate electromagnetic energy at the same frequency as the driving wave according to the well-understood laws of electricity and magnetism.

The instantaneous rate of energy emission from an accelerating charge is given by

$$\Phi = \left(\frac{q^2}{6\pi\epsilon_0 c^3}\right)a^2, \qquad (12.9)$$

where a is the acceleration, and q is the magnitude of the electric charge. This can be written in terms of the effective dipole moment of the oscillating charge collection and averaged over one cycle by solving the equation of motion of a harmonic oscillator:

$$m\ddot{x} = kx. \qquad (12.10)$$

The solution to this equation is of the form $x = x_0 \cos \omega_0 t$, where $\omega_0 = (k/m)^{1/2}$. Since the acceleration is $\ddot{x}$, Eq. 12.9 can be written

$$\Phi_{av} = \frac{p_0^2 \omega_0^4}{12\pi\epsilon_0 c^3}, \qquad (12.11)$$

where $p_0 = qx_0$ is the dipole moment of the charge distribution. But when the numbers corresponding to atmospheric scattering are used to evaluate this expression, the average energy loss per unit time calculated by Eq. 12.11 is much smaller than the observed. There must be other effects present that introduce additional loss mechanisms.

Atmospheric molecules are not free oscillators in the sense used above. There are "friction" effects—an excited molecule can lose energy by collision with another molecule before it has a chance to reradiate, or it can reradiate at a lower frequency. Simultaneously, charge distributions absorbing electromagnetic radiation are being subject to a driving force, $\mathbf{F} = q\mathbf{E} = qE_0 \cos \omega t$, where the driving frequency ω is in general different from the free-oscillator frequency ω_0. The actual equation of motion, then would be

$$m\ddot{x} = -kx - h\dot{x} \pm qE_0 \cos \omega t, \quad (12.12)$$

where k is the restoring constant, and h is the "frictional" coefficient (friction is a velocity dependent term). Changing variables such that the dipole moment appears explicitly is useful, since it is the dipole moment p_0 that we need for the evaluation of Eq. 12.11. Let $\tau = m/h$ and $p = qx$; then we have

$$\ddot{p} + \frac{1}{\tau}\dot{p} + \omega_0^2 p = \pm \frac{q^2}{m} E_0 \cos \omega t, \quad (12.13)$$

Equation 12.13 is a rather complicated differential equation, the solution to which has two parts—a transient part and a steady-state part. We are not interested in transient effects; that is, we do not need to know what happens to the charge distribution just as the electromagnetic wave impinges. Rather, we are interested in the behavior after steady oscillation has begun. There are various techniques for finding the steady-state solution to this equation. The rotating-vector method yields this result for p_0:

$$p_0 = \frac{(q^2/m)E_0}{[(\omega_0^2 - \omega^2)^2 + \omega^2/\tau^2]^{1/2}}. \quad (12.14)$$

In a real situation $\tau \approx 10^{-4}$ s, and $\omega \approx 10^{-15}$ s, so Eq. 12.14 can be written

$$p_0 = \frac{(q^2/m)E_0}{|\omega_0^2 - \omega^2|}, \quad (12.15)$$

which gives for $(\Phi)_{av}$

$$(\Phi)_{av} = \frac{q^2\omega^4 E_0^2}{12\pi\epsilon_0 c^3 m^2 (\omega_0^2 - \omega^2)^2}. \quad (12.16)$$

We can define the scattering cross section σ_s

to be the average energy scattered per unit time divided by the average incident energy per unit time. This is a kind of probability for scattering; the higher the cross section, the higher the probability that scattering will occur. The average energy incident on the scattering center per unit time is given by the Poynting vector. From electromagnetic theory this quantity has the value $\frac{1}{2}cE_0^2$, so the scattering cross section is given by

$$\sigma_s = \frac{(\Phi)_{av}}{\frac{1}{2}cE_0^2}. \quad (12.17)$$

If we assume $\omega_0/\omega \gg 1$, as it is in general, substitution of Eq. 12.13 into Eq. 12.17 and simplification yield

$$\sigma_s = \left(\frac{8}{3}\right)\pi r_e^2 \left(\frac{\omega}{\omega_0}\right)^4, \quad (12.18)$$

where $r_e = q^2/(4\pi\epsilon_0 mc^2)$. (This is often called the classical radius of the electron.)

Now we know why the sky is blue and sunsets are red! These conclusions are evident from the frequency dependence of the scattering cross section. The frequency of blue light is about 10^{15} Hz, while for red light it is about 2×10^{14} Hz. Therefore the atomic and molecular scatterers in the atmosphere are $[10^{15}/(2 \times 10^{14})]^4$, or about 625 times as effective in scattering blue light as in scattering red. The sky appears blue then because a substantial amount of blue light is removed from the normal-incidence solar beam and bounced around in the atmosphere, much of it eventually finding its way to the surface of the earth, but coming from all directions. Sunsets are red because the light beam from the setting sun travels through the maximum amount of atmosphere to get to the observer. This long path length provides ample opportunity for the scattering centers to do their work removing the blue component from the solar beam.

Example 12.4   Compare X-band radar to red light in terms of atmospheric scattering.

Using Eq. 12.18 the ratio of cross section is given by

$$\frac{\sigma_{\text{radar}}}{\sigma_{\text{red light}}} = \left(\frac{9 \times 10^9}{2 \times 10^{14}}\right)^4 = 4.1 \times 10^{-18}.$$

We can conclude that X-band radar is even less likely, by a very large fraction, than red light to be scattered by the atmosphere.

Radiation scattered by clouds, water vapor, dust, and haze in the atmosphere accounts for about 31 percent of the total energy loss from the incident solar radiation. An additional 15 percent is absorbed via the frictional coefficient h in Eq. 12.12. This frictional effect appears as heating of the atmosphere and is almost as important as reflection in the global energy balance.

We started this discussion of scattering by assuming that classical physics could be used in the description of the phenomenon. Was this assumption justified? As far as the average energy change per unit time is concerned, the classical calculations seem to agree rather well with measurement. But what we have not done, and cannot do without quantum mech-

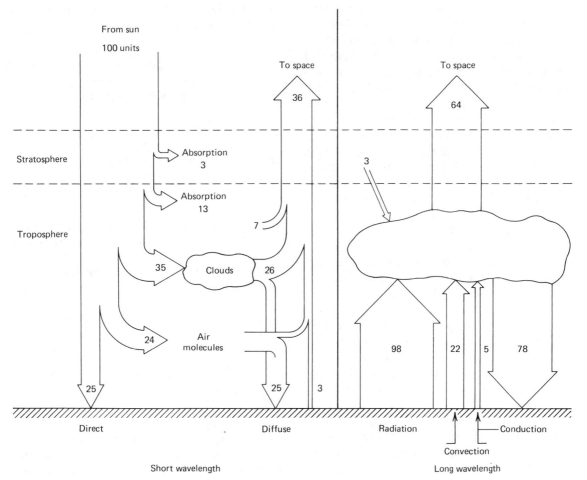

Figure 12.8 The energy balance of the earth and atmosphere.

anics, is calculate the detailed behavior of a scattering event. For example, we have not calculated the angular dependence of scattered radiation with respect to the incident beam. Nor have we attempted to calculate the percentage of the incident radiation lost by collisional transfer or radiation at a lower frequency. These details depend on the nature of the scatterer, whether it be an atom, molecule, or particle. As a consequence, only quantum mechanics can give an appropriate description.

The two mechanisms absorption and scattering represent the only means by which the atmosphere can alter an incident energy flux. Solar energy remains the only source of energy of any significance for the earth. The quantitative energy balance of the earth in terms of 100 units of incident solar radiation is summarized in Figure 12.8. According to the diagram, three units of incident short-wavelength radiation are absorbed in the upper atmosphere and 14 units in the lower atmosphere. Of the remainder, 35 units interact with the cloud cover. Of these, 26 are reflected, 2 are absorbed, and 7 are reradiated into the atmosphere. Twenty-four units of the primary beam interact with the atmosphere along with the 7 units from the clouds; 6 units are reflected and lost, and 25 units reach the surface of the earth. Of this, diffuse scattered short-wavelength radiation, 3 units are reflected, and the remainder are absorbed.

In order for an energy balance to be maintained, the earth must rid itself of the 47 units of short-wavelength radiation absorbed from the sun. One way of doing so is by radiating into the atmosphere. But because the temperature of the earth is relatively low, the radiation is concentrated in the long-wavelength region—the infrared.

The troposphere radiates isotropically at IR wavelengths; 78 units reach the surface of the earth. The surface then must rid itself of a total of 125 units. A total of 27 units are transferred by convection and conduction; 98 units must be radiated.

We can use Eq. 8.7 to calculate the effective temperature of the earth's surface, since this is the equation representing heat transfer by radiation.

Example 12.5 We have $\dot{q} = \epsilon\sigma T^4$, and $\dot{q}$ is the rate of energy radiation, or 98/100 of the rate of energy reception E_R.

$$E_R = \frac{\text{area of earth's disk} \times 1400 \text{ W/m}^2}{\text{surface area of earth}},$$

$$\dot{q} = 0.98 \, E_R = 343 \text{ W/m}^2.$$

For a blackbody, which we must assume, $\epsilon\sigma = 5.67 \times 10^{-8} \text{ W/m}^2 \text{ K}^4$. Therefore,

$$T^4 = \frac{343}{5.67 \times 10^{-8}},$$

$$\boxed{T = 278.9 \text{ K, or } 5.89° \text{ C.}}$$

Since the average temperature of the earth is about 15° C, this calculation slightly underestimates the amount of energy radiated by the earth. However, if the calculation were based solely on short-wavelength radiation (see Example 12.1), we would obtain a significantly lower temperature characteristic of the stratosphere.

The heat-transfer question between the earth and space is more complicated than we have indicated. It is part of a larger problem that connects oceans, continents and the atmosphere.

12.3 THE EARTH'S HEAT ENGINES

Air Currents

The air–ocean interface is the transfer point for a substantial amount of energy and matter. This interface is part of a global heat machine that drives air and ocean currents, transferring heat energy to the atmosphere and to the

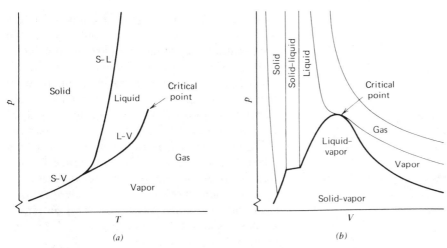

Figure 12.9 Typical phase diagrams for a real substance: (*a*) *p–T*, (*b*) *p–V*.

poles. The magnitude of this transfer is enormous. There are two mechanisms for the transfer of energy at this interface: sensible heat and latent heat of vaporization. Water requires 1 cal/g to increase in temperature by 1°C. If dry air were placed in contact with the ocean, water vapor would be evaporated into the air, the amount depending on the temperature. At equilibrium at the interface, the air is saturated with water vapor; that is, it has as much water vapor as it can have. The properties of mixtures of liquids and their vapors are more difficult to study in thermodynamics than those of simple systems.

This difficulty arises because an ideal gas— the convenient system that we have used for calculations and model-building—cannot change phase! But real gases do liquefy and solidify, and the behavior of a gas–liquid or liquid–solid mixture is difficult to predict analytically. With reference to Figure 12.9, the dark curves in both the *p–T* and *p–V* diagrams are called *saturation curves*. (Refer back to Chapter 4 for details.) Along these lines the substance exists in equilibrium in the two phases indicated to the left and right of the lines. As the temperature of a saturated mix-

ture is increased at constant pressure (along a horizontal line to the right in a *p–V* diagram), the mixture becomes unsaturated, and can accept more water vapor and concomitant latent heat to restore the equilibrium saturation condition. The amount of latent heat required to heat 1 g of air 1°C and to maintain saturation varies as a function of temperature because the water content of the air varies; the heat required to raise 1 g of saturated air $dT°C$ is given by

$$dQ = (c_p)_{\text{dry air}} m_{\text{air}} + m_w c_{\text{pw}} \, dT + l \, dm - p \, dV, \tag{12.19}$$

where m_w is the mass of water in the air at the initial temperature T, dm is the additional amount of water required to maintain saturation at the higher temperature $T + dT$, l is the heat of vaporization of water (540 cal/g), and $p \, dV$ is the work done by the expanding water vapor. This equation is not easily solved analytically, but an answer may be obtained for a given set of conditions by the use of tables.[6] The results of such calculation are

[6]See, for example, "Vapor Pressure of Water Below 100° C," *Handbook of Chemistry and Physics*, Chemical Rubber Company.

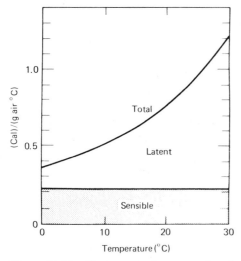

Figure 12.10 Heat required to warm 1 g of saturated air 1°C (sensible heat) and maintain it saturated with water vapor (latent heat).

shown in Figure 12.10. Because of variation in heat absorption as a function of temperature, oceans act as effective temperature-regulating devices.

Example 12.6 Find the quantity of water vapor present in a volume that contains 1 g of saturated air at 20°C at 760 mmHg.

At 20°C the vapor pressure of water is 17.535 mmHg. The dry air pressure must then be 760.000 − 17.535 = 742.465. The volume occupied by 1 g is given by

$$V = \frac{1}{\rho(p/p_0)} = \frac{760.000}{1.293 \times 10^{-3} \times 742.465}$$
$$= 791.7 \text{ cm}^3.$$

Use $pV = nRT$:

$$(17.535 \text{ mmHg}) \times 1.33$$
$$\times (10^2 \text{ Pa/mmHg}) \times 791.7 \times 10^{-6}$$
$$= n \times 8.35 \times 293.3,$$

$$n = 7.575 \times 10^{-4} \text{ mol} = 1.36 \times 10^{-2} \text{ g}.$$

Since land areas in general contain far less water than the oceans, you would expect the temperature-regulating properties of latent heat transfer to be less effective in those areas. Indeed, they are; land areas have much larger temperature variations and much higher summer temperatures than ocean regions. And, of course, certain arid regions near 30° latitude being very low in water content have very high summer temperatures. This difference in heat transfer over land and ocean areas results in a heat gradient between these regions and a resulting air circulation pattern. This pattern is masked usually by a stronger driving force, the poleward circulation.

The poles receive less solar heat than the tropics because of the latitude variation of the incident flux and because of the higher reflectivity of the polar ice. As a consequence, there is a differential heating of the air such that there is a constant poleward flow. This flow, however, is interrupted by two effects. Because of the rotation of the earth the air that would flow back from the poles to the equator along the surface is deflected to blow from East to West. The deflection of moving objects in rotating systems is known as the **Coriolis effect** and was studied mathematically by Gaspard Coriolis in about 1840. It is interesting to note that George Hadley predicted the Coriolis effect on atmospheric circulation in 1735. The other effect, which Hadley did not account for, is the cooling of the tropical air before it reaches the pole. This cooling comes about because of radiative transfer in the atmosphere. By the time the tropical air reaches about 30° latitude, it has cooled sufficiently that it begins to drop. As it drops it is heated by compression and flows along the surface both toward the equator and the pole (see the three-cell circulation model in Figure 12.11). This downward flow at 30° latitude accounts for the almost constant high pressure in this region and also is instrumental in the formation of desert regions at this latitude.

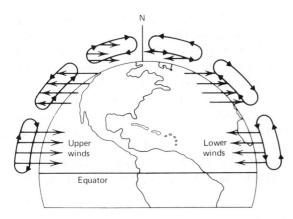

Figure 12.11 The three-cell circulation model of the earth's air currents.

The water vapor accepted by the atmosphere in the transfer of latent heat is quite important in terms of the global energy balance. The presence of water vapor in the air reduces the rate of temperature decrease as a function of altitude because of condensation of the water. This condensation forms clouds, which, as we have learned, have a significant effect on both the albedo and the infrared absorption of the atmosphere. Lastly, water vapor changes the specific volume of air; moist air is less dense than dry air and as a consequence would be more likely to participate in up-currents, low-pressure systems. We see that water vapor and the air–ocean interface are very important to the global energy balance; any mechanism that tends to change either—for example, production of large quantities of water vapor via fossil fuel combustion or large area oil spills on the ocean surface that inhibit evaporation—is likely to have major climatic effects.

Ocean Currents

The air–ocean interface is primarily responsible for driving the atmospheric currents; at the same time, the atmospheric currents are primarily responsible for driving the ocean currents. These currents transport not only water, but also heat energy and nutrients in lateral and vertical motions. They are of great importance in the overall energy-balance picture.

If the earth were all ocean, the circulation of ocean currents would follow the atmospheric currents exactly. As the wind blows across the ocean surface it produces waves. The motion of an individual water droplet is not, in general, in the direction of the wind, but rather in a circular path, much like the bobbing up and down of a cork in the water. However, as a wave "breaks," water drops are sprayed in the wind direction; by this technique there is a general transfer of matter along the direction of wind currents. The earth is not entirely water, however; the presence of the continents alters the idealized ocean-current structure. The result is a system of closed looping currents called **gyres**, shown for a hypothetical elliptical ocean in Figure 12.12.

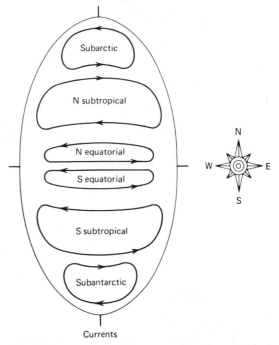

Figure 12.12 Surface currents for a hypothetical elliptical ocean.

Because of the latitudinal variation in insolation the poleward branches of each gyre tend to be cooler than the equatorward sides. As a consequence the symmetry of the hypothetical currents is broken. In addition, the difference in heat and water-vapor transfer between various regions of the ocean leave these areas with varying amounts of salinity. A more saline ocean region will have a net inflow of water in order that an equilibrium mixture should be reestablished. The centers of the subtropical gyres tend to be high in surface salinity because of the high water evaporation from these regions. Consequently, there is a net inflow of water superimposed on the natural circulation into these areas.

Salinity also affects the relationship between density and temperature for water. As we see in Figure 12.13, when the salinity is above 24 parts per thousand (written as 24‰), the temperature of maximum density falls below the freezing point. Seawater, then, always expands

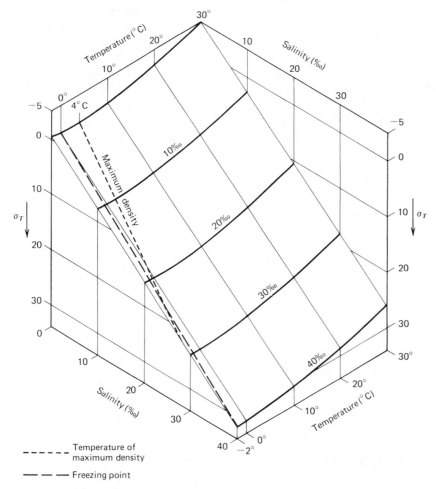

Figure 12.13 "Density" σ_T of water (see text) as a function of salinity and temperature. (From Peter K. Weyl, *Oceanography*. New York: John Wiley & Sons, 1970.)

as the temperature is raised (pure water has a maximum density at 4° C). Note that the "density" σ_T is defined by

$$\sigma_T = (\rho - 1) \times 1000. \qquad (12.20)$$

This way of quoting densities is useful to oceanographers, since it very clearly displays the relative salt content of the water. For example, at 35‰ salinity, $\rho = 1.028$, and $\sigma_T = 28$. For pure water $\rho = 1.00$, $\sigma_T = 0$.

The circulation of ocean currents is a difficult problem, even in a hypothetical elliptical ocean. In the real ocean, because of the irregularity of land mass distribution, detailed predictions are impossible. However, the general features of the six-gyre scheme seem to be reproduced in actual current circulations. Any effect that tends to modify the present circulation system would have substantial effects not only on the climate of coastal regions, but also on the fish population, which

depends on nutrient currents brought up from colder, deeper levels. A major modification would no doubt have a profound effect on the global climate.

For some time now scientists have been attempting to devise mathematical models for the air and ocean environments that would enable them to predict the behavior of those environments. These attempts have met with mixed degrees of success. The difficulty stems from the large number of interactions that must be taken into account (see Figure 12.14). In the last 10 years, however, since the advent of the fast, large-memory, integrated-circuit digital computers, advances in modeling have been made. There is still a great deal of work to be done before the two major goals of climate modeling are achieved: determination of the large-scale, long-term dispersive and storage characteristics of the combined atmospheric–hydrospheric system and the

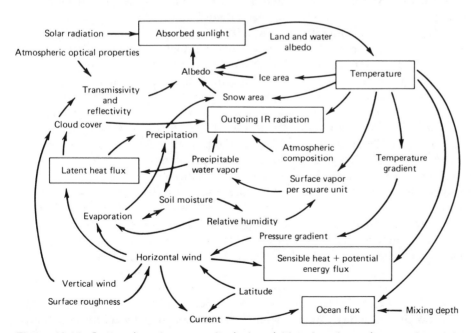

Figure 12.14 Interactions between physical variables that determine weather and climate. (From John R. Herman and Richard A. Goldberg, *Sun, Weather and Climate*. Washington D.C.: NASA, 1978.)

assessment of secondary interactions that would significantly influence the structure and variability of the combined system.

12.4 MANKIND'S EFFECTS

Mankind's interaction with the environment, particularly in search of energy, produces several undesirable by-products that can influence the global energy balance: heat, particulates, and various gases. We should examine the extent to which these by-products are being produced to determine whether they pose an immediate or a more long-range problem for the earth and to determine if there are any means by which these problems, if such exist, can be alleviated.

Heat Production

The precise amount of heat currently being produced on the earth is not known; remember that all energy produced in any form eventually becomes heat. Several independent estimates have been made that are in remarkable agreement; about 5×10^{12} J/s for the current rate of production. This is really a very small number compared to the amount of energy absorbed by the earth, only about 0.00025 of the total rate of energy absorption. Clearly this will be insignificant in determining the earth's temperature, if it is reasonably well distributed over the surface of the earth. It is not, in fact; the Los Angeles basin is estimated to be producing energy equivalent to 5 percent of its incident solar flux now. So, in this small area, heat production may be influencing climate; it certainly will be in the future, if energy production continues to increase as it has in the past.

Until recently, it was believed that electric generating capacity should increase by 10 percent every year to keep up with the consumer demand. This corresponds to a doubling time of about 7 yr. A more realistic figure would be a 4 percent per year increase, although this leads to a doubling time of only 17.5 yr, and so is almost equally unreasonable. What are the consequences in terms of the global climate of 4 percent/yr heat load increase?

It has been estimated that a temperature increase of 1° C would cause substantial changes in the boundaries between plant communities; a rise of 3° C could melt the ice caps, thus submerging most of the major cities of the world and, of course, all of Florida. At a 4 percent annual rate of increase a 1° rise would be accomplished in only about 150 yr. The 3° rise would require substantially longer, assuming the 4 percent rate remained constant. Of course, fossil fuels, and uranium and thorium reserves as well, would probably be exhausted before the 3° rise could be observed. But within that period of time nuclear fusion could be made economically feasible and the melting of the pole caps would no longer be science fiction but a real possibility.

The 150-yr figure quoted above is not so long a period of time that we should not be worrying about the problem here and now. How can we avoid the consequences of the trends that we seem to be following at present? There will always be a segment of society that will opt for the *status quo*, let the future take care of itself, "I'm all right, Jack." Scientists belonging to this school might plan to air condition a part of the earth for habitation and agriculture, while letting the rest attain a very high temperature to radiate heat away more efficiently. This is certainly a possible course of action; there is probably no technical reason why it could not be done, although a myriad of social and political problems come to mind. Such problems are likely to arise, however, in any solution.

For example, if one could determine the optimum population that the earth could support at a desirable standard of living without suffering a deterioration of recuperative processes or energy balances, and then initiate policies to achieve that steady-state condition and at the same time develop means to utilize solar energy directly, the excess heat problem would no longer exist. Of course, the social and political problems of putting this plan into use are formidable, but perhaps worth the effort in the long run. For even if the air-conditioned cities, ultrahot-desert plan were used, other consequences of energy generating by combustion would still cause difficulties in maintaining the proper energy balance.

The concentration of several trace constituents of the atmosphere up to about the top of the stratosphere is shown in Figure 12.15. We discuss these as well as other pollutants

resulting from mankind's activities, since we know they have a substantial effect on the earth's energy balance.

Particulate Production

Compared to nature, society's global particulate production is minor (refer to Table 12.2). Out of a total of more than 4×10^9 tons of particulate matter estimated to be in the atmosphere in 1968, only 0.7×10^9 tons, or about 17 percent, can be attributed to man-made sources. The majority of these particles are converted gaseous pollutants, that is, gas molecules that have clumped together to form

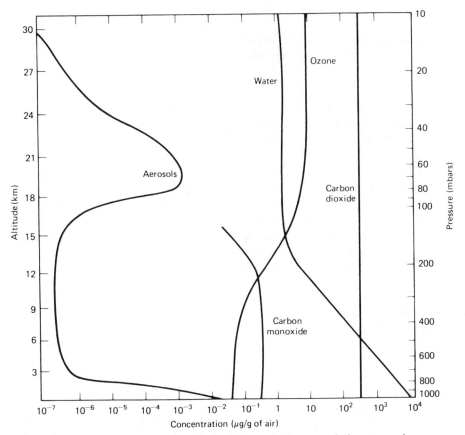

Figure 12.15 Concentration of certain trace constituents of the atmosphere as a function of altitude.

TABLE 12.2 Estimated Mass of Global Emissions of Atmospheric Particulates, 1968[a]

Source	$d < 500$ nm	All Sizes
Seasalt	500	1000
Converted natural sulfate	335	420
Natural windblown dust	250	500
Converted man-made sulfate	200	220
Converted natural hydrocarbons	75	75
Converted natural nitrates	60	75
Converted man-made nitrates	35	40
Man-made particles	30	135
Volcanoes	25	?
Converted man-made hydrocarbons	15	15
Forest fires	5	35
Meteoric debris	0	10
Total	1530	2525 +

[a]In 10^6 tonnes.
Source: Reprinted by permission from James T. Peterson and Christian E. Junge, "Sources of Particulate Matter," in *Man's Impact on Climate.* Cambridge: MIT Press, 1970.

a particle. There is some evidence that man-made particulates are beginning to penetrate the stratosphere. The Cl to Br ratio in the stratosphere is about 1/20 of its sea-level value, indicating an overabundance of bromine. This bromine may come from the lead compounds exhausted by automobiles.

Fuel combustion in stationary sources and industrial processes accounts for the majority of the particulate emissions. Stationary combustion sources emit particulates directly in the form of ash or soot and indirectly in the form of oxides of sulfer and nitrogen that form particles by clumping. We know the effect of particulates on the global climiate; they can also have important effects in the local regime (see Chapter 13).

The types of particulates emitted in industrial processes are generally different from those given off by stationary combustion sources. Manufacture of fertilizer, storing and milling of grain, pulp and paper milling, iron and steel refining, and organic and inorganic

chemical manufacturing all produce substantial amounts of particulate matter of various types. For the global climate, the effects of these aerosols are the same as those from combustion; but for the local regime, the biological effects may be quite different. We shall consider some of these effects later. Since the major amount of man-made particulates in the atmosphere comes from the agglutination of gas molecules, we should study in some detail the emission of these gases.

Production of Gases

By far the most important man-made gaseous pollutant[7] is carbon dioxide, CO_2 (refer again to Table 12.1). Since the International Geophysical Year in 1958, careful measurements of atmospheric CO_2 have been

[7]Carbon dioxide is ordinarily not considered to be a pollutant, because it does not enter into photochemical smog reactions. But because of its significance in the global energy balance, it certainly should be so considered.

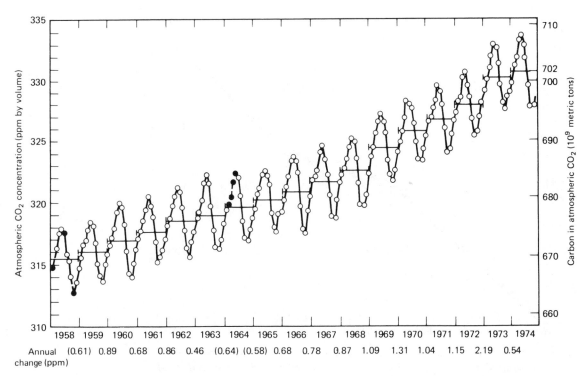

Figure 12.16 Monthly average values of CO_2 concentration at Mauna Loa Observatory, Hawaii. [Reprinted, with permission, from C. F. Baes, Jr., H. E. Goeller, J. S. Olson and R. M. Rotty, "Carbon Dioxide and Climate: The Uncontrolled Experiment," *American Scientist* **65**, 310 (1977), journal of Sigma Xi, The Scientific Research Society.]

made. Figure 12.16 is a plot of the CO_2 content versus time for a period of about 16 yr. We can see that there are not only seasonal variations but also a distinct increasing trend. The data do not extend over a long enough period of time to determine whether the increase itself is linear or whether it has variations as well. However, in Figure 12.17 we show an estimate of CO_2 production from fossil fuel combustion from about 1860. There has clearly been an exponential increase of a more or less constant slope over this time period. Breaks in the curve can be attributed to the two world wars and the depression.

It is clear that the production of CO_2 by the combustion of fossil fuels will continue for many years and that the rate of production will continue to increase in the near future. Since CO_2 has a significant role in the global energy balance, it is important to determine the sinks and reservoirs of CO_2 on the earth. We need to know to what extent these reservoirs are becoming saturated, at what point the atmospheric content of CO_2 will begin to increase substantially, and what alternatives we have, if any.

There are basically only two CO_2 sinks available: organic material and the oceans. Living plants and the organic material on which they feed absorb a great deal of CO_2 in the photosynthesis process and the nitrogen-fixation reactions. An increase in atmospheric CO_2 would result in an increase in the rate of photosynthesis, assuming growing things are

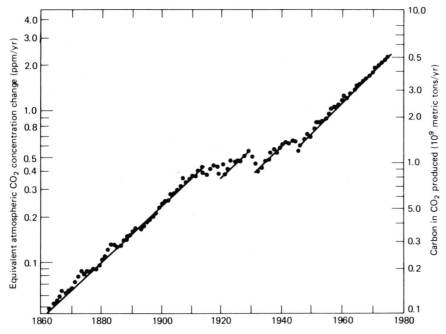

Figure 12.17 Annual global CO_2 production estimate from about 1860. [Reprinted, with permission, from C. F. Baes, Jr., H. E. Goeller, J. S. Olson and R. M. Rotty, "Carbon Dioxide and Climate: The Uncontrolled Experiment," *American Scientist* **65**, 310 (1977), journal of Sigma Xi, The Scientific Research Society.]

not nutrient- or water-limited. At the same time a reduction in forested areas, for example, may have the effect of increasing the atmospheric CO_2 load, with an accompanying increase in surface temperature. Recent calculations using empirically derived models indicate that the biosphere could not absorb a significant amount of the CO_2 expected to be produced in the next 150 years (see Figure 12.18).

Carbon dioxide is stored in the oceans as carbonate and bicarbonate ions, but only in a relatively thin surface layer no more than about 60 m thick. There is some mixing, but measurements are inconclusive on how rapidly various layers mix. The amount of CO_2 that can be absorbed in water depends on the water temperature in an inverse relationship. This, according to our understanding of the green-

house effect, would provide an undesirable positive feedback mechanism. It is likely the amount of CO_2 taken up by the oceans is in direct proportion to the partial pressure of the CO_2 in the atmosphere, so that an increase in atmospheric CO_2 would be followed by a proportionate increase in CO_2 stored in the oceans.

It is not clear that the same relationship is true for biochemically stored CO_2; nor is it clear whether the oceans or uptake by living organisms dominates as the overall most efficient storage mechanism. If current trends are sustained, that is, half the released CO_2 remains in the atmosphere, it will be possible to multiply the amount of atmospheric CO_2 by more than a factor of 4 in the next century. The implications of this increase are not entirely predictable, as we know; but certainly,

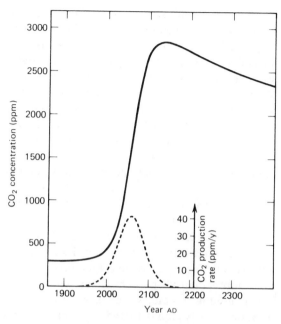

Figure 12.18 CO_2 concentration as a function of time assuming all recoverable fossil fuel is burned; CO_2 production rate is the lower, bell-shaped curve. [From U. Siegenthaler and H. Oeschger, "Predicting Future Atmospheric CO_2 Levels," *Science* **199**, 388 (1978). Copyright © 1978 by the American Association for the Advancement of Science.]

the global energy balance will show deviations.

Another combustion by-product that is ordinarily not thought of as a pollutant is water. We know how important water vapor is in the radiative balance of the atmosphere and in the formation of clouds, so the production of large amounts of water vapor by man should concern us. Water vapor is produced not only as a combustion by-product in the oxidation of hydrocarbons—petroleum, but is also inserted into the atmosphere by wet cooling towers used in electrical generating plants. Water vapor production by jet aircraft is really quite remarkable: about 1.25 kg of water vapor is produced for each kilogram of fuel consumed; and for the SST the fuel consumption

is in excess of 90,000 kg/hr. Evaporative cooling towers are likewise substantial water sources. A 1000-MWe plant requires about 10^7 gal of water per day, enough for a community of about 100,000 people.

For the purposes of discussing the effects of water vapor we must again separate the atmosphere into the troposphere and the stratosphere, since it is in the former region that weather phenomena are formed and in the latter region that chemical, photochemical, and radiative interactions are important. What is more, the concentration of water vapor in these two regions is quite different, the mechanisms for obtaining water vapor in the two regions are different, and under normal conditions there is very little exchange of water vapor between them.

Exchange of water vapor between the troposphere and the stratosphere is inhibited by an "ice trap" at the tropopause. Referring to Figure 12.19, we see there is a steady decrease in atmospheric temperature up to the tropopause. The trapping effect results from the temperature dependence of the partial pressure of water vapor. You will recall from Dalton's law that in a mixture of noninteracting gases, the total pressure is the sum of the pressures of the individual gases, each behaving as if the others did not exist. The partial pressure of water vapor as a function of temperature can be calculated approximately using the relationships developed for ideal gases in Chapter 3;

$$T_1 p_1^{(1-\gamma)/\gamma} = T_2 p_2^{(1-\gamma)/\gamma}. \qquad (12.21)$$

The quantity γ, the ratio of the constant-pressure and constant-volume heat capacities, is not easily calculated for polyatomic molecules such as H_2O and must be determined empirically. For water vapor in the temperature region around $0°C$ it is equal to about 1.05. The consequence of the temperature dependence of partial pressures is that as a given mixture of air and water vapor is cooled by rising in the

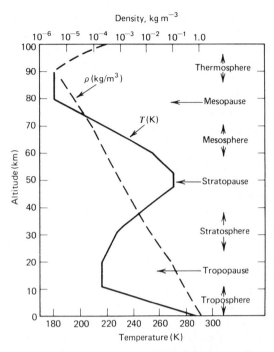

Figure 12.19 Temperature and density profile according to the U.S. Standard Atmosphere (1962) and the IUGG nomenclature (1960).

atmosphere, the quantity of water in the vapor form must decrease to maintain the appropriate partial pressure. Since the temperatures and pressures involved are beyond the triple point, the water vapor freezes out as ice. Hence, the tropopause is often referred to as an ice trap. Atmospheric temperatures increase above the tropopause; water-vapor concentrations above this boundary average around $2\,\mu$g of H_2O to 1 g of air. There is some evidence that water vapor in the stratosphere increases with altitude, presumably being produced by the destruction of methane by reactions with atomic oxygen.

The tropopause ice trap can be bypassed by natural and man-made phenomena. Major volcanic eruptions known to eject substantial quantities of particulates into the stratosphere also deposit massive quantities of water and

trace gases well above the ice-trap region. Thermonuclear weapons testing in the stratosphere has also been observed to bypass the normal cold trap. Water vapor in the lower stratosphere appears to have a residence time of more than 2 yr; in the upper regions it may be as high as 50 yr. It is possible that massive bypasses of the ice trap in the past two decades are still having an effect on the water-vapor concentration in the stratosphere. Measurements have not been made systematically, and so conclusions cannot be formed. However, observations of other phenomena may be significant.

The air at 10,000 m altitude is very dry—that is, the amount of water vapor per liter of air is very small. Yet the *relative humidity* is very high, more than 99 percent, because the very small amount of water vapor there has a partial pressure nearly equal to the maximum it can have for the temperature at that altitude. As a consequence, any additional water deposited at that altitude is likely to trigger a precipitation of water, usually as ice. The contrails of high-flying jet aircraft are a familiar example of this process. Since the introduction of commercial jet aircraft in 1958, the number of flights of such planes in the upper troposphere has increased almost exponentially. One may indeed ask if these flights have influenced the ice content of this region—are there more clouds now than before 1958? While the data are not necessarily conclusive, observational evidence seems to point to an effect.

A permanent increase in cloudiness has to have an effect on the earth's albedo. An increased albedo would produce a loss of solar radiation at the earth's surface, which in turn would produce a net loss in evaporation from the surface. This would reduce the temperature-moderating effect of the water vapor and tend to make the atmosphere hotter, thus increasing the water-vapor loading capability, leading to a lesser cloud-cover production. If this reasoning is correct, additional water vapor in the troposphere has a negative feedback effect, not undesirable. The effects of additional water vapor in the stratosphere, other than ice formation, are not clear. If water vapor reacts chemically with ozone as it does with methane and other trace constituents, serious problems could result from large water depositions. Ozone is probably the most important constituent of the stratosphere, as we have seen, being one of the major filters for short-wavelength electromagnetic radiation, which would be injurious if incident upon the earth. Clearly, experimental work is needed to investigate the distribution of water vapor in the upper atmosphere and gain an understanding of its influence on the chemical and physical processes that are part of the earth's heat engines before its natural abundance is significantly increased by contributions from man-made sources. For that matter, the same could be said for all the trace constituents of the atmosphere.

Carbon monoxide is the second most prevalent gas emitted in man-made processes, principally by the internal combustion engine used in automobiles (see Table 12.3). The production and control of CO emissions from auto engines have already been discussed in

TABLE 12.3 Global CO Emissions in the United States in 1968

Source	CO Emission ($\times 10^6$ tonnes/yr)			Percent of Total		
Technological Sources						
Fuel combustion in mobile sources	63.8			67.5		
Motor vehicles		59.2			62.7	
Gasoline			59.0			62.5
Diesel			0.2			0.2
Aircraft		2.4			2.5	
Vessels		0.3			0.3	
Railroads		0.1			0.1	
Non-highway use of motor fuels		1.8			1.9	
Fuel combustion in stationary sources	1.9			2.0		
Coal		0.8			0.8	
Fuel oil		0.1			0.1	
Natural gas		[a]			[a]	
Wood		1.0			1.0	
Industrial processes	11.2			11.9		
Solid waste combustion	7.8			8.3		
Miscellaneous: man-made fires, coal refuse, etc.	9.7			10.3		
Cigarette smoke		<0.01			[a]	
Technological sources (subtotal)	94.4			100.0		
Natural Sources						
Forest fires	7.2					
Total from all sources	101.6					

Source: I. Jaffe, "Global Balance of Carbon Monoxide, in *Global Effects of Environmental Pollution*, ed. by S. Fred Singer, copyright © by D. Reidel Publishing Co., Dordrecht, Holland.
[a]Negligible.

TABLE 12.4   Estimated Global Sulfur Compound Emissions, 1968

Compound	Source	Estimated Emissions ($\times 10^6$ tonnes/yr)	Emissions as Sulfur ($\times 10^6$ tonnes/yr)
SO_2	Coal combustion	102	51
	Petroleum refining	6	3
	Petroleum combustion	23	11
	Smelting operations	16	8
H_2S	Industrial emissions	3	3
	Marine emissions	30	30
	Terrestrial emissions	70	70
SO_4	Marine emissions	130	44
	Total emissions		220

Source: E. Robinson and R. C. Robbins, "Gaseous Atmospheric Pollutants from Urban and Natural Sources", in *Global Effects of Environmental Pollution*, ed. by S. Fred Singer, copyright © by D. Reidel Publishing Co., Dordrecht, Holland.

Chapter 4. We are concerned here with the global implications of CO production. A small amount of CO is produced in nature; the background concentration of CO in "clean" air of the lower atmosphere seems to be about 0.1 ± 0.09 pm. This figure is subject to con- siderable variability depending upon the exact location of the sample, the time of day, the season, and so on. Levels as high as 150 ppm have been observed in Los Angeles freeway traffic during rush hours.

The precise mechanism for the removal of

TABLE 12.5   Estimated Annual Global Emissions of Nitrogen Compounds, 1968

Compound	Source	Source Magnitude ($\times 10^6$ tonnes/yr)	Estimated Emissions ($\times 10^6$ tonnes/yr)	Emissions as Nitrogen ($\times 10^6$ tonnes/yr)
NO_2	Coal combustion	3,074	26.9	8.2
	Petroleum refining	$11,317 \times 10^6$(bbl)	0.7	0.2
	Gasoline combustion	379	7.5	2.3
	Other oil combustion	894	14.1	4.3
	Natural gas combustion	20.56×10^{12}(ft^3)	2.1	0.6
	Other combustion	1,290	1.6	0.5
Total NO_2			52.9	16.1
NH_3	Combustion		4.2	3.5
NO_2	Biological action		500	150
NH_3	Biological action		5,900	4,900
N_2O	Biological action		650	410

Source: E. Robinson and R. C. Robbins, "Gaseous Atmospheric Pollutants from Urband Natural Sources," in *Global Effects of Environmental Pollution*, ed. by Fred Singer, copyright © by D. Reidel Publishing Co., Dordrecht, Holland.

CO from the atmosphere is unknown, although several possible sinks have been explored. Since the background level of CO seems to have remained constant over the same period of time the CO_2 level has increased, active CO removal mechanisms surely exist. CO is relatively inert chemically at the level of concentration normally found in the atmosphere, and photoinduced reactions involving CO are rather infrequent—CO is mostly transparent to solar radiation, although it may be reformed to CO_2 by a collision with atomic oxygen and a third body, such collisions being rare. What is more, chemical reactions in the lower atmosphere are very slow; for example, the reactions

$$CO + O_2 \rightarrow CO_2 + O \qquad (12.22)$$

and

$$CO + H_2O \rightarrow CO_2 + H_2 \qquad (12.23)$$

require 2.1×10^4 and 2.3×10^4 J/mol, respectively, constituting a rather high energy barrier.

Other possible CO sinks include biological interactions with soil bacteria or vegetation, oceanic absorption, and absorption by surfaces. Each of these mechanisms is a possible CO sink, but the extent to which each contributes to the global CO balance is not known.

Aside from the obvious health hazard of high concentrations of CO, this particular gaseous emission from combustion sources seems to be relatively inert in terms of the global energy balance. It removes very little solar energy via photoinduced reactions and is not a critical ingredient in any major environmental system, *so far as is known*. The fact that the exact sink for CO is not known is somewhat disquieting, since major environmental changes involving other problems, heat production for example, could unwittingly destroy the CO sink, thereby creating a problem of potentially major proportions.

The other gaseous pollutants are emitted in large quantities in combustion processes: oxides of sulfur and oxides of nitrogen. These compounds are very important in the formation of photochemical smog, but are not large factors in determining the global energy balance, except in one respect. Oxides of sulfur in the presence of water vapor readily form sulfuric acid, which is very hydroscopic. As a consequence sulfur-containing molecules often act as nucleating centers for the formation of raindrops, and rain is often acidic. Oxides of nitrogen readily form ammonium radicals in the atmosphere and behave much as the sulfur-containing molecules. In fact, many raindrops contain ammonium sulfate. Most of these processes happen in the troposphere; residence times for these compounds are fairly short—10 days at most. There is an ambient background concentration of sulfur and nitrogen compounds in the parts per billion range. So even though vast quantities of these compounds are emitted in man-made activities and natural processes (see Tables 12.4 and 12.5), they do not appear to be as effective in the global regime as other previously discussed gases.

Stratospheric Ozone

We have mentioned the importance of the minute quantities of ozone in the upper atmosphere. Indeed, life as we know it may be possible on the surface of the earth because of the absorption of the shorter-wavelength ultraviolet portion of the solar flux by ozone. We should certainly make efforts to discover if any of mankind's activities have a potentially adverse effect on the concentration of this species.

The possibility of deleterious interactions between supersonic aircraft emission and ozone led the Department of Transportation, after the cancellation by Congress of federal funding of the SST project, to establish the Climatic Impact Assessment Program (CIAP).

The CIAP effort has produced some valuable data and even a few surprises. Of major importance was the discovery that water-vapor production by SSTs is probably not a serious problem; but at the same time it was found that the oxides of nitrogen may prove to be an even more serious threat to the stratosphere than had been imagined. Before we look at these interactions in detail, let us examine the natural equilibrium situation for ozone and the other constituents of the stratosphere.

Ozone is formed in the stratosphere by the combination of O_2 and atomic oxygen in the presence of a third body, usually on the surface of a particulate. The atomic oxygen results from the photolytic dissociation of oxygen. Absorption by oxygen takes place primarily in the visible and ultraviolet regions of the spectrum; the major part of the ozone absorption takes place only in the ultraviolet region.

Ozone is chemically very reactive; as a consequence, it is found normally in the less than 1-ppm concentration range in the dense atmosphere near the surface of the earth. Local concentrations associated with severe smog may temporarily exceed this range by 2 or 3 orders of magnitude.

It is the great chemical activity of ozone that prevents a large buildup, even in the stratosphere. One of the most rapid chemical reactions in this region is that between nitric oxide and ozone:

$$NO + O_3 \rightarrow NO_2 + O_2. \qquad (12.24)$$

The nitric oxide is quickly regenerated:

$$NO_2 + O \rightarrow NO + O_2, \qquad (12.25)$$

so NO becomes a catalyst in a chain reaction that has the net result of ozone destruction. There would normally be an equilibrium between production and destruction; the equilibrium concentrations would be altitude-dependent, peaking at about 20 km and falling off linearly above that.

The reactions of Eqs. 12.24 and 12.25 do not contribute significantly to ozone destruction ordinarily, since the oxides of nitrogen are not normally very abundant.[8] There are, however, two man-made mechanisms for greatly enhancing the stratospheric NO_x abundance: nuclear weapons testing (or conflict) and supersonic aircraft.

Thermonuclear weapons detonations produce locally very high temperatures that are most effective in disassociating oxygen and nitrogen. A wide variety of nitrogen species are produced; as equilibrium is achieved in the middle stratosphere, substantial amounts of NO_x are deposited. These are effective in the catalytic destruction of ozone. Each NO_x molecule may destroy thousands of ozone molecules before it is itself destroyed or removed from the region.

There is some evidence that weapons testing has indeed produced an ozone depletion. Both the United States and the Soviet Union conducted extensive nuclear tests in 1961 and 1962 just prior to signing the atmospheric test ban treaty. The ozone concentration failed to peak in 1963, as it should have by following the 11-yr solar cycle. Model calculations based on ^{14}C production in nuclear weapons tests (assuming NO_x production is proportional to ^{14}C production) yielded an estimated 3 to 6 percent ozone reduction.

The ozone concentration appears to be "normal" at the present time, although there are many fluctuations on various time scales that make unambiguous determinations difficult. The implication of the nuclear test results seems to be that widespread nuclear conflict could have a very significant ozone-depletion effect. This would be particularly exacerbated if weapons were detonated in the upper stratosphere to destroy satellites or to disrupt communications by causing the various charged-particle layers to disperse.

---

[8] NO_x production in the stratosphere is enhanced by the charged-particle bombardment of the solar wind. This probably explains the correlation between ozone *maximum* concentrations and sunspot *minima* in the 11-yr solar cycle.

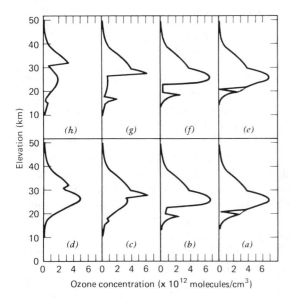

Figure 12.20 Effect of SSTs on stratospheric ozone. See text for details. [From H. Johnston, "Reduction of Stratospheric Ozone by Nitrogen Oxide Catalysts from Supersonic Transport Exhaust," *Science* **173**, 517 (1974). Copyright © 1974 by the American Association for the Advancement of Science.]

Generation of NO_x is also a major environmental effect of SSTs. A detailed computer-model study based on an assumed fleet of 500 SSTs each emitting 350 ppm NO concluded that the ozone reduction could be as much as a factor of 2. Figure 12.20 shows the results of this study. In Figures 12.20*a* and 12.20*e* the solid curve represents the normal ozone vertical distributions at 45° latitude. In 12.20*a* through 12.20*d* the accumulated 2-yr effect of the emissions of the 500 SST fleet uniformly distributed at various heights is shown: 12.20*a* = 20 to 21 km, 12.20*b* = 19 to 23 km, 12.20*c* = 17 to 25 km, and 12.20*d* = 15 to 31 km. For 12.20*e* through 12.20*h* 10 times the additional NO_x is assumed (a 5000 SST fleet); the altitude span for the uniform distribution is as for 12.20*a* through 12.20*e*.

A consensus of other groups studying the NO_x–O_3 problem indicates an increase of 50 percent in the stratospheric NO_x concentration would reduce the ozone level by between 7 and 12 percent. One group with a more sophisticated three-dimensional model estimates a 500 SST fleet in the northern hemisphere would not only cause a 16 percent reduction in ozone level there, but would also cause an 8 percent drop in the southern hemisphere!

The threat to stratospheric ozone posed by nuclear weapons and SSTs can be handled; at least, if we assume rationality on the part of our leaders, it can be. We need not have nuclear detonations in the atmosphere; we need not have SSTs. There is an argument that goes "Well, we *have* SSTs, so we must rely on our technology to find a way to reduce undesirable emissions." That is correct, of course; but technology has found ways to reduce emissions from power plants and automobiles—are we ready to commit ourselves to using these methods?

The final threat (so far) to stratospheric ozone comes from the **fluorocarbons**, or halomethanes as they are sometimes called, used as propellants in spray cans. Because of the widespread use of spray cans it may be much more difficult to control this threat.

The fluorocarbons, principally Feon-11 ($CFCl_3$) and Freon-12 (CF_2Cl_2), have been widely used since the 1930s as propellants because they are essentially chemically inert and nonflammable.[9] Until recently, there was little evidence of any health-related problems with the Freons.[10] Because of the chemical

[9]Freon is the brand name of fluorinated hydrocarbons manufactured by Du Pont; other trade names are Genetron (Allied Chemical Company), Isotron (Pennwalt Corporation), and Ucon (Union Carbide Chemicals Company). Other chemicals used as propellants are propane, butane, isobutane, CO_2, methylene chloride, N_2O, and vinyl chloride.

[10]J. W. Clayton, "Fluorocarbon Toxicity and Biological Action," *Fluorine Chem. Rev.* **201**, (1967).

inertness and the vast amount being produced currently ($\sim 500,000$ metric tons of Freon-11 and -12 in 1972), measurable amounts of this material are being found around the globe.

An average concentration of 61 parts per trillion (ppt) was found at points along a path between Los Angeles and the Antarctic. Concentrations between 50 and 150 ppt have been found at various other northern and southern hemispheric locations. So far, concentrations this high have been measured only in the troposphere, but it is only a question of time—perhaps a few years—before diffusion will bring the halomethanes to an altitude where they will be exposed to short-wavelength ultraviolet radiation.

The Freons photolytically dissociate when exposed to radiation between 170 and 220 nm, that which is available in the stratosphere. This dissociation produces free chlorine and fluorine atoms, which catalytically destroy ozone in much the same manner as does NO_x:

$$Cl + O_3 \rightarrow ClO + O_2,$$
$$ClO + O \rightarrow Cl + O_2. \qquad (12.26)$$

This process can continue many hundreds of times before the chlorine becomes bound or diffuses down. How much of an ozone decrease can the fluorocarbons produce?

Of course, only the results of model calculations are available, but they suggest it may already be too late to keep significant environmental effects from occurring. Figure 12.21 shows the results of one computer study, plotting ozone decrease as a function of time for various scenarios. The effect on the ozone concentration is large, but does this guarantee an effect on our lifestyles?

It has been estimated that any decrease in ozone concentration would cause a corresponding increase in skin cancer incidence. Thus, a 5 percent ozone depletion (10 percent

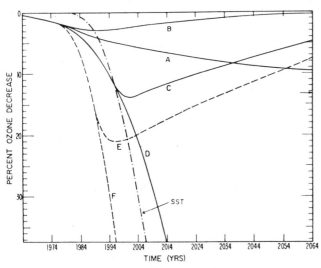

Figure 12.21 Predicted ozone decrease for various fluorocarbon production schemes. [From S. C. Wofsy, M. B. McElroy and N. D. Sze, "Freon Consumption: Implications for Atmospheric Ozone," *Science* **187**, 535 (1975). Copyright © 1975 by the American Association for the Advancement of Science.]

increase UV flux) could cause between 20,000 and 60,000 additional skin cancers. Skin aging would also be accelerated by an increase in ultraviolet.

Many more subtle effects might appear as a consequence of ozone reduction. The ocean phytoplankton, which produce a large portion of the earth's oxygen, could be adversly affected. Some plants, notably vegetables, appear to suffer from retarded growth when exposed to increased UV. Prolonged exposure can also produce mutations. Insects can see a portion of the ultraviolet region. Changes in the overall solar spectrum could affect their perception of skylight, flower colors, and sexual markings, although the role that sight plays is not entirely clear.

Some researchers have suggested that ozone depletion would lead to an average temperature change for the earth; whether it would be an increase or a decrease would depend on the many considerations we have examined earlier. Other climatologists feel that the stratosphere is not important in the regulation of climate. This question is only beginning to be studied.

There seems to be a clear and present danger to stratospheric ozone. What is being done? There is a nuclear test ban treaty. Not all nations have signed it; two of those that have not—the People's Republic of China and France—continue to test in the atmosphere. Two versions of SSTs are flying—the Russian and the Anglo-French Concorde.

World production of the halomethanes by 1979 had dropped to pre-1973 levels, primarily as a result of the U.S. ban on their use as aerosol spray propellants. Production in other countries continues to increase. The position of producers of fluorocarbons in these countries is that adverse effects have not been unequivocably proven. That hardly seems to be the point any more—indeed, we should be mounting a concentrated search for other possible stratospheric degraders. The discovery of the effects of the fluoromethanes on ozone has prompted one researcher, Michael McElroy of Harvard, to ask, "What the hell else has slipped by?"[11]

The interrelationships between the various parts of the earth's heat engines are very complex. We cannot say with certainty that even in the absence of humans the earth's system would be in stable equilibrium. Our mathematical models do not have enough sophistication to take into account all the variables. We know that human activities, especially in the past few decades, have had a measurable effect on the earth; for example, the CO_2 content has increased noticeably. The upper stratosphere is a very sensitive region of our atmosphere because in it gas concentrations are small and very important photo-induced chemical reactions occur. Thermonuclear tests in the stratosphere and large quantities of particulates and gaseous emissions from high-flying aircraft, volcanic eruptions, and artificial gases can seriously change the energy balance of this fragile region. We should understand the chemical and physical processes of the atmosphere a great deal better than we do before we allow the magnitude of these disruptions to increase, as it is likely to do at the present rate of growth of energy consumption.

SUMMARY

There are three kinds of equilibrium: *stable*, *unstable*, and *neutral*. Which kind of equilibrium (if any) is the earth in? Are there restoring forces?

There are fluctuations and variations in the sun's solar output. The amount of solar energy actually received by the earth depends on its albedo. Ice cover can affect the albedo profoundly.

[11]As quoted by A. L. Hammond and T. H. Maugh II, *Science* **186**, 335 (1974).

Solar radiation may be either *absorbed* or *scattered* by the earth's atmosphere. Absorption is frequency-dependent, leading to the *greenhouse Effect*.

Absorption is an atomic process; it can therefore be studied in detail only with the use of *quantum mechanics.* Treating atomic scatterers as simple electric dipoles yields a result consistent with observation—the sky is blue!

By using energy balancing between the sun, earth, and atmosphere, the temperature of the earth can be calculated.

Evaporation of water from the oceans is a very important aspect of the *earth's heat engines.* Ocean currents carry heat to the poles.

Heat production on the earth's surface could overwhelm the natural restoring forces in certain high-population-density areas. Particulates are produced in great abundance by nature, but not usually over large cities.

Carbon dioxide production will affect the global climate in the next few hundred years, if nothing changes. Other gases may not have the same effect, but other problems could occur. Ozone destruction by NO_x and the Freons could be disastrous in the next few decades.

REFERENCES

Sun, Weather, Climate

Bryson, Reid A., "'All Other Factors Being Constant...'—Theories of Global Climate Change," in *Man's Impact on Environment*, Thomas R. Detwyler (Ed.). New York: McGraw-Hill, 1971.

Eddy, John A. "The Maunder Minimum," *Science* **192**, 1189 (1976).

Fletcher, Joseph O., "Polar Ice and the Global Climate Machine," *Bull. Atomic Sci.,* December (1970), p. 40.

Hays, J. D., John Imbrie, and N. J. Shackleton, "Variations in the Earth's Orbit-Pacemaker of the Ice Ages," *Science* **194**, 1121 (1976).

Herman, John R., and Richard A. Goldberg, *Sun, Weather and Climate.* Washington, D.C.: National Aeronautics and Space Administration, 1978.

Langley, Richmond W., *Elements of Meteorology.* New York: John Wiley, 1970.

McCormac, Billy M., and Thomas A. Seliga (Eds.), "Solar–Terrestrial Influences on Weather and Climate." Dordrecht: D. Reidel, 1979.

Newell, Reginald E., "The Earth's Climatic History," *Tech. Rev.,* December (1974), p. 31.

Wilcox, John M. "Solar Structure and Terrestrial Weather," *Science* **192**, 745 (1976).

Earth and Oceans

Garland, George D., *Introduction to Geophysics.* Philadelphia: W. B. Saunders, 1971.

Munk, W. "The Circulation of the Oceans," *Sci. Amer.,* September (1955).

Newell, Reginald E., "Climate and the Ocean," *Amer. Sci.,* **67**, 405 (1979).

Sarter, J. Doyne, "Clouds and Precipitation," *Phys. Today,* October (1972), p. 32.

Stacey, Frank D., *Physics of the Earth.* New York: John Wiley, 1969.

Stewart, R. W., "The Atmosphere and the Ocean," *Sci. Amer.,* September (1969), p. 76.

Weyl, Peter K., *Oceanography.* New York: John Wiley, 1970.

Heat, Particulates, CO_2

Baes, C. F., Jr., H. E. Goeller, J. S. Olson, and R. M. Rotty, "Carbon Dioxide and Climate: The Uncontrolled Experiment," *Amer. Sci.* **65**, 310 (1977).

Bosck, William L., "Meteorological Consequences of Atmospheric Krypton-85," *Science* **193**, 195 (1976).

Broecker, W. S., T. Takahashi, H. J. Simpson, and T. -H. Peng, "Fate of Fossil Fuel Carbon Dioxide and the Global Carbon Budget," *Science* **206**, 409 (1979).

Bryson, Reid A., and Brian M. Goodman, "Volcanic Activity and Climate Changes," *Science* **207**, 1041 (1980).

Cole, Lamont C., "Thermal Pollution," in *Man's Impact on Environment*, Thomas R. Detwyler (Ed.). New York: McGraw-Hill, 1971.

Matthews, W. H., W. W. Kellog, and G. D. Robinson (Eds.), *Man's Impact on the Climate*. Cambridge, Mass.: MIT Press, 1971.

Newell, Reginald E., "The Global Circulation of Atmospheric Pollutants," *Sci. Amer.*, January (1971), p. 32.

Siegenthaler, U., and H. Oeschger, "Predicting Future Atmospheric Carbon Dioxide Levels," *Science* **199**, 388 (1978).

Singer, S. Fred (Ed.), *Global Effects of Environmental Pollution*. Dordrecht: D. Reidel, 1970.

Stommel, Henry, and Elizabeth Stommel, "The Year Without a Summer," *Sci. Amer.*, June (1979), p. 176.

Toon, Owen B., and James B. Pollack, "Atmospheric Aerosols and Climate," *Amer. Sci.*, **68**, 268 (1980).

Wang, W. C, Y. L. Yang, A. A. Lacis, T. E. Mo, and J. E. Hanson, "Greenhouse Effects Due to Man-Made Perturbations of Trace Gases," *Science* **194**, 685 (1976).

———, *Man's Impact on the Global Environment*, Report of the Study of Critical Environmental Problems (SCEP). Cambridge, Mass.: MIT Press, 1970.

Ozone

Alyea, Fred N., Derek M. Cunnold, and Ronald G. Prinn, "Stratospheric Ozone Destruction by Aircraft-Induced Nitrogen Oxides," *Science* **188**, 117 (1975).

Cutchis, Pythageras, "Stratospheric Ozone Depletion and Solar Ultraviolet Radiation on Earth," *Science* **184**, 13 (1974).

Lubkin, Gloria B., "Fluorocarbons and the Stratosphere," *Phys. Today*, October (1975), p. 34.

Maugh II, Thomas H., "The Threat to Ozone is Real, Increasing," *Science* **206**, 1167 (1979).

Wofsy, Steven C., Michael B. McElroy, and Nien Dak Sze, "Freon Consumption: Implications for Atmospheric Ozone," *Science* **187**, 535 (1975).

GROUP PROJECTS

Project 1. Survey a reasonable number of people to find out how many are concerned enough about the fluorocarbon problem to switch from spray deodorants to other forms. What about other household spray products? Is the ozone problem understood by the average citizen?

Project 2. Estimate the heat load generated by your college or university. Include building heat, electrical appliances, and people. What fraction of the solar flux is it? Could this affect the local temperature?

EXERCISES

Exercise 1. In Equation 12.1 what percentage change in α is required to produce a 1 percent change in T?

Exercise 2. If the solar flux were suddenly to drop by 1.6 percent, calculate the annual reduction in energy absorbed by the earth. Calculate the length of time for the earth to radiate that amount of energy.

Answer: 1.44×10^{16} J/yr, about 147 hr.

Exercise 3. Assume an average decay energy of 2 MeV for the radioisotopes in the earth. Use Figure 12.4 to calculate the total radioactivity in the earth. Assume some average values of Q to account for the difference between oceans and continents. Compare this value to the presumed natural abundance of uranium and thorium.

Exercise 4. Calculate the total daily heat energy available at the earth's surface resulting from radioactive decay, and compare it to the daily solar energy received by the earth.

Exercise 5. Use the average heat flux and the temperature gradient of the earth to estimate the thermal conductivity near the earth's surface.

Exercise 6. Assume a 1 percent reduction in solar insolation worldwide for one year. Assume that in order to maintain the global balance, the earth retrieves this lost energy by converting ocean water to ice, thus liberating its heat of fusion. How much ice would this create (in cubic kilometers)? Assume the Antarctic continent to be circular with a radius of 1800 km and that the nearby oceans freeze to a depth of 5 m. How many additional degrees of latitude will be covered by half this ice (the other half being in the Arctic)?

Answer: 2.93×10^3 km, $10.2°$ of latitude.

Exercise 7. Assume a one percent increase in solar insolation world-wide, and that in order to maintain the global temperature balance the earth increases its average albedo by converting margin lands to deserts. What increase in desert area is required? What other mechanisms could the earth use to accomplish the same goal?

Exercise 8. Calculate the percentage absorption of 279-eV photons from the sun, assuming absorption by helium only.

Exercise 9. The calculation of Example 12.3 assumed no water vapor in the stratosphere. If μ_w/ρ_w at 400 nm is $3\mu_{air}/\rho_{air}$, how much of an effect does the addition of 10^{-12} g/cm^3 H_2O have on the absorption of 400-nm photons?

Answer: $I/I_0 = 0.861$.

Exercise 10. Use Eq. 12.18 to design a lighting system for penetrating fog. What assumptions do you have to make to do that? What biological effects are you neglecting, if any?

Exercise 11.
(a) The potential energy of a harmonic oscillator is given by $PE = \frac{1}{2}kx^2$. The kinetic energy is given by $KE = \frac{1}{2}m\dot{x}^2$. Show that the total energy $(KE + PE)$ is given by

$$W = \frac{1}{2}m\omega_0^2 x_0^2.$$

(Note: $x = x_0 \cos \omega_0 t$).

(b) Show that

$$\Phi_{av} = \frac{8\pi^2}{3}\left(\frac{r_e}{\lambda_0}\right)\frac{W}{\tau_0},$$

where $r_e = e^2/(4\pi\epsilon_0 mc^2)$ is the "classical radius of the electron," and $\lambda_0 = c\tau_0 = 2\pi c/\omega_0$.

(c) Justify this statement:

$$(\Phi)_{av} = -\frac{dW}{dt}.$$

(d) Hence, show that

$$W = W_0 \exp\left(\frac{-t}{\tau_r}\right),$$

where

$$\tau_r = \frac{3}{8\pi^2}\frac{\lambda_0}{r_e}\tau_0.$$

(e) The equation in (d) indicates an oscillating electron loses energy by radiation exponentially. After a time τ_r the initial energy of excitation W_0 is reduced a factor of $1/e$. A wave train emitted by the electron would have a length $l = c\tau_r$. Show that for visible light this is of the order of 300 cm.

(f) Light from two independent sources will not interfere coherently in a double-slit experiment. What does that tell you about the result in (e)?

Exercise 12. Using Figure 12.8, calculate the final percentage of radiation absorbed by the earth, if the solar flux interacts with the cloud cover by 44 percent. Assume all other interactions scale as shown. Estimate the surface temperature of the earth.

Exercise 13. Repeat Example 12.6 for 21°C. Use Equation 12.19 to find the heat required to raise 1 g of air from 20° to 21°C, maintaining saturation.

Exercise 14. A certain parcel of air saturated with water vapor at 1 atm has a total mass of 1 g at a temperature of 25°C.

(a) How many moles of air and water vapor are there in the parcel?
(b) If the temperature rises to 26°C and saturation is maintained by contact with the ocean, how many moles of each are in the parcel?
(c) What total heat input is required to accomplish (b)?

Answer: (a) 0.0691 mol air, 0.00174 mol water. (b) 0.06906 mol air, 0.00184 mol water. (c) 6.097 J.

Exercise 15. Assume the oceans absorb an amount of solar energy in proportion to their area. Use a result of Exercise 14 to estimate the total amount of water evaporated from the oceans daily.

Exercise 16. The heat released by dissolving m mol of salt in 1000 g of water at 298 K is given by

$$\Delta H = 3.861m + 1.992m^{3/2} - 3.038m^2 + 1.019m^{5/2} \text{ kJ.}$$

Find the heat required for 1000 g to *reduce* the salinity of seawater from 30‰ to 20‰.

Exercise 17. An SST traveling at 400 m/s produces about 110,000 kg of water per hour at an altitude of 30 km. Assume a cylindrical exhaust plume having a diameter equal to the wingspan. How much excess water, beyond the amount required for saturation in the exhaust stream, is produced per hour?

Exercise 18. The CO_2 production curve in Figure 12.18 is of the form

$$P(t) = \frac{d}{dt}\left(\frac{A}{1 + B\exp(-Ct)}\right).$$

Estimate the parameters A, B, and C from the figure. Would a function of this form fit the petroleum production curve of Figure 2.6?

CHAPTER THIRTEEN

LOCAL CLIMATIC EFFECTS

In Chapter 12, we examined the large-scale effects of environmental pollution. In this chapter we study much smaller-scale phenomena. These range from the effects of a single power plant on the surrounding countryside to the combined effect of a large number of automobiles in a metropolitan area. We also look at the details of the formation of such environmental disturbances and consider several techniques for eliminating or reducing their production. It would be of considerable interest to include a detailed study of water pollution as well. But this could take us rather far afield from physical impacts toward biological impacts. These important aspects of energy production are discussed in the ample literature of the subject.

13.1 MICROCLIMATE CHANGES

Microclimate is the climate of a small geographic area—from a few hundred square meters to a few hundred square kilometers. The overall climate of any region is that forced upon it by the global weather forces, but modulating this can be significant departures caused by purely local occurrences. It is these local effects that we wish to study. The first of these is the effect of gathering together a large

number of people to live or work in a small region.

The Climate of Cities

Life in the city is different from life in the country. Aside from the sociological differences implies by this truism, we also mean to indicate that there are climatic differences between the city and the country. These differences have been known for years, although their causes are only now beginning to be understood. Several urban–rural climatic differences are noted in Table 13.1.

Of course, the existence of the city is the reason for the differences; but what characteristics of cities make them so different from rural areas as to cause climatic changes? There are several factors. First of all, the city is a region of high population density; the by-products of human activities are amplified in this concentration. What is more, many of the activities that are most injurious to the environment take place generally only in the city or metropolitan area—heavy manufacturing, refining, large central station energy production, and so on. As a consequence, the urban air is generally heavily laden with aerosols and gaseous contaminants. These contaminants not only reinforce the green-

TABLE 13.1 Climatic Differences in Cities

Element	Comparison with Rural Environs
Temperature	
Annual mean	0.5 to 0.8°C higher
Winter minima	1.1 to 1.6°C higher
Relative humidity	
Annual mean	6% lower
Winter	2% lower
Summer	8% lower
Dust particles	10 times more
Cloudiness	
Clouds	5 to 10% more
Fog, winter	100% more
Fog, summer	30% more
Radiation	
Total on horizontal surface	15 to 20% less
Ultraviolet, winter	30% less
Ultraviolet, summer	5% less
Wind speed	
Annual mean	20 to 30% lower
Extreme gusts	10 to 20% lower
Calms	5 to 20% more
Precipitation	
Amounts	5 to 10% more
Days with <0.2 in.	10% more

Source: J. T. Peterson, *The Climate of Cities: A Survey of Recent Literature.* Washington D.C.: National Air Pollution Control Administration, 1969.

house effect over the city, but also they absorb heat.

This heat absorption also tends to cause a stable air mass to form over the city, hindering dispersion of the pollutants, in effect acting as a positive feedback mechanism—the more pollutants there are, the more there can be.

The city is a compact mass of buildings and pavement; these materials are more like rocks than soil or vegetation in terms of their heat conduction. Stone absorbs heat about three times slower than soil does but conducts heat three times faster, so that for equal volumes and equal exposure times to the solar flux the stone will have stored more heat but have a lower temperature. At night the stonelike material of the city releases heat by radiation three times slower than the surrounding soil; as a consequence the city tends to be substantially warmer at night than neighboring rural areas. The same effect occurs during the day, but is less pronounced. The fact that metropolitan areas are warmer in general than their surroundings has led to their being called "heat islands."

The geometry of city construction also leads to climatic effects. Cities usually consist of buildings substantially taller than the surrounding terrain. This high density of tall buildings serves as a very effective solar flux trap; the building surfaces absorb and reflect the sunlight. The reflected sunlight usually impinges on another building surface; very little escapes. This is to be contrasted with open fields in which solar radiation reflected from the soil is almost always directed back up to the atmosphere, as shown in Figure 13.1. In addition, the tall buildings serve as effective wind breaks, introducing turbulence and reducing cooling by evaporation.

Indeed, the water content of city surfaces is quite different from their rural environs. Rain is quickly ducted away from the city by gutters and sewers, and even snow is usually quickly cleared away. The result is that little water remains on or near the city surface to be evaporated, thereby cooling the city.

Example 13.1 Compare heat absorption by evaporation in cities and countryside.

Assume 95 percent runoff in the city and 60 percent soil absorption in the country. For a given quantity of rain, say *M* kg, the ratio of the energy absorbed by a city to that by the country would be

$$\frac{(M \times 10^3 \, g) \times (540 \, cal/g) \times (1 - 0.95)}{(M \times 10^3 \, g) \times (540 \, cal/g) \times (1 - 0.6)} = \frac{0.05}{0.4}$$

$$= 0.125$$

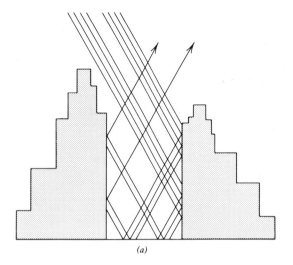

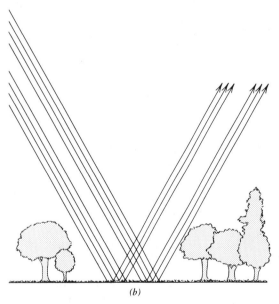

Figure 13.1 (*a*) Cities absorb incident sunlight relatively efficiently, while (*b*) in the countryside there is little absorption.

The city loses only 12.5 percent of the heat to water evaporation compared to an equal area of countryside.

Finally, the city is a gigantic heat machine, summer and winter. In the summer, of course,

heat extracted from buildings by air conditioners (plus the work input!) is rejected to the city atmosphere. In the winter the poorly insulated buildings pour heat out of doors. We can estimate the effect of heat production on local temperature. We generalize Eq. 8.7 to include a solar heat term $\dot{Q}_s$ plus a human-generated heat term $\dot{H}$:

$$\dot{Q}_s + \dot{H} = \epsilon\sigma T_H^4, \qquad (13.1)$$

where T_H is the new equilibrium temperature that we wish to calculate. The $\dot{Q}_s$ term was calculated in Section 12.2 to be 343 W/m^2, and by taking a ratio of Eq. 13.1 with the same equation without H,

$$\frac{343 + \dot{H}}{343} = \frac{T_H^4}{T^4}. \qquad (13.2)$$

Thus,

$$T_H = T \sqrt[4]{1 + \frac{\dot{H}}{343}}. \qquad (13.3)$$

The radical can be expanded in a Taylor's series. The temperature difference $\Delta T = (T_H - T)$ is given by

$$\Delta T \approx \frac{T}{4}\left(\frac{\dot{H}}{343}\right). \qquad (13.4)$$

With this result the approximate effect of a given heat production rate can be estimated.

All these characteristics combine to alter substantially the climate in *and near* urban centers as compared to their rural surroundings. These alterations are not necessarily bad; cities have warmer winters with more frost-free days, lower humidity, and lower wind speeds than rural areas. On the other hand, they also have more smog, more precipitation, and more fog (although there is less very dense fog).

Most of us have had the experience of approaching a large city either by car or by plane and recognizing the metropolitan area first as a dark smudge on the horizon. Upon closer ap-

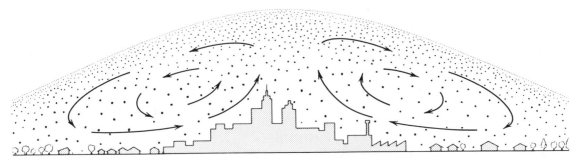

Figure 13.2 Cities generate "dust domes" by breaking up natural wind patterns, generating heat, and trapping pollutants.

proach the city seems to be enveloped in a dusty, murky dome. This "haze hood," or **dust dome** as it is often called, is characteristic of most large cities these days (see Figure 13.2). It is the result of the interaction of the heat island effect, the emission of pollutants, and the modification of winds by the city. The dust dome starts to accumulate on Monday morning; automobiles and factories begin to pour heat and pollutants into the atmosphere. At night the particulates suspended over the city begin to cool by radiation more rapidly than the surrounding air, especially those at the top of the dome. These particles serve as condensation nuclei for fog. The fog blanket drops down over the city and retards further radiative cooling. The fog layer also keeps the particles from being convected upward and out of the dome. The city's influence on the winds of the region also affects the particulate distribution. By Tuesday morning the fog–particulate layer is still there and serves as an efficient trap for another day's output of pollutants. This process will continue until either a strong wind or a heavy rain removes the dust accumulation, or a weekend comes with sufficient natural circulation to allow the dust dome to drain off.

There seems to be little that can be done to relieve this problem short of drastically altering the concepts of cities as we know them, or somehow doing away with all particulate

emissions. Neither solution is particularly practical at present. For other reasons as well, altering the city structure by reducing the density of heat-producing devices (cars, factories, people), by lowering the city's vertical profile, by changing the water content by introducing large green areas, and by using a wider variety of construction materials could be very beneficial. Obviously, a solution of this type would be difficult to implement for existing cities; but one can hope that plans for expansion of present urban areas or construction of "new towns" will be done with these principles in mind, as they often are in Europe.

Urban characteristics combine to produce another phenomenon typical of large cities: fog. Fogs are ordinarily produced by two mechanisms: radiative cooling of the surface air layer below the dew point and layering of cool dry air and warmer moist air. We cannot use thermodynamics to describe the formation of water droplets from water vapor—an ideal gas does not even liquefy! Fog, rain, and clouds all require nucleation centers, particulates usually, which promote the accumulation of water into a droplet. In cities there is a large abundance of particulate pollutants to promote droplet formation. Also the alteration of normal wind patterns by the city geometry retards mixing and dispersal, making fog formation more likely.

Studies indicate that cities do indeed have

more days and nights of fog than neighboring rural areas; however, they do not have as much very dense fog—this notwithstanding what you may have heard about London's "pea soupers." The extra warmth of the city increases the water-holding ability of the surface air, reducing the severity of water droplet formation. In fact, the worst fogs are to be found in valleys, where normal air circulation is stagnated, the ground is moist, and radiative cooling proceeds rapidly at night. California's central San Joaquin Valley presents such a situation in the winter, where dense fogs may last more than 15 days before being blown away by strong off-the-ocean winds. At least this particular weather phenomenon of the largely rural valley cannot be traced to characteristics of the urban centers of California. There are places where climatic effects can be observed in rural areas downwind of population concentrations.

Metropolitan Influences on Rural Climate

The La Porte Anomaly The U.S. Weather Bureau not only maintains its own well-equipped observation stations throughout the country but also uses privately owned and maintained substations at many hundreds of locations to supplement its data. Records from these stations have generally been scrupulously kept for about the past 60 years. An examination of the records of such stations in the northern Indiana area near the Gary–Chicago industrial complex has revealed a curious weather phenomenon (these stations are shown in Figure 13.3). The weather records for the small town of La Porte, Indiana, seem to be significantly different from those of stations only a few miles away. Among other discrepancies there appeared to be more precipitation, more hail, and more thunderstorms at La Porte (see Figures 13.4 and 13.5).

At first the La Porte anomaly was thought to be caused by observer bias, since more than one observer had been involved. But from

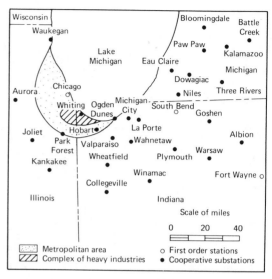

Figure 13.3 Weather stations in the area around the southern tip of Lake Michigan.

1927 to 1965, a single person reported the data from La Porte; detailed studies discredit the observer bias idea and conclude that the climatic observations at La Porte are real. The most likely cause of the observed phenomena is the large industrial complex to the west coupled with the close proximity of La Porte to Lake Michigan to the north.

This industrial complex has a number of steel refineries that exhaust large amounts of particulates and water vapor to the atmosphere; in addition, there is substantial heat generated by the metropolitan area. The particulates probably act as nuclei for condensation either as rain or hail, and the heat island may be a factor in the development of thunderstorms. The apparent trend toward the equivalence of weather records at La Porte and its neighbors beginning around 1960 may have been caused by a change in the method of steel production using the basic oxygen furnace. By 1965 this technique, which requires only 7 percent of the time required in the open hearth process, was producing 25 percent of the U.S. steel output. The precise interaction

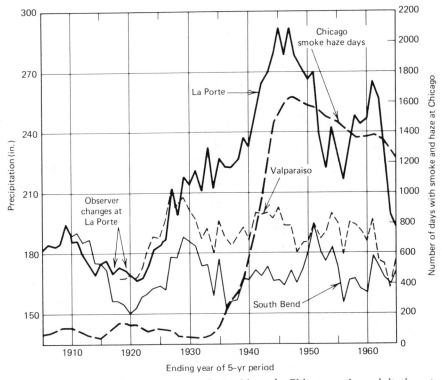

Figure 13.4 Correlation between smoke and haze in Chicago and precipitation at La Porte, Indiana as a function of time. [From Stanley A. Chagnon, Jr., *American Meteorological Society* **49**, 4 (1968).]

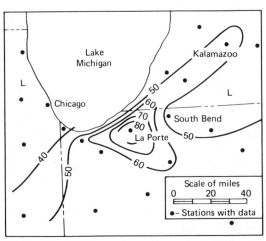

Figure 13.5 Lines of equal number of severe thunderstorms downwind of Chicago. [From Stanley A. Chagnon, Jr., *American Meteorological Society* **49**, 4 (1968).]

mechanisms between urban pollutants and downwind climatic effects are not known. However, these interactions are certainly real, as is amply demonstrated by another recent phenomenon.

Acid Rain It has been known for several decades that the rain immediately downwind of coal-burning plants is acidic. However, in the 1950s it was discovered that the precipitation (rain, snow, and dustfall) at considerable distances from sources of air pollution is also acidic. This was in Scandinavia, where later measurements have shown an even more pronounced acidity. In Figure 13.6 we show lines of average pH in Scandinavia deduced from measurements taken in 1957 and 1970.

Recall that the pH of a solution is a measure

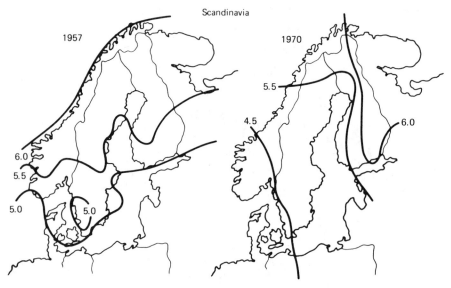

Scandinavia

1957

1970

Average pH from 12 monthly samples

Figure 13.6 pH of rainfall in Scandinavia in 1957 and 1970. (Reprinted with permission from Gene E. Lickens, "Acid Precipitation," *Chemical Engineering News,* November 22, 1976, p. 31. Copyright © 1976 by the American Chemical Society.)

of the hydrogen-ion content: The pH is the negative logarithm of the hydrogen-ion concentration expressed in moles per liter. For pure neutral water the pH should be exactly 7.0. Because of the presence of carbon dioxide in the normal (unpolluted) air, rain should have a pH of 5.7. The 1957 values in Scandinavia were close to this value. But by 1970 and subsequently, the hydrogen-ion concentration had increased markedly. Note that a change in pH from 5 to 4 represents a factor of 10 increase in hydrogen-ion concentration. The most acidic rain observed to date fell in Scotland in 1974; it had a pH of 2.7, about equivalent to household vinegar!

A similar situation has been found in North America. Both the northeastern United States and southeastern Canada have experienced marked increases in acid precipitation in the years from 1955 to 1972, as indicated in Figure 13.7. Acid rain is believed to have a number of adverse effects:

- Decrease in agricultural and forest yields.
- Decreases in freshwater fish production.
- Depletion of nutrients from soils or aquatic ecosystems.
- Inactivation of important microorganisms.
- Corrosion or deterioration of some exposed materials.

Effects of this nature have been noted. In lakes above about 600 m elevation in the Adirondacks, a significant depletion of the fish population has been correlated with pH. This is illustrated in Figure 13.8: note that no fish have been found in lakes that have a pH lower than about 4.6. This may be because certain microorganisms, for example plankton, are destroyed by the acidity of the water. Plankton are important links in the freshwater fish food chain.

Significant effects have also been observed in Canadian lakes. In this case, the airborne pollutants, most of which originate in the United

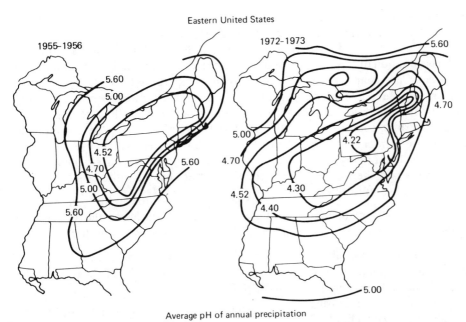

Figure 13.7 pH of rainfall in the eastern part of the United States in 1955 and 1972. (Reprinted with permission from Gene E. Lickens, "Acid Precipitation," *Chemical Engineering News*, November 22, 1976, p. 31. Copyright © 1976 by the American Chemical Society.)

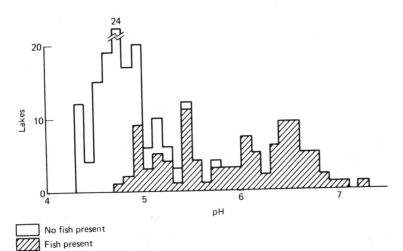

Figure 13.8 Frequency distribution of pH and fish populations in Adirondack mountain lakes above 610 m, 1975. [From C. L. Shofield, "Acid Precipitation: Effects on Fish," *Ambio* **5**, 229 (1976).]

States, have an effect over a large region. So perhaps the terminology *microclimate* change is not entirely appropriate!

The mechanism for the production of this high level of acidity in rain water seems clear. Oxides of sulfur and nitrogen are produced in great quantities in urban areas where there are concentrations of power plants burning coal and oil, industrial plants such as smelters, home heating, and automobiles. These oxides are rapidly converted to sulfates and nitrates and thence into sulfuric and nitric acids in the moist atmosphere.

Example 13.2 Estimate the amount of sulfates formed daily in the Chicago region.

The Chicago region has a population of about 6×10^6, or about 3 percent of the population of the United States. It should have, then, about 3 percent of the electric power generating capability, or about 20,000 MW. But about 40 percent of this is nuclear, so about 12,000 MW must be coal fired. But there must be at least 5000 MW equivalent of other coal-burning industries in this area. So we estimate 17,000 MW, or a daily energy production of $17,000 \times 10^6 \times 24 \times 60 \times 60$, or 1.47×10^{15} J. Under proper operating conditions there will be about 1000 kg of sulfur emitted per 10^{11} J of energy produced. We can thus estimate that at least 15,000 tonnes of sulfur is emitted in the Chicago region; most of this will come down as acid rain.

The pressures on the government to relax clean air standards by allowing the burning of higher-sulfur coal will not improve the acid precipitation problem. We should be attacking this problem at its source—the coal itself. Not only must we be good international neighbors, but it is also in our own enlightened self-interest to reduce the acidity of rainfall in our forests and recreational regions.

Before we discuss some of the mechanisms that promote air pollution, let us examine one more example of a microclimate change induced by anthropogenic (human) causes.

Power Plants We know already of the effects of the oxides of sulfur and nitrogen from power plants in terms of acid rain. But there is another, more localized effect attributable to power plants: increased humidity. Power plants that use spray ponds or wet cooling towers (see Chapter 8) will discharge considerable amounts of water to the surrounding countryside. This water will cause fogs and icing in the winter and increased humidity in general. The effect can be quite large.

Example **13.3** Estimate the amount of water evaporated to dissipate the waste heat from a 1000-MW coal-fired plant daily.

A coal-fired plant will be about 40 percent efficient, so that 1500 MW of heat must be dissipated; or

$$1500 \times 10^6 \times 24 \times 60 \times 60 = 1.3 \times 10^{14} \text{ J}.$$

Each gram of water evaporated requires 540 cal or 2.26×10^3 J of heat. So $(1.3 \times 10^{14})/(2.26 \times 10^3)$, or 5.75×10^4 tonnes of water per day is required.

This kind of persistent high humidity can have permanent effects on the downwind area: deterioration of wooden structures, changes in plant growth, and leaching of minerals from the soil. If a trend toward dry cooling towers develops for new power plants, then these problems can be avoided. In dry towers the heat is dissipated by radiation and convection rather than by evaporation. But if cooling water continues to be used in this fashion, then many communities located near large power plants can be expect to encounter problems in the next decade.

13.2 PRIMARY AIR POLLUTANTS

The substances usually associated with air pollution are carbon monoxide, hydrocarbons, sulfur oxides, nitrogen oxides, and particulates. Several other substances are produced

Table 13.2 Origin of Air Pollutants

	CO	HC	SO$_x$	NO$_x$	Particulates
	\multicolumn{5}{c}{Percentage of Total}				
Transportation	58	52	—	51	3
Forest fires and controlled burning	19	—	—	1	9
Solvent evaporation	—	27	—	—	—
Industrial processes	11	14	20	1	51
Solid waste disposal	8	4	1	2	5
Stationary combustion sources (power plants)	2	2	78	44	26
Miscellaneous	4	1	1	1	6

secondarily, that is, are derived from these primary pollutants. As discussed in Chapter 12, there are a number of natural sources for these pollutants, so that even in the absence of humankind there would still be a residual background level of undesirable substances. However, we should note that, except for people living on the slopes of volcanoes, this natural background of pollution would be very small and unlikely to produce the kinds of effects noted in our metropolitan concentrations.

The origins of these pollutants are given in Table 13.2. Clearly, transportation and stationary combustion sources account for the lion's share of the emissions. This is what we must have expected from our study of the internal combustion engine in Chapter 4 and the magnitude of electrical energy generation in Chapter 2.

In this section we examine each of these polluting substances to see what its effect on the environment is. In the next section we deal with secondary pollutants.

Carbon Monoxide

Carbon monoxide is not particularly harmful to plants, although reduced nitrogen fixation has been reported in some plants that have been exposed to high concentrations of CO for prolonged periods of time. In air-breathing animals, however, CO can be quite harmful and even toxic. Carbon monoxide is about 210 times more readily absorbed by the blood than oxygen, so that if both gases are present, CO is preferentially acquired by hemoglobin molecules. The compound carboxyhemoglobin (COHb) is formed; COHb ties up hemoglobin molecules so that oxygen transport from the lungs to the remainder of the body, normally as oxyhemoglobin (OHb), is impaired. This forces the heart and lungs to work harder, and if a sufficient amount of COHb is formed, coma and death result. The symptoms in humans of various levels of COHb are indicated in Table 13.3.

As a general rule of thumb, it has been found that for every 1 ppm (1.15 mg/m^3) CO with which the body comes into equilibrium, about 0.16 percent of the body's supply of hemoglobin will be combined as COHb. The process of reaching equilibrium at a given concentration level is approximately exponential and can therefore be written as

$$N = N_0(1 - e^{-mt}), \qquad (13.5)$$

where N_0 is the equilibrium concentration of COHb, m is a constant that depends on the concentration, and t is time. For a concentration of about 77 ppm, m has a value of 0.3, so that about 8 hr is required to produce 90 percent of the saturation value of 12.4 percent COHb.

TABLE 13.3 Symptoms of COHb in Humans

Percentage of COHb[a]	Symptom
< 1	No effects known
2.5	Nonsmokers suffer impairment of time-interval discrimination
3	Nonsmokers notice impairment of visual acuity
5	Consistent impairment in cognitive and psychomotor performance
10	Significant reduction in oxygen transport
15	Headaches, dizziness, and lassitude
35	Flickering before the eyes, ringing in the ears, nausea, vomiting, palpitations, muscular weakness, apathy
40	Coma, death

[a]Percentages are approximate and vary with individuals.

It takes from 3 to 4 hr to reduce a saturated bloodstream by a factor of 2, more or less independently of the level of saturation. Note that females clear faster than males, presumably because of the smaller amount of blood.

Expired air from a nonsmoker normally has 2 to 3 ppm CO. Smokers, however, are generally believed to have a continuous COHb content of about 5 percent compared to 0.5 percent for nonsmokers. This estimate may be too low; during smoking, it may go as high as 15 percent. Chronic smokers live in a pollution level of their own making equivalent to about 30 ppm of CO. Note that the ambient background is about 0.15 ppm, exclusive of the sources listed in Table 13.2.

Prior to the enforcement of emissions regulations on automobiles the CO levels on urban freeways during rush hours were very high, varying from 30 ppm to as high as 300 ppm. Even now, a smoker forced to crawl along in heavy traffic is being exposed to a very high CO risk.

Hydrocarbons

The word hydrocarbon (HC) refers to a very large class of chemicals having a molecular structure consisting only of hydrogen and carbon atoms. This definition is too narrow for us, because there are a number of compounds, formed as secondary pollution products, which contain elements other than hydrogen and carbon, notably nitrogen and oxygen. Since such a large variety of compounds exist in the atmosphere, it has been difficult to make quantitative statements concerning concentrations and effects of individual HCs. However, more than half of all HC molecules in the atmosphere are methane. So it has become the practice to specify both the methane concentration and the nonmethane hydrocarbon concentration—essentially everything else.

The alkanes, methane and its family, are relatively inert in terms of health effects (in small quantities) and in terms of the production of secondary pollutants. However, many of the other nonmethane HCs present health risks, even in the absence of photochemical reactions. These include the aldehyde, benzene, ketone, and ethylene families. These compounds are known to cause skin irritation, pulmonary dysfunction, eye irritation, and, in the case of benzene, cancer at concentrations less than 25 ppm.

Most of the information about the health effects of specific HCs has been obtained from animal studies, since it is difficult to isolate the effect of a single HC in the presence of so many others in a polluted atmosphere. However, it is believed that the lowest level of HC concentration at which health effects would be observed is about 130 μg/m^3, which corresponds to about 0.2 ppm. We discuss HCs further in the section on secondary air pollutants.

Total Suspended Particulates

Much of the particulate matter emitted by power plants and industrial processes comes to earth as dustfall. Dustfall particles are usually those having diameters larger than about 10 μm, although there is no clear distinction between dustfall and suspended particulates. Figure 13.9 illustrates the relative sizes of particles of various types. Dustfall has been of concern for some time, because it is so visible, figuratively as well as literally. However, the larger particles represented by dustfall are of much less concern from the point of view of public health than the smaller suspended particulates.

All particulates are eventually brought to earth, either by gravitational action or by precipitation. We can use some simple physics to predict settling times for particles with diameters greater than about 1.3 μm. For smaller particulates the atmosphere does not seem to be continuous; these are small enough to be influenced by the molecular nature of the air. For larger particles the total force is the vector sum of the buoyant force, gravitational force, and frictional drag forces:

$$\mathbf{F} = \mathbf{F}_b + \mathbf{F}_g + \mathbf{F}_d. \qquad (13.6)$$

The buoyant force is the weight of the displaced medium:

$$F_b = \left(\frac{4}{3}\right) \pi r^3 \rho_g g, \qquad (13.7)$$

where r is the radius of the particles (presumed to be spherical), and ρ_g is the density of air. This assumes that there is no vertical air velocity.

The drag force is given by

$$F_d = 3\pi\mu \, dv, \qquad (13.8)$$

where μ is the viscosity of the air, and v is the velocity of the particle. When these forces add to zero, the net force on the particle will be zero, and it will fall with a constant, terminal velocity v_t:

$$v_t = \frac{d^2 g(\rho_p - \rho_g)}{18\mu}, \qquad (13.9)$$

where ρ_p is the density of the particle. We can use this to calculate the time required for these particles to settle.

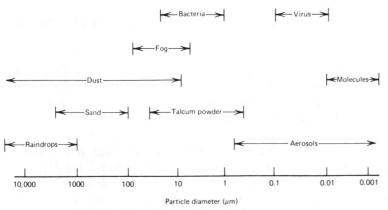

Figure 13.9 Relative sizes of several particulates.

Example 13.4   Particles having $d = 15\ \mu$m and $\rho_p = 1$ g/cm^3 are carried to a height of 200 m by a smokestack. There is a horizontal wind velocity of 4 m/s. How far downwind will these particles settle?

If we assume the terminal velocity is acquired very quickly, then

$$v_t = \frac{(15 \times 10^{-6})^2\ \text{m}^2 \times (9.8\ \text{m/s}^2) \times (1\ \text{g/cm}^3)}{18 \times (1.83 \times 10^{-4}\ \text{g/cm s})}$$

$$\times \frac{(0.001\ \text{kg/g}) \times (10^6\ \text{cm}^3/\text{m}^3)}{(0.001\ \text{kg/g}) \times (100\ \text{cm/m})}$$

$$(\rho_g \ll \rho_p),$$

$$\boxed{v_t = 6.69 \times 10^{-3}\ \text{m/s}.}$$

At this speed a 200-m fall would require $200/(6.69 \times 10^{-3}) = 2.99 \times 10^4$ s. In this period of time the particle would be transported a distance of $4 \times 2.99 \times 10^4 = 59.8$ km.

Apparently, particulates will be dispersed over a very large distance from the source. We could expect health effects to be quite noticeable.

The human health effects of inhaled particulates depend upon the chemical nature of the particle, on the distribution of the sizes of the particulates in the air, and on the efficiency with which the chemical agents of the particles are extracted in the region of respiratory deposition.

The various chemical elements produce a wide variety of health effects in humans, from no effect to toxicity. Many of the trace metals found in the environment are systemic poisons. Some elements produce lung disorders, such as silicosis. It is generally accepted that particle size is a major factor in determining the toxic effects of airborne particulates. These effects usually increase with decreasing particle diameter.

Extraction of toxic chemicals from particulates proceeds more efficiently if the particu-

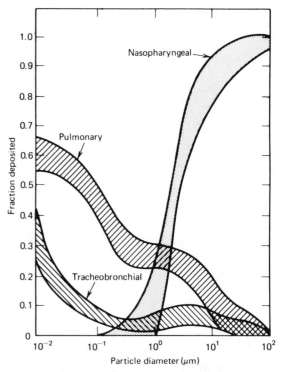

Figure 13.10 Fractional deposition of particulates as a function of particle diameter in the nose (nasopharyngeal), throat (tracheobronchial), and lungs (pulmonary) systems. [From David F. S. Natusch, "Urban Aerosol Toxicity: The Influence of Particle Size," *Science* **186**, 695 (1974). Copyright © 1974 by the American Association for the Advancement of Science.]

lates are lodged in the lung than if they are stopped higher up in the respiratory system. Figure 13.10 indicates that large particles are stopped effectively in the nose and trachea, but that the smallest particulates are lodged primarily in the lung, where they are the most damaging.

Studies in Buffalo and Nashville have shown increased death rates in older individuals at mean annual concentrations greater than 100 μg/cm^3. A study of British school children indicates increases in respiratory illness for smoke levels above 120 μg/cm^3. Excess deaths

TABLE 13.4 Definition of Pollutant Standard Index (PSI) Values

PSI Index Value	Air Quality Level	Pollutant Level					Health Effect	General Health Effects	Cautionary Statements
		TSP (24 hr, $\mu g/m^3$)	SO$_2$ (24 hr, $\mu g/m^3$)	CO (8 hr, mg/m^3)	O$_3$ (1 hr, $\mu g/m^3$)	NO$_2$ (1 hr, $\mu g/m^3$)			
500 400	Significant Emergency	1000 875	2620 2100	57.5 46.0	1200 1000	3750 3000		Premature death of ill and elderly. Healthy people will experience adverse symptoms that affect their normal activity.	All persons should remain indoors, keeping windows and doors closed. All persons should minimize physical exertion and avoid traffic.
300	Warning	625	1600	34.0	800	2260	Hazardous	Premature onset of certain diseases in addition to significant aggravation of symptoms and decreased exercise tolerance in healthy persons.	Elderly and persons with existing diseases should stay indoors and avoid physical exertion. General population should avoid outdoor activity.
200	Alert	375	800	17.0	400[a]	1130	Very unhealthful	Significant aggravation of symptoms and decreased exercise tolerance in persons with heart or lung disease, with widespread symptoms in the healthy population.	Elderly and persons with existing heart or lung disease should stay indoors and reduce physical activity.
100	National Ambient Air Quality Standards (NAAQS)	260	365	10.0	240		Unhealthful[b]	Mild aggravation of symptoms in susceptible persons, with irritation symptoms in the healthy population.	Persons with existing heart or respiratory ailments should reduce physical exertion and outdoor activity.
50	50% of NAAQS	75[c]	80[c]	5.0	120		Moderate[b]		
0		0	0	0	0		Good[b]		

[a] 400 $\mu g/m^3$ was used instead of the O$_3$ Alert level of 200 $\mu g/m^3$.
[b] No Index values reported at concentration levels below those specified by "Alert level" criteria.
[c] Annual primary NAAQS.

and a considerable increase in illness were noted in London and New York at smoke levels of about 750 $\mu g/cm^3$. In all these cases SO_2 was also present, although not measured. Since there may be synergistic action between SO_2 and particulates, we should be careful in interpreting these data.

Particles suspended in the air scatter and absorb sunlight, produce haze, and reduce visibility. For a typical urban area in the United States with a mean annual particulate concentration of about 100 $\mu g/m^3$, the total sunlight, including direct and reflected light, is reduced about 5 percent for every doubling of particulate concentration. This reduction is frequency-dependent, and is even larger in the ultraviolet portion of the spectrum.

Buffalo, New York, is one of the few areas of the country that suffers from significant particulate pollution. In the last year for which data are available, 1977, Buffalo had 21 days with a Pollution Standard Index (PSI) between 100 and 200 and almost 300 days with a PSI between 50 and 100. The definition of PSI is given in Table 13.4. It is likely that residents of Buffalo suffer from a continual reduction in visibility that may become even worse during severe episodes.

Oxides of Sulfur

Sulfur dioxide, SO_2, and sulfur trioxide, SO_3, are emitted in the ratio of about 30 to 1 in the combustion of fossil fuels. Both convert primarily to sulfuric acid in the presence of water vapor in the atmosphere.

Sulfuric acid and SO_2 have adverse health effects, producing construction and irritation of the bronchial passages. It is somewhat difficult to separate the effects of these two chemicals, since they almost always appear together in a polluted atmosphere. Sulfuric acid tends to be formed as an acid mist, requiring nucleation centers. As a consequence SO_2 and particulates appear to have a syner-

gistic effect; that is, the total health effect is greater than the sum of the individual effects.

This synergistic action may come from either one or both of two mechanisms: adsorption of sulfates on the particulate surface; the particulate itself is a liquid sulfate. Many particulates are liquid droplets, notwithstanding the usual identification of the word *particle* with a solid. Sulfates, formed from SO_2 dissolving in water vapor, form the primary nucleation centers for the formation of mists and droplets in the atmosphere. A long, visible, white plume from a power plant is there *only* because of the excessive amount of SO_2 being emitted to form nucleating centers for water vapor. A "clean" power plant plume will quickly dissipate and will not trail off across the countryside.

Because of the combined action of SO_2 and particulates it is difficult to single out SO_2 effects quantitatively. Instead, we note only that at a PSI index between 50 and 100 increased frequency and severity of respiratory ailments have been noted in children and older people, with some deaths reported of those already suffering from bronchitis. Also excess cases of lung cancer have been reported for individuals suffering chronic exposure at this level. At a PSI index greater than 200 a significant increase in the daily death rate has been reported. Clearly, SO_2–particulate pollution has severe direct health effects—this is in addition to any acid-rain effects, discussed earlier.

Oxides of Nitrogen

Both nitric oxide (NO) and nitrogen dioxide (NO_2), among the several oxides of nitrogen, are emitted as a result of high-temperature combustion in air; NO_2 is usually less than 0.5 percent. Nitric oxide slowly is oxidized to NO_2 in the atmosphere, although the process is more rapid in the presence of other pollutants and light.

Nitric oxide and NO_2 are present in a normal, unpolluted atmosphere at about 2 and $8\,\mu g/m^3$, respectively. Nitric oxide is not known to have any adverse health effects at the concentrations normally found, even in polluted atmospheres. Nitrogen dioxide, on the other hand, has severe health effects and at high enough concentrations can be fatal. Both NO and NO_2 affect plant life, reducing growth and yield. These oxides also affect materials, causing fading of dyes and deterioration of cotton and nylon fibers.

The threshold for smelling NO_2 is about $225\,\mu g/m^3$. Prolonged exposure to NO_2 concentrations of the order of $150\,\mu g/m^3$ has caused bronchial constriction and swelling and increased incidence of respiratory ailments.

Nitrogen dioxide absorbs sunlight preferentially in the blue end of the spectrum, giving the transmitted light a reddish tinge. Note that NO itself is colorless, but the NO_2 is a brown-orange color. The total effect as far as the observer is concerned is the now familiar orangish-brown smudge characteristic of polluted cities.

The oxides of nitrogen are not as important as air pollution problem in terms of their direct effects, but rather in terms of the secondary pollution constituents they produce. These are a variety of substances known collectively as **oxidants**, which we discuss in the next section.

Indoor Air Pollution

Before discussing secondary air pollutants, we should mention an aspect of air pollution that has only recently become important: indoor pollution. With the newfound concern over energy conservation, many residents have taken steps to make their homes more airtight to reduce infiltration and consequent heating and humidification costs, as we discussed in Chapter 11. As a consequence many homes have fewer than 0.2 air exchanges per hour, and pollution from inside the home can become severe enough to present a health hazard. In fact, in a poorly ventilated home and air quality can be worse than the federal standards for outdoor air!

There are primarily four pollutants of concern: radon gas, CO, particulates, and organics. Radon gas, in the form of ^{222}Rn, is a noble gas resulting from the radioactive decay of uranium. Radon is radioactive, being an α-emitter. This isotope has a half-life of only 3.8 days, but its decay products are also radioactive, also by α-emission. These decay particles tend to attach themselves to household particulates, dust, and the like and become inhaled. As we shall see in Chapter 14, inhaled radiation presents a far more serious risk than exterior radiation. Epidemiological studies show increased lung cancer risk in uranium miners who must inhale radon-laden dust. The chemically inert gas radon is a radioactive daughter of one of the daughters of uranium decay. Radon is not likely to be a problem in homes where there is little concrete or brick construction. If ventilation in a brick or stone home is sufficiently restricted, the radiation background level can be increased by as much as a factor of three by the collection of radon from decays of uranium in the building materials. In such cases, care should be taken to provide adequate ventilation, even at the cost of increased heat and humidification expenses.

Carbon monoxide comes primarily from wood fires, which have become very popular and even practical in the past few years. Inadequate ventilation of wood stoves and fireplaces, especially when coupled with heavy smoking indoors, can significantly raise the CO level in the house. Also we should mention that the trend to "do it yourself" energy-efficiency adaptations has led to some problems involving alterations to furnaces; these changes have the potential to generate lethal quantities of CO. A good thing to keep in mind is the old adage: "When all else fails, read the instructions." Only, do it *before* everything fails!

The particulates referred to above as being potential pollution problems indoors are primarily asbestos fibers. Asbestos has been used routinely for many years as an insulating material in homes and schools. It has also been known for many years that asbestos is linked with many respiratory diseases and lung cancer. There is also a synergistic effect between asbestos and smoking. Asbestos workers who smoked were found to have 92 times the excess cancer risk of those who did not smoke. The nearest Environmental Protection Agency office will provide information to enable those interested to determine whether or not there is a significant asbestos hazard in their home or school.

Finally, there are a large number of organic substances that can be of concern. The largest worry is **formaldehyde**. This organic compound is incorporated in particleboard, chipboard, and urea-formaldehyde (UF) foam insulation. It is commonly present in homes, particularly in mobile homes. In concentrations of less than 1 ppm it can cause watery eyes; nose, throat, and lung irritation' dizziness; skin rashes; headache; fatigue, and general body aches. These symptoms are most likely to occur in warm and humid weather after UF foam has been installed or in a new mobile home. However, they can also occur in the winter in a well-sealed home having few air exchanges per day. There are currently no national standards for formaldehyde concentrations; Wisconsin requires concentrations less than 0.02 ppm in new mobile homes. In that state warning labels are also required for houses constructed with UF foam. But, unless you live in Wisconsin, *caveat emptor*.

13.3 SECONDARY AIR POLLUTION

We have discussed thus far the direct effects of most of the pollutants emitted in industrial processes, including power plants and automobiles. However, many of these primary pollutants interact with each other, with sunlight, and in the presence of special meteorological features to form secondary pollutants that are every bit as bad as, perhaps worse than, the primary substances. Let us examine first the meteorology that enters into these processes.

Temperature Inversions

Natural sources of air pollution—volcanoes, sea spray, and dust storms, among others—put a far greater amount of material into the air than does the sum total of human activities. Then why do we seem to have such a problem? Probably for three reasons: we put different materials into the air, we concentrate our pollution sources in urban areas, and on some days we cannot rely on natural effects to carry away, dilute, and disperse our pollution. One of the principal culprits in this last difficulty is the **temperature inversion**.

The temperature of the atmosphere normally decreases by a uniform amount as a function of increasing elevation. We can discover what this amount is. Consider a column of fluid, as shown in Figure 13.11. In order for a given volume of it to be in equilibrium, it must be true that

$$\Delta p = -g\rho \, \Delta z, \qquad (13.10)$$

where ρ is the density of the fluid, and g is the acceleration of gravity. This is called the *hydrostatic equation*.

Temperature does not explicitly appear in this equation, but since we are interested in the atmosphere, which is a gas, there must be a relationship between temperature, pressure, and volume. We assume that relationship to be the same as for an ideal gas. Remember that

$$\rho = \frac{m}{V} \qquad (13.11)$$

and

$$m = nM, \qquad (13.12)$$

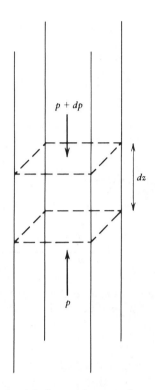

Figure 13.11 A column of atmosphere with a small volume having a pressure difference dp between the top and bottom surfaces.

where n is the number of moles contained in the volume of interest and M is the kilogram-molecular weight. Using Eqs. 13.11 and 13.12 in the ideal gas law, we have

$$p = \frac{\rho R T}{M}. \qquad (13.13)$$

By solving Eq. 13.13 for the density and substituting in the hydrostatic equation, we have a result that explicitly displays the temperature dependence:

$$dp = -\frac{gM}{RT} \, dz, \qquad (13.14)$$

where we are now using differential notation, instead of finite differences.

However, we have not found the rate of change of temperature with respect to elevation, which would be dT/dz. To do that we must make an assumption: namely, that as this volume of air changes elevation, it does so adiabatically. That being the case, recall from Chapter 3 that

$$Tp^{(1-\gamma)/\gamma} = \text{constant} \qquad (13.15)$$

$$dTp^{(1-\gamma)/\gamma} + \left(\frac{1-\gamma}{\gamma}\right) Tp^{(1-2\gamma)/\gamma} \, dp = 0 \qquad (13.16)$$

A little algebra on Eq. 13.16 will produce the result that

$$\frac{\gamma \, dT}{T} = (\gamma - 1) \frac{dp}{p}. \qquad (13.17)$$

And combining this result with Eq. 13.14 yields the required rate of change:

$$-\frac{dT}{dz} = \left(\frac{\gamma - 1}{\gamma}\right)\left(\frac{gM}{R}\right). \qquad (13.18)$$

Meteorologists refer to this as the **adiabatic lapse** rate. For dry air, $\gamma = 1.41$, $M = 28.96$, so $-(dT/dz) = (9.9 \times 10^{-3})°C/m$, or just about $1°C/100$ m. This is the rate at which dry air would decrease in temperature as a function of elevation, up to the tropopause, at which point, as we found in Chapter 12, the temperature begins to rise.

For wet air the situation is more complicated since temperature changes will induce relative humidity changes, which are accompanied by release or absorption of heat energy by phase change. Since these effects depend on the initial temperature and relative humidity, it is not possible to quote a simple constant for the lapse rate for wet air. Suffice it to say that it may be considerably different from the $1°C/100$ m for dry air.

The lapse rate can tell us about the **stability** of the atmosphere. By this we mean whether or not a given volume, called a *parcel*, of air will rise, fall, or remain in equilibrium. The adiabatic lapse rate is the "model" rate of change for temperature. In a real atmosphere, for a variety of reasons, the actual lapse rate may be larger or smaller than the adiabatic rate. Consider a situation where the actual lapse rate is larger, say, 1.2°C/100 m.[1] This means the temperature of the atmosphere is decreasing faster than it would if the lapse rate were equal to the adiabatic rate. Therefore, a parcel of air initially at equilibrium that is displaced upward a distance Δz and not mixed with its surroundings would have a temperature change given by

$$\Delta T = -9.9 \times 10^{-3} \Delta z, \qquad (13.19)$$

whereas the surroundings would have a temperature change of

$$\Delta T' = -12 \times 10^{-3} \Delta z. \qquad (13.20)$$

The parcel of air would find itself surrounded by cooler air, so that it would continue to rise. This condition is referred to as an *unstable atmosphere*. Obviously, this is a desirable condition as far as pollution is concerned. Under this condition pollutants would be carried upward to be mixed with large volumes of air and dispersed over a wide area.

If the actual lapse rate of the atmosphere were smaller than 1°C/100 m, the reverse would occur. A parcel of air would fall, because it would be more dense than its surroundings. This is referred to as a *stable atmosphere*, and is clearly undesirable from the point of view of dispersing pollutants.

Lapse rates are often plotted on a z versus T plot, as in Figure 13.12. The adiabatic rate is shown as the solid line. Any line with a slope less than the adiabatic line results in an unstable atmosphere; any line with a greater slope gives a stable atmosphere. Note carefully which slope is being referred to here: dz/dT!

Example 13.5 If the prevailing lapse rate is +0.75°C/100 m, and a parcel of air is emitted from a smokestack at an elevation of 25 m, having a temperature 1°C warmer than the surrounding air, how high will the air rise?

[1]By this we mean $-(dT/dz) = 1.2°C/100\,m$.

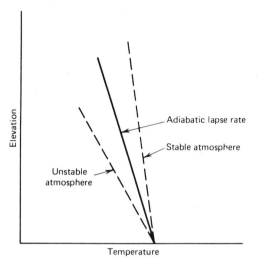

Figure 13.12 An elevation versus temperature plot. Unstable and stable lapse rates are shown relative to the adiabatic lapse rate.

This situation is illustrated in Figure 13.13. The air temperature at 25 m is $T_0 - 0.75 \times 0.25 = T_0 - 0.19$, where T_0 is the ground temperature. The air released from the smokestack has a temperature of $T_0 + 0.81$. The released air will rise, following the adiabatic lapse rate, until its temperature equals that of its surroundings:

$$(T_0 + 0.81) - 1 \times z = (T_0 - 0.19) - 0.75 \times z,$$

$$T_0 + 0.81 - T_0 + 0.19 = z - 0.75z,$$

$$1.00 = 0.25z,$$

$$z = 400 \text{ m above the smokestack.}$$

Thus the two lapse rate lines in Figure 13.13 cross at 425 m.

From the above example we conclude that a lapse rate less than 1°C/100 m results in a stagnation of vertical air currents at some particular elevation. It is possible for the lapse rate to have an opposite sign, indicating an *increase* in temperature with increasing ele-

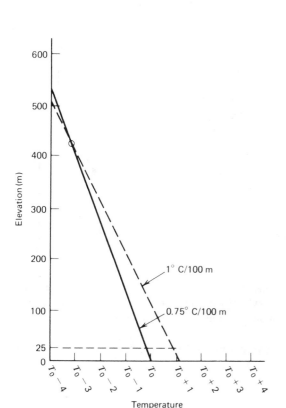

Figure 13.13 Diagram for Example 13.5.

the air immediately in contact with it. Then the surface air slowly cools by conduction. When the sun comes up, the earth warms and begins to warm the surface layer of air. This process is illustrated in Figure 13.14. If the atmosphere is heavily polluted, there may be enough solar energy absorption so that the warming does not completely eliminate the inversion. This type of phenomenon is particularly common in winter; the 1952 London episode was from December 4 to 9.

Another type of temperature inversion is that created along with slow-moving high-pressure systems. Gradual subsidence of air aloft creates adiabatic warming at the upper levels. Such inversions are very common in the late autumn and lead to "Indian Summers." Temperature inversions have caused most of the "smog" episodes in this and other countries. When an inversion occurs over a metropolitan area, pollutants produced in that region cannot disperse. Rather, they are trapped in the inversion region. Sulfur oxides and particulates produced from coal burning, in particular, can accumulate to very high levels, as they did in London in 1952 when 4000 people died or earlier in Donora, Pennsylvania, in 1948 when 40 people died. There have been many other episodes dating back several hundred years, in fact.

Today, we have found yet another way to insult our environment by means of tem-

vation. This is called a **temperature inversion**, and it too results in vertical air current stagnation.

Inversions are common at night when there are clear skies. Radiative cooling of the warm earth brings it to a temperature below that of

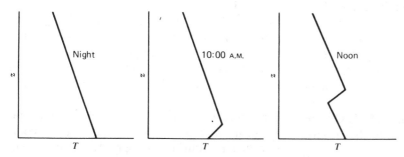

Figure 13.14 The development of a radiation inversion. Earth warming occurs more rapidly in the early morning than air warming.

perature inversions. This is the effect produced when automobile effluents are stagnated over a city and are acted upon by light.

Photochemical Reactions

The secondary pollutants that are the major sources of health problems in metropolitan areas are the consequence of a number of complex chemical reactions, not yet thoroughly understood, that are triggered by sunlight. There are two important ingredients in this photochemical soup: ozone and several of the nonmethane hydrocarbons. Collectively these are referred to as oxidants. As used in this context an **oxidant** is any substance capable of causing an oxidation reaction to occur. There are many hydrocarbon oxidants, but since they almost always occur in conjunction with ozone, it is convenient from the standpoints of both measurement and standard setting to refer to oxidant levels in terms of only the ozone concentration.

Ozone is produced in the lower atmosphere by the reaction of an oxygen molecule with an oxygen free radical liberated in the interaction of sunlight with nitrogen dioxide:

$$NO_2 + h\nu \rightarrow NO + O\cdot, \qquad (13.21)$$

$$O_2 + O\cdot \rightarrow O_3. \qquad (13.22)$$

However, if there were no other pollutants in the atmosphere these reactions would rather quickly be negated by the recombination of the O_3 and NO, as in Figure 13.15a. The net effect would be a slight warming of the atmosphere by the absorption of the photon of sunlight. In the presence of hydrocarbons, as shown in Figure 13.15b, the ozone produced in the first step can be preserved, and an organic nitrogen compound can be formed. Additional ozone can be formed by the reduction of HCO-type compounds by NO, which would also return the NO_2.

Several of the organic nitrates in the polluted atmosphere are very strong oxidants.

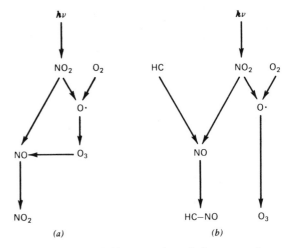

Figure 13.15 (a) Net warming of the atmosphere by NO_2. (b) Production of organic nitrogen compounds by sunlight in the atmosphere.

These belong to the *peroxyacetyl nitrate* family (PAN). These have structural formulas of the form

$$
\begin{array}{c}
O \\
\parallel \\
RCOONO_2,
\end{array}
\qquad (13.23)
$$

where normally $R = CH_3$, but it could be any of the other alkane roots. The PAN family members are somewhat unstable and are always unpredictable when being studied in the laboratory in large quantities. They are powerful oxidants, damaging to human tissue and plant life.

Oxidant production is linked to NO_x production. Measurements show this; Figure 13.16 plots the buildup of NO, NO_2, nonmethane hydrocarbons, and oxidant as a function of time in a metropolitan area. The NO and NO_2 peaks are correlated with the morning and evening rush hours. Oxidant formation continues soon after sunrise and builds to a peak near noon. If the temperature inversion exists in the region, the oxidant will not completely clear during the late afternoon; it will be strengthened by the evening rush hour, so long as sufficient daylight exists.

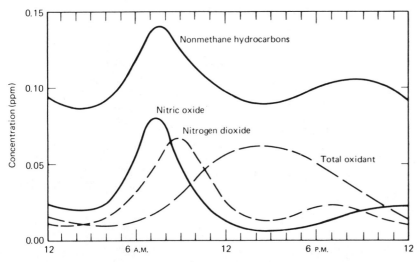

Figure 13.16 Time-dependent buildup of air pollutants in an urban environment.

Oxidant levels higher than 0.3 ppm have been measured in urban areas. Studies have found that concentrations of the order of 0.05 ppm (100 μg/m^3) produce leaf injury to some plant species. Levels above 0.1 ppm produce eye irritation, and levels above 0.25 aggravate respiratory ailments.

Photochemical air pollution is almost entirely the consequence of automobile usage. There is a very close correlation between emissions and "smog" episodes; this was well documented even before federal air pollution standards were introduced in 1970.

13.4   AIR POLLUTION CONTROL

Air pollution has been a fact of civilized life for several centuries. Photochemical air pollution, to be sure, is a rather recent phenomenon. In 1969, in order to cope with growing environmental problems, Congress passed the National Environmental Policy Act and in 1970 the Amendments to the 1955 Clean Air Act.[2]

[2]These amendments are usually referred to as "the Clean Air Act of 1970."

With these pieces of legislation, air pollution control finally took a major step forward.

This legislation has brought about significant improvements in air quality for the "criteria pollutants." The primary standards, that is, the levels at which human health effects are important, as well as the secondary standards, that is, the levels at which human welfare effects are important, have been set (see Table 13.5).

Hazardous air pollutants not controlled by the National Ambient Air Quality Standards (NAAQS) are defined as any substance that may cause or contribute to mortality or irreversible or incapacitating illness. The list includes vinyl chloride, beryllium, mercury, asbestos, and benzene. The EPA has proposed standards for these materials that have provoked controversy and, in the case of vinyl chloride, court action. The question turns on the idea of the existence of a threshold level beneath which there are no visible health effects. Naturally, this threshold, if it exists, would be at very small concentrations; this fact contributes to the difficulty in establishing

TABLE 13.5 National Ambient Air Quality Standards (NAAQS)

Pollutant	Averaging Time	Primary Standard Levels	Secondary Standard Levels
Particulate matter	Annual (geometric mean)	75 μg/m^3	60 μg/m^3
	24 hr[a]	260 μg/m^3	150 μg/m^3
Sulfur oxides	Annual (arithmetic mean)	80 μg/m^3 (0.03 ppm)	—
	24 hr[a]	365 μg/m^3 (0.14 ppm)	—
	3 hr[a]	—	1300 μg/m^3
Carbon monoxide	8 hr[a]	10 mg/m^3 (9 ppm)	10 mg/m^3 (9 ppm)
	1 hr[a]	40 mg/m^3 (35 ppm)[b]	40 mg/m^3 (35 ppm)
Nitrogen dioxide	Annual (arithmetic mean)	100 μg/m^3 (0.05 ppm)	100 μg/m^3 (0.05 ppm)
Ozone	1 hr[a]	235 μg/m^3 (0.12 ppm)[c]	235 μg/m^3 (0.12 ppm)
Hydrocarbons (nonmethane)[d]	3 hr (6 to 9 A.M.)	160 μg/m^3 (0.24 ppm)	160 μg/m^3 (0.24 ppm)

Source: U.S. Environmental Protection Agency.
[a]Not to be exceeded more than once per year.
[b]Revision to 25 ppm proposed by EPA in August 1979.
[c]Revised from 0.08 ppm in January 1979.
[d]A nonhealth related standard used as a guide for ozone control.

whether or not the threshold exists. We return to the very serious question of thresholds in Chapter 14, when we discuss the health effects of radioactive substances.

The designation of clean air areas and the desire to maintain (if not improve) the air quality in those areas have led the EPA to require rather more severe emissions standards for SO$_x$. After some successful opposition by industry to EPA's initial proposed standards, a final set was announced in May of 1979. This calls for from 70 to 90 percent removal of SO$_2$ from stack gases based on a monthly average. The variable amount depends on the initial sulfur content of the coal. Also, a maximum amount of 1.2 lb/MBtu (5.16 × 10^{-4} kg/MJ) of sulfur is to be averaged over a month, rather than 24 hr. However, the primary standard of

80 μg/m^3 for any area still cannot be exceeded on an annual average basis.

These regulations are tough, but the levels proposed are not impossible to achieve. With firm enforcement and a willingness on the part of the public to pay for the gains that will be made, there is no reason that the air quality hoped for in the Clean Air Amendments of 1970 cannot be achieved by 1990. Certainly, the technology for doing it exists.

Control Techniques

The power sources of airborne pollutants are automobiles and coal-burning power plants. We discussed in Chapter 4 several means that have been used to minimize auto emissions. In this section we consider techniques for removing particulates and SO_2 from power plant effluents, since these are the only pollutants that can be controlled effectively.

Several techniques for controlling particulates have been used for some time. These include cyclone separators, filters, and electrostatic precipitators (ESP). Wet scrubbing systems also remove particulates, but they are more often used primarily for SO_2 removal.

Cyclone action is illustrated in Figure 13.17. Particulates are carried to the container walls by "centrifugal" action. Upon striking the walls the particles fall to the bottom of the chamber, where they are collected. Of course, many are reincorporated into the gas stream and exit along with the gas. This type of device has high efficiency for heavier particulates and decreasing efficiency as the size and mass of the particles decrease. Particulates with diameters less than $5 \mu m$ are removed with less than 50 percent efficiency. Cyclones are designed with the aid of a number of empirically determined relationships between the

various dimensions of the cyclone and the inlet gas velocity and pressure.

Cyclones are rather simple to maintain and inexpensive to operate, the only cost being that associated with the removal of the captured solid matter. But because of the inefficiency with which they remove very small particulates, and because of the rather severe health hazard that these impose, cyclones must be used in tandem with one of the other devices.

Electrostatic precipitators can have high efficiencies for all but the smallest particulates. They are rather expensive to install, but maintenance and operating costs are low. A typical ESP system is illustrated schematically in Figure 13.18. A pulsed DC voltage, obtained by full-wave or half-wave rectification of the stepped-up line voltage is applied to a positive collecting plate and a negative central conductor. This creates a pulsating electric field in the space between the conductors. The resulting corona discharge liberates electrons from the central wire. These electrons may become attached to gas molecules, which are then adsorbed or absorbed by particulates in the gas. The particulate, now having a net negative charge, migrates toward the positive collecting plate.

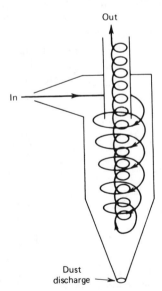

Figure 13.17 Schematic diagram of a cyclone separator for particulate removal from exhaust streams.

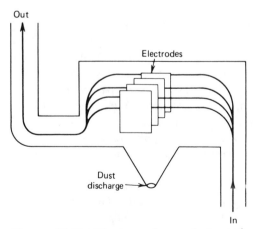

Figure 13.18 Diagram of an electrostatic precipitator system for exhaust cleanup.

A high migration velocity results in a high efficiency of collection. Very small particulates will in general have large migration velocities, but nevertheless they are not collected with high efficiency in an ESP. This is a consequence of the fact that the probability for a very small particulate ($<0.5\ \mu$m in diameter) to capture an ion from the gas is very small; it is inversely dependent of the particulate diameter.

Example 13.6 Determine the migration velocity of a 30-μm radius particulate having $\rho_p = 1.5\ \text{g/cm}^3$ as it strikes the collecting plate. Assume that the uniform electric field is 10,000 V/m, and the distance of travel is 15 cm.

Equate the electrical force with the accelerating force:

$$qE = ma.$$

The acceleration is then qE/m and the equations of linear motion give $v^2 = v_0^2 + 2as$. Thus,

$$v^2 = 0 + \frac{2 \times (1.6 \times 10^{-19}) \times 10^4 \times 15}{4/3 \times \pi \times (0.3 \times 10^{-4})^3 \times 1.5},$$

$$\boxed{v = 0.52\ \text{cm/s.}}$$

This assumes a singly charged ion is picked up by the particulate.

Filter systems, shown in Figure 13.19, are generally used in industrial processes such as cement making and carbon black, clay, and pharmaceutical production. This is because the fibers of the filter may seriously degrade if the effluent is too hot, as it might be in a power plant. Several techniques are used to clean the filterbags; a reverse pulse of air can be used. Alternatively, automatic knockers or shakers can be employed.

Efficiencies of filter systems are the highest of any type. Even for particulates smaller than 0.3 μm in diameter, efficiencies of the order of 98 percent can be routinely obtained. However, filters are more expensive to main-

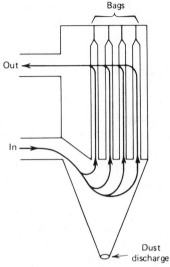

Figure 13.19 An air-bag effluent cleaning arrangement.

tain and operate, since the fabric must be changed periodically, and a large filter area is required to process a large effluent rate.

Wet scrubbers can be used both as particulate and SO_2 control systems. A typical installation is diagramed schematically in Figure 13.20. The scrubbing fluid, usually water, is atomized under high pressure at the point where the dirty gas enters the system. Collection of particles by wet scrubbing relies on

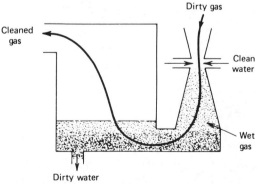

Figure 13.20 Gas cleaning by water scrubbing.

impaction of the particulates against the water droplets. This is usually effective for particles having diameters greater than about 0.3 μm. Hybrid systems in which ESP and wet scrubbing are combined can be more effective against the smallest particulates.

Wet scrubbing requires higher operating energy, discharges wet gases, and produces a wet product. Meeting additional environmental constraints on mist discharge is more difficult because of this. However, wet scrubbing is one of the few proven techniques for SO_2 removal.

SO_2 removal can be accomplished either by wet scrubbing with a solution containing sodium, magnesium, or calcium atoms or by converting the SO_2 gas to particulates and then filtering or applying ESP. Conversion to sulfate or sulfite particulates can be done by injecting an alkali, such as limestone, into the boiler or by spraying an alkali solution into the gas space. Currently, about 70 percent of the SO_2 removal in this country is done using this technique followed by a filtration scheme, as discussed above.

It may not be possible to meet the New Source Performance Standards without full scrubbing, something that the industry has sought to avoid. However, the full potential of dry spraying has not been realized, and work continues in this field.

Air Pollution Trends

In the few years since the NAAQS were established, there have been some noticeable changes in urban air quality. Figure 13.21 indicates the air quality for 25 Standard Metropolitan Statistical Areas (SMSAs) for 1974 through 1977. While the trend is in the direction of improvement, it has not been very large. Even worse, the Department of Energy predicts that, with the projected increase in fossil fuel burning as a consequence of the "energy crisis," air quality can be expected to deteriorate, *even if all new power plants are in*

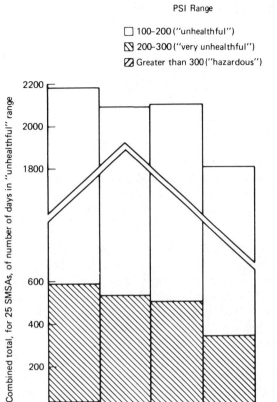

Figure 13.21 Total PSI for 25 Standard Metropolitan Statistical Areas (SMSAs) as a function of time between 1974 and 1977, showing the noticeable improvement. [From *Environmental Quality, The 10th Report of the Council on Environmental Quality.* Washington D.C.: U.S. Government Printing Office, 1979.]

compliance with the NSPS. This is indicated in Table 13.6.

The trend indicated in Figure 13.21 is misleading, as well. If the trend for ozone by itself is examined, as shown in Figure 13.22, we see that while the nation as a whole has made some gains, the worst area, Los Angeles, has made only minor headway in reducing the number of days per year in which the standard

TABLE 13.6 Predicted Emissions from Increased Coal Combustion

	Emission During the Year (10^9 kg)		
	1975	1985	2000
TSP	14.1	8.6	12.0
SO_2	24.0	26.1	28.6
NO_x	15.8	19.9	25.9

Source: Annual Environmental Analysis Report.
Washington, D.C.: U.S. Department of Energy, 1977.

was violated. This should not be a surprise,
since ozone formation in the lower atmosphere
comes primarily from automobiles through
NO_x emission. And while NO_x emission has
been reduced, it has not been controlled.
Effective and economic means for doing so in
automobiles have not yet been found.

The outlook for significantly cleaner air in
1990 is not very good, even if we rigorously
enforce the current emission standards.
Adopting more stringent standards in the
future may be difficult to do. There can be no
doubt that urban air, even rural air in some
places, is simply unhealthful and that an incremental increase in energy production by
fossil fuel burning, dE/E, will result in a corresponding incremental increase in morbidity
and mortality, dM/M. That is,

$$\frac{dM}{M} = k \frac{dE}{E}. \qquad (13.24)$$

A very good argument can be made that k in
Eq. 13.24 is actually negative. That is, an increase in energy production will cause a
decrease in sickness and death because of
increased heating availability and improved
general lifestyle.

Since good arguments can be made in favor

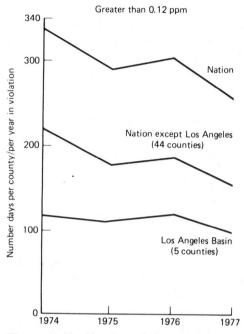

Figure 13.22 Number of days per county
per year of ozone violations in the United
States and in Los Angeles as a function of
time. [From *Environmental Quality, The
10th Report of the Council on Environmental
Quality.* Washington D.C.: U.S. Government
Printing Office, 1979.]

of either sign for k, perhaps a better, more
general relationship would be

$$\frac{dM}{m} = k \left(\frac{dE}{E}\right)^2 - k' \frac{dE}{E}. \qquad (13.25)$$

Now, whether dM/M increases or decreases
with dE/E depends on the magnitudes of k and
k'. It is certainly beyond the scope of this
textbook to try to determine those parameters.
Answers to this question and similar ones will
have to be obtained in the next decade. The
health hazards of air pollution are **not** under
control, and they will not be reduced in the
future without vigorous intercession.

SUMMARY

The climate in the around cities is affected to a great extent by the existence of the city. Cities absorb heat and produce pollutants. The temperature difference between a city and its environs can be more than 3°C.

Large cities can also influence climate a large distance downwind, by sending pollutants to interact with the prevailing winds.

Sometimes this interaction produces *acid precipitation*, which may cross national boundaries.

The primary pollutants are *carbon monoxide, hydrocarbons, total suspended particulates, oxides of sulfur*, and *oxides of nitrogen*. These pollutants are harmful to health in sufficient concentrations. Maximum levels have been set by the Environmental Protection Agency.

Smog attacks are a combination of chemistry and a meteorological phenomenon: the *temperature inversion*. This is the inversion in sign of the *lapse rate*.

The chemistry involves *photochemical reactions* that produce *ozone* and a family of organic compounds known as *PAN*.

Air pollution, especially from power plants, can be controlled with a variety of postcombustion devices.

REFERENCES

Babich, Harvey, Devian Lee Davis and Guenther Stotzky, "Acid Precipitation," *Environment*, May (1980), p. 7.

Bethea, Robert M., *Air Pollution Control Technology*. New York: Van Nostrand Reinhold, 1978.

Bodkin, L. D., "Carbon Monoxide and Smog," *Environment* **16**, (4), 34 (1974).

Cleveland, William S., and T. E. Graedel, "Photochemical Air Pollution in the Northeastern United States," *Science* **204**, 1273 (1979).

de Nevers, Noel, "Enforcing the Clean Air Act of 1970," *Sci. Amer.* **228**, 14 (June 1973).

Dobbins, Richard A., *Atmospheric Motion and Air Pollution*. New York: John Wiley, 1979.

Grad, Frank P., et al., *The Automobile and the Regulations of Its Impact on the Environment*. Norman: The University of Oklahoma Press, 1975.

Hesketh, Howard E., *Air Pollution Control*. Ann Arbor, Mich.: Ann Arbor Science Publishers, 1979.

Hodges, Laurent, *Environmental Pollution*. New York: Holt, Rinehart and Winston, 1973.

Husar, R. B., J. P. Lodge, and D. J. Moore (Eds.), *Sulfur in the Atmosphere*. New York: Pergamon Press, 1977.

Lowry, William P., "The Climate of Cities," *Sci. Amer.* **217**, 15 (August 1967).

Lynn, David A., *Air Pollution: Threat and Response*. Reading, Mass.: Addison-Wesley, 1976.

Natusch, David F. S., and John R. Wallace, "Urban Aerosol Toxicity: The Influence of Particle Size," *Science* **186**, 695 (1974).

Perera, Frederica, and A. Karim Ahmed, *Respirable Particles*. Cambridge: Ballinger, 1979.

Shaw, David T. (Ed.), *Fundamentals of Aerosol Science*. New York: John Wiley, 1978.

Sittig, Marshall, *Environmental Sources and Emissions Handbook*. Park Ridge: Noyes Data Corporation, 1975.

Tomany, James P., *Air Pollution: the Emissions, the Regulations and the Costs*. New York: American Elsevier, 1975.

———, "Effect of Acid Rain," Hearing before the Committee on Energy and Natural Resources of the U.S. Senate, May 28, 1980, Publication No. 96-126.

GROUP PROJECTS

Project 1. Estimate the level of CO in a cafeteria or similar eating place during a lunch hour. Take into account the number of people, the number of cigarettes and cigars smoked, and the number of air exchanges, if possible. Estimate the health effects for nonsmokers.

Project 2. Visit a nearby fossil-fuel-burning plant. Determine the daily SO_x, NO_x, and TSP emissions. Is there a preferential wind direction? Estimate concentration exposures for residents at various distances from this source.

Project 3. If there is an Air Quality Control Office in your community, determine your annual air quality record. What are the worst seasons? worst pollutants? highest concentrations? Has there been improvement since the office began to keep records?

EXERCISES

Exercise 1. The albedo of a certain city is 0.2; a nearby country region of the same area has an albedo of 0.4. The product mc_p for the city is one-third that for the country. What can you conclude about the temperature of the two regions at sunset?

Exercise 2. What heat production would be required to raise the temperature of your local area 0.01°C? What global heat production rate would be required to raise the overall temperature 1°C? What effects do you neglect in these calculations?

Exercise 3. Would you expect a city viewed during the day from a satellite orbit to appear brighter or darker than its surrounding countryside? Explain.

Exercise 4. Following Example 13.2, estimate the amount of nitrates formed over the Chicago region daily.

Exercise 5. The air surrounding a 1000-MW coal-fired power plant is 20°C at 30 percent humidity. If the waste heat is dissipated in a wet cooling tower, with how much air should the tower effluent be mixed in order that the humidity not rise about 40 percent?

Answer: $3.3 \times 10^{10} \, m^3$.

Exercise 6. A city has an area of 750,000 ha; there are 4,000,000 automobiles in the city. Estimate the amount of water exhausted into the atmosphere above the city daily from automobile operation.

Exercise 7. A nonsmoker enters a house where the CO level is 35 ppm. How long will it take for that person to notice some impairment in visual acuity?

Exercise 8. A cigarette smoker commutes to and from work. If 1 hr of driving at an average CO concentration of 100 ppm is required, what COHb level could be expected at the end of that hour?

Answer: 4.15% COHb.

Exercise 9. If the SO_2 concentration in your classroom were 0.01 ppm, what amount of sulfur in grams would be suspended in it?

Exercise 10. The number of respiratory and heart disease deaths from SO_2–particulate pollution per day can be approximated by an equation:

$$N = 150.3 + 0.25c_p + 20.7c_s + 1.25\delta,$$

where c_p is the particulate concentration in micrograms per cubic centimeter, c_s is the SO_2 concentration in micrograms per cubic centimeter, and δ is the difference between the ambient temperature and 18°C. How many deaths are to be expected in the absence of pollution? How could this equation be adjusted to account for the difference in susceptibility of older individuals?[3]

Exercise 11. If you lived 100 km downwind from a power plant that produced particulates of $\rho_p = 1$ g/cm³ and the average horizontal wind speed were 6 m/s, what diameter particles would be likely to fall on you? Would you be better off living 500 km downwind? Assume stack height is 200 m.

Answer: At 100 km, $d \approx 2 \times 10^{-3}$ cm; at 500 km, $d \approx 2 \times 10^{-4}$ cm.

Exercise 12. Find an expression for the pressure as a function of elevation.

Exercise 13. The prevailing lapse rate is 1.2°C/100 m except that between 1 and 1.5 km it is −0.5°C/100 m. Describe the motion of a parcel of air released at the surface, initially in equilibrium.

Exercise 14. A 100-MW power plant burns 5 percent sulfur coal continuously. Assume the power plant to be at the center of a circular city having a radius of 10 km and that there is a temperature inversion at 500 m that effectively traps all the effluent. Assume that this effluent is mixed thoroughly over the city. What is the SO_x concentration over the city after 24 hr?

Exercise 15. Smoke rises from a smokestack at 100 m with a temperature 20°C higher than its surroundings. How high will it rise, if the prevailing lapse rate is +0.5°C/100 m? What if it were −0.5°C/100 m?

Answer: 4100 m, 1366 m.

Exercise 16. Smoke is observed to rise from the ground to a height of 900 m. If the smoke temperature difference with its surroundings at ground level is 15°C, what must be the prevailing lapse rate?

Exercise 17. Smoke is emitted from a smokestack at 25 m with the same temperature as that of the surrounding air. If the prevailing lapse rate is −7°C/100 m, describe what happens to the smoke.

Exercise 18. A 40 percent efficient 1000-MW power plant is to burn 4.5 percent sulfur coal. What percentage of the SO_2 must be removed from the effluent in order for the plant to be in compliance?

Exercise 19. What is the maximum percentage-surfur coal a new 40 percent efficient 100-MW plant can burn and still remain in compliance? How would this change if the efficiency dropped to 35 percent?

Exercise 20. A particular cyclone processes 100,000 m³/hr of effluent containing 100 g/m³ particulates at an average efficiency of 0.7. What is the mass of solid matter removed each day? If $\rho_p = 1.5$, what size storage bin is required for 1 week's storage? Assume a 60 percent packing efficiency.

Answer: 1.68×10^8 g, 1306.7 m³.

Exercise 21. If the horizontal gas motion in the ESP of Example 13.6 is 4 m/s, what must be the minimum spacing of the collecting surfaces (length = 3 m) to ensure that 50 percent of the ionized particulates of the example are collected?

[3]See T. A. Hodgson, Jr., "Short-Term Effects of Air Pollution on Mortality in New York City," *Environ. Sci. Tech.* **4**, 589 (1970).

CHAPTER FOURTEEN   ENVIRONMENTAL RADIATION

Perhaps the most controversial aspect of nuclear power is the production of radioactive by-products during the fission process. The average citizen is alarmed at the prospect, remote or not, of being bombarded by invisible rays that can induce cancer and against which there is little protection. In this chapter we discuss the nature of radiation, both man-made and naturally occurring, and the effects of radiation on people. We conclude by attempting to indicate how we could go about making rational decisions, when many around us are not.

14.1   THE INTERACTION OF RADIATION WITH MATTER

Radiation is a general term; there are in fact several different types of radiation that interact differently with matter. Some types are normally found in nature; some are not. We should start off this discussion by identifying these various types of radiation.

Nature of Radiation

First of all, note that some types of radiation are particles, and some are massless photons of electromagnetic energy. Some of the particles have an electric charge; some have none.

There are several particles normally found in nature (by normally, we mean in the absence of human actions). Alpha particles are helium nuclei; they are usually emitted by heavy, radioactive nuclei beyond lead in the atomic table. Beta particles, both plus and minus, are emitted by unstable nuclei throughout the atomic chart. The properties of these particles are summarized in Table 14.1.

Cosmic rays are also particles that originate outside the earth's atmosphere. These may be protons or heavier particles. They normally do not penetrate very far into the atmosphere but, rather, collide with a nucleus and produce a shower of secondary particles of inter-mediate mass, called **mesons**. There are several types of mesons, charged and uncharged, that have been produced in very-high-energy collisions induced by protons from large accelerators. Cosmic rays usually produce only two types that are likely to penetrate to the earth's surface: pi-mesons (pions) and mu-mesons (muons). Mesons are unstable, decaying eventually into electrons and uncharged massless entities called neutrinos. Uncharged mesons decay differently. Mesons produced in cosmic rays probably do not constitute a health hazard, even though we are constantly being showered with them.

Photons go by a variety of names. Never-

TABLE 14.1 Particle Properties

Particle	Mass (MeV)	Charge (Electron Charges)	Lifetime (s)
Electron	0.511	±1	∞
Muon	105.659	±1	2.197×10^{-6}
Pion	139.569	±1	2.603×10^{-8}
	134.964	0	0.828×10^{-16}
Proton	938.280	±1	∞
Neutron	939.573	0	918
Alpha	3726.40	+2	∞

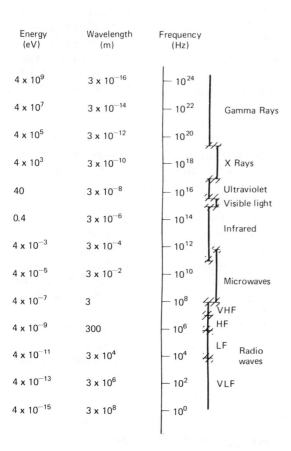

theless, they are all photons, differing only in energy (therefore frequency, remember Eq. 5.21), their place of origin, and their method of detection. Figure 14.1 gives the electromagnetic spectrum. Note that the highest-energy photons are called gamma rays and the lowest-energy photons radio waves. All photons originate from the acceleration of electric charge. In the case of gamma rays these charges are in nuclei. Nuclear binding energies are large on the atomic scale, so nuclear vibrations produce high-energy photons. Electrons that are outside a nucleus can also produce photons. If tightly bound inner electrons are involved, X rays are emitted. Vibrations of valence electrons may produce photons corresponding to ultraviolet (UV), visible, or infrared (IR) light. The accelerations of charges in circuits or electrical discharges in the atmosphere produce even lower-energy photons: radio waves. These are further classified by wavelength. Microwaves, for example, are radio waves with wavelengths between about 10 cm and 1 mm.

All of these "radiations" are basically the same: electromagnetic waves. The names are a convenience to us. There is no sharp distinction between an X ray and an UV photon. Certainly, a measuring device could not tell how a given 5-eV photon originated!

Figure 14.1 The electromagnetic spectrum. There are no sharp boundaries between the various classifications.

Certain types of radiation that are not normally found in nature are important because of their biological effects. Fission fragments are found in the high-level radioactive wastes from nuclear reactor spent-fuel rods or in the fallout from nuclear weapons. These are nuclei resulting from ^{236}U fission (or ^{240}Pu). These nuclei may have significant biological effects because of their chemical identities as well as their inevitable radioactivity.

Certain light, charged ions, such as protons, are also important because of their method of energy loss in matter. Protons are emitted by some radioactive elements not normally found in nature but produced in reactors. They also may be accelerated by machines into high-energy beams for the purpose of irradiation of either tissue in cancer therapy or materials for radiation-damage studies.

Neutrons are unstable with a half-life of 636 s, so they are not found in large quantities in nature. They are obviously produced in very large quantities in a nuclear reactor and in a nuclear explosion. Neutrons have also been used in cancer therapy. Radiation damage studies using neutrons are important to determine the "survivability" of communications and weapons systems in a nuclear attack.

Energy Loss Mechanisms

The various types of radiation interact with matter in quite different ways, depending on the charge state, mass, and energy of the radiation. Since so much of the material we want to cover in this chapter depends on these fundamental interactions, we look at them in detail.

Charged Particles Electrons, protons, fission fragments, and so forth interact with matter through the Coulomb force, principally with atomic electrons. For sufficiently large incident particle energies, every interaction will free an electron from an atom and leave a positive ion.

TABLE 14.2 Effective Composition of Soft Tissue

Element	Charge	Atoms/cm^3
Hydrogen	1	5.98×10^{22}
Carbon	6	9.03×10^{21}
Nitrogen	7	1.29×10^{21}
Oxygen	8	2.45×10^{22}

Obviously, the incident particle energy must be greater than the electron binding energy for this to be the case. These binding energies range from a few electron volts for valence electrons to many thousand electron volts for the K-shell electrons of heavy elements. In this chapter we are interested principally in the interaction of radiation with living tissue, which can be taken to be the mixture of light elements given in Table 14.2. The methods we develop will, of course, be applicable to any type of material.

The amount of energy lost in a particle–electron collision depends on the incident particle charge, mass, and energy. The exact dependence is complicated and does not concern us here. It has been calculated and tabulated for a variety of situations and can be referred to, if the need should arise.[1]

Since a charged particle in matter will make many thousands of collisions before coming to rest, the rate of energy loss, called the **stopping power**, is usually of interest in determining radiation effects. It may be given in terms of keV/cm or any unit equivalent to energy per unit length. From an experimental point of view, it is often easier to measure thickness by measuring area and mass. So as in Chapter 7, we define the *areal density* ξ by

$$\xi = \rho x, \qquad (14.1)$$

where ρ is the density. Figure 14.2 is a plot of

[1]J. E. Turner, in *Studies in Penetration of Charged Particles in Matter*, National Academy of Science, National Research Council Report 1133 (1964).

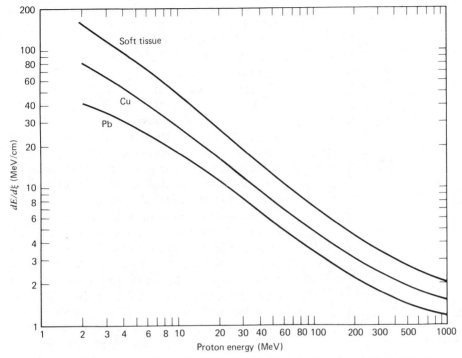

Figure 14.2 The rate of energy loss for protons in soft tissue, copper, and lead as a function of proton energy.

the stopping power for protons in several materials.

Example 14.1 What is the thickness of an aluminum foil having an areal density of $2.4 \, mg/cm^2$?

From Eq. 14.1

$$x = \frac{\xi}{\rho}.$$

Since ρ for aluminum is $2.7 \, g/cm^3$,

$$x = \frac{2.4 \times 10^{-3}}{2.7},$$

$$\boxed{x = 0.89 \times 10^{-3} \, cm.}$$

It is worth noting that the stopping power is not a constant, but increases nonlinearly with decreasing energy. This effect is shown in Figure 14.3 for 400-MeV protons. Note in particular the very large increase in stopping power near the end of the range. This comes about because a slow-moving particle is able to transfer a great deal more energy to atomic electrons than a rapidly moving one. Also near the end of the range, each collision removes a large fraction of the total remaining energy, so that random processes tend to make the actual range uncertain. This will cause **straggling** to the range, so that for a given energy particle it is not possible to predict the exact range. The approximate range of a particle having an initial energy E_i would be given by integrating the stopping power:

$$R = \int_{E_i}^{0} \left(\frac{dE}{dx}\right)^{-1} dE. \qquad (14.2)$$

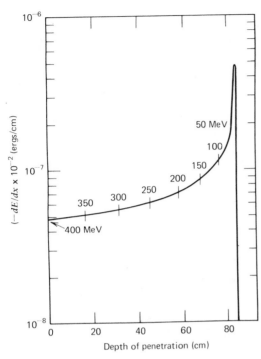

Figure 14.3 The rate of energy deposition for protons in tissue as a function of penetration depth, showing the Bragg peak at the end of the range.

Some corrections are required at very low energies, but again tables of ranges have been compiled and are useful for most applications.

The large peak in the stopping power at the end of the range is called the **Bragg peak**. This phenomenon has been exploited in cancer therapy, where it is desired to deliver a large dose of energy to a deep-seated tumor without destroying the intervening tissue or, at least, minimizing the damage to it. An even larger effect can be obtained with pions, since there is not only a Bragg peak associated with the change in stopping power, but also, when the pion stops, it is usually absorbed by a nucleus, giving up all of its mass energy (see Table 14.1) and producing a **spallation** or shattering of the nucleus that absorbed it. Pion therapy is in its infancy, as pion beams are not easily obtained

in large numbers from most accelerators capable of producing them.

An additional effect must be taken into account when the incident charged particle is an electron. Because of the small electron mass, any type of collision will cause a significant momentum transfer from the incident electron. This will usually cause a direction change, which is an acceleration. As mentioned in Chapter 12, accelerating charged particles must radiate electromagnetic energy. This radiation caused by acceleration is called **Bremmsstrahlung**, or braking radiation. It can account for a significant fraction of the total electron energy. Photon beams used in cancer therapy and nuclear research are generated by stopping a high-energy electron beam in a heavy metal target. This is the same technique used to generate X rays, except that in the case of X rays the electron beam is less energetic.

Because electrons are lighter than most other charged particles they are not as effective in producing ions in matter. Figure 14.4 gives the range of electrons in tissue as a

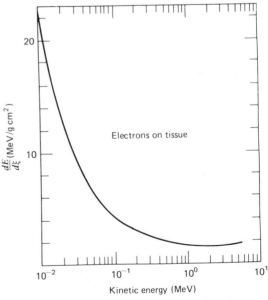

Figure 14.4 The rate of energy loss for electrons in tissue as a function of electron energy.

function of energy. For comparison note that a 10-MeV proton has a range of 0.12 cm in soft tissue. It is convenient to define the **linear energy transfer** (LET) of a charged particle as the average energy in keV transferred to the material per unit of track length, usually expressed in micrometers. The LET varies with particle and energy, but is a good relative measure of the amount of radiation damage that can be expected. Average LETs for several particles and energies are listed in Table 14.3.

Neutrons Neutrons have no electric charge, and therefore must interact with matter through a mechanism other than the Coulomb force. As we saw in Chapter 7, neutrons can be characterized by speed. *Thermal neutrons* have energies less than about 0.05 eV, *slow neutrons* up to about 0.1 keV. Above that, neutrons are called *fast*. Fast neutrons lose energy primarily by direct collisions with nuclei. Applying conservation of momentum and energy will convince you that if the neutron makes a collision with a nucleus having a mass more than a few times the neutron mass, the amount of energy transferred to the nucleus is very small. However, in tissue there are many hydrogen atoms, tissue being very nearly equal to water in terms of density and

stopping power. A neutron could transfer very nearly all of its energy to a hydrogen nucleus in a head-on collision. For other nuclei, the amount of energy transferred in each collision would be smaller on average, but the net result would be the same: high-LET charged particles produced in the tissue.

Neutrons also can induce nuclear reactions. The reactions $^{16}O(n, \alpha)^{13}C$ and $^{16}O(n, p)^{16}N$ have high cross sections for fast neutrons and are likely to contribute to the overall damage effect of such neutrons. Of course, similar reactions can occur with charged particles; but they contribute an insignificant amount of energy, compared with the direct ionization produced by the particles themselves.

Neutrons can also produce gamma rays by means of inelastic scattering from nuclei. This can contribute as much as 20 percent of the total energy deposition. Thermal neutrons cannot in general produce high-LET secondaries, as is the case of fast neutrons. Thermal energies are so low that they often do not exceed the molecular binding energies of even the hydrogen compounds. However, these neutrons can induce atomic excitations and molecular rotation–vibration excitations resulting in heating. In addition, thermal neutrons can be readily absorbed by some nuclei, usually producing a radioactive product. However, the absorption cross sections for nuclei likely to be found in tissue are usually small.

Photons The interaction of photons with matter is strikingly different from that of charged particles. A photon usually gives up a large fraction, perhaps all of its energy, in a single collision; charged particles lose a very small fraction in each encounter. Photons may travel several meters, perhaps kilometers, before making a collision; charged particles make a collision every few atoms.

The loss of photons from a beam passing through matter is reminiscent of an exponential change process. The rate of change of the

TABLE 14.3 Average LET Values

Particle	Energy (MeV)	Average LET (keV/μm)	Tissue Penetration (μm)
Electron	0.01	2.3	1
	0.1	0.42	180
	1.0	0.25	5000
Proton	0.1	90	3
	2	16	80
	5	8	350
	100	4	1400
Alpha	0.1	260	1
	5	95	35

intensity of the beam with respect to position at any point in the material depends on the intensity at that position and the type of material;

$$-\frac{dI}{dx} = \mu I. \qquad (14.3)$$

This equation is analogous to Eq. 1.6; the minus sign indicates the intensity becomes smaller as penetration into the material increases. The factor μ is called the *linear attenuation coefficient*. It is not a constant but, rather, depends on material and energy in a complex fashion. A mass attenuation coefficient μ/ρ is often defined, similarly to the attenuation coefficient we used in Chapter 12. The complex dependence of these coefficients on energy is a consequence of the fact that photon interactions with matter are themselves very complex.

Example 14.2. The mass attenuation coefficient for 1-MeV photons in lead is 0.04 cm²/g. What thickness in millimeters is required to reduce the photon beam intensity of 1-MeV photons by 90 percent?

From Eq. 14.3,

$I = I_0 e^{-(\mu/\rho)x}, \quad \ln 0.5 = -0.04x,$

$x = 17.33 \text{ g/cm}^2, \quad \rho \text{ for lead} = 11.3 \text{ g/cm}^3.$

For 50 percent reduction,

$$\boxed{x = 15.33 \text{ mm.}}$$

For 90 percent reduction,

$$\boxed{x = 50.9 \text{ mm.}}$$

This testifies to the tremendous penetrating power of high-energy photons!

When a photon does interact, the interaction may be one of five types: nuclear scattering (elastic or inelastic), photoelectric effect, Compton scattering, pair production, or atomic or molecular excitation.

Elastic nuclear scattering changes only the direction of travel of the photon. The only energy change is that required to conserve energy. Inelastic nuclear scattering does result in the transfer of energy to the nucleus. This excess energy is usually emitted as another photon, most often in a different direction. Except for the small fraction of nuclear excitations that result in atomic electrons being emitted, nuclear photon scattering serves only to extract photons from the beam, rather than to deposit energy in the vicinity of the collision.

The photoelectric effect, diagramed schematically in Figure 14.5, is the total absorption of the energy of a photon by a bound atomic

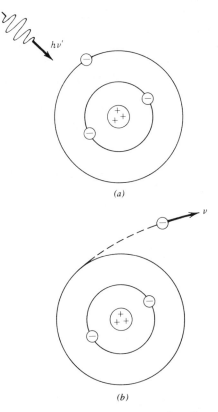

Figure 14.5 The photoelectric effect. (*a*) A photon of energy $h\nu$ is completely absorbed by an atom, (*b*) which then ejects an electron.

electron. The electron is usually freed by this absorption, creating an ion pair. One or more electron energy-level shifts occur in the atom to replace the freed electron, generating additional photons in the process. These may be X rays, UV, IR, or visible light photons, depending on the energy levels involved in the shifts. Clearly, the energies of these photons will be less than that of the photon that initiated the event.

The Compton effect is a collision between a photon and an atomic electron in which energy and momentum are conserved. The energy and momentum of a photon are given in terms of its wavelength λ by

$$E = \frac{hc}{\lambda} \tag{14.4}$$

and

$$p = \frac{h}{\lambda}. \tag{14.5}$$

From this it may be shown that

$$\lambda' - \lambda = \frac{h}{mc}(1 - \cos \theta), \tag{14.6}$$

where m is the electron mass, λ' is the wavelength after the collision, and θ is the scattering angle, as shown in Figure 14.6.

Pair production (shown in Figure 14.7) is a process that has no analogue in classical physics. A photon of sufficiently high energy may spontaneously disappear (in the presence of another body to conserve momentum) and convert all its electromagnetic energy into the mass energy of an electron–positron pair. The minimum amount of energy required is simply twice the electron mass energy: 1.02 MeV. For photon energies higher than this, the excess over 1.02 MeV is shared by the electron and positron as kinetic energy.

The three processes—photoelectric effect, Compton scattering, and pair production—are energy dependent. We show their relative importance in Figure 14.8 for photons on soft

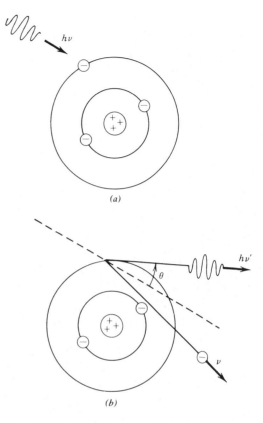

Figure 14.6 The Compton effect. (*a*) A photon of energy $h\nu$ scatters from an atomic electron. (*b*) After the scattering the photon has a reduced energy $h\nu'$, and the electron is knocked away with a speed v.

tissue. For other substances, the attenuation coefficients will usually be larger for photon energies below 1 MeV and above about 5 MeV.

Photons having energies below about 5 eV cannot interact by the methods described above. The binding energies of electrons in atoms usually exceed this value. However, these low-energy photons can interact by causing atomic or molecular excitations. The photon energy is usually completely absorbed by the atom or moelcule, which is then raised to an excited state. The excited state decays by

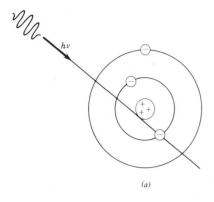

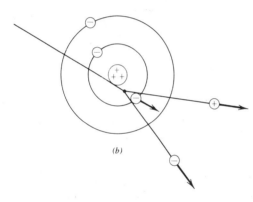

Figure 14.7 Pair production. (*a*) A photon of energy $h\nu > 1.02$ MeV spontaneously transforms (*b*) into an electron–positron pair in the neighborhood of an atomic electron.

emission of one or more photons; these may also be absorbed by neighboring atoms in a similar fashion. Eventually the original photon energy degrades to thermal vibrations in the absorbing substance. Microwaves have energies well below that required to produce ionization. So far as is known, microwave radiation produces only heating in biological tissue, although many other effects have been discussed. No hard experimental evidence has been found to verify claims of additional microwave effects, particularly those for low levels of radiation.

It is generally true that the net result of the impaction of high-energy radiation of any type on matter is the creation of high-velocity charged particles, electrons or positive ions, in the materials. These charged secondaries are primarily responsible for the radiation damage, as they affect a larger number of atoms than does the primary radiation. The time scale for the creation of charged secondaries is very short, as we see in this example.

Example 14.3 A 2-MeV proton impinges on tissue. Assume it comes uniformly to rest. How long does this require?

Use the equations of uniform motion:

$$v^2 = v_0^2 + 2as,$$

$$v = v_0 + at.$$

This assumes the acceleration is constant. The time required is thus

$$t = \frac{(v - v_0)2s}{(v^2 - v_0^2)} \quad \text{and} \quad v_0 = \left(\frac{2E}{m}\right)^{1/2}.$$

From Fig. 14.2,

$$s \approx \frac{2\text{ MeV}}{150\text{ MeV/cm}} = 1.3 \times 10^{-4}\text{ m},$$

and

$$\boxed{t = 8.2 \times 10^{-12}\text{ s.}}$$

This is a very short time period; so short that all other motions, chemical reactions, and so on are effectively frozen. It is therefore possible to think of the incoming radiation from a single particle as occurring during a very short pulse of time.

Radiation Chemistry

Ion pairs last about 10^{-10} s, which is long compared to the time required to form them. They interact with the molecules of the sub-

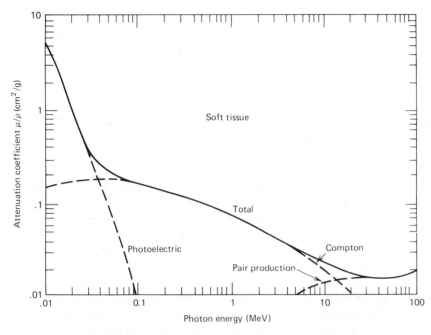

Figure 14.8 The relative importance of the photoelectric effect, the Compton effect, and pair production in the interaction of photons with soft tissue as a function of photon energy.

stance in several ways, but the end result is almost always the same: **free radicals**. Free radicals are electrically neutral atoms or molecules that have an unpaired electron in their outer or valence shells. They are always extremely reactive. A typical chain of reactions would be

$$A \rightarrow A^+ + e^-, \quad (14.7)$$

$$EF + e^- \rightarrow EF^-, \quad (14.8)$$

$$EF^- \rightarrow E^- + F\cdot, \quad (14.9)$$

$$F\cdot + O_2 \rightarrow (FO_2)\cdot. \quad (14.10)$$

This final peroxyl radical is extremely reactive and can cause structural alterations on the macromolecules of biological significance.

Since tissue is primarily water, $H\cdot$ and $OH\cdot$ radicals are abundantly formed by irradiation (with radiation of any kind). The first of these

is a potent reducing agent, even more potent than the $OH\cdot$ is as an oxidizing agent. However, there are more ways to form oxidizing agents, so irradiation in tissue produces an overall oxidizing effect. Two $OH\cdot$ radicals can and often do combine to form hydrogen peroxide, H_2O_2, which is also a powerful oxidizer.

Up to this point, the physics and biophysics of the interaction of radiation with matter are relatively well understood. The picture begins to cloud here, because the number of possible ways to form radicals, their relative probability of formation, the reaction rates, and so forth, are not well known. We have also not been very quantitative so far in the discussion of radiation effects. Since we wish to do so, it is time to introduce units of measure that will enable quantitative comparisons to be made.

14.2   RADIATION UNITS

Since we are interested primarily in the biological effects of radiation that are caused by the deposition of energy in tissue, you might think that we do not need any special units, that joules or perhaps joules per kilogram would suffice. Unfortunately, there are some valid reasons that radiation effects must be treated somewhat differently than simple energy transfers from one substance to another. The Fifteenth General Conference of Weights and Measures in 1975 adopted a new set of terms for dealing with radiation and its energy deposition. Unfortunately, there was already in existence a set of terms that has been in very widespread usage for many years. The conference recommended that the new terms be phased into use in a 10-yr period, namely, by 1985. But recent reviews of the American radiation protection literature indicate these new terms are having a difficult time catching on. As a consequence, we define both the new and the old radiation units, but we try to present information in the new units as much as possible. Data taken from the literature, however, will be expressed in whatever units were used by the original authors.

Activity

We have already encountered a unit of **activity**, the *curie* (Ci). This unit was devised many years ago to provide a measure of the strength of radioactive sources. One curie is defined to be a quantity of radioactive material that has a decay rate of 3.7×10^{10} disintegrations per second. Unfortunately, the "quantity" is not well defined. Does this refer to volume or mass? As a remedy and to make the unit of activity independent of specific substances[2] or arbitrary values, the unit of activity is now defined as one disintegration

[2]The curie was originally defined as the number of disintegrations of 1 g of radium.

per second, the *becquerel* (Bq). Note that the becquerel is an SI unit; all the SI prefixes can be used as appropriate scaling factors.

Exposure

X Rays were in use before radioactivity was understood. It became convenient in the early days of radiation physics to use the ionizing effect of X rays in air as a quantitative measure of radiation exposure. Air was convenient because it has approximately the same effective atomic number as tissue, and so might be expected to have a similar response to X rays. Accordingly, the unit of **exposure** was defined as the *roentgen* (R). One roentgen of radiation liberates 2.58×10^{-4} C of charge of one sign in 1 kg of air. The conference decided to abolish the term roentgen, as no need was felt for an additional name. Expressing exposure in terms of coulombs per kilogram does not conflict with any other use for that particular combination of SI units and carries all the intelligence required. Nevertheless, the roentgen and milliroentgen are used a great deal today in research laboratories, and, in the opinion of many, will require a rather longer period of time than 10 yr to eliminate.

Absorbed Dose

Neither the becquerel nor coulombs per kilogram is an appropriate measure of the biological effect of radiation, because neither indicates the amount of energy actually absorbed by tissue. Consequently, a measure of **absorbed dose** has been defined. In the older units, 1 rad is 0.01 J/kg of material. (The original definition was 1 rad = 100 ergs/g, but cgs units should never be used in a definition.) The conference adopted the *gray* (Gy) as the unit of absorbed dose, with 1 Gy = 1 J/kg. This is a more appropriate measure, since being exposed to 1 C/kg of ionizing radiation does not mean that a person would absorb a large amount of energy; indeed, if the radiation were in the form of alpha particles and if the radioactive substance were not inhaled, it could be that the exposed individual would not receive any dose at all, However, it is not simple to calculate the absorbed dose, even when everything is known about the radioactive source.

Example 14.4 An individual ingests 10 mg of ^{55}Fe. What dose does this person receive in a 10-yr period?

Iron-55 decays by electron capture and subsequent X-ray emission with a total decay energy of 0.22 MeV per decay. To find the number of atoms that decay in 10 yr, note that the half-life of ^{55}Fe is 2.9 yr. From this $m = 0.239$ and mass $= 10 \exp(-0.239 \times 10)$, mass $= 0.916$ mg. Thus, $(10.0 - 0.916) = 9.084$ mg has decayed. This is

$$\frac{9.084 \times 10^{-3}}{55} \times 6.022 \times 10^{23} \text{ atoms}$$
$$= 1.0849 \times 10^{20} \text{ atoms}$$

for a total of

$$1.0949 \times 10^{20} \times 0.22 \times 10^6 \text{ eV} = 2.3867 \times 10^{25} \text{ eV}$$
$$\text{or} \quad 1.49 \times 10^6 \text{ J}.$$

Now to find the energy per unit mass, we need to know not the mass of the person but, rather, the mass of tissue that received the radiation. It is unlikely the radiation would have been received equally throughout the body—at least in this case. Let us assume the mass to be 25 kg; this is about one-third the average male body mass. The total dose then would be

$$\frac{1.49 \times 10^6}{25} = 5.96 \times 10^4 \text{ Gy} \quad (5.96 \times 10^2 \text{ rads}),$$

a very large amount of absorbed dose. It illustrates that what seems to be a small quantity of material can be very dangerous.

Although the definition of absorbed dose gives us an acceptable measure of radiation dose in materials, it does not provide exactly what we need. This is a consequence of the fact that radiations with different LETs will produce different effects for the same total deposited energy.

This implies that the density of energy deposition along the track of the incident radiation determines the extent of damage caused by that energy. This is a function of the detailed microscopic interaction of the charged secondaries and uncharged radicals produced by the primary radiation. We examine this in more detail later in this chapter when we talk about the effects of low levels of radiation.

Dose Equivalent

But for now, we need a measure of radiation effects that includes the variation in LET for different radiations. For this we define the **dose equivalent** to be

$$H = D \times Q \times N, \quad (14.11)$$

where D is the dose in grays, and Q is a quality factor that depends on the type of radiation (see Table 14.4). The factor N depends upon the type of tissue involved. Dose equivalent is measured today in the new units *sieverts* (Sv). Dose equivalent is far more commonly quoted in the older units *rems*; the definition of rem follows from Eq. 14.11 when D is measured in rads.

TABLE 14.4 Quality Factors[a]

Average LET in Water (keV/μm)	Q
3.5	1
3.5 to 7.0	1 to 2
7.0 to 23	2 to 5
23 to 53	5 to 10
53 to 175	10 to 20

[a]X Rays, electrons, and positrons of any energy have $Q = 1$.

TABLE 14.5 Biological Weighting Factors

Organ	$1/N$
Total body	1.0
Gonads	0.25
Breast	0.15
Red marrow	0.12
Lung	0.12
Thyroid	0.03
Bone	0.03

Some misunderstanding has arisen about the difference between grays and sieverts. If Q and N are considered to be dimensionless quantities, then both the gray and the sievert have the units of joules per kilogram. Clearly, only the gray represents the actual energy deposited per unit mass; the sievert represents the effective energy received by the tissue per unit mass and can be a very much larger quantity.

The factor N can be thought of as a weighting factor that takes into account the sensitivity of the tissue involved. Certain types of tissue such as bone are rather insensitive, while others, such as sperm cells, are rather more sensitive. Table 14.5 lists the weighting factors recommended by the International Committee on Radiation Protection for several organs.

From Table 14.4 we see that high-LET radiation is much more damaging in tissue than low-LET radiation depositing the same dose. Note that 5-MeV alpha particles have a Q of about 15. This is the reason that inhaling or ingesting heavy radionuclei, such as plutonium, is far more dangerous than receiving an exterior exposure to any kind of radioactivity.

Neutrons are not themselves ionizing, but produce charged secondaries that are. The LET for neutrons can be defined similarly to that for charged particles. For thermal neutrons (i.e., for energies less than about 0.1 eV) Q is 1; for fast neutrons Q is about 10.

The current situation with radiation units is summarized in Table 14.6. The appropriate unit for radiation effects is the sievert, although because of its newness its use is not very widespread. In fact, the General Conference did not adopt it in 1975; it is expected to do so shortly. Some countries have already converted to the new units, although in Britain the centisievert (equivalent to the rem) has been adopted as the unit of dose equivalent. This is

TABLE 14.6 Radiation Units

	New	Old	Definition	Equivalent Unit[a]
Activity	Becquerel (Bq)		1 disintegration per second	t^{-1}
		Curie (Ci)	3.7×10^{10} disintegrations per second	t^{-1}
Exposure	Coulomb/kilogram		—	q/m
		Roentgen	2.58×10^{-4} C/kg	q/m
Absorbed Dose	Gray (Gy)		1 J/kg	l^2/t^2
		Rad	0.01 J/kg	l^2/t^2
Dose Equivalent	Sievert (Sv)		$H = D \times Q \times N$	l^2/t^2
		Rem	$H = D \times Q \times N$ (D = rads)	l^2/t^2

[a]Equivalent unit is the unit expressed in terms of the basic measures: t for time, q for charge, l for length, and m for mass.

an error that becomes apparent when smaller quantities are required; you cannot say *milli-centisievert* ("millirem") without breaking the SI rules!

New ways of doing things come hard, particularly in an "old" profession with well-established protocols. Perhaps authors of textbooks owe it to their constituency to use the most up-to-date terminology to foster its use and eventual adoption.

14.3 THE RADIATION BACKGROUND

We are continually bombarded by radiation: from the sun, from naturally occurring radionuclides, from man-made radionuclides produced by reactors, and fallout from nuclear weapons testing. In addition, we voluntarily subject ourselves to additional radiation when we have X rays taken, when we have radiation therapy, when we watch television, and under several other circumstances, many of which we discuss in this chapter.

Solar Flux

The sun produces both ultraviolet and infrared radiation in addition to visible light. The IR radiation does not carry enough energy to be ionizing, but it can and does cause heating of the skin tissue. Ultraviolet can cause ionization in some atoms, but it can more readily cause the severing of molecular bonds. Consequently, UV can be very damaging to skin. It can cause severe burns, and prolonged exposure can cause skin cancer. It has been established that in societies in which exposure to the sun is excessive and where skin pigmentation is light, skin cancers are far more prevalent than in societies where sun exposure is not as typical.

Since UV is absorbed by the atmosphere, UV intensities will be higher at higher elevations, as anyone who has exposed unprotected skin to the sun for a few hours at the top of a mountain will tell you. Residents of cities at higher elevations will therefore be exposed to a higher average UV level than others. This exposure, coupled with the higher cosmic-ray flux (see below), leads to higher overall rates of radiation-induced problems for residents of cities such as Denver and Mexico City, as verified by epidemiological studies.

However, such studies cannot isolate mechanisms; they see the totality of causes by looking at only the effects. In a city like Denver one might easily guess that there would be a higher abundance of naturally occurring radioisotopes, as well, because of all the mining activity in the Rocky Mountains.

Naturally Occurring Radioisotopes

A number of radioactive elements were created at the time the earth was formed. These are listed in Table 14.7. There are also a number of decay products, called "daughters" of ^{238}U and ^{232}Th; notable among these is ^{226}Ra, with a half-life of 1622 years.

There is also a continuous production of shorter-lived isotopes by cosmic-ray bombardment of the atmosphere. Cosmic rays, like UV radiation, are attenuated by the atmosphere. You would expect to find higher levels of cosmic-ray flux at higher elevations, and such is the case. Denver (1610 m elevation) has an average cosmic-ray flux equivalent to 2×10^{-3} C/kg hr, while Boston (6.5 m elevation) has the equivalent of 0.9×10^{-3} C/kg hr. The important radionuclides produced by this means are listed in Table 14.8. Of these ^{14}C is perhaps the most significant, since it becomes incorporated into the food chain very easily and has a rather long half-life.

Many of these naturally occurring radioisotopes are distributed more or less uniformly around the globe; others can be concentrated by several natural mechanisms in certain kinds of soils or in certain plants. Thus, areas in Brazil (Minas Gerais, Espirito Santo) and in

TABLE 14.7 Primeval Radionuclides

Isotope	Abundance	Half-Life (yr)	Radiation
Uranium-238	4×10^{-6} g/g[a]	4.5×10^9	Alpha
Thorium-232	12×10^{-6} g/g	1.4×10^{10}	Alpha, gamma
Potassium-40	3×10^{-4} g/g	1.3×10^9	Beta, gamma
Vanadium-50	0.2 ppm[b]	5×10^{14}	Gamma
Rubidium-87	75 ppm	4.7×10^{10}	Beta
Indium-115	0.1 ppm	6×10^{14}	Beta
Lanthanum-138	0.01 ppm	1.1×10^{11}	Beta, gamma
Samarium-147	1 ppm	1.2×10^{11}	Alpha
Lutetium-176	0.01 ppm	2.1×10^{10}	Beta, gamma

[a]Gram per gram of soil.
[b]Parts per million.

India (Kerala), where there are alluvial deposits of monazite sands, have large radiation backgrounds caused by the high content of thorium and rare earths in the monazite.

These heavy radionuclides are particularly dangerous in the body, since they are deposited primarily in the bones, have a long biological half-life (a long residence time in the body), and are alpha-emitters. The lighter common radionuclides, such as ^{14}C and ^{40}K, are much more biologically active and so have a much shorter biological half-life. The data for several elements (regardless of isotopic nature) are shown in Table 14.9.

The net effect of the primeval and cosmic-ray-produced radionuclides is the establishment of a radiation background continuously present, continuously exposing all individuals. This background is not uniform but depends on altitude, geographic location, diet, and type

TABLE 14.8 Some of Radionuclides Produced by Cosmic Rays

Isotope	Half-Life	Concentration (disintegrations/min m³)[a]
Tritium-3	12.3 yr	10^1
Beryllium-7	53 days	1
Beryllium-10	2.7×10^6 yr	10^{-7}
Carbon-14	5760 yr	4
Sodium-22	2.6 yr	10^{-4}
Silicon-32	700 yr	2×10^{-6}
Phosphorus-32	14.3 days	2×10^{-2}
Phosphorus-33	25 days	1.5×10^{-2}
Sulfur-35	87 days	1.5×10^{-2}
Chlorine-36	3×10^5 yr	3×10^{-8}

[a]Disintegrations per minute per cubic meter of air in the lower troposphere.

TABLE 14.9 Biological Behavior of Certain Elements

Element	Critical Organ	Mass of Organ (kg)	Biological Half-Life (days)	Fraction of Full Dose in Critical Organ
H	Whole body	70	19	1.0
C	Fat	10	35	0.6
Na	Whole body	70	19	1.0
K	Muscle	30	37	0.92
Sr	Bone	7	4,000	0.7
I	Thyroid	0.2	120	0.2
Cs	Muscle	30	17	0.45
Ba	Bone	7	200	0.96
Ra	Bone	7	20,000	0.99
Th	Bone	7	40,000	0.82
U	Kidneys	0.3	30	0.065
Pu	Bone	7	43,000	0.75

of dwelling among other things. In the United States, the background level varies between about 0.85 mGy/yr (85 mrad/yr) along the Atlantic Coast to as much as 1.8 mGy/yr (180 mrad/yr) in the Rocky Mountains. Note that this is an average; it could be from 10 to 20 percent higher for individuals who live in stone or concrete buildings, who must travel by air a great deal, or who work with radioactive materials.

Doses are higher for those who live in stone or concrete buildings because of the higher concentrations of uranium, thorium, and their daughter nuclides, especially radon gas, in these materials. While a 20 percent annual increase may seem small, it must be remembered that these individuals may be exposed for a large number of years. The cumulative effect of such exposure could be quite large.

Example 14.5 Assume the average annual U.S. background radiation dose to be 1.0 mGy/yr and that the average individual has a cancer risk of 1800 in 1,000,000. What risk of cancer would this person have over a 30-yr period, if the radiation background were 20 percent higher? (Assume 120 cancer deaths per million people for each 0.01 Gy)

Over 30 yr the average dose would be $1.0 \times 10^{-3} \times 30 = 3.0 \times 10^{-2}$ Gy. At 2 mGy/yr, it would be 3.6×10^{-2} Gy or an increase of 0.6×10^{-2} Gy. This could be expected to result in 72 cancers per 1,000,000. Therefore this individual's risk has increased to 1872 in 1,000,000, or a 4 percent increase in risk.

Man-Made Radiation

In addition to the naturally occurring sources of radiation, we expose ourselves, willingly and unwillingly, to a large variety of radiation sources generated by our own activities. Without a doubt the most important of these is medical X-ray diagnosis. The dose received from an X ray can vary widely, depending on the type of film used, the region of the body X-rayed, the age and quality of the equipment used, and the skill of the radiological technicean. Exposures have been estimated to vary from 10 mR (2.4×10^{-7} C/kg) to 3000 mR (7.2×10^{-4} C/kg) with doses from 100 μGy (10 mrad) to 30 mGy (3 rads). An individual can easily receive a dose from a single X ray equal to the entire dose from a year's exposure to the background radiation.

Fortunately, this country has abandoned the program of mass chest X ray that had been in operation for many years in the 1940s and 1950s. A few states require X-ray technicians to have training and certification, but most do not. Most physicians and dentists receive little radiological training. While older equipment that is not well shielded and does not produce a tightly collimated beam is being replaced, in many localities such old equipment is still in use. As a general rule every person should insist that only necessary X rays be taken.

Another important source of radiation during the past 20 years is fallout from nuclear weapons tests. Most of the nuclear powers have agreed not to test in the atmosphere, but China and France have not agreed and continue to test, albeit not with the frequency of 20 years ago. Up to the signing of the test ban

treaty in 1962, about 511 megatons of nuclear weapons (in terms of the equivalent TNT explosive power) had been exploded in the atmosphere. Of this, about 193 megatons were fission weapons, the remainder being fusion weapons (H-bombs).

More than 200 radioisotopes are produced when a fission device explodes in the atmosphere. Most of these have very short half-lives and will decay before falling to earth. Some have long enough half-lives to be health hazards; many were listed in Table 7.13. Prominent among these are ^{131}I, ^{90}Sr, and ^{137}Cs. These have half-lives of 8 days, 28 yr and 30 yr, respectively, and all three are biologically active. Strontium behaves chemically as if it were calcium; similarly, cesium appears to the body as if it were potassium. Figure 14.9 shows the ^{131}I level in milk in the United States

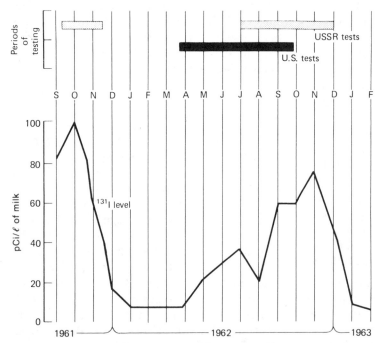

Figure 14.9 The average level of ^{131}I in milk in the United States related to periods of weapons testing. (From C. L. Comer, *Fallout from Weapons Tests*. Washington D.C.: U.S. Atomic Energy Commission, 1966.)

from late 1961 to early 1963, the period of intense U.S. and Soviet atmospheric nuclear tests. This figure gives monthly averages for the entire country; there were wide variations from one section to another, with some monthly averages as high as 26 Bq/ℓ (700 pCi/ℓ). Daily measurements were often three times this value in some localities.

These three radionuclides find their way into the body by ingestion of food products that have become contaminated, usually milk from cows that have fed on grasses on which the radioactive debris from weapons tests has fallen. Because of the biological activity of these species (see Table 14.9), they are particularly dangerous and can result in very large doses to sensitive organs.

Example 14.6 An average individual in New York City in 1963 had a daily intake of ^{90}Sr of about 0.94 Bq. What dose to the bone did such an individual receive in that year?

Strontium-90 decays by beta emission to ^{90}Y, which decays rather quickly to stable ^{90}Zr, as shown in Figure 14.10. The beta energies indicated in the figure are the end point or maximum energy of each decay. The average beta energy will be 0.3 or 0.4 of this value, depending on what kind of decay it is (i.e., what

nuclear quantum-number changes are involved). Assume $0.4E_{max}$ (worst case); the gammas will not all be stopped in the body. Assume 10 percent of the gamma energy is absorbed in the bone. The total energy absorbed per decay is then

$$(0.54 + 2.27) \times 4 + 1.75 \times 0.1 \text{ MeV} = 1.299 \text{ MeV}$$
$$\text{or} \quad 2.078 \times 10^{-13} \text{ J}.$$

Each day there is 0.94 Bq of ^{90}Sr activity ingested of which, according to Table 14.9, 70 percent is absorbed by the bone, or 0.658 decays/s. This 5.685×10^4 decays/day. So for each day $5.685 \times 10^4 \times 2.078 \times 10^{-13}$ J is absorbed by the bone. Over a period of a year there will be then

$$D \sum_{n=1}^{365} n$$

total absorbed, where D is the amount absorbed each day. The dose in grays is the total absorbed energy in joules divided by the mass of the organ. Therefore the dose received in a year would be

$$\boxed{0.79 \text{ mGy.}}$$

Fallout, medical X rays, nuclear power plants, even television sets because of the X-ray emissions—all add to the radiation background that we receive. The dose varies considerably for each person but is between 1 and 3 mGy (100 to 300 mrad) per year for most people. The next question is, what does this radiation do when it impinges upon the human body?

14.4 RADIATION EFFECTS

The effects of radiation in tissue are generally divided into two categories: **somatic** and **genetic**. Somatic effects occur in the individual receiving the radiation; genetic effects occur in that person's progeny. We could also divide

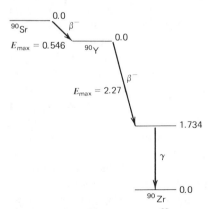

Figure 14.10 The decay of ^{90}Sr and its daughter ^{90}Y.

radiation effects into two categories depending on dose rate: **acute** and **chronic**. Acute doses are large amounts given in short time periods. Chronic doses are usually small amounts given over long periods of time. Somatic and genetic effects are relatively easy to describe. Acute effects are also straightforward; but chronic low-level effects are an area of controversy. In this section we discuss all the easy parts of radiation effects and then present some of the facts leading up to the low-level radiation controversy.

Somatic Effects

There is no controversy about the result when tissue is exposed to high levels of radiation. There is some uncertainty about the precise microscopic mechanisms, since the conversion of incident radiation to energy deposition is complicated, as we saw earlier. One mechanism could be that the large numbers of free radicals that result indirectly from the interaction of ionizing radiation with tissue are responsible for large numbers of cell deaths by causing the plasma membranes of the cells to rupture.

The cell, indicated in general form in Figure 14.11, consists largely of a complex substance called *cytoplasm*. This material is bounded by the plasma membrane and itself surrounds the cell *nucleus*. The cell membrane allows the passage of nutrients and wastes selectively. The cytoplasm uses the materials to generate the energy required by the cell for sustenance and for self-duplication. The genetic character of a cell is determined by *chromosomes*, which are found in the cell nucleus. Several other specialized types of structures are found in the cell as indicated in Figure 14.11.

Cell membranes are formed from layers of phospholipids sandwiched between proteins. These lipids are vigorously attacked by the free radicals OH· and O· and by H_2O_2, which we know are found abundantly in irradiated tissue. When a cell membrane is breached, the

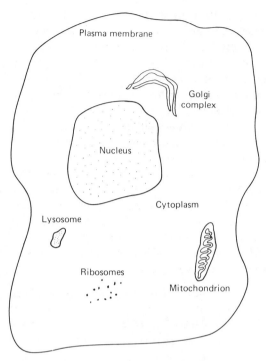

Figure 14.11 A typical cell, consisting of a nucleus surrounded by cytoplasm enclosed by a plasma membrane. Many of the other structures shown can be found in most cells.

cytoplasm spills out, and the cell is unable to function. This mechanism for cell destruction would be enhanced in the presence of large quantities of oxygen. Indeed, cancer treatment by radiation therapy does seem to be enhanced when the cancer is located in tissue having a higher-than-average water content, as in the gastrointestinal tract, compared to cases where the cancer is located in low-water tissue, such as bone or lung.

Direct damage can also result in cell death. Radiation tracks near a deoxyribonucleic acid (DNA) molecule in one of the chromosomes can cause ionization of one or more atoms or radicals from the DNA string, or even complete rupture of the helical structure of the DNA molecule. DNA has been determined to be the genetic blueprint, the carrier of information that enables cells to reproduce themselves properly to carry out their functions in the body. If the DNA molecule affected cannot repair itself and if it is an essential molecule in the reproductive process, then the cell will not be able to divide. It will simply die. The damage to a DNA molecule may not inhibit reproduction, but it may change the genetic information carried by the molecule. After cell division, a *mutation* is formed. This mutation may be fatal to the cell, as most mutations are—the cell simply cannot perform its duty any longer. The mutation may have no effect at all. Finally, the mutation may be cancerous; it may grow at a significantly higher rate than do normal cells and eventually lead to death of the organism.

If a large enough number of cells die, the organ that those cells make up must be affected. Its performance will deteriorate, and conceivably that organ could also die. Large enough doses of radiation can and have resulted in death. But the effects of radiation are not always so dramatic, although they may be just as fatal. Radiation acts as a poison—it is *abscopal*. The effect may not occur at the location of the event. A large radiation dose to the hand can result in a tumor in another body organ. Like poisons, radiation damage can be specific to one organ or tissue, or it can be general. The effects of radiation exposure can be cumulative. Some organs are much more sensitive than others, as we have seen. Unlike most poisons, radiation damage can act at high speed. And it can destroy the body's immune system, so that the individual becomes more susceptible to diseases like pneumonia.

Energy is deposited randomly along the path of ionizing radiation. The ions and radicals created interact randomly with surrounding molecules. As a consequence, the same amount of energy deposited at different times or in different persons or in slightly differing locations will not necessarily produce the same effect. It is not necessarily possible to say that a particular dose will always result in a parti-

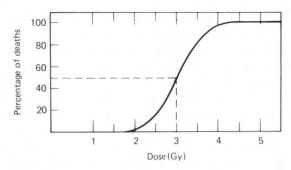

Figure 14.12 Percentage of human deaths within 30 days as a function of dose. The dashed line indicates the $LD_{50}(30)$ dose.

cular effect. If the percentage of the time a particular effect is observed when a particular dose is administered is plotted as a function of dose, a curve of the type of Figure 14.12 results. This is a plot of the percentage of human deaths within 30 days as a function of dose in grays. From this we see that the dose that produced death within 30 days 50 percent of the time is about 3 Gy. This is called the $LD_{50}(30)$ dose. The $LD_{50}(30)$ for rats exposed to X rays is about 7.4 Gy; for other animals the dose varies considerably (see Table 14.10).

TABLE 14.10 $LD_{50}(30)$ for Several Animals

Animal	$LD_{50}(30)$ (Gy)
Goldfish	20.0
Tortoise	15.0
Sparrow	8.0
Rabbit	8.0
Rat	7.5
Hamster	7.0
Frog	7.0
Chicken	6.0
Mouse	6.0
Monkey	5.5
Guinea pig	4.0
Dog	3.5

The response of the human body to large doses of radiation varies with the magnitude of the dose. Table 14.11 indicates the major symptoms of large doses, the Acute Radiation Syndrome.

Doses larger than about 100 Gy appear to affect the central nervous system. The symptoms reported in the few instances where accidental exposures have resulted in doses of this order of magnitude are consistent with central nervous system damage or failure. Large doses like this are the result of an exposure to massive amounts of radiation, as can be seen from a simple calculation.

Example 14.7 An individual receives a 100-Gy dose from a 10-s exposure to a ^{60}Co source. If each photon loses about 40 percent of its energy in the person's body, how many photons did the individual intercept in the 10 s?

The decay scheme for ^{60}Co is shown in Figure 14.13. The most probable decay is a beta to the 2.50-MeV state in ^{60}Ni. This state decays by successive emissions of 1.17- and 1.33-MeV gamma rays. Assuming isotropy of decay, each gamma-ray pair would deposit on average $0.4 \times 2.50 = 1.00 \text{ MeV} = 1.6 \times 10^{-13}$ J. For a 75-kg person, each pair of gammas

TABLE 14.11 Symptoms of Acute Radiation Syndrome

Time after Exposure	Central Nervous System Syndrome ($> \sim 100$ Gy)	Gastrointestinal Tract Syndrome (9–100 Gy)	Hematopoietic Syndrome (3–9 Gy)	Sublethal Dose (<3 Gy)
1 to 2 days	Hyperexcitability, incoordination, respiratory distress, intermittent stupor, death	Nausea and vomiting		
1 to 2 weeks		Nausea, vomiting, diarrhea, fever, inflammation of throat, prostration, dehydration, emaciation leading to death		
3 to 6 weeks			General malaise, loss of appetite, loss of hair, hemorrhage, pallor, diarrhea, fever, inflammation of throat, emaciation leading to death in most victims	Loss of appetite, loss of hair, inflammation of throat, pallor, hemorrhage, diarrhea
1 to 10 yr			Increased risk of leukemia and other cancers	

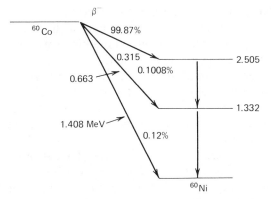

Figure 14.13 The beta decay of ^{60}Co and the subsequent gamma decay of the ^{60}Ni daughter.

results in a dose of

$$\frac{1.6 \times 10^{-13}}{75} = 2.13 \times 10^{-15} \text{ Gy}.$$

A 100-Gy dose is therefore

$$\frac{100}{2.13 \times 10^{-15}} = 4.688 \times 10^{16} \text{ gamma pairs}$$

$$\text{or} \quad 9.37 \times 10^{16} \text{ photons}.$$

For absorbed doses between about 9 and 100 Gy death usually results from bacterial poisoning, the bacteria coming from the intestine after the intestinal walls have been destroyed by the action of the radiation. For this reason, this effect is often called the gastrointestinal syndrome. Note that there is no sharp dividing line between this syndrome and the one discussed above. The result is the same: irreversible damage leading almost exclusively to death.

In the dose range between 3 and 9 Gy, death results more often than not. All individuals receiving such a dose show characteristic alterations in blood cells and blood-forming tissue) hence, the name of the effect, the *hematopoietic sysndrome.*

For doses below about 0.5 Gy, the percentage of exposed individuals who die in 30 days is very small. However, there always seems to

be an increased risk of leukemia and other cancers as well as risk from other diseases in radiation survivors.

A number of experiments with animals have shown that radiation exposure produces life-shortening; that is, the exposed animals died earlier than a group of control animals that did not receive the radiation. This is indicated schematically in Figure 14.14. Here, line A represents the normal aging process of an animal in which the natural "injuries" of life accumulate to the point when either a terminal illness sets in or one or more essential body organs fail. Curve B represents the history of an animal receiving an acute (sublethal) radiation dose at time t_B. A certain amount of repair of the associated injury occurs up to time t_B', then the normal aging pattern continues. However, the individual dies earlier, at t_M. Such life-shortening has been observed in many animals, but it is not possible to do controlled "experiments" on humans to determine the extent of the effect.

The induction and growth of cancerous cells is not well understood. It is believed that cancerous cells are produced continuously in the human body by a variety of mechanisms including radiation, chemicals, and spontaneous

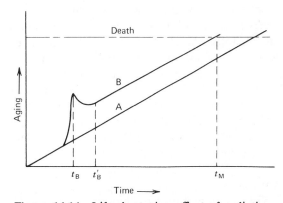

Figure 14.14 Life-shortening effect of radiation. An acute dose is absorbed at t_B; after t_B' aging continues at the same rate, but death will result earlier.

formation. Usually, these cells either die or are destroyed by the body's immune system. In some individuals, possibly by inherited susceptibility, cancerous cells establish a foothold and develop into a tumor. Late-appearing cancers are now believed to be one of the most severe effects of radiation; originally, genetic effects were though to be more important.

Genetic Effects

The genetic information in human cells—the inheritable traits—is contained in two sets of 23 chromosomes that are found in cell nuclei. Human sex cells, sperm and eggs, each contain one set of chromosomes. The genetic data for each trait are believed to be localized at a specific gene site on a chromosome. The chromosomes are made up of DNA molecules, which serve as the genetic data banks. When a cell reproduces itself, a process called *replication*, the DNA molecules provide the blueprints for the construction of additional DNA, so that the chromosomes are replicated. Half of the new total of 46 pairs goes to each of the two halves of the new cells, as shown in Figure 14.15. This process is called **mitosis**. Sex cells contain only 23 chromosomes, not 23 pairs. When they divide, the process is somewhat more complex and is termed **meiosis**.

When a cell divides, its nucleus contains a significantly larger-than-normal amount of DNA. As a consequence, dividing cells are rather more sensitive to radiation damage than quiescent cells. Chromosomes can be damaged in several different ways: breakage, cross linkage, alteration of particular gene sites, and alteration of one or more DNA structures. Any one of these effects could result in the death of the cell, since it could disrupt the process of mitosis.

Genetic alteration from external causes such as radiation need not produce a fatal change. Some genetic changes, called **mutations**, can produce viable cells that can transmit the change from generation to generation. Sometimes those changes are benign, such as eye coloration, skin pigmentation (albinoism), skin flaps between digits, and the like. But frequently mutations are harmful or ultimately fatal, such as Down's syndrome (mongolism), hemophilia, Mediterranian syndrome, sickle-cell anemia, and Tay-Sachs' disease. There are many others.

Some mutations are *recessive*. That is, both parents must carry at least one of this particular type of altered chromosome in order for the mutation to be expressed in the offspring. A dominant trait will always be expressed if either parent carries the alteration, assuming the offspring carries that alteration as well.

The last caveat was necessary because of the way genetic information is transmitted from one generation to the next. The standard example of Mendelian genetics is that of eye coloration. Brown eyes are dominant, blue eyes are recessive. If one parent's chromosome pair that contains the eye-coloration gene site can be denoted Bb (one Brown, one blue) and the other parent's chromosome pair is Bb as well, then the offspring could have the combinations shown in Figure 14.16. Out of four offspring, three could have brown eyes, and one could have blue. But these are *average* figures, averaged over many such combinations. Since the actual sharing of genetic traits is a random process, based on the available genes from each parent, it is possible for all offspring to have blue eyes or for all offspring to have brown eyes.

Similarly, undesirable traits can likewise be passed along, even if they are not expressed in the particular carrier of the trait. These undesirable genes are slowly removed from the population over a period of many generations, particularly if they result in fatal changes. But since many of the traits are recessive, they may take a great deal of time to remove. This burden of undesirable genetic traits is often referred to as the *genetic load*. It might appear

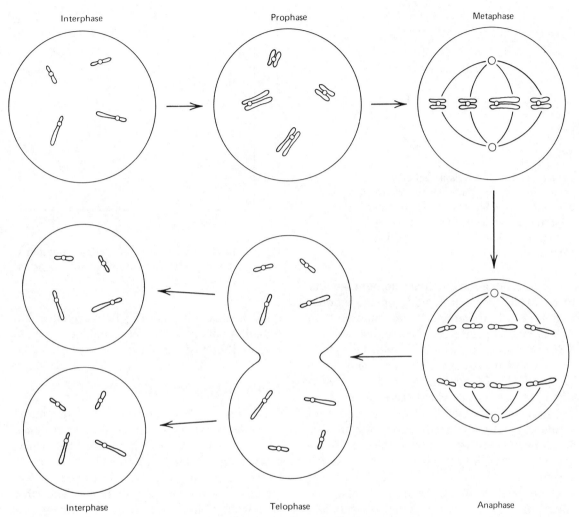

Figure 14.15 Mitosis—the process of cell division in which genetic information is duplicated in each new cell.

that, in time, all undesirable genes would be removed from the total genetic pool; but this is not the case, since there are always mutations being produced. These come about from incidental ionizing radiation, mutagenic chemicals, or spontaneous effects that are not clearly understood. Eventually, an equilibrium is reached between mutation production and removal by natural selection.

The rate of new mutation production can be estimated directly by examining offspring who display a dominant trait that was not present in the parents. Clearly, this must represent the production of a mutation in that generation. For dwarfism this rate has been estimated to be 1 in 12,000 children—which is the same as 1 in 24,000 genes. Most "direct" observations of this type lead to mutation rates between 1 in 10,000 and 1 in 1,000,000.

Ionizing radiation can increase the fre-

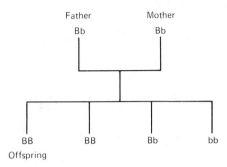

Figure 14.16 Mendelian genetics. If each parent contributes a brown-eye gene (B) and a blue-eye gene (b), four offspring could have the combinations shown. Only one (bb) would be blue-eyed.

quency of mutation by large amounts—from 10 to 10,000 times, depending on the amount and type of radiation. Also important in determining the effects of radiation are the types of cells being irradiated and the mitotic condition of those cells. Cells that are dividing are particularly sensitive; sex cells are sensitive. Muscle cells, which have no nucleus, and nerve cells, which have nuclei but do not divide, are not very sensitive to radiation. The developing fetus is particularly sensitive to ionizing radiation, especially in the first trimester.

There is little doubt about the mutagenic effects of relatively large amounts of radiation. Figure 14.17 is a plot of percent lethal mutations observed in fruit-fly irradiation as a function of dose. The effect is linear down to quite small doses. It is possible from data such as these in other animals and from epidemiological studies in humans to estimate the quantitative effects of radiation for a given result. For example, it is generally agreed that 100 deaths from cancer per million people will result per 0.01 Gy dose of radiation. (See Table 14.12.) This rule of thumb seems to hold for relatively large doses—above perhaps 1 Gy. The question and the source of controversy is:

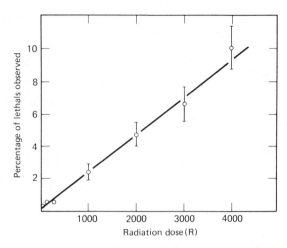

Figure 14.17 Percentage of lethal mutations observed in fruit flies as a function of radiation dose in roentgens.

does this linear relationship between effect and radiation hold down to very small doses, in the neighborhood of 0.01 Gy?

Low Levels of Radiation

For many years it was believed that radiation effects exhibited a threshold at very small doses, similar to curve a in Figure 14.18. Curve b represents a linear relationship, and curve c would result if very low doses were more effective at producing radiation effects than somewhat higher doses. There is evidence for the support of each of these three low-level radiation effects behaviors.

Evidence for a threshold comes from the effects of fractionating doses. Fractionation is the process of delivering a certain dose by dividing it into a number of smaller doses delivered over a longer period of time. Studies with animals show that, while a single large dose may be fatal, the same total dose delivered in several smaller batches may have very little effect. The implication is that somatic or genetic repair mechanisms exist, which enable the cells affected to heal the damage

TABLE 14.12 Incidence of Cancer

Type of Cancer	Radiation-Induced Deaths per Million per 10 μSv[a]	Spontaneous Deaths per Million per Year[b]
Leukemia	15–25	69
Lung	25–50	410
Breast (female)	60	160
Bone	2–5	9
Thyroid	5–15	5
Others	25	1117
Total	120	1770

[a]*Sources and Effects of Ionizing Radiation.* New York: United Nations Committee on the Effects of Atomic Radiation, 1977.
[b]*Cancer Facts and Figures.* New York: American Cancer Society, 1977.

caused by the initial radiation. But it is also clear that this repair cannot be complete. The $LD_{50}(30)$ dose changes for animals with a past history of irradiation.

Because it is generally believed that repair mechanisms are not complete, most professionals in the radiation biology field adhere to the linear hypothesis. A number of studies tend to indicate linearity, but definitive measurements at very low doses have not been made.

The major objection to curve c in Figure 14.18 has been that there is no known mechanism for enhancing radiation effects at very low doses. Some time ago an experiment[3] was reported that gave very interesting results. The time required for the rupture of cell membranes in an aqueous solution containing ^{22}Na was measured as a function of dose rate. It was found that the time required went down as the dose rate was increased. However, the total dose received to produce cell-membrane rupture calculated from these data clearly implies that a smaller dose is required to cause cell membrane rupture at *small* dose rates than at *large* dose rates.

[3]A. Petkau, "Effect of ^{22}Na$^+$ on a Phospholipid Membrane," *Health Phys.* **22**, 239 (1972).

This is argued to be reasonable if it is assumed that it is primarily the free radicals that are responsible for cell membrane damage. The free radicals are produced within a small cylindrical volume surrounding the incident radiation track. If there are many such radicals produced in this cylinder, then the chances are high that many or even most will recombine before escaping the cylinder. The

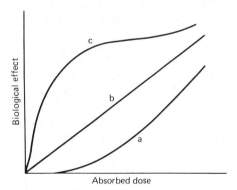

Figure 14.18 Hypotheses of the biological effect of radiation as a function of absorbed dose for very low doses. Curve a implies a threshold below which effects are negligible. Curve b is a linear relationship, and curve c implies an enhanced effect at low doses.

only cells affected would be those within the cylinder. But if only a few radicals are produced, the chances are very good that they will not recombine within the cylinder and will diffuse away, affecting a larger number of cells. This argument has not been verified.

In fact, it is very difficult to verify low-level radiation effects. First of all, it is unethical to perform potentially hazardous or fatal experiments on humans. Administering controlled doses of radiation, even at low levels, would certainly fall into this category. It is then necessary to do experiments on animals; but extrapolating from animals to humans is a tricky business. We know that insects are more radiation-resistant than mice, which are more resistant than monkeys, and so on. Second, very large numbers of animals are required.

Example 14.8 The induction of a particular cancer in mice at high radiation doses is known to be predicted by

$$C = AR,$$

where C is the number of cancers per 100,000 mice, R is the dose in grays, and $A = 1000$. The spontaneous cancer induction rate S is known to be 5 per 100,000. How many animals must be irradiated at 1 mGy to make a statistically significant test?

At 1 mGy the equation predicts 1 cancer per 100,00 mice. The spontaneous production would produce 5. Can you "see" an excess of 1 cancer over the spontaneous rate of 5? No. The difference between the two should be statistically significant, say 10 percent. That is,

$$\frac{\sqrt{\delta}}{\delta} = 0.1,$$

where δ is the difference. Since

$$\delta = N \times 6 - N \times 5,$$

where N is the number of hundred thousands of mice,

$$0.1 = \frac{\sqrt{(N \times 6 - N \times 5)}}{(N \times 6 - N - 5)} = \frac{\sqrt{N}}{N},$$

or

$$\boxed{N = \frac{1}{0.01} = 100.}$$

In other words, 10,000,000 mice must be studied—a heroic effort!

Because the direct study of low level effects is so difficult, the international bodies responsible for setting radiation standards have decided that, even if the linear hypothesis overestimates effects at low levels, it is better to assume linearity than thresholds for the public safety.

14.5 RADIATION STANDARDS

The biological effects of ionizing radiation were not fully appreciated by early workers in the field. As a consequence, many died untimely deaths. Today, while there is general agreement on the need to establish "safe" levels of radiation emission to which radiation workers and the general public are exposed, there is not a similar agreement on what constitutes a "safe" level.

Current Standards

The International Committee on Radiation Radiation Protection (ICRP) was formed in 1928 to set standards for radiation workers. It was believed then, and still is, that although radiation workers accept a higher exposure risk than the general public, they should not be exposed to radiation levels so high that their collective gene pool would become adversely affected. The Environmental Protection Agency sets standards in the United States, but usually accepts the recommendations of the ICRP. The National Academy of Sciences has been asked by the EPA to develop the criteria on which standards could be based. A committee of the academy has periodically issued reports, "Biological Effects of Ionizing Radiation" (BEIR), the latest one having been issued in 1979. This report, BEIR-III, reaffirms the standards originally set by BEIR-I in 1972. Namely, that there is no level of radiation below which it is safe; in essence, the linear hypothesis is recognized. Accordingly, radiation workers were allowed a *maximum permissible dose* (MPD) of 5 rads (50 mGy) in 1 yr, with the MPD for any week not to exceed 0.1 rad (1 mGy) and the MPD for any quarter year not to exceed 3 rads (30 mGy). There are also MPDs for particular organs and maximum possible concentrations (MPC) for certain radioisotopes in air and in water.[4]

The EPA standards for the general public are that the dose from radiation sources *other than natural background and medical procedures* be no more than 5 rads (50 mGy) per generation (30 yr). This is equivalent to 170 mrad (1.7 mGy) per year. In addition, the ICRP recommends that radiation doses to the general public be "as low as reasonably achievable," the **ALARA** criterion.

Since this MPD does not count medical or background exposure, there are only a few sources of radiation that could contribute: occupational exposure, fallout from weapons testing, nuclear power, and exposure in the workplace to persons not engaged in radiological occupations. The recommendations of the ICRP on how these various sources should contribute to the allowed 5 rads are listed in Table 14.13. Note that 1.5 rad is "reserved" for future developments.

The maximum dosage from sources such as nuclear would be 2 rads per 30 yr (20 mGy), or 67 mrad/yr (670 μGy). To accomplish this, the Nuclear Regulatory Commission requires that emissions from light water reactors should be as follows:

[4]See, for example, *Los Alamos Handbook of Radiation Monitoring*. Washington, D.C.: ASAEC, 1970.

TABLE 14.13 ICRP Recommendations for Public Radiation Dose

Source	Amount (rads)	(mGy)
Occupational exposure	1.0	10
Exposure of adult workers not directly engaged in radiation work	0.5	5
General exposure (nuclear power)	2.0	20
Reserve	1.5	15
Total	5.0	50

1. No more than 10 mrem (100 μSv) dose equivalent integrated over a year's time at any point on the site boundary from noble gas emissions.

2. No more than 5 mrem (50 μSv) dose equivalent calculated to any person over a year's time as the result of the discharge of liquid waste.

3. No more than 5 mrem (50 μSv) dose equivalent to any organ as a result of the discharge of long-lived activities ($t_{1/2} = 8$ days or more).

With these standards (assuming they are not violated), any individual will receive no more than 15 mrem (150 μSv) dose equivalent from nuclear power, which is substantially less than the 67 mrem recommended by the ICRP and is about 10 percent of the average natural radiation background.

This is not to say that large radiation releases cannot happen from LWRs, even during routine operation. And, of course, in cases of nonroutine operation, such as the Three Mile Island accident, large releases can occur. Even in this particular case, however, it has been estimated that an individual standing at the north gate of the plant 24 hours a day for the three days immediately following the accident would have received at most 90 mrem

(900 μSv) of whole-body dose equivalent. While this may seem to be a large amount of radiation, measured in terms of annual MPD, it should be placed in the proper perspective.

Example 14.9 How much time would be required for a person living in Denver to receive 900 μSv more dose equivalent than a person living in Boston?

In an earlier discussion, we noted that the annual natural background dose in Denver could be as high as 1.8 mGy and in Boston about 0.85 mGy. The Denver person would then receive 0.95 mGy/yr more radiation. Therefore somewhat less than 1 yr would be required for a person living in Denver to receive the additional radiation equivalent to the amount received by the three-day observer at Three Mile Island. (Of course, this assumes no medical exposures for the Denver inhabitant! This also assumes that for low-LET radiation 1 Gy dose produces 1 Sv dose equivalent.)

Another perspective is gained by comparing the Three Mile Island activity released to that resulting from some natural process. The total Three Mile Island activity has been estimated at 9×10^{16} Bq, while the total activity emitted during the Mt. St. Helens eruption on May 18, 1980 has been estimated to be about 1.1×10^{17} Bq.[5] It should be noted as well that the Three Mile Island activity was principally in the form of radioactive xenon gas, while that from Mt. St. Helens was chiefly radium, thorium, polonium, lead, and potassium. These elements are much more biologically active than xenon and are therefore potentially far more hazardous.

Of course, the standards set up by the ICRP and adopted by the EPA are arbitrary. They do not take into account synergistic effects between ionizing radiation and inheritable traits, such as asthma. Several studies have shown

asthma sufferers to have much higher cancer risk from radiation than others. There may be other such combinations that are not as well known.

Since there is no recognised safe lower limit for radiation doses, the assessment of any value, even 170 mrad/yr—which may be smaller than the natural background, as being tolerable suggests that the ALARA criterion has been applied and that this figure represents reasonability. Some workers in this area believe it is being overly reasonable to the point of being ridiculously low. Others feel it is too high by an order of magnitude. It seems to come down to a question of how much does it cost to lower the risk, and how much benefit do you get from it?

Benefit versus Risk

It may seem crass to be asking what a life is worth, but that *is* what is being asked. We answer this question perhaps unconsciously virtually every day, when we make decisions about which brand of automobile tire to buy, whether or not to have a medical checkup, and so forth. Similarly, when we as a society choose not to develop strict SO_2 removal regulations for coal-burning plants because of excessive costs, we implicitly place a cash value on lives and conclude the value is too high.

We are continually faced with hazards; how we deal with those hazards determines the extent of the risk we take. For example, crossing the surface of the Atlantic ocean certainly presents a hazard. However, the risk in crossing depends on the choice of transport: a rowboat yields high risk, while an ocean linear presents relatively little risk. Radiation from nuclear power plants represents a hazard; the amount of risk an individual assumes because of the existence of nuclear power depends on a large number of factors, which in general can be controlled.

It is generally true that a law of diminishing

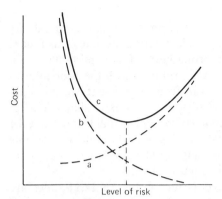

Figure 14.19 The cost of risk control to society versus level of risk.

returns applies to all controllable risk: risk reduction becomes progressively more expensive, curve a on Figure 14.19. This is a plot of cost to society of risk control versus level of risk. This tells us the obvious, that it becomes more difficult and expensive to control risk as the level of risk becomes progressively higher. Curve b in this figure is the cost to society of the hazard from which the risk is being controlled, plotted technically as a function of level of risk but in reality as a function of the amount of control. This tells us that at low levels of risk, but with little expenditure for control, the cost to society of this hazard is high. As the level of risk increases and the expenditure for control increases, the cost of the hazard decreases. Curve c is a sum of these two curves; this sum displays à minimum, as indicated. This level of risk must be the optimum; it is the level at which the slopes of the a and b curves are equal but opposite in sign. In economic terms, this is the level at which the marginal increase in expenditure for controls just equals the marginal decrease in cost to society of the hazard. A further incremental increase in controls expenditure will result in a smaller incremental societal savings.

In theory, this is fine. In practice there are two difficulties: estimating the risk and calculating the cost to society of each level of risk. Risk estimation has been done for many

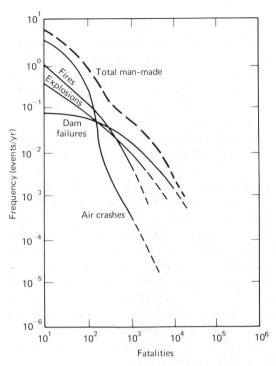

Figure 14.20 A frequency versus fatalities plot for several man-made hazards. [From *Reactor Safety Study: An Assessment of Accident Risks in U.S. Commercial Nuclear Power Plants* (WASH-1400). Washington D.C.: U.S. Atomic Energy Commission, 1974.]

years for various kinds of occupations, events, and individual actions. A large number of studies have been made with general agreement.[6] Figure 14.20 is a plot of the frequency of various events as a function of the number of fatalities per event. This figure illustrates several points. Risk involves not only a probability but also a magnitude. If an event has a low probability of occurring, it will be perceived to have a low risk. But if the low-probability event also has the possibility of causing a large number of fatalities, the perception of risk will be higher.

[6]See, for example, C. Starr, R. Rudman, and C. Whipple, "Philosophical Basis for Risk Analysis," *Ann. Rev. Energy* **1**, 629 (1976) and the references cited therein.

The curves all relate to accidents of one kind or another for which frequency and fatality statistics are reasonably available. At the high-fatality end of some of these curves the data become somewhat sketchy, so the curves are dashed. For other types of hazards the risks are not so well established. In particular, the risks from nuclear power plants are currently being hotly debated. But there are no statistics on nuclear power plant failures leading to general population radiation exposure, since there have been so few. There are no known deaths as a result of the Three Mile Island accident, and it will be impossible to assign blame to TMI for deaths occurring in the population that might reasonably assumed to have ben exposed. This is because the dose to the local polulation is generally agreed to be so small that the number of cancer inductions and other radiation-caused diseases would be so small as to be statistically indistinguishable from the spontaneous rate.

So how do we know what the frequency of nuclear power plant failure is? There are two methods. First, there are more than 90 nuclear power plants operating in this country and more than 170 operating in the rest of the world. These have operated for an average of more than 8 yr apiece with only one failure—resulting in no fatalities. This would lead you to believe the failure rate is of the order of 1 in 2520 reactor-years of operation. This could be very misleading, however; the actual rate could be much higher or much lower than this—the statistics are not good enough to be certain. (We do *not* advocate a higher failure rate to improve the statistics!)

A second method for estimating failure rate would be to analyze the power plant in terms of the expected failure rates of the components, the safety systems, and so on. The techniques used today are called **event-tree** and **fault-tree analysis**. Event-tree analysis begins with a particular postulated event, for example, valve closure failure, and then traces forward in time in a sequence along all possible paths that derive from it, noting the probability of failure at each junction, as shown in Figure 14.21. The probability of any particular consequence is the product of all the probabilities along the path from the event to the consequence. Fault-tree analysis is similar except that the anlaysis proceeds backward in time from a postulated event through all possible connections that could have led to the event, as shown in Figure 14.22.

In the event tree of Figure 14.21 there are four **lines of assurance** (LOA). Each one will either succeed (S) or fail (F) with an estimated probability given at each juncture. We have estimated the probability of smoke detector success to be 0.7; of course, there are many circumstances under which it could be higher or lower. Similarly, the probability that the resident could raise an alarm will depend on whether or not the smoke detector worked. The third LOA is an alarm raised by a neighbor; we have estimated this LOA success at 0.5. Finally, quick response by the fire department will determine whether or not the house can be saved. The consequences are abbreviated RS—residents saved, HS—house saved, HD—house destroyed, and RK—residents killed. It is interesting to note that, with the optimistic success probabilities estimated here, the total probability of saving the home is only barely over 0.5!

The fault tree of Figure 14.22 is an attempt to find all sequences of occurrences that could lead up to the event. Each occurrence in Figure 14.22 is labeled with an estimate of the probability that occurrence would happen. Such probabilities must be derived from statistical analyses of the various component failures or situations, and of course they are subject to considerable variation. In this illustration we see that the most likely cause of a house fire is the ignition of flammable materials on the stove. It is useful to examine such diagrams because additional LOAs can sometimes be suggested. For example, the existence of CO_2 fire extinguishers in the kitchen could

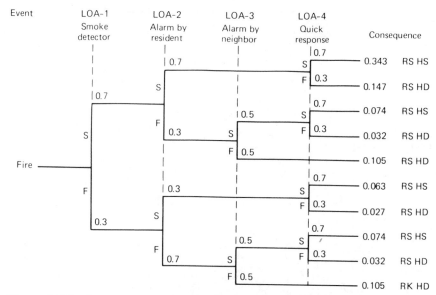

Figure 14.21 An event-tree analysis of a house fire showing several lines of assurance (LOAs) and resulting consequences. RS = residents saved, HS = house saved, RK = residents killed, and HD = house destroyed.

significantly decrease the probability of damage for a fire originating on the stove.

The simple ideas we have applied to house fires could also be applied to the much more complex nuclear reactor system. There have been many safety studies published in the past 20 years or so; indeed, the field is awash with them. The most recent one,[7] referred to as the Rasmussen report after the MIT physicist who chaired the study, used fault-tree and event-tree analysis. This study deduced the frequency versus fatality curve of Figure 14.23; it is rather considerably lower than any of the corresponding curves for other human activities, as shown in Figure 14.20. In effect, the conclusion of this study was that the likelihood of a member of the general public being killed by a nuclear power plant was about the same as that of being killed by a meteorite.

Naturally, this report was severely criticized by the opponents of nuclear power. The criti-

cism has been leveled at both the risk estimates and the fatality estimates. The event-tree, fault-tree method itself is only as satisfactory as the data that go into it, and herein lies the problem. It is very difficult to obtain failure rates for all the components in the reactor. In addition, human error cannot be easily quantified and reduced to a probability. So the final risk values are likely to have error bars. How large? Perhaps an order of magnitude could be justified; but even with two orders of magnitude the risk from nuclear power would still be smaller than that from other hazards.

The other criticism had to do with the level of fatalities estimated by the group preparing the report. The Americal Physical Society, in studying the same problem,[8] arrived at much higher estimates of fatalities by taking into account delayed deaths from cancers and other late-appearing diseases and by using somewhat

[7]WASH-1400, *Reactor Safety Study*, AEC, 1975.

[8]*Rev. Modern Phys.* **47**, S1 (1975).

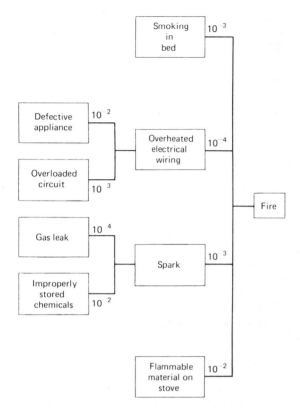

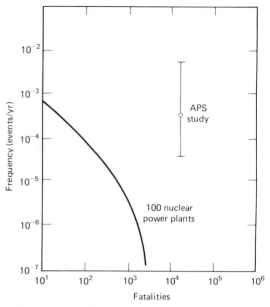

Figure 14.23 Frequency versus fatalities plot for 100 nuclear power plants. The solid curve is that published in WASH-1400; the point is that deduced by the American Physical Society study.

Figure 14.22 A fault-tree analysis in which several events that could cause a house fire and their probabilities are indicated.

different values for failure probabilities. This estimate is given as the data point with error bars in Figure 14.23. This value is more than three orders of magnitude greater than the Rasmussen study in terms of fatalities, but has about the same estimated frequency of occurrence.

The Rasmussen report has been a very valuable addition to our library of nuclear power safety literature; it is the most comprehensive application of event-tree and fault-tree analysis yet. No doubt future studies will refine this technique and make better predictions on risk and level of fatalities for nuclear power.

The conclusions from the recent debate on nuclear safety seem to be that, while nuclear power poses a threat that is probably no greater than that posed by comparable energy-producing plants, there is room for improvement. But we should be careful to remember curve c of Figure 14.19. Do we really want to continue to invest in risk controls, when the return in benefits becomes smaller and smaller?

The real difficulty lies not in the determination of the value of life, for several estimations have been made, all in general agreement, but in the determination of risk and how much the risk level is reduced by any particular control. Even though nuclear power is probably better understood than any of the other hazards to which we expose ourselves, the understanding can always be improved, particularly in the area of multiple component failure and human error leading to radiation release.

In the past decade the interaction of ionizing radiation with human tissue has been studied extensively. More is known about this interaction and the ensuing consequences than is known about the much larger effects of the many chemical carcinogens in the biosphere. Radiation standards appear today to be adequate. Periodic reviews by the ICRP will ensure that any new evidence for enhanced low-level effects or significant long-delayed disease increase can be incorporated quickly into the standards.

Whether we wish to admit it or not, the development of a standard of any kind is always a balance between risks and benefits. We tolerate bad air in our major cities because we do not believe the benefits of cleaning it are worth the costs. We accept inadequate fire protection in older high-rise hotels built before

satisfactory municipal codes were passed, for the same reason. There are innumerable such examples; to insist that there should be no trade-offs with radiation is irrational. We list some of the benefits and risks of radiation in Table 14.14. Most of these risks are voluntary—no one forces you to buy a microwave oven. However, the risk from nuclear power is not voluntary.

As a consequence, we must always make certain the risk we take with a nuclear power plant does not exceed the risk from any other technology that provides us with the same benefit. Nuclear power, as an industry, does not today present an excessive risk. An individual plant, because of poor construction or operation, may present such a risk. The operating characteristics of every nuclear power plant in the country are on file with the Nuclear Regulatory Commission. Those with poor records can be easily found.

TABLE 14.14 Radiation Axiology

Source	Benefits	Risks
Diagnostic X rays	Improved health	Genetic mutations, leukemia, other cancers, life-shortening
Industrial radiography	Inspection of critical metal assemblies by X rays (aircraft wings, etc.)	As above
Color TV	Education, entertainment, information	As above
Nuclear power	"Cheap electricity," reduced air pollution, pure water (desalination), increased overall health (better home heating)	As above
Radioisotopes	Improved medical diagnosis and treatment	As above
Ultraviolet	Germicide, industrial processing	Skin and eye damage
Radar	Navigation, defense	Cataracts, genetic mutations
Microwave ovens	Convenience, improved food service (hospitals, etc.)	As above

SUMMARY

Several types of atomic particles go under the name *radiation*. Some have mass but no charge, some have mass and charge, and some have no mass and no charge.

The various kinds of particles lose energy in matter in very different ways. Charged particles lose energy continuously along their paths. *Alpha particles* lose energy much more rapidly than do *electrons*. The rate of energy loss is referred to as the *linear energy transfer (LET)*.

Neutrons lose energy by direct nuclear collisions. This produces high-speed charged secondary particles that have high LETs.

Photons lose all or most of their energy in a single collision, so an LET cannot be defined. Several different processes can be involved.

In tissue the charged secondaries resulting from incoming radiation produce ions and *free radicals*. In tissue free radicals are powerful oxidizing agents.

The units used in the radiation biology field are changing throughout the world, although slowly in the United States. Both sets of units are used in this textbook; the new units are preferred.

There is a continuous *radiation background* to which we are all exposed. This consists of ultraviolet radiation from the sun as well as cosmic rays and natural radioactivity from the earth.

X rays, even dental plates, can result in doses much larger than from the average annual background.

Nuclear weapons testing in the atmosphere produced very large amounts of radioactivity, some of which is still with μs.

Radiation can produce *somatic* or *genetic effects. Acute* somatic effects are reasonably well understood, although the sequence of events on the microscopic cell level is not known.

Genetic effects are not as well understood. *Mutations* can certainly be produced by radiation, but since most mutations result in the death of the cell when it attempts to *replicate*, few mutations are passed to succeeding generations.

The effects of very low levels of radiation over a long period of time are not well known. The assumption is made that no threshold for radiation effects exists. It is difficult to verify this because of the large numbers of laboratory animals that would be needed.

Current radiation standards allow no more than 5 rads (50 mGy) per person per 30-yr period. However, any radiation source should follow the *ALARA criterion*: as low as reasonably achievable.

There must be trade-off between *benefit and risk*. Risk can be studied by event-tree and fault-tree analysis. Using this technique, nuclear power can be considered to be following ALARA.

REFERENCES

General

Casarett, Alsion P., *Radiation Biology.* Englewood Cliffs, N.J.: Prentice-Hall, 1968.

Cohen, Bernard L., and I-Sing Lee, "A Catalog of Risks,"*Health Phys.* **36**, 707 (1979).

Eisenbud, Merrill, *Environmental Radioactivity.* New York: McGraw-Hill, 1963.

Lewis, Harold W., "The Safety of Fission Reactors," *Sci. Amer.*, March (1980), p. 53.

Lieberman, Joseph A., "Ionizing-Radiation Standards for Population Exposure," *Phys. Today*, November (1971), p.32.

McBride, J. P., R. E. Moore, J. P. Wither-spoon, and R. E. Blanco, "Radiological Impact of Airborne Effluents of Coal and Nuclear Plants," *Science* **202**, 1045 (1978).

Morgan, Karl Z., "Never Do Harm," *Environment*, January/February (1971), p. 28.

Morgan, Karl Z., "Cancer and Low Level Ionizing Radiation," *Bull. Atomic Sci.*, September (1978), p. 30.

Morgan, Karl Z., and J. E. Turner (Eds.), *Principles of Radiation Protection*. New York: John Wiley, 1967.

O'Donnell, E. P., and J. J. Mauro, "A Cost-Benefit Comparison of Nuclear and Non-nuclear Health and Safety Protective Measures and Regulations," *Nucl. Safety* **20**, 525 (1979).

Robertson, J. Craig, *A Guide to Radiation Protection*. New York: John Wiley, 1976.

Starr, Chauncey, Richard Rudman, and Chris Whipple, "Philosophical Basis for Risk Analysis," *Ann. Rev. Energy* **1**, 629 (1976).

Steneck, Nicholas H., Harold J. Cook, Arthur J. Vander, and Gordon L. Kane, "The Origins of U.S. Safety Standards for Microwave Radiation," *Science* **208**, 1230 (1980).

Wallace, Bruce, and Th. Dobzhansky, *Radiation, Genes and Man*. New York: Holt, Rinehart and Winston, 1963.

Journals

Annals of the ICRP, Pergamon Press, New York.

Applied Health Physics and Notes, Nuclear Technology Publishing, Ashford, England.

Health Physics, Pergamon Press, New York.

Radiation and Environmental Biophysics, Springer-Verlag, Berlin.

Radiation Effects, Gordon-Breach, London.

Radiation Protection Dosimetry, Nuclear Technology Publishing, Ashford, England.

GROUP PROJECTS

Project 1. Determine the status of X-ray technician certification in your state. Interview X-ray technicians in the nearest hospital or clinic to determine the average dose given during a typical X ray. Do the same for a complete dental X-ray set. Estimate the total average radiation dosage for an individual in your community.

Project 2. Most universities and colleges have a significant amount of radioactive materials on hand, typically in biology, physics, and chemistry departments. Discover the type, location, and amount of such material in your institution. Estimate the dose a student *not connected with the use of these radionuclides* would be likely to receive.

EXERCISES

Exercise 1. Show that the maximum amount of energy a particle of mass M and energy K can transfer to an atomic electron is given by

$$E_{max} = 2K \left(\frac{m}{M} \right),$$

where m is the electron mass.

Exercise 2. A flat lead foil has four unequal straight sides. The corners lie on grid points $(0.60, 0.71)$, $(0.69, 1.21)$, $(2.22, 0.55)$, and $(2.36, 1.38)$ with distances measured in centimeters. The mass of the foil is 120 mg. Find its areal density:

Answer: $170 \, mg/cm^2$.

Exercise 3. Find the density of the tissue equivalent given in Table 14.2.

Exercise 4.   Show that the thickness of a certain substance required to cause a charged particle of energy E_1 to lose ΔE of energy is given by

$$t = R(E_1) - R(E_1 - \Delta E),$$

where $R(E)$ is the range of that particle with energy E.

Exercise 5.   Show that the acceleration of a charged particle is given by

$$a = \left(\frac{1}{m}\right)\left(\frac{dE}{dx}\right).$$

Use this, along with Figure 14.3, to make a qualitative argument that the actual ion production time in tissue is a great deal smaller than the number found in Example 14.3.

Exercise 6.   A plot of the mass attenuation coefficient versus photon energy for lead has sharp peaks, called "edges," as shown in Figure 14.24. Note that these edges occur in an energy regime where the photoelectric effect is expected to dominate. What is the physical process causing these edges?

Exercise 7.   What is the maximum Compton scattering angle a 2.0-Mev photon can scatter at and still have enough energy left to create an electron–positron pair?

Answer: 60.7°.

Exercise 8.   Estimate the thickness of tissue required to stop a 10-MeV proton, and compare with the thickness required to reduce the intensity of a 10-MeV gamma-ray beam by 99.9 percent.

Exercise 9.   How many becquerels of activity are represented by a 100-μg quantity of freshly prepared ^{239}Pu?

Exercise 10.   A student proposes to use heavy gloves to handle a 370-MBq source of ^{90}Sr that is contained in a glass test tube. Is this safe? What if it were 370 MBq of ^{22}Na in a glass test tube?

Exercise 11.   If 100 μg of ^{239}Pu were inhaled by a person and remained in the body for 30 yr, how many grays of dose would the individual have from this material?

Answer: 7.2 Gy (720 rads).

Exercise 12.   What is the maximum amount in grams of ^{90}Sr that may be ingested by an individual, if the total dose from it is to be less than 1 mGy/yr?

Exercise 13.   As a general rule of thumb, 1 megaton of fission yield in the atmosphere will eventually result in 1.15 Bq/g of calcium of ^{90}Sr activity in the bones of children. Up to 1963 about 193 megatons equivalent of fission weapons was tested in the atmosphere. What dose in grays did children receive from the absorbed ^{90}Sr over the period from 1963 to 1983?

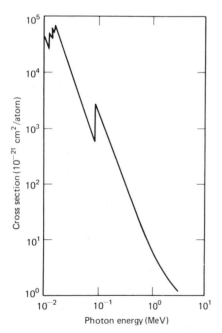

Figure 14.24   Drawing for Exercise 6.

Exercise 14. What activity of alpha-emitter would produce the same dose equivalent in the lungs as 1 kBq of ^{90}Sr in the bones? (Assume the alpha energy to be 4 MeV.)

Exercise 15. Natural sources of radiation contribute a dose to each person of about 1.15 mGy/yr, made up from terrestrial radiation ($\sim 600 \,\mu$Gy), cosmic rays ($\sim 300 \,\mu$Gy at sea level), and radioisotopes internal to the body ($\sim 250 \,\mu$Gy). Of the sources internal to the body, ^{40}K is the major contributor.

(a) Using the following data, calculate the dose due to ^{40}K. A 70-kg man has about 150 g of potassium in his body. One in 8500 potassium nuclei is ^{40}K. ^{40}K has a half-life of 1.3×10^9 yr. In the ^{40}K decay about 0.6 MeV of the 1.3 MeV decay energy is absorbed by the body.
(b) Compare the dose received in 1 yr from ^{40}K with the current NRC guideline for nuclear reactor doses to the general population.

Answer: (a) 280 μGy.

Exercise 16. Assume the individual in Example 14.7 was standing 3 m from the ^{60}Co source when the 10-s exposure occurred. What was the strength of the source in becquerels?

Exercise 17. The decay scheme of ^{131}I is shown in Figure 14.25. Estimate the thyroid

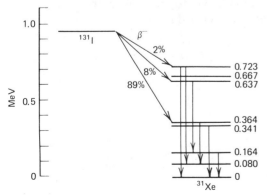

Figure 14.25 The decay scheme of ^{131}I.

dose received by a child who drank a liter of milk per day between September 1961 and January 1962.

Exercise 18. The average background levels at New Haven, Connecticut, and Cheyenne, Wyoming, are 730 μGy and 1.64 mGy, respectively. How many excess cancers would be expected per year in Cheyenne as a consequence of the background radiation difference? Would this number be statistically significant?

Exercise 19. At some distant future time a nuclear exchange between two countries results in an overall increase of the radiation background by 30 percent. How many excess cancers per year could be expected worldwide? in the United States? in your city?

Exercise 20. A person lives in Denver for 10 yr. In order to reduce his lifetime dose, he decides to move to Charlotte, North Carolina, where the average background is 0.86 mGy/yr. How long should he live here in order that his lifetime dose would be that obtained from a 1.0-mGy average background?

Answer: 57.1 yr.

Exercise 21. Mrs. Smith lives in Atlantic City, New Jersey. She is told that a western climate would improve her health; so she spends three months each year in Flagstaff, Arizona. In terms of radiation exposure has her health improved?

Exercise 22. In an isolated community with a large number of people it is found that 75 percent have brown eyes (assume 50% BB and 50% Bb) and 25 percent have blue. What is the likelihood that any single offspring will have blue eyes?

Exercise 23. In an isolated community of a large number of people it is found that children having brown eyes are born 89 percent of the time. What is the distribution of brown eyes (assume 50% Bb and 50% BB) and blue eyes in the parents?

Exercise 24. If the routine radiation released from a nuclear power plant could be evenly divided over the entire U.S. population, the resulting dose would be of the order of 10^{-7} Gy/person per year. For 200 reactors over a 30-yr period, how many excess cancers would be expected in the United States? A more realistic estimate is that the total dose would be absorbed by only about 1 percent of the population. How would this estimate change the result?

Exercise 25. In a coal-burning plant with a good scrubber, about 2.5 percent of the mass of burned coal is released as fly ash. This contains about 4 parts in 10^{13} of radium. Estimate the daily release of radioactivity from a 1000-MW coal plant. If a resident at the plant boundary receives about 50 mGy dose for every 18 GBq emitted, what dose does such a resident receive from a coal-burning plant in the course of a year?

Answer: 3.12 mGy (0.31 rad).

Exercise 26. At what airborne Pu concentration will the inhaled dose be 1.7 mGy in the course of a year? Assume an average breathing rate of 6 ℓ/min and that 1 percent of the Pu inhaled remains in the lung.

Exercise 27. What is the excess risk that a radiation worker who receives the maximum permissible dose each year for 30 yr will develop cancer?

Exercise 28. Four hundred thousand people live immediately downwind of a 2000-MW nuclear reactor complex that emits no more than the allowed level of airborne radioactivity. What is the excess cancer risk for each over the period of a year? How many *total* cancers would be expected from this group? How many are attributable to the nuclear power plant?

Exercise 29. Develop an event tree and a fault tree for brake failure. Use LOAs of emergency brake, roadway guard rail, and seat belt. Assume consequences that include passenger injury and vehicle damage. Estimate the various probabilities. What is the total passenger injury probability?

Exercise 30. How would the total probability for saving the house in Figure 14.21 change if the probability of failure of the smoke detector were 1.0 (i.e., no detector)?

Answer: HS = 0.455.

Exercise 31. In Figure 14.21, how small does the probability of success for alarm by resident in the event of successful smoke-detector operation have to be before the total probability of saving the house drops below 0.5?

GLOSSARY

Absolute scale A temperature scale for which zero represents the lowest achievable temperature; the degree size is the same as the Celsius degree.

Absolute zero The lowest achievable temperature. $0°$ absolute $= -273.16°$ C.

Absorption refrigeration A system in which the working fluid is dissolved into a solution with another fluid, compressed as a liquid, then expanded as a gas, extracting heat in the process.

Absorptivity A frequency-dependent number between zero and one that is a measure of the ability of an object to absorb radiation.

AC—Alternating current Sinusoidally time-varying current.

Acceptor An impurity atom that accepts an electron from the crystal.

Acute dose A large radiation dose of short duration.

Adiabat A path on a $p-V$ diagram representing a process during which there is no energy transfer by means of temperature difference; that is, $Q = 0$.

Adiabatic lapse rate The calculated rate of change of temperature with respect to elevation; about $-0.99°$ C/100 m.

Adiabatic process A process for which $Q = 0$.

Adsorption refrigeration A system in which water vapor is removed from the air by surface adsorption.

AEC—Atomic Energy Commission The first civilian agency to have authority over nuclear power.

Aerodynamic drag The resistance to forward motion provided by the flow of air across the surfaces of the moving object. It is proportional to the square of the velocity of motion of the vehicle relative to the air.

Albedo The ratio of the light reflected from to the light incident upon an object.

Alpha ray (α) The nucleus of a helium atom; a component of radioactivity.

Anaerobic process A process that normally takes place only in the absence of air.

Anode That terminal of an electrical cell that supplies electrons to the external circuit.

Antineutrino The antiparticle of the neutrino; emitted in beta decay: $n \rightarrow p + \bar{\nu}$.

Approach The difference between the theoretical cooling limit and the actual cooling limit of a cooling tower.

Areal density The thickness of an object measured in terms of the density times the thickness.

Atomic number The designator for the number of protons in the nucleus; written as a lower left subscript: $_6$C.

Available work The maximum amount of work that can be done by a particular process.

Band gap The energy spacing between allowed energy bands of a solid, usually the valence and conduction bands.

Barn A non-SI unit for area used in nuclear physics and engineering: 1 barn $= 10^{-24}$ cm^2.

Battery A collection of electrical cells wired in series, parallel, or some combination thereof.

Becker nozzle A uranium enrichment process in effect similar to a centrifuge.

Becquerel (Bq) A measure of activity; 1 Bq = 1 disintegration per second.

BEPS—Building Energy Performance Standards A set of standards to be promulgated by the Department of Housing and Urban Affairs that will require that new houses be significantly more energy-efficient than in the past.

Bergius process A coal liquefaction process.

Bernoulli's law The relationship between pressure and velocity in steady nonviscous flow:

$$p + \tfrac{1}{2}\rho V^2 = \text{const.}$$

Beta-alumina A particular crystalline form of alumina (AlO_2).

Beta ray (β) A positive or negative electron; a component of radioactivity.

Binding energy The difference between the sum of the masses of the individual nucleons and the actual mass of a nucleus.

Bioconversion A technique for converting natural materials into usable fuels by means of living organisms.

BOF—Basic oxygen furnace A steel production furnace that uses oxygen blown through the molten iron to produce heat by reactions with carbon and silicon in the mix.

Blackbody radiation Radiation having the frequency distribution characteristic of a blackbody: Eq. 6.17.

Bragg peak The large peak in the energy versus penetration curve for charged particles in matter near the end of the range.

Brayton cycle The idealization of high-temperature gas turbine operation.

Breeding The process whereby a fissile nucleus is produced from a nonfissile nucleus in a reactor by neutron absorption.

Bremmstrahlung "Braking radiation"; electromagnetic radiation emitted by an charged particle when accelerating.

Btu—British Thermal Unit The amount of energy required to raise the temperature of one pound of water one degree Fahrenheit.

BWR—Boiling-water reactor A nuclear power reactor in which the cooling water is allowed to boil.

Calorie The amount of energy required to raise the temperature of one gram of water one degree Celsius.

CANDU The Canadian natural-uranium, deuterium-moderated power reactor.

Capacitance The ratio of the electrical charge of an object to its potential; measured in *farads*.

Carboxyhemoglobin Molecule formed by the attachment of a carbon monoxide molecule to a hemoglobin molecule.

Carnot cycle A conceptual cycle consisting of alternating adiabats and isotherms.

Catalytic converter As applied to automobiles, a device for removing carbon monoxide and hydrocarbons catalytically from exhaust.

Cathode That terminal of an electrical cell that accepts electrons from the external circuit.

Cell The basic electrochemical device that converts the potential energy stored in chemical form to free electrons and ions in a electrolyte.

Cell nucleus That portion of a biological cell that carries the genetic information required for replication.

Celsius scale A temperature based on $0° =$ ice-water equilibrium and $100° =$ steam–water equilibrium.

Centrifuge As applied to nuclear power, a device for separating ^{235}U from natural uranium.

Cetane rating The rating of diesel fuel that measures its ability to ignite appropriately; equivalent to octane rating for gasoline.

Chemical energy The electrical potential energy that results from chemical bonding between interacting species.

China syndrome Tongue-in-cheek reference to the fact that a reactor core, if molten, could melt its way into the ground beneath the reactor.

Chromosome The carrier of genetic information in the cell nucleus, composed of DNA molecules.

Chronic dose A low-level, sustained radiation dose.

CIAP—Climatic Impact Assessment Program A program of the Department of Commerce to assess the impact of supersonic flight on the stratosphere.

Coal gas A low-heating-value gas produced by heating coal in air.

Cold cranking power The amount of current that can be delivered by a battery at $0°$ F with the battery potential being maintained above 7.2 V for a 12-V battery and 3.6 V for a 6-V battery.

Combined cycle Two or more heat-conversion cycles operating in tandem; that is, the exhaust heat from one is the input heat for another; usually different types of cycles.

Compression ratio The ratio of the volume inside the cylinder at the full downward extension of the piston to that at the full upward extension of the piston.

Compton effect The elastic scattering of a photon by an electron.

Conduction Heat transfer through a substance by means of a temperature difference (alternatively, electrical current flow).

Conduction band The band of available energy states that valence electrons may scatter into to provide electrical conduction.

Constant quality curves Lines on the p–V or T–S diagram that connect points having the same vapor fraction.

Convection Heat transfer from the surface of a body by means of conduction into a moving layer of fluid.

Conversion ratio A measure of the breeding effectiveness of a reactor; should be greater than 1.0 for successful breeding.

Cooling range The amount of cooling of the inlet water by a cooling tower.

Coolness storage The use of a reservoir that can be heated by room air to reduce the temperature of the room air.

Coriolis effect The force that acts on a body as a consequence of its rotation.

Corona discharge The breakdown of the air in the electric field gradient of a high-voltage transmission line and subsequent electrical current discharge.

Criteria pollutants CO, total suspended particulates, SO_2, NO_x, and oxidant.

Critical The condition in which a reactor just sustains the neutron chain reaction.

Critical isotherm The isotherm of the highest temperature for which saturated liquid or vapor can be obtained.

Critical temperature As applied to superconductors, the highest temperature at which superconductivity is observed in a particular material.

Curie (Ci) An older measure of activity; $1 \text{ Ci} = 3.7 \times 10^{10}$ disintegrations per second.

Cut-in speed The minimum wind speed at which a given horizontal wind turbine will generate power.

Cyclone separator A device for removing particulates from an effluent stream by creating a vortex motion in the gas.

Darrieus rotor A vertical "eggbeater"-type wind machine.

DC—Direct current A current that does not change in direction as a function of time.

Deadweight tonnes (dwt) The cargo capacity of a ship.

Depletion region A region of a crystal that is free of mobile electric charges.

Dielectric losses Power losses caused by the leakage currents in the insulating medium between conductors.

Diesel cycle The idealization of diesel engine operation.

Diode An electrical device that permits current in one direction, but not the other.

Dirigible Literally, steerable; the name usually given to lighter-than-air craft other than balloons.

DNA—Deoxyribonucleic acid A large macromolecule consisting of sequences of amino acids arranged in a double helical spiral.

DOE—Department of Energy The most recent government agency to have authority over nuclear power research.

Donor An impurity atom that donates an electron to the crystal.

Dopant An impurity deposited in a semiconductor to change its electrical properties.

Doppler effect The name given to the broadening of neutron absorption resonances as the temperature of the fuel increases.

Doubling time The time required for a substance increasing in number at an exponential rate to double.

Dust dome The cloud of pollutants trapped over a city by a temperature inversion.

ECCS—Emergency core cooling system A high-pressure water spray that is used to cool the core of a reactor in the event of a primary coolant failure.

ECE—External Combustion Engine An engine for which the fuel combustion takes place outside the region containing the working fluid.

EER—Energy efficiency ratio As applied to air conditioners, the ratio of the cooling capacity in British thermal units per hour to the electrical power input in watts.

Effective value As applied to AC circuits, the peak value of the current or voltage divided by $2^{1/2}$.

Efficacy Used by lighting engineers to denote for a source of light the luminous flux in lumens divided by the electrical power input in watts.

EGR—Exhaust gas recirculation A system to reduce NO_x formation in an engine by ducting exhaust back in with the incoming air to lower the flame-front temperature.

Electrolyte A medium between the anode and cathode of an electrical cell that conducts ions easily but electrons poorly.

Electron The smallest unit of matter known; $m_e = 9.1 \times 10^{-31}$ kg.

emf—Electromotive force An electrical potential difference capable of causing a current.

EMI—Electromagnetic interference Electrical noise over a wide range of frequencies including the TV bands caused primarily by corona discharge.

Emissivity A frequency-dependent number between zero and one that is a measure of the ability of an object to emit radiation.

Enthalpy A thermodynamic variable defined by $H = U + pV$.

Entombment The process of decommissioning a reactor in which the volatile radioactivity is removed and the remainder sealed in place.

Entropy Defined in terms of the entropy difference between two states:

$$S_2 - S_1 \equiv \int_1^2 \frac{dQ}{T}.$$

Equilibrium In the thermodynamic sense, a condition during which all the thermodynamic variables have well-determined values.

ERDA—Energy Research and Development Agency An interim agency formed from the AEC to oversee nuclear power research; superseded by the DOE.

ESP—Electrostatic precipitator A device for removing particulates from an effluent by corona discharge.

Ethanol A potable alcohol: CH_3CH_3O.

eV—Electron volt An amount of energy equal to 1.602×10^{-16} J.

Event tree A method of accident analysis in which a postulated event is tracked through all pathways leading to a variety of consequences.

Excess Reactivity The amount of reactivity in a reactor over that required for criticality.

Exponential growth A process in which the rate of change of a substance with respect to time is proportional to the amount present.

Fahrenheit scale Temperature scale for which $32° =$ ice–water equilibrium and 1 degree is 5/9 of a Celsius degree.

Faraday An amount of electrical charge equal to 96,493 C.

Faraday's law The emf induced in a circuit is directly proportional to the rate of change of magnetic flux through the circuit.

Fast fission factor The contribution to the neutron flux by fissions induced by fast neutrons.

Fast reactor A reactor without moderation; fissions are produced primarily by fast neutrons.

Fault tree A method of accident analysis in which a postulated accident is tracked forward in time to find all possible failure modes that could cause the accident.

Fermi A non-SI unit of measure used in nuclear physics; 1 fermi $= 10^{-13}$ cm.

Fertile material Nuclei that by neutron absorption and possibly subsequent radioactive decay can be used to create fissile nuclei.

Fill Material placed in a cooling tower to increase the exposure of the water to the air.

First law Conservation of energy as applied to thermodynamic system: $dQ = dU + dW$.

Fischer–Tropsch process A coal-liquefaction process.

Fissile material Nuclei that will undergo fission when a neutron is absorbed.

Fluorocarbon A chemical compound in which the radical is a combination of fluorine and carbon; Freon is a typical fluorocarbon.

Four-factor formula The multiplication factor for an infinite reactor.

Free energy The Gibbs function of a thermodynamic system.

Free radical An electrically neutral, highly reactive

atom or molecule having a vacancy in the valence shell.

Fuel cell An electrical cell in which the fuel material is supplied externally.

Furl speed The wind speed at which a wind machine must be shut down in order to prevent damage.

Fusion The process of bringing two nuclei together to produce products that have less mass than the original nuclei.

Gamma ray (γ) High-energy photon of electromagnetic energy.

Gaseous diffusion The separation of ^{235}U by diffusion of UF_6 gas through porous barriers.

Gasification The process of converting coal to a usable gaseous fuel.

Gasohol A mixture of alcohol (10 to 20 percent) and gasoline.

GCFBR—Gas-cooled fast-breeder reactor A breeder reactor that uses helium as a coolant.

Genetic effect An effect passed on to the next or subsequent generations.

Geosynchronous orbit An orbit that places a satellite continuously over the same point on the surface of the earth.

Gibbs function Defined for a thermodynamic system by $G = H - TS$.

Gray (Gy) A unit of absorbed dose; 1 Gy = 1 J/kg.

Greenhouse effect The trapping of radiation by a material that passes short wavelengths but is opaque to long wavelengths.

Ground state The state of lowest energy for an atomic or molecular system; the state ordinarily occupied by that system.

GVT—Gravity vacuum transport A mass-transit method employing a train in an evacuated tube following a path curving down, then up to the next station, using gravity to increase the downward speed and inertia to carry it upward.

Gyre The large-scale ocean current system.

Head The height of the water behind a water storage dam.

Heat capacity The ratio dQ/dT for a substance.

Heat pump A device that operates as a refrigerator in the winter, cooling the outside and exhausting warm air into the house; in the summer it cools the house, exhausting the hot air out of doors.

Heat quality The measure of a heat source to provide available work; directly proportional to the temperature at which the heat is extracted.

Heliostat A mirror that may be positioned to reflect sunlight to a central receiver throughout the day.

Hole A missing electron in a covalent bond; acts as a net positive charge.

Hot dry rock A geothermal resource consisting of heated rock that may be used to heat water pumped into it.

HTGR—High-temperature gas-cooled reactor A gas-cooled, graphite-moderated, nonbreeder reactor.

Hydrates Chemical compounds that contain one or more molecules of water as an integral part of their cystalline structure.

Hydraulic fracturing The process by which the area of hot, dry rock used to heat water is increased by the pressure of the water being pumped from above.

Hydride A chemical compound of hydrogen and another element or radical.

Hydroelectricity Electricity produced by turbogenerators turned by water flow, either from a dam or in a stream.

Hydrostatic equation The relationship between pressure change and elevation change: $dp = -g\rho\, dz$.

Hydrostorage Storing potential energy by pumping water up to a higher elevation, then letting the water turn a turbogenerator when it flows back down to the original elevation.

ICE—Internal combustion engine An engine in which the combustion of fuel takes place inside the region containing the working fluid.

Ideal gas A gas of noninteracting molecules that obeys this equation of state: $pV = nRT$.

Illuminance A measure of light intensity; measured in lumens per square meter.

Impedance The net effect of resistance R and reactance X: $Z^2 = R^2 + X^2$.

Inductance The ratio of the emf developed in a circuit to the rate of change of current through that circuit; measured in henrys.

Insulator A material having a high resistivity as a consequence of a filled valence band and a large energy gap.

Internal energy A measure of the energy a substance has by virtue of its temperature.

Isobaric process A process occurring at a constant pressure.

Isothermal process A process occurring at a constant temperature.

Isotope From Greek, "having the same place": nuclei that have the same atomic number but differing numbers of neutrons are isotopes.

Isovolumnic process A process occurring at a constant volume.

Joule's law The conversion of electrical energy to heat energy. The rate at which thermal energy is generated is given by $P = I^2 R$.

Kelvin The degree name for the absolute scale: 5° absolute = 5 K.

Kerogen A solid form of petroleum; often erroneously called oil shale.

Laminar flow Fluid flow in which there is no component of velocity of the fluid normal to the direction of motion; the Reynolds number for such flow is usually less than about 2000.

Lapse rate The rate of change of temperature with elevation.

Laser A source of coherent light; typically ruby or He–Ne.

Lawson criterion For D + T fusion reaction to occur the product of plasma density and confinement time must be at least 10^{14}.

LD$_{50}$(30) The dose required to produce death in 50 percent of the test individuals within 30 days.

LET—Linear energy transfer The rate of energy transfer from an incident particle of radiation to the medium in which it travels; usually measured in terms of keV/μm.

Linear attenuation coefficient The coefficient of proportionality between the rate of intensity change per unit length and intensity.

Liquefaction The process of converting coal into usable liquid hydrocarbon fuels.

Liquid-dominated reservoir A geothermal resource in which the thermal energy is available primarily as a hot brine.

LMFBR—Liquid-metal fast breeder reactor A breeder reactor without moderation using liquid sodium as a coolant.

LOA—Line of assurance In an event tree, a control or safety measure.

LOCA—Loss of coolant accident The worst accident postulated to occur in a power reactor: one of the two major coolant pipes breaks.

LOT—Load on top A procedure for cleaning oil tanks in a tanker by skimming off oil into a central containment tank.

Luminous flux A measure of the intensity of a light source that takes into account the sensitivity of the eye to various wavelengths.

Lurgi process A particular coal-gasification process.

Magma The hot, semiplastic material under the earth's crust and mantle.

Magnetic levitation A technique for supporting high-speed ground vehicles away from the track; both attractive and repulsive techniques are used.

Majority carrier The charge carrier having the same sign as the dopant atom: donors produce electrons, acceptors produce holes.

Manhattan project The code name given to the project to develop the atomic bomb during World War II.

Mass attenuation coefficient A means of expressing the absorption of radiation by a substance.

MCU—Modular combustion unit A boiler for burning trash to produce steam.

MEC cycle—Munters Environmental Control A system of refrigeration using adsorption of water vapor from the air.

Meiosis The process by which sex cells replicate.

Meissner effect The expulsion of magnetic field lines from a superconductor as the critical temperature is reached.

Meson One of a group of intermediate-mass unstable particles.

Metastable state A nuclear, atomic, or molecular state having a relatively long lifetime.

Methanol A toxic alcohol: CH_3OH.

Methyl fuel A mixture of methanol and higher alcohols.

MHD—Magnetohydrodynamics The direct conversion of thermal energy to electricity by means of hot, ionized gas in a magnetic field.

Microballoon A small, hollow, glass sphere used to enclose the fuel in a laser-fusion system.

Microclimate The climate of a small region.

Minority carrier The charge carrier having the sign opposite to that of the dopant atom: donors produce hole minority carriers; acceptors produce electron minority carriers.

Mitosis The process by which cells other than sex cells replicate.

Moderation The process of reducing neutron energies to values low enough to be of use in causing fission.

Mothballing The process of decommissioning a power reactor in which the volatile radioactive material is removed, and the remainder of the facility is kept beyond casual reach.

MPD—Maximum permissible dose The maximum radiation dose allowed; doses to various body organs as well as whole-body doses are defined.

MSBR—Molten-salt breeder reactor A breeder reactor using thermal neutrons in which the fuel is a molten mixture circulated through the system.

Multiplication factor The factor by which the number of first-generation neutrons in a reactor is increased in the second generation.

Mutation A change in some inheritable trait in the next generation.

NAAQS—National Ambient Air Quality Standards Standards promulgated by the Environmental Protection Agency on the criteria pollutants.

Natural gas Gaseous hydrocarbons, principally methane, occurring naturally in the earth, often along with oil.

Neutral equilibrium A state in which a perturbation creates neither a restoring force nor a force in a direction away from equilibrium.

Neutrino A massless, uncharged particle emitted during the beta decay of protons in the nucleus: $^{22}Na \rightarrow {}^{22}Ne + \beta + \nu$.

Neutron A neutral particle found in the nucleus of most atoms having about the same mass as the proton: $m_n = 1.00867$ u.

NGL—Natural gas liquids Heavier hydrocarbons, principally propane, that are liquid at standard temperature and pressure, found along with natural gas.

NRC—Nuclear Regulatory Commission That agency responsible for the licensing and operation of nuclear power reactors.

NSPS—New Source Performance Standards Standards of performance for new sources of criteria pollutants.

Nucleon The collective term given to particles normally found in the nucleus, namely, protons and neutrons.

Nucleus That part of the atom containing virtually all the mass and the positive electric charge.

Octane rating A measure of gasoline that predicts its ability to function in high-compression engines without preignition.

Ohm's law The empirical relationship between current, potential difference, and resistance in an electrical circuit: $V = IR$.

OTEC—Ocean thermal-energy conversion A system to produce electricity using the temperature difference between the ocean surface and bottom to drive a Rankine-cycle device.

Otto cycle The idealization of ICE operation.

Oxidant A substance that oxidizes other substances.

Oxyhemoglobin A molecule consisting of one oxygen atom and one hemoglobin molecule.

Pair production The process in which a photon having at least 1.02 MeV energy is converted into an electron–positron pair.

PAN—Peroxyacetyl nitrate Secondary air pollutant caused by action of NO_2, sunlight, and hydrocarbons.

People mover General name for a variety of devices, moving sidewalks, and the like, designed to carry small groups of passengers short distances at low speeds.

Permafrost Soil that is frozen the year around.

Permeability As applied to magnetic fields, the ratio of the magnetic field produced by a current loop to the current; for free space it has the value 1.26×10^{-6} H/m.

Permittivity For a capacitive circuit, the ratio of the charge per unit area of the circuit to the electric field intensity; for free space it has the value 8.85×10^{-12} F/m.

Petroleum The general term for a wide variety of complex hydrocarbons, usually liquid.

pH A measure of the acidity or alkalinity of a liquid. Neutrality is a pH of 7.

Phase change The change of substance from one form to another, that is, from solid to liquid, and so forth.

Photoelectric effect The release of electrons from a material by the absorption of light.

Photolysis The use of radiation to split water molecules in order to produce hydrogen.

Photon A massless particle of electromagnetic energy.

Pitchblende A uranium ore.

Plasma The state of matter in which all atoms are free and ionized.

Plutonium A man-made element having atomic number 94; undergoes fission readily and can be "bred" in reactors.

p–n junction The region of close contact between p-type and n-type materials; a depletion region is formed at this junction.

Polyphase systems Electrical transmission systems in which multiple windings in the generators create multiple pairs of transmission circuits.

Positron A positively charged electron.

Power factor Defined in terms of the reactance X and impedance Z of an AC circuit: power factor $= \cos \tan^{-1} |X|/R$.

Power tower A central tower containing a receiver for light reflected by a field of heliostats around the tower.

Preheating Heating air prior to use in combustion to incease the efficiency of the combustion process.

Proton A positively charged nucleon; $m_p = 1.00782$ u.

PSI—Pollution Standard Index A means of classifying health hazards in terms of the level of concentration of the criteria pollutants.

Pumped hydrostorage A technique for storing energy by using excess electricity to pump water to a high reservoir from which it is drained to generate electricity during periods of peak demand.

Purex process The plutonium–uranium chemical extraction process used to separate these elements from each other and from other radioactive materials in spent nuclear reactor fuel elements.

PWR—Pressurized-water reactor A nuclear power reactor in which the cooling water is kept under high pressure and not allowed to boil.

Q An amount of energy equal to 10^{18} Btu.

Quad An amount of energy equal to 10^{15} Btu.

Quantization The process of selecting specific values for physical variables.

Quantum mechanics The mathematics that must be used to study atomic and molecular systems.

Quench zone In an ICE cylinder, the region near the cylinder wall in which fuel combustion is quenched.

rad An old unit of absorbed dose; 1 rad $= 0.01$ J/kg.

Radiation Heat transfer by the emission of long-wavelength electromagnetic waves from a substance. Also refers to the emissions from radioactive nuclei.

Radiation inversion A temperature inversion caused by radiative cooling of the earth with subsequent cooling of the air in contact with it.

Radioactivity Emissions from unstable nuclei; usually α, β, or γ rays.

Rank The classification of coal by energy content.

Rankine cycle The idealization of turbine power systems.

Rankine scale A temperature scale for which zero represents the lowest achievable temperature; the degree size is the same as the Fahrenheit degree. $0°$R $= -459.67°$F.

Rated speed The speed for which a given wind turbine is designed to produce maximum power.

RDF—Refuse-derived fuel The light fraction of municipal trash.

Reactance The AC analogue of resistance. Inductive reactance $X_L = 2\pi f L$; capacitive reactance $X_C = (2\pi f C)^{-1}$.

Reactor poison A substance that absorbs neutrons lowering the overall efficiency of operation of a power reactor.

rem An old unit of dose equivalent; rems = dose (rads) × quality factor × weighting factor.

Replication The process in which a cell divides to create a copy of itself.

Reserves That part of a mineral resource that is known and can be recovered.

Resistivity A measure of the ease with which an electrical current passes through a substance; varies from material to material.

Resource The total amount of a mineral presumed to exist.

Reversible A process is reversible in the thermodynamic sense if its direction can be reversed, that is, a compression becomes an expansion, and so on, without any change in the state variables or the heat or work done, except a change in direction.

Reynolds number A dimensionless constant $(\rho v D/\mu)$ that is used to characterize fluid flow.

Roentgen (R) An old unit of exposure; $1\,R = 2.58 \times 10^{-4}$ C/kg.

Rolling resistance The resistance of tires to motion; caused principally by tread slippage, air resistance, and internal flexing.

R-value The "resistance" of a substance to the conduction of heat; the inverse of the conduction coefficient U, measured in British thermal units per hour per foot per degree Fahrenheit.

Saturation current The minority current in a diode.

Saturation curve The curve connecting points on a p–V or T–S diagram that represent the substance in a saturated state.

Savonius rotor A vertical wind machine consisting of two opposing semicircular cylindrical "sails."

Scram The rapid insertion of control rods in a reactor to reduce the fission rate dramatically.

Second law A statement that describes the capability of thermal processes to transform heat into work.

Semiconductor A material that has a resistivity between that of an insulator and of a conductor; examples are germanium and silicon.

Separation factor A measure of the efficacy of a process that separates ^{235}U from the other naturally occurring uranium isotopes.

Sievert (Sv) New unit of dose equivalent; $Sv =$ dose (Gy) × quality factor × weighting factor.

Sigmoid curve An S-shaped curve; characteristic of physical systems that grow initially at an exponential rate, then asymptotically approach a constant.

Skin effect The tendency of AC to concentrate on the surface of conductors rather than to distribute evenly throughout the conductor volume.

SMOG—"Smoke + fog" The term given to various types of air pollution.

SNAP—Systems nuclear auxiliary power Power packs using radioactive materials to provide the energy, usually as heat.

SNG—Synthetic natural gas Methane produced from coal.

Solar constant The amount of sunlight available at the top of the earth's atmosphere; about $1.4\,kW/m^2$.

Somatic effect An effect on the individual receiving the injury.

Spallation A shattering of the nucleus of an atom.

Specific heat The ratio dQ/dT per unit mass of substance.

Stable atmosphere A prevailing lapse rate such that an air parcel would tend not to rise.

Stable equilibrium A state in which a perturbation creates a restoring force.

Steam reforming The reaction of methane with steam to produce carbon dioxide and hydrogen; Eq. 6.9.

Stirling engine An ECE developed in the early nineteenth century having a theoretical efficiency equal to that of the Carnot engine.

Stoichiometry The condition in which the reactants in a chemical reaction exist in the correct amounts so that the reaction proceeds completely with no reactants in excess.

Stopping power dE/dx for a particle of a specific energy.

Straggling The nonuniformity in energy loss rate near the end of the range of a charged particle.

Stratified charge A method of providing the fuel to an engine in two distinct regions of fuel concentration: a rich region to promote burning and a lean region to minimize emissions.

Stratosphere The region of the atmosphere above roughly 15 km.

Subsidence inversion A temperature inversion caused by a downward motion of upper air with subsequent heating of that air.

Superconductivity The ability of some metals to carry currents without Joule's law losses, that is, with zero resistance.

Superheat A condition in which the working fluid is an unsaturated vapor at that particular p, V, and T.

SWU—Separative work unit A measure of the work required to perform uranium isotope separation.

Syncrude Synthetic crude oil; usually made from coal.

Synthesis gas A high-quality gas produced by the heating of coal in the presence of steam.

TACV—Tracked air-cushion vehicle A class of high-speed ground vehicles that are supported away from a guideway by a cushion of air.

Tailing cycle A heat conversion cycle using the exhaust heat from a conventional Rankine cycle.

Tailings The remainder after some process; in particular, the remainder after the uranium isotope-separation process.

Temperature inversion A condition in which the lapse rate has the opposite sign to the normal rate.

Thermal conductivity The coefficient of proportionality between the rate of heat transfer and the product of area and temperature gradient in an object.

Thermal neutrons Neutrons that have energies comparable to the thermal vibration energies of the reactor fuel; $E_n < 0.1\,\text{eV}$.

Thermal relaxation time The time required for the temperature difference between the interior and exterior of a house to drop to $1/e$ of its initial value.

Thermodynamics The study of the transformations of heat and work on a macroscopic scale.

Thermonuclear ignition The instant when the fusion of fuel material begins in a fusion reactor.

Throttling process A process in which the working fluid is made to pass through a constriction.

Tip speed The velocity of the tips of the blades of a wind turbine.

Tokamak A particular type of fusion reactor involving magnetic confinement and ohmic heating.

Topping cycle A heat conversion cycle having a high-temperature input and exhausting heat to a conventional Rankine cycle.

Transformer A device that steps up or down the applied AC voltage.

Transmission line Conducting cables that carry large currents usually at high voltages from primary generators to major distribution centers.

Tropopause The small region that separates the troposphere from the stratopshere.

Troposphere The region of the atmosphere up to roughly 15 km; weather is a tropospheric phenomenon for the most part.

Turbine A mechanical device for converting the internal energy of a fluid (heat, in the case of steam, and kinetic energy, in the case of flowing water) into rotational motion of a shaft.

Turbulent flow Fluid flow in which there is a randomly varying normal component of velocity; characterized by Reynolds numbers larger than about 2000.

u—Unified atomic mass unit An amount of mass equal to 1/12 of the carbon atom;

$$1\,\text{u} = 1.6598 \times 10^{-27}\,\text{kg}.$$

ULCC—Ultra large crude carrier An oil tanker having a capacity greater than about 300,000 dwt.

Unstable atmosphere A prevailing lapse rate such that an air parcel would tend to rise.

Unstable equilibrium A state in which a perturbation creates a force in a direction away from equilibrium.

Valence band The energy band available to the valence electrons of the atoms of a solid.

Vapor dome The shape formed by the liquid and vapor saturation curves in the p–V and T–S diagrams.

Vapor fraction That fraction of a substance that is in the vapor state at some point intermediate between the liquid and vapor saturation curves.

Viscosity The "friction" of fluid flow; the coefficient of proportionality between viscous force and the product of area and fluid velocity gradient.

VLCC—Very large crude carrier An oil tanker having a capacity between about 200,000 and 300,000 dwt.

Wankel engine A rotary engine developed by Felix Wankel.

Water gas shift The reaction of carbon monoxide with steam to produce carbon dioxide and hydrogen.

Water-splitting cycle A sequence of chemical reactions in which the net effect is the breakup of water into hydrogen and oxygen.

Wet-bulb temperature The lowest temperature that can be achieved by evaporating water into the air.

Wet scrubber A device for removing particulates or SO_2 from an effluent that uses a spray of solution as the removal agent.

Working fluid The fluid used in a cyclic device to transform heat into work or vice versa.

WWI—Water wall incinerator A trash incinerator in which steam is formed in water pipes lining the combustion chamber.

Xenon override Additional reactivity built into a reactor to compensate for neutron absorption by xenon fission products.

X ray The photon emitted during transition between electrons to inner shells of medium to heavy nuclei.

Zircalloy A zirconium alloy from which nuclear fuel rods are fabricated.

ZPG—Zero population growth The condition in which birth rate plus net immigration just equals death rate.

APPENDIX A   UNITS OF MEASURE

Length

1 meter(m) *Definition*: 1,650,763.73 wavelengths of the $2p_{10}5d_5$ line of krypton-86 in vacuum. Formerly 1 m was 10^{-7} of the quadrant of the earth through Paris. Thus girth of the earth is $\simeq 4 \times 10^7$ m.

1 foot (ft) $\equiv 0.3048$ m.

1 yard (yd) $\equiv 3$ ft $\equiv 0.9144$ m.

1 mile $\equiv 1760$ yd $\equiv 1.609344$ km.

Area

1 square meter	(m^2)	$= 1$ m $\times$ 1 m.
1 hectare	(ha)	$= 100$ m $\times$ 100 m $= 10^4\,m^2$.
1 square foot	(ft^2)	$= 0.092\,903\,04\,m^2$.
1 square yard	(yd^2)	$= 0.836\,127\,3\,m^2$.
1 acre		$\equiv 4840(yd^2) = 0.404\,685\,61$ ha.
1 square mile	$(mile^2)$	$\equiv 640$ acre $= 258.999$ ha.

Volume

1 cubic meter (m^3) $= 1$ m $\times$ 1 m $\times$ 1 m.

1 liter (ℓ) $\equiv 10^{-3}\,m^3$. Formerly the liter was the volume of 1 kg of water at 4°C at 760 mmHg pressure. Thus the density of water is very close to 1 kg/ℓ or 1 metric ton/m^3.

1 pint (USA)   (pt) $= 0.47318\,\ell$.

1 gallon (USA)(gal) $\equiv 8$ pt (US) $= 3.78541\,\ell$ (1 USA pt of water has a mass of 1 lb).

1 barrel (of oil)(bbl) $\equiv 42$ gal $= 159.0\,\ell$.

1 acre-foot $\equiv 1233.48\,m^3$ (Lakes, dams, and oceans).

Mass

1 kilogram (kg)	$\equiv$ mass of platinum–iridium prototype kilogram kept at BIPM Sevres, France.
1 gram (g)	$\equiv 10^{-3}$ kg.
1 metric ton(tonne)	$\equiv 10^3$ kg $\equiv 10^6$ g.
1 pound (lb)	$\equiv 0.453\,592\,37$ kg.
1 ton (short ton, USA)	$\equiv 2000$ lb $= 0.907185$ metric ton.

Time

1 second(s)	*Definition*: 1 s is the duration of 9,192,770 cycles of radiation of the ground-state hyperfine splitting of cesium-133.
1 minute(min)	$\equiv 60$ s.
1 hour (hr)	$\equiv 60$ min $= 3600$ s.
1 day (solar day)	$\equiv 24$ hr $= 86,400$ s.
1 year (solar year)	$= 3.156 \times 10^7$ s.

Force

1 newton (N)	That force which imparts an acceleration of 1 m/s^2 to a mass of 1 kg.
1 kilogram force (weight)(kgf)	$\equiv 9.806\,65$ N ($g \equiv 9.80665$ m/s^2).
1 pound force (weight) (lbf)	$= 4.44822$ N.

Pressure

1 pascal (Pa)	$\equiv 1$ N/m^2.
1 bar	$\equiv 10^5$ Pa $= 10^6$ dyne cm^2.
1 millibar (mbar)	$\equiv 10^2$ Pa.
1 kilogram weight/m^2	$= 9.807$ Pa.
1 pound weight/square inch(psi)	$= 6.895 \times 10^3$ Pa.
1 torr	$\equiv$ pressure of 1 mm of Mercury $= 1.333 \times 10^2$ Pa.
1 atmosphere	$\equiv 760$ torr $\equiv$ pressure of 76 cm of mercury $= 1.01325 \times 10^5$ Pa (as the name indicates, atmospheric pressure is about 1 "atmosphere" at sea level).

Energy

1 joule (J)	$\equiv$ work done by a force of 1 N moving through a distance of 1 m $= $ N/m $=$ kg m^2/s^2.
1 kilowatt-hour (kWh)	$= 3.600 \times 10^6$ J.
1 calorie (cal)	$= 4.186$ J.
1 British thermal unit(Btu)	$= 1.05506 \times 10^3$ J (the Btu is the energy to raise 1 lb of water 1° F).
1 kilocalorie (kcal)	$\equiv 1$ Calorie (Cal) $= 4.186 \times 10^3$ J. The calorie is the energy to raise 1 g of water 1 K, the kilocalorie or Calorie (used by dietitions) is the energy required to raise 1 kg of water 1 K.
1 therm	$\equiv 10^5$ Btu $= 1.05506 \times 10^8$ J $= 1.05506 \times 10^2$ MJ.
1 foot-pound weight	$= 1.35\,582$ J(energy required to raise 1 lb through 1 ft).

Power

1 watt   (W) $\equiv$ energy of 1 J expended in 1 s.

1 horse power(hp) $\equiv$ 550 foot-pounds weight per second $\equiv$ 0.7457 kW.

Temperature

1 kelvin   (K) = 1/273.16 of the absolute thermodynamic temperature of the triple point of water.

T (Celsius)   ($^\circ$C) = $(T - 273.15)$, T = absolute temperature in kelvins.

T (Fahrenheit)($^\circ$F) = $32 + \frac{9}{5}T(^\circ C)$.

APPENDIX B THERMODYNAMIC PROPERTIES OF WATER AND AMMONIA

The tables in this appendix are reprinted from the *ASHRAE Fundamentals Handbook*, 1981.

TABLE B.1 Thermodynamic Properties of Water at Saturation

Temp °C	Absolute Pressure kPa p	Specific Volume, m³/kg Sat. Solid v_i	Specific Volume Evap. v_{ig}	Specific Volume Sat. Vapor v_g	Enthalpy, kJ/kg h_i	Enthalpy Evap. h_{ig}	Enthalpy Sat. Vapor h_g	Entropy, kJ/kg·K Sat. Solid s_i	Entropy Evap. s_{ig}	Entropy Sat. Vapor s_g	Temp °C
−60	0.00108	0.001082	90942.00	90942.00	−446.40	2836.27	2389.87	−1.6854	13.3065	11.6211	−60
−59	0.00124	0.001082	79858.69	79858.69	−444.74	2836.46	2391.72	−1.6776	13.2452	11.5677	−59
−58	0.00141	0.001082	70212.37	70212.37	−443.06	2836.64	2393.57	−1.6698	13.1243	11.5147	−58
−57	0.00161	0.001082	61805.35	61805.35	−441.38	2836.81	2395.43	−1.6620	13.1243	11.4623	−57
−56	0.00184	0.001082	54469.39	54469.39	−439.69	2836.97	2397.28	−1.6542	13.0646	11.4104	−56
−55	0.00209	0.001082	48061.05	48061.05	−438.00	2837.13	2399.13	−1.6464	13.0054	11.3590	−55
−54	0.00238	0.001082	42455.57	42455.57	−436.29	2837.27	2400.98	−1.6386	12.9468	11.3082	−54
−53	0.00271	0.001083	37546.09	37546.09	−434.59	2837.42	2402.83	−1.6308	12.8886	11.2578	−53
−52	0.00307	0.001083	33242.14	33242.14	−432.87	2837.55	2404.68	−1.6230	12.8309	11.2079	−52
−51	0.00348	0.001083	29464.67	29464.67	−431.14	2837.68	2406.53	−1.6153	12.7738	11.1585	−51
−50	0.00394	0.001083	26145.01	26145.01	−429.41	2837.80	2408.39	−1.6075	12.7170	11.1096	−50
−49	0.00445	0.001083	23223.70	23223.69	−427.67	2837.91	2410.24	−1.5997	12.6608	11.0611	−49
−48	0.00503	0.001083	20651.69	20651.68	−425.93	2838.02	2412.09	−1.5919	12.6051	11.0131	−48
−47	0.00568	0.001083	18383.51	18383.50	−424.17	2838.12	2413.94	−1.5842	12.5498	10.9656	−47
−46	0.00640	0.001083	16381.36	16381.35	−422.41	2838.21	2415.79	−1.5764	12.4949	10.9285	−46
−45	0.00721	0.001084	14612.36	14612.35	−420.65	2838.29	2417.65	−1.5686	12.4405	10.8719	−45
−44	0.00811	0.001084	13047.66	13047.65	−418.87	2838.37	2419.50	−1.5609	12.3866	10.8257	−44
−43	0.00911	0.001084	11661.85	11661.85	−417.09	2838.44	2421.35	−1.5531	12.3330	10.7799	−43
−42	0.01022	0.001084	10433.85	10433.85	−415.30	2838.50	2423.20	−1.5453	12.2799	10.7346	−42
−41	0.01147	0.001084	9344.25	9344.25	−413.50	2838.55	2425.05	−1.5376	12.2273	106897	−41
−40	0.01285	0.001084	8376.33	8376.33	−411.70	2838.60	2426.90	−1.5298	12.1750	10.6452	−40
−39	0.01438	0.001085	7515.87	7515.86	−409.88	2838.64	2428.76	−1.5221	12.1232	10.6011	−39
−38	0.01608	0.001085	6750.36	6750.36	−408.07	2838.67	2430.61	−1.5143	12.0718	10.5575	−38
−37	0.01796	0.001085	6068.17	6068.16	−406.24	2838.70	2432.46	−1.5066	12.0208	10.5142	−37
−36	0.02005	0.001085	5459.82	5459.82	−404.40	2838.71	2434.31	−1.4988	11.9702	10.4713	−36
−35	0.02235	0.001085	4917.10	4917.10	−402.56	2838.73	2436.16	−1.4911	11.9199	10.4289	−35
−34	0.02490	0.001085	4432.37	4432.36	−400.72	2838.73	2438.01	−1.4833	11.8701	10.3868	−34
−33	0.02772	0.001085	3998.71	3998.71	−398.86	2838.72	2439.86	−1.4756	11.8207	10.3451	−33
−32	0.03082	0.001086	3610.71	3610.71	−397.00	2838.71	2441.72	−1.4678	11.7716	10.3037	−32
−31	0.03425	0.001086	3263.20	3263.20	−395.12	2838.69	2443.57	−1.4601	11.7229	10.2628	−31
−30	0.03802	0.001086	2951.64	2951.64	−393.25	2838.66	2445.42	−1.4524	11.6746	10.2222	−30
−29	0.04217	0.001086	2672.03	2672.03	−391.36	2838.63	2447.27	−1.4446	11.6266	10.1820	−29

T											T
−28	0.04673	0.001086	2420.89	2420.89	−389.47	2838.59	2449.12	−1.4369	11.5790	10.1421	−28
−27	0.05175	0.001086	2195.23	2195.23	−387.57	2838.53	2450.97	−1.4291	11.5318	10.1026	−27
−26	0.05725	0.001087	1992.15	1992.15	−385.66	2838.48	2452.82	−1.4214	11.4849	10.0634	−26
−25	0.06329	0.001087	1809.35	1809.35	−383.74	2838.41	2454.67	−1.4137	11.4383	10.0246	−25
−24	0.06991	0.001087	1644.59	1644.59	−381.82	2838.34	2456.52	−1.4059	11.3921	9.9862	−24
−23	0.07716	0.001087	1495.98	1495.98	−379.89	2838.26	2458.37	−1.3982	11.3462	9.9480	−23
−22	0.08510	0.001087	1361.94	1361.94	−377.95	2838.17	2460.22	−1.3905	11.3007	9.9102	−22
−21	0.09378	0.001087	1240.77	1240.77	−376.01	2838.07	2462.06	−1.3828	11.2555	9.8728	−21
−20	0.10326	0.001087	1131.27	1131.27	−374.06	2837.97	2463.91	−1.3750	11.2106	9.8356	−20
−19	0.11362	0.001088	1032.18	1032.18	−372.10	2837.86	2465.76	−1.3673	11.1661	9.7988	−19
−18	0.12492	0.001088	942.47	942.46	−370.13	2837.74	2467.61	−1.3596	11.1218	9.7623	−18
−17	0.13725	0.001088	861.18	861.17	−368.15	2837.61	2469.46	−1.3518	11.0779	9.7261	−17
−16	0.15068	0.001088	787.49	787.48	−366.17	2837.47	2471.30	−1.3441	11.0343	9.6902	−16
−15	0.16530	0.001088	720.59	720.59	−364.18	2837.33	2473.15	−1.3364	10.9910	9.6546	−15
−14	0.18122	0.001088	659.86	659.86	−362.18	2837.18	2474.99	−1.3287	10.9480	9.6193	−14
−13	0.19852	0.001089	604.65	604.65	−360.18	2837.02	2476.84	−1.3210	10.9053	9.5844	−13
−12	0.21732	0.001089	554.45	554.45	−358.17	2836.85	2478.68	−1.3132	10.8629	9.5497	−12
−11	0.23775	0.001089	508.75	508.75	−356.15	2836.68	2480.53	−1.3055	10.8208	9.5153	−11
−10	0.25991	0.001089	467.14	467.14	−354.12	2836.49	2482.37	−1.2978	10.7790	9.4812	−10
−9	0.28395	0.001089	429.21	492.21	−352.08	2836.30	2484.22	−1.2901	10.7375	9.4474	−9
−8	0.30999	0.001090	394.64	394.64	−350.04	2836.10	2486.06	−1.2824	10.6962	9.4139	−8
−7	0.33821	0.001090	363.07	363.07	−347.99	2835.89	2487.90	−1.2746	10.6552	9.3806	−7
−6	0.36874	0.001090	334.25	334.25	−345.93	2835.68	2489.74	−1.2669	10.6145	9.3476	−6
−5	0.40178	0.001090	307.91	307.91	−343.87	2835.45	2491.58	−1.2592	10.5741	9.3149	−5
−4	0.43748	0.001090	283.83	283.83	−341.80	2835.22	2493.42	−1.2515	10.5340	9.2825	−4
−3	0.47606	0.001090	261.79	261.79	−339.72	2834.98	2495.26	−1.2438	10.4941	9.2503	−3
−2	0.51773	0.001091	241.60	241.60	−337.63	2834.72	2497.10	−1.2361	10.4544	9.2184	−2
−1	0.56268	0.001091	223.11	223.11	−335.53	2834.47	2498.93	−1.2284	10.4151	9.1867	−1
0	0.61117	0.001091	206.16	206.16	−333.43	2834.20	2500.77	−1.2206	10.3760	9.1553	0
0	0.6112	0.001000	206.143	206.141	−0.04	2500.81	2500.77	−0.0001	9.1554	9.1553	0
1	0.6571	0.001000	192.456	192.455	4.18	2498.43	2502.61	0.0153	9.1134	9.1286	1
2	0.7060	0.001000	179.770	179.769	8.39	2496.05	2504.44	0.0306	9.0716	9.1022	2
3	0.7581	0.001000	168.027	168.026	12.60	2493.68	2506.28	0.0459	9.0301	9.0761	3
4	0.8135	0.001000	157.138	157.137	16.81	2491.31	2508.12	0.0611	8.9890	9.0501	4
5	0.8725	0.001000	147.033	147.032	21.02	2488.94	2509.95	0.0762	8.9482	9.0244	5
6	0.9353	0.001000	137.654	137.653	25.22	2486.57	2511.79	0.0913	8.9076	8.9990	6
7	1.0020	0.001000	128.948	128.947	29.42	2484.20	2513.62	0.1064	8.8674	8.9738	7
8	1.0729	0.001000	120.851	120.850	33.62	2481.84	2515.46	0.1213	8.8274	8.9488	8
9	1.1481	0.001000	113.327	113.326	37.82	2479.47	2517.29	0.1362	8.7878	8.9240	9
10	1.2280	0.001000	106.329	106.328	42.01	2477.11	2519.12	0.1511	8.7484	8.8995	10
11	1.3128	0.001000	99.808	99.807	46.21	2474.74	2520.95	0.1659	8.7093	8.8751	11
12	1.4026	0.001001	93.741	93.740	50.40	2472.38	2522.78	0.1806	8.6705	8.8510	12
13	1.4979	0.001001	88.085	88.084	54.59	2470.02	2524.61	0.1953	8.6319	8.8272	13
14	1.5987	0.001001	82.813	82.812	58.78	2467.66	2526.44	0.2099	8.5936	8.8035	14
15	1.7055	0.001001	77.896	77.895	62.97	2465.30	2528.26	0.2244	8.5556	8.7800	15

The MSBR is a thermal breeder—it does not use the fast-neutron spectrum required by the LMFBR and GCFBR. In order to obtain the $\eta = 2$ requirement for breeding, ^{233}U is used as the fissile material. The fertile material in this reactor is ^{232}Th; the breeding reaction is as follows:

$$^{232}Th + n \rightarrow {}^{233}Th \quad (\tau_{1/2} \times 22 \text{ min})$$
$$\quad \hookrightarrow {}^{233}Pa + \beta^- + \bar{\nu} \quad (\tau_{1/2} = 27.4 \text{ days})$$
$$\quad \hookrightarrow {}^{233}U + \beta^- + \bar{\nu}. \quad (7.26)$$

Thorium is a rather abundant heavy element; it is common in granites and shales. The mass-232 isotope is the only one that occurs naturally. The MSBR employs a fuel composed of ^{232}Th and ^{233}U dissolved in molten lithium and beryllium fluorides. This viscous liquid passes through a moderating graphite matrix within which most of the fissions occur. It also circulates around the outside of the core region, forming a blanket within which most of the breeding occurs.

The major advantage of the MSBR is that it eliminates fuel pins entirely and makes on-site fuel reprocessing practical. The first advantage may become a crucial point in the development of commercial breeder capability because of the great difficulty that has been experienced in testing current fuel-pin assemblies under conditions similar to those of the LMFBR or GCFBR neutron and heat environment. The on-site reprocessing is also an extremely desirable feature, although adequate safety

TABLE 7.8 Characteristics of a 1000-MWe MSBR

Reactor thermal power (MWt)	2250
Overall plant thermal efficiency (percent)	44
Fuel salt inlet and outlet temperatures (°C)	566, 704
Coolant salt inlet and outlet temperatures (°C)	454, 621
Throttle steam conditions	240 atm, 538° C
Core height/diameter (m)	4.0/4.3
Radial blanket thickness (m)	0.5
Graphite reflector thickness (m)	0.8
Number of core elements	1412
Size of core elements (cm)	$10.2 \times 10.2 \times 396$
Salt volume fraction in core (percent)	13
Salt volume fraction in undermoderated zones (percent)	37 and 100
Salt volume fraction in reflector (percent)	< 1
Average core power density (W/cm³)	22
Maximum thermal neutron flux (neutrons/cm² s)	8.3×10^{14}
Graphite damage flux (> 50 keV) at point of maximum damage (neutrons/cm² s) (neutrons/cm² s)	3.3×10^{14}
Estimated graphite life (yr[a])	4
Total salt volume in primary system (ℓ)	48,700
Thorium inventory (kg)	68,000
Fissile fuel inventory of reactor system and processing plant (kg)	1470
Breeding ratio	1.07
Fissile fuel yield (percent/yr)	3.6
Fuel doubling time (exponential) (yr)	19

Reprinted, with permission, from A. Perry and A. Weinberg *Annual Review of Nuclear Science* 22, 318 (1972), copyright © 1972 by Annual Reviews Inc.
[a]Based on 80 percent plant factor and a fluence of 3×10^{22} neutrons/cm² (> 50 keV).

48	8.1097	7.4318	0.6778	2587.73	2386.77	200.97	13.215	13.214	0.001011	11.1754	48
49	8.0921	7.4013	0.6908	2589.50	2384.36	205.15	12.606	12.605	0.001012	11.7502	49
50	8.0747	7.3709	0.7038	2591.27	2381.94	209.33	12.029	12.028	0.001012	12.3503	50
51	8.0574	7.3407	0.7167	2593.04	2379.52	213.51	11.483	11.482	0.001013	12.9764	51
52	8.0403	7.3107	0.7296	2594.80	2377.10	217.70	10.965	10.964	0.001013	13.6293	52
53	8.0232	7.2808	0.7424	2596.56	2374.68	221.88	10.474	10.473	0.001014	14.3108	53
54	8.0064	7.5212	0.7552	2598.32	2372.25	226.06	10.008	10.007	0.001014	15.0205	54
55	7.9897	7.2217	0.7680	2600.07	2369.83	230.25	9.566	9.565	0.001015	15.7601	55
56	7.9730	7.1923	0.7807	2601.82	2367.39	234.43	9.146	9.145	0.001015	16.5311	56
57	7.9566	7.1632	0.7934	2603.57	2364.96	238.61	8.748	8.747	0.001016	17.3337	57
58	7.9402	7.1342	0.8061	2605.32	2362.32	242.80	8.370	8.369	0.001016	18.1691	58
59	7.9240	7.1053	0.8187	2607.06	2360.08	246.98	8.010	8.009	0.001017	19.0393	59
60	7.9079	7.0767	0.8313	2608.80	2357.63	251.17	7.6686	7.6676	0.001017	19.994	60
61	7.8920	7.0482	0.8438	2610.54	2355.18	255.36	7.3436	7.3426	0.001018	20.886	61
62	7.8761	7.0198	0.8563	2612.28	2352.73	259.54	7.0345	7.0335	0.001018	21.865	62
63	7.8604	6.9916	0.8688	2614.01	2350.28	263.73	6.7406	6.7395	0.001019	22.883	63
64	7.8448	6.9636	0.8812	2615.74	2347.82	267.92	6.4607	6.4597	0.001019	23.941	64
65	7.8293	6.9357	0.8936	2617.46	2345.36	272.11	6.1943	6.1933	0.001020	25.040	65
66	7.8140	6.9080	0.9060	2619.18	2342.89	276.29	5.9407	5.9397	0.001021	26.181	66
67	7.7987	6.8804	0.9183	2620.90	2340.42	280.48	5.6991	5.6980	0.001021	27.366	67
68	7.7836	6.8530	0.9306	2622.62	2337.95	284.67	5.4688	5.4678	0.001022	28.597	68
69	7.7686	6.8257	0.9429	2624.33	2335.47	288.86	5.2493	5.2483	0.001022	29.874	69
70	7.7537	6.7986	0.9551	2626.04	2332.99	293.06	5.0401	5.0391	0.001023	31.199	70
71	7.7389	6.7716	0.9673	2627.75	2330.50	297.25	4.8405	4.8394	0.001023	32.573	71
72	7.7242	6.7448	0.9794	2629.45	2328.01	301.44	4.6501	4.6491	0.001024	33.998	72
73	7.7096	6.7181	0.9916	2631.15	2325.51	305.63	4.4684	4.4673	0.001025	35.476	73
74	7.6952	6.6915	1.0037	2632.84	2323.02	309.83	4.2950	4.2940	0.001025	37.006	74
75	7.6808	6.6651	1.0157	2634.53	2320.51	314.02	4.1292	4.1282	0.001026	38.594	75
76	7.6666	6.6388	1.0278	2636.22	2318.01	318.22	3.9711	3.9701	0.001026	40.237	76
77	7.6525	6.6127	1.0398	2637.90	2315.49	322.41	3.8200	3.8189	0.001027	41.939	77
78	7.6384	6.5867	1.0517	2639.58	2312.98	326.61	3.6754	3.6744	0.001028	43.702	78
79	7.6245	6.5608	1.0636	2641.26	2310.46	330.80	3.5374	3.5364	0.001028	45.525	79
80	7.6106	6.5351	1.0755	2642.93	2307.93	335.00	3.4053	3.4043	0.001029	47.414	80
81	7.5969	6.5095	1.0874	2644.60	2305.40	339.20	3.2790	3.2780	0.001030	49.367	81
82	7.5833	6.4840	1.0993	2646.26	2302.86	343.40	3.1582	3.1572	0.001030	51.386	82
83	7.5697	6.4587	1.1111	2647.92	2300.32	347.60	3.0426	3.0416	0.001031	53.475	83
84	7.5563	6.4335	1.1228	2649.58	2297.78	351.80	2.9320	2.9309	0.001032	55.634	84
85	7.5430	6.4084	1.1346	2651.23	2295.22	356.00	2.8260	2.8249	0.001032	57.866	85
86	7.5297	6.3834	1.1463	2652.88	2292.67	360.21	2.7245	2.7234	0.001033	60.173	86
87	7.5165	6.3586	1.1580	2654.52	2290.11	364.41	2.6273	2.6262	0.001034	62.554	87
88	7.5035	6.3338	1.1696	2656.16	2287.54	368.62	2.5340	2.5330	0.001035	65.017	88
89	7.4905	6.3093	1.1812	2657.79	2284.97	372.82	2.4447	2.4437	0.001035	67.558	89
90	7.4776	6.2848	1.1928	2659.42	2282.39	377.03	2.3591	2.3581	0.001036	70.182	90
91	7.4648	6.2604	1.2044	2661.04	2279.80	381.24	2.2770	2.2760	0.001037	72.890	91

(Table B.1, continued)

Temp °C	Absolute Pressure kPa p	Specific Volume, m³/kg Sat. Solid v_i	Evap. v_{ig}	Sat. Vapor v_g	Enthalpy, kJ/kg h_i	Evap. h_{ig}	Sat. Vapor h_g	Entropy, kJ/kg·K Sat. Solid s_i	Evap. s_{ig}	Sat. Vapor s_g	Temp °C
92	75.685	0.001037	2.1972	2.1983	385.45	2277.22	2662.66	1.2159	6.2362	7.4521	92
93	78.567	0.001038	2.1217	2.1227	389.66	2274.62	2664.28	1.2274	6.2121	7.4395	93
94	81.543	0.001039	2.0491	2.0502	393.87	2272.02	2665.89	1.2389	6.1881	7.4270	94
95	84.609	0.001040	1.9796	1.9806	398.08	2269.41	2667.49	1.2504	6.1642	7.4145	95
96	87.771	0.001040	1.9127	1.9138	402.29	2266.80	2669.09	1.2618	6.1404	7.4022	96
97	91.033	0.001041	1.8485	1.8496	406.51	2264.18	2670.69	1.2732	6.1167	7.3899	97
98	94.394	0.001042	1.7869	1.7879	410.72	2261.55	2672.28	1.2845	6.0932	7.3777	98
99	97.853	0.001043	1.7277	1.7287	414.94	2258.92	2673.86	1.2959	6.0697	7.3656	99
100	101.420	0.001043	1.6708	1.6718	419.16	2256.28	2675.44	1.3072	6.0464	7.3536	100
101	105.095	0.001044	1.6160	1.6171	423.38	2253.64	2677.01	1.3185	6.0231	7.3416	101
102	108.877	0.001045	1.5634	1.5645	427.60	2250.99	2678.58	1.3297	6.0000	7.3297	102
103	112.773	0.001046	1.5128	1.5139	431.82	2248.33	2680.15	1.3410	5.9770	7.3179	103
104	116.782	0.001047	1.4642	1.4652	436.04	2245.66	2681.70	1.3521	5.9541	7.3062	104
105	120.908	0.001047	1.4173	1.4184	440.27	2242.99	2683.26	1.3633	5.9313	7.2946	105
106	125.155	0.001048	1.3723	1.3733	444.49	2240.31	2684.80	1.3745	5.9085	7.2830	106
107	129.524	0.001049	1.3289	1.3300	448.72	2237.63	2686.34	1.3856	5.8859	7.2715	107
108	134.015	0.001050	1.2872	1.2882	452.95	2234.93	2687.88	1.3967	5.8634	7.2601	108
109	138.635	0.001051	1.2470	1.2480	457.18	2232.23	2689.41	1.4078	5.8410	7.2488	109
110	143.390	0.001052	1.2082	1.2093	461.41	2229.52	2690.93	1.4188	5.8187	7.2375	110
111	148.271	0.001052	1.1709	1.1720	465.64	2226.81	2692.45	1.4298	5.7965	7.2263	111
112	153.289	0.001053	1.1350	1.1361	469.88	2224.09	2693.96	1.4408	5.7744	7.2152	112
113	158.447	0.001054	1.1004	1.1014	474.11	2221.35	2695.47	1.4518	5.7523	7.2041	113
114	163.749	0.001055	1.0670	1.0680	478.35	2218.62	2696.97	1.4628	5.7304	7.1931	114
115	169.192	0.001056	1.0348	1.0359	482.59	2215.87	2698.46	1.4737	5.7085	7.1822	115
116	174.786	0.001057	1.0038	1.0048	486.83	2213.12	2699.94	1.4846	5.6868	7.1714	116
117	180.530	0.001058	0.9738	0.9749	491.07	2210.35	2701.42	1.4954	5.6652	7.1606	117
118	186.420	0.001059	0.9450	0.9460	495.32	2207.58	2702.90	1.5063	5.6436	7.1499	118
119	192.476	0.001059	0.9171	0.9181	499.56	2204.80	2704.36	1.5171	5.6221	7.1392	119
120	198.688	0.001060	0.8902	0.8912	503.81	2202.02	2705.83	1.5279	5.6007	7.1286	120
122	211.603	0.001062	0.8392	0.8402	512.31	2196.42	2708.73	1.5494	5.5582	7.1076	122
124	225.198	0.001064	0.7916	0.7926	520.82	2190.78	2711.60	1.5709	5.5160	7.0869	124
126	239.496	0.001066	0.7472	0.7482	529.33	2185.11	2714.44	1.5922	5.4741	7.0664	126
128	254.518	0.001068	0.7057	0.7068	537.86	2179.40	2717.26	1.6135	5.4326	7.0461	128
130	270.306	0.001070	0.6670	0.6680	546.39	2173.66	2720.04	1.6347	5.3914	7.0261	130
132	286.871	0.001072	0.6308	0.6318	554.93	2167.87	2722.80	1.6558	5.3505	7.0063	132
134	304.251	0.001074	0.5968	0.5979	563.48	2162.05	2725.53	1.6768	5.3099	6.9867	134

136	322.479	0.001076	0.5651	0.5662	572.03	2156.18	2728.22	1.6977	5.2696	6.9673	136
138	341.568	0.001078	0.5354	0.5364	580.60	2150.28	2730.88	1.7185	5.2296	6.9481	138
140	361.572	0.001080	0.5074	0.5085	589.18	2144.33	2733.51	1.7393	5.1899	6.9292	140
142	382.503	0.001082	0.4813	0.4823	597.76	2138.34	2736.10	1.7599	5.1505	6.9104	142
144	404.392	0.001084	0.4567	0.4578	606.36	2132.31	2738.66	1.7805	5.1113	6.8918	144
146	427.306	0.001086	0.4336	0.4347	614.97	2126.22	2741.19	1.8011	5.0724	6.8734	146
148	451.222	0.001088	0.4119	0.4129	623.58	2120.10	2743.68	1.8215	5.0338	6.8553	148
150	476.207	0.001091	0.3914	0.3925	632.21	2113.92	2746.13	1.8419	4.9954	6.8372	150
152	502.292	0.001093	0.3722	0.3733	640.85	2107.70	2748.55	1.8622	4.9572	6.8194	152
154	529.499	0.001095	0.3541	0.3552	649.50	2101.43	2750.93	1.8824	4.9193	6.8017	154
156	557.882	0.001097	0.3370	0.3381	658.16	2095.11	2753.26	1.9025	4.8817	6.7842	156
158	587.472	0.001100	0.3209	0.3220	666.83	2088.73	2755.56	1.9226	4.8442	6.7669	158
160	618.283	0.001102	0.3058	0.3069	675.52	2082.31	2757.82	1.9427	4.8070	6.7497	160
162	650.382	0.001104	0.2914	0.2925	684.22	2075.82	2760.04	1.9626	4.7700	6.7326	162
164	683.792	0.001107	0.2779	0.2790	692.93	2069.29	2762.21	1.9825	4.7332	6.7157	164
166	718.546	0.001109	0.2651	0.2662	701.65	2062.70	2764.35	2.0023	4.6967	6.6990	166
168	754.675	0.001112	0.2530	0.2541	710.39	2056.05	2766.44	2.0221	4.6603	6.6824	168
170	792.245	0.001114	0.2415	0.2427	719.14	2049.34	2768.48	2.0418	4.6241	6.6659	170
172	831.293	0.001117	0.2307	0.2318	727.91	2042.57	2770.47	2.0614	4.5881	6.6496	172
174	871.852	0.001119	0.2204	0.2216	736.69	2035.74	2772.42	2.0810	4.5523	6.6333	174
176	913.902	0.001122	0.2107	0.2118	745.48	2028.85	2774.33	2.1005	4.5167	6.6173	176
178	957.586	0.001125	0.2015	0.2026	754.29	2021.89	2776.18	2.1200	4.4813	6.6013	178
180	1002.899	0.001127	0.1928	0.1939	763.12	2014.87	2777.99	2.1394	4.4460	6.5854	180
182	1049.859	0.001130	0.1845	0.1856	771.96	2007.78	2779.74	2.1588	4.4109	6.5696	182
184	1098.548	0.001133	0.1766	0.1777	780.82	2000.63	2781.45	2.1781	4.3759	6.5540	184
186	1149.005	0.001136	0.1691	0.1702	789.69	1993.40	2783.10	2.1973	4.3411	6.5384	186
188	1201.247	0.001139	0.1620	0.1631	798.59	1986.11	2784.69	2.2165	4.3065	6.5230	188
190	1255.367	0.001141	0.1552	0.1564	807.50	1978.74	2786.23	2.2356	4.2720	6.5076	190
192	1311.304	0.001144	0.1488	0.1500	816.42	1971.30	2787.72	2.2547	4.2376	6.4924	192
194	1369.253	0.001147	0.1427	0.1439	825.37	1963.78	2789.15	2.2738	4.2034	6.4772	194
196	1429.196	0.001150	0.1369	0.1380	834.34	1956.18	2790.52	2.2928	4.1693	6.4621	196
198	1491.103	0.001153	0.1313	0.1325	843.32	1948.51	2791.84	2.3118	4.1353	6.4470	198
200	1555.099	0.001157	0.1261	0.1272	852.33	1940.76	2793.09	2.3307	4.1014	6.4321	200

TABLE B.2 Thermodynamic Properties of Ammonia as a Liquid and Saturated Vapor

Temp K	Pressure MPa	Volume Vapor m³/kg	Density Liquid kg/m³	Enthalpy Liquid kJ/kg	Enthalpy Vapor kJ/kg	Entropy Liquid kJ/kg·k	Entropy Vapor kJ/kg·K
**195.48	0.006075	15.648	733.86	−1110.11	380.09	4.2032	11.8265
200	0.008646	11.237	728.85	−1088.77	388.51	4.3111	11.6976
205	0.012512	7.9469	723.25	−1066.17	397.68	4.4228	11.5635
210	0.017746	5.7290	717.54	−1044.12	406.68	4.5290	11.4375
215	0.024706	4.2037	711.72	−1022.33	415.50	4.6315	11.3190
220	0.033811	3.1351	705.80	−1000.59	424.12	4.7314	11.2072
222	0.038159	2.8000	703.41	−991.89	427.50	4.7707	11.1643
224	0.042959	2.5065	701.00	−983.18	430.85	4.8097	11.1223
226	0.048248	2.2488	698.58	−974.45	434.16	4.8485	11.0812
228	0.054061	2.0220	696.16	−965.71	437.43	4.8870	11.0410
230	0.060439	1.8219	693.72	−956.95	440.66	4.9252	11.0017
232	0.067420	1.6450	691.27	−948.17	443.85	4.9631	10.9632
234	0.075048	1.4882	688.82	−939.38	447.00	5.0009	10.9256
236	0.083366	1.3490	686.36	−930.56	450.11	5.0383	10.8887
238	0.092420	1.2251	683.89	−921.72	453.18	5.0756	10.8525
239.82	0.101325	1.1241	681.64	−913.67	455.92	5.1092	10.8203
240	0.10226	1.1145	681.41	−912.86	456.20	5.1126	10.8171
242	0.11293	1.0157	678.92	−903.98	459.17	5.1494	10.7823
244	0.12448	0.92726	676.43	−895.07	462.10	5.1859	10.7483
246	0.13696	0.84790	673.92	−886.15	464.99	5.2223	10.7149
248	0.15044	0.77657	671.40	−877.20	467.82	5.2584	10.6821
250	0.16496	0.71234	668.88	−868.23	470.61	5.2944	10.6499
252	0.18058	0.65441	666.34	−859.24	473.35	5.3301	10.6184
254	0.19736	0.60206	663.79	−850.23	476.04	5.3656	10.5874
256	0.21536	0.55468	661.23	−841.20	478.68	5.4009	10.5569
258	0.23465	0.51174	658.65	−832.14	481.27	5.4361	10.5270
260	0.25529	0.47274	656.06	−823.06	483.80	5.4710	10.4976
262	0.27733	0.43728	653.46	−813.95	486.28	5.5058	10.4687
264	0.30086	0.40498	650.84	−804.82	488.71	5.5403	10.4403
266	0.32593	0.37553	648.20	−795.67	491.07	5.5747	10.4124
268	0.35262	0.34863	645.55	−786.50	493.39	5.6089	10.3849
270	0.38100	0.32402	642.88	−777.29	495.64	5.6430	10.3578
272	0.41113	0.30148	640.19	−768.07	497.84	5.6768	10.3312
274	0.44310	0.28081	637.48	−758.81	449.97	5.7106	10.3049
276	0.47698	0.26183	634.75	−749.54	502.04	5.7441	10.2791
278	0.51284	0.24438	632.00	−740.23	504.05	5.7775	10.2536
280	0.55077	0.22831	629.22	−730.90	505.99	5.8107	10.2284
282	0.59083	0.21350	626.43	−721.54	507.87	5.8438	10.2036
284	0.63312	0.19984	623.61	−712.15	509.68	5.8768	10.1792
286	0.67771	0.18721	620.77	−702.73	511.42	5.9095	10.1550
288	0.72469	0.17553	617.91	−693.28	513.08	5.9422	10.1311
290	0.77413	0.16472	615.02	−683.81	514.68	5.9747	10.1076
292	0.82613	0.15470	612.11	−674.30	516.19	6.0071	10.0843
294	0.88077	0.14540	609.17	−664.76	517.64	6.0393	10.0612
296	0.93813	0.13676	606.20	−655.19	519.00	6.0715	10.0384
298	0.99830	0.12873	603.21	−645.59	520.28	6.1035	10.0158

(Table B.2, continued)

Temp K	Pressure MPa	Volume Vapor m³/kg	Density Liquid kg/m³	Enthalpy Liquid kJ/kg	Enthalpy Vapor kJ/kg	Entropy Liquid kJ/kg · K	Entropy Vapor kJ/kg · K
300	1.0614	0.12126	600.19	−635.95	521.47	6.1354	9.9935
305	1.2324	0.10472	592.50	−611.70	524.08	6.2145	9.9384
310	1.4235	0.09079	584.63	−587.23	526.10	6.2931	9.8845
315	1.6362	0.07898	576.55	−562.51	527.51	6.3710	9.8314
320	1.8721	0.06893	568.24	−537.53	528.24	6.4484	9.7789
325	2.1327	0.06033	559.70	−512.25	528.25	6.5253	9.7268
330	2.4196	0.05293	550.89	−486.65	527.48	6.6019	9.6750
335	2.7344	0.04653	541.79	−460.68	525.86	6.6783	9.6232
340	3.0789	0.04099	532.36	−434.30	523.31	6.7546	9.5711
345	3.4549	0.03615	522.56	−407.44	519.75	6.8309	9.5184
350	3.8641	0.03191	512.34	−380.02	515.07	6.9075	9.4650
355	4.3085	0.02819	501.62	−351.97	509.13	6.9846	9.4103
360	4.7902	0.02489	490.33	−323.15	501.79	7.0625	9.3541
363	5.3112	0.02197	478.35	−293.42	492.81	7.1416	9.2957
370	5.8740	0.01936	465.54	−262.58	481.93	7.2222	9.2345
375	6.4811	0.01701	451.69	−230.38	468.76	7.3051	9.1696
380	7.1352	0.01489	436.52	−196.46	452.74	7.3911	9.0996
385	7.8395	0.01294	419.59	−160.25	433.01	7.4814	9.0224
390	8.5977	0.01113	400.21	−120.88	408.14	7.5783	8.9348
395	9.4144	0.009410	377.04	−76.642	375.49	7.6856	8.8303
400	10.2956	0.007689	346.94	−23.456	328.97	7.8133	8.6943
*405.4	11.304	0.00426	235.	142.7	142.7	8.216	8.216

**Triple point
*Critical point

INDEX

CONVERSION FACTORS

Multiply	By	To Obtain
Feet	0.305	Meters
Yards	0.915	Meters
Miles	1.609	Kilometers
Acres	4046	Square meters
Square miles	459	Hectares
Cubic feet	28.3	Liters
Pounds	0.454	Kilograms
Tons	0.907	Tonnes
Megaelectron Volts (MeV)	1.6×10^{13}	Joules
British thermal units (Btu)	1054	Joules
Kilowatt-hours (kWh)	3.6×10^{6}	Joules
British thermal units per pound (Btu/lb)	2.324×10^{3}	Joules per kilogram
Horsepower	746	Watts
Pounds per square inch (lb/in^2)	6.8948	Kilopascals
Atmospheres	101	Kilopascals

FUNDAMENTAL CONSTANTS

Name	Symbol	Value
Avogadro's number	N_0	6.02×10^{23} molecules/mol
Boltzmann constant	k	1.38×10^{-23} J/K
Electron rest mass	m_e	9.11×10^{-31} kg
Elementary charge	e	1.60×10^{-19} C
Faraday constant	F	9.65×10^{4} C/mol
Planck constant	h	6.63×10^{-34} J s
Proton rest mass	m_p	1.67×10^{-27} kg
Speed of light	c	3.00×10^{8} m/s
Stefan–Boltzmann constant	σ	5.67×10^{-8} W/m^2 K^4
Universal gas constant	R	8.314 J/mol K
		1.987 cal/mol K

USEFUL NUMBERS

Density of air at STP	1.293 kg/m^3
Density of moist air (20°C)	1.200 kg/m^3
Earth's surface heat flux	$6.28 \, \mu\text{J/cm}^2 \text{ s}$
Land area of the earth	$1.5 \times 10^{14} \text{ m}^2$
Mean molecular weight of dry air	28.97
Specific heat of dry air	1000 J/kg K
Standard volume of an ideal gas	$2.24 \times 10^{-2} \text{ m}^3/\text{mol}$
Sun's output	$3.9 \times 10^{26} \text{ W}$
Volume of the oceans	$1.37 \times 10^{18} \text{ m}^3$

ENERGY CONTENTS[a]

1 barrel of oil	6.1×10^9 J
1 ft^3 of natural gas	1.1×10^6 J
1 gallon of oil	1.5×10^8 J
1 liter of diesel fuel	3.8×10^7 J
1 liter of gasoline	3.4×10^7 J
1 tonne of coal	2.6×10^{10} J

[a]All values are approximate.